ASTRONOMY AND ASTROPHYSICS LIBRARY

For further volumes:
http://www.springer.com/series/848

To my father

to my father

Preface

Over the last decade, stellar interferometry has developed from a specialist tool to a mainstream observing technique. The user community has expanded well beyond the experts, attracting scientists whose research benefits from milliarcsecond angular resolution. As a result, the number of scientific publications has grown exponentially, showing the same trend as in radio interferometry some 30 years earlier.

Stellar interferometry has become part of the astronomer's toolbox, complementing single telescope observations by providing unique capabilities that will advance astronomical research.

While there is a large number of publications dealing with individual topics of interferometric observations and technical developments – all requiring a good level of understanding of the underlying physical principles – there is no text introducing these principles, deriving the relevant properties for interferometry and relating them to interferometric observations.

This book provides this information both for the astronomer using interferometry, but not being an interferometrist per se, and for the student starting in this field either to prepare astronomical research or to develop instruments. Rather than attempting to detail technical developments that are constantly evolving, the physical ideas behind the concept of interferometric observations are analysed, and the fundamental limitations are discussed. Numerical examples are given so that the basic performance of interferometers can be assessed.

Having spent the better part of the last 13 years with the Very Large Telescope Interferometer (VLTI), the material in this book is based on my experience with planning and developing this facility, and, thus, it is based on the work of the VLTI team. It is my pleasure to acknowledge this invaluable contribution. The kind support and the patience of the Springer-Verlag are also appreciated. In addition, I would like to thank Michael Dobers, Georg Junker and Gautam Vasisht for reading parts of the text, and Bruno Leibundgut for going through the whole manuscript, providing valuable advice.

This book would have never been completed without the unfailing support of my family. I am indebted to my wife Birgit, to Paula and to Jasper.

Garching
July 2010

Andreas Glindemann

Contents

List of Symbols, Notations and Abbreviations

Symbols

F	Focal length
k	Wavenumber $k = 2\pi/\lambda$
λ	Wavelength
l_c	Coherence length, $l_c = \lambda^2/\Delta\lambda$, with $\Delta\lambda$ the spectral bandwidth
n	Refractive index
ν	Frequency
r_0	Fried parameter

Abbreviations, Operators and Functions

FWHM	Full width at half maximum
OPD	Optical path difference
OTF	Optical transfer function
PSF	Point-spread function
$\mathcal{F}_x(.)$	Fourier transform of a function of x
$E\{.\}$	Ensemble average
$< . >$	Time average
∇	Gradient or nabla operator $(\frac{\partial}{\partial x}, \frac{\partial}{\partial y}, \ldots)$
$. \cdot .$	Scalar product
$. * .$	Convolution
$\delta(x)$	δ-function with $\int \delta(x)\mathrm{d}x = 1$, $\delta(0) \rightarrow \infty$ and $\delta(x) = 0$ if $x \neq 0$
$\mathrm{sinc}(x)$	sinc-function with $\mathrm{sinc}(x) = \sin(x)/x$
$\mathrm{circ}(x)$	circ-function with $\mathrm{circ}(x) = 1$ if $x \leq 1$, and $\mathrm{circ}(x) = 0$ elsewhere
$\mathrm{Besinc}(x)$	Besinc-function with $\mathrm{Besinc}(x) = 2\,\mathrm{J}_1(x)/x$, and $\mathrm{J}_1(x)$ the first order Bessel function

Coordinates

Note: The prime denotes object and entrance pupil, the lack of it image and exit pupil

$\boldsymbol{r}$	Three-dimensional spatial vector
$\boldsymbol{x}' = (x', y')$	Coordinate vector in the source plane
$\boldsymbol{\alpha}'$	Angular coordinate vector in the source plane, $\boldsymbol{\alpha}' = \boldsymbol{x}'/z_0$, or relative coordinate in the sky, $\boldsymbol{\alpha}' = (\Delta RA, \Delta\delta)$
$\boldsymbol{\alpha}$	Angular coordinate vector in the image plane, $\boldsymbol{\alpha} = \boldsymbol{x}/z_1$
$\boldsymbol{B}', \boldsymbol{B}$	Baseline vector, $\boldsymbol{B} = \boldsymbol{\xi}_1 - \boldsymbol{\xi}_2$
$\boldsymbol{f}_\xi$	Spatial frequency vector, $\boldsymbol{f}_\xi = \boldsymbol{\xi}/\lambda$
$\boldsymbol{R}', \boldsymbol{R}$	Spatial frequency vector as coordinate difference, $\boldsymbol{R} = \boldsymbol{f}_{\xi,1} - \boldsymbol{f}_{\xi,2} = \frac{\boldsymbol{\xi}_1 - \boldsymbol{\xi}_2}{\lambda}$, $\boldsymbol{R} = (u, v)$ is the coordinate vector in the *uv-plane*
$\boldsymbol{R}'_B, \boldsymbol{R}_B$	Baseline vector in the uv-plane, $\boldsymbol{R}_B = \boldsymbol{B}/\lambda$
$\boldsymbol{\xi} = (\xi, \zeta)$	Coordinate vector in the aperture plane
$\boldsymbol{x} = (x, y)$	Coordinate vector in the image plane
z_0	Distance between source plane and aperture plane
z_1	Distance between aperture plane and image plane

Amplitudes

$v(\boldsymbol{r}, t)$	Optical disturbance, dimensionless, proportional to the electric field vector of the electromagnetic wave
$v(\boldsymbol{\xi}, t), v(\boldsymbol{x}, t), v(\boldsymbol{\alpha}, t)$	Optical disturbance as a function of $\boldsymbol{\xi}$, $\boldsymbol{x}$ or $\boldsymbol{\alpha}$
$V(\boldsymbol{r}, \nu)$	Spectral amplitude in Hz^{-1}, $v(\boldsymbol{r}, t) = \mathcal{F}_\nu(V(\boldsymbol{r}, \nu))$
$V(\boldsymbol{\xi}, \nu), V(\boldsymbol{x}, \nu), V(\boldsymbol{\alpha}, \nu)$	Spectral amplitude as a function of $\boldsymbol{\xi}$, $\boldsymbol{x}$ or $\boldsymbol{\alpha}$
$V(\boldsymbol{r})$	Monochromatic amplitude, dimensionless $V(\boldsymbol{r}) = \int V(\boldsymbol{r}, \nu)\delta(\nu - \nu_0)\mathrm{d}\nu$

Intensities

$I(\boldsymbol{r})$	Intensity, dimensionless, also as function of $\boldsymbol{\xi}$, $\boldsymbol{x}$ or $\boldsymbol{\alpha}$ $I(\boldsymbol{r})$ is proportional to the *flux* in $\mathrm{W\,m^{-2}}$, $I(\boldsymbol{r}) =< v(\boldsymbol{r}, t)v^*(\boldsymbol{r}, t) >$
$I(\boldsymbol{r}, \nu)$	Spectral intensity in Hz^{-1}, also as function of $\boldsymbol{\xi}$, $\boldsymbol{x}$ or $\boldsymbol{\alpha}$, $I(\boldsymbol{r}, \nu)$ is proportional to the *flux density* in $\mathrm{W\,m^{-2}\,Hz^{-1}}$ $I(\boldsymbol{r}) = \int I(\boldsymbol{r}, \nu)\mathrm{d}\nu$
$G(\nu)$	Spectrum in Hz^{-1}
$g(\tau)$	Fourier transform of the centred spectrum $G(\nu - \nu_0)$

$G_B(\boldsymbol{R})$ Spectrum, dimensionless, as a function of spatial frequency $\boldsymbol{R}$ for baseline $\boldsymbol{B}$, with $\boldsymbol{R} = \frac{\boldsymbol{B}}{c}\nu$

$I_b(\boldsymbol{\alpha}')$ Source brightness distribution, dimensionless

$I_b(\boldsymbol{\alpha})$ Source brightness distribution, projected into the image plane $\boldsymbol{\alpha}$

$I(\tau)$ Intensity of the fringe pattern as a function of τ (instead of $\boldsymbol{\alpha}$) using
$k\boldsymbol{\alpha} \cdot \boldsymbol{B} = 2\pi\nu\tau$, with OPD $= \boldsymbol{\alpha} \cdot \boldsymbol{B}$ and $\tau =$ OPD$/c$

$\hat{I}(\nu)$ Fourier transform of $I(\tau)$, in Hz^{-1}

Coherence Functions

$\Gamma(\boldsymbol{x}_1, \boldsymbol{x}_2, \tau)$ Mutual coherence function (MCF), dimensionless

$\phi_{\mathrm{MCF}}(\boldsymbol{x}_1, \boldsymbol{x}_2, \tau)$ Phase of the MCF
$\Gamma(\boldsymbol{x}_1, \boldsymbol{x}_2, \tau) =< v(\boldsymbol{r}, t+\tau)v^*(\boldsymbol{r}, t) >$
$I(\boldsymbol{x}) = \Gamma(\boldsymbol{x}, \boldsymbol{x}, 0)$
The MCF and the following coherence functions are also written as functions of $\boldsymbol{x}'$, $\boldsymbol{\xi}$, $\boldsymbol{\alpha}'$ or $\boldsymbol{\alpha}$.

$\hat{\Gamma}(\boldsymbol{x}_1, \boldsymbol{x}_2, \nu)$ Mutual spectral density function (MSDF) in Hz^{-1}

$\hat{\phi}(\boldsymbol{x}_1, \boldsymbol{x}_2, \nu)$ Phase of MSDF
$\hat{\Gamma}(\boldsymbol{x}_1, \boldsymbol{x}_2, \nu) = \lim_{T\to\infty} \frac{1}{2T} E\{V(\boldsymbol{x}_1, \nu)V^*(\boldsymbol{x}_2, \nu)\}$
$\Gamma(\boldsymbol{x}_1, \boldsymbol{x}_2, \tau) = \mathcal{F}_\nu(\hat{\Gamma}(\boldsymbol{x}_1, \boldsymbol{x}_2, \nu))$
$I(\boldsymbol{x}, \nu) = \hat{\Gamma}(\boldsymbol{x}, \boldsymbol{x}, \nu)$

$\hat{\Gamma}(\boldsymbol{\xi}_1 - \boldsymbol{\xi}_2, \nu)$ MSDF in the aperture plane as a function of coordinate difference only in case of an incoherent source

$\Gamma(\boldsymbol{\xi}_1 - \boldsymbol{\xi}_2, \tau)$ MCF in case of an incoherent source

$\mu_{\nu_0}(\boldsymbol{\xi}_1 - \boldsymbol{\xi}_2)$ Source visibility function in quasi-monochromatic approximation $\mu_{\nu_0}(\boldsymbol{\xi}_1 - \boldsymbol{\xi}_2) = \frac{\Gamma_{\mathrm{qm}}(\boldsymbol{\xi}_1 - \boldsymbol{\xi}_2, 0)}{\sqrt{I(\boldsymbol{\xi}_1)I(\boldsymbol{\xi}_2)}}$

$\phi_{\nu_0}(\boldsymbol{\xi}_1 - \boldsymbol{\xi}_2)$ Phase of $\mu_{\nu_0}(\boldsymbol{\xi}_1 - \boldsymbol{\xi}_2)$

$\mu(\boldsymbol{R})$ Source visibility function as a function of spatial frequency, $\boldsymbol{R} = \boldsymbol{f}_{\xi,1} - \boldsymbol{f}_{\xi,2} = \frac{\boldsymbol{\xi}_1 - \boldsymbol{\xi}_2}{\lambda}$

$\phi(\boldsymbol{R})$ Phase of $\mu(\boldsymbol{R})$

$\mathcal{V}$ Fringe visibility, $\mathcal{V} = (I_{\max} - I_{\min})/(I_{\max} + I_{\min})$, in the absence of any disturbances it is $\mathcal{V} = |\mu(\boldsymbol{R})|$

Atmospheric Turbulence

C_n^2 Structure constant of atmospheric turbulence, in $\mathrm{m}^{-2/3}$

$n(\boldsymbol{r})$ Random refractive index in atmospheric turbulence

$D_n(\boldsymbol{r})$ Structure function of refractive index, dimensionless
$D_n(\boldsymbol{r}) =< |n(\boldsymbol{r}') - n(\boldsymbol{r}' + \boldsymbol{r})|^2 >$

$\Gamma_n(\boldsymbol{r})$	Correlation function of refractive index $\Gamma_n(\boldsymbol{r}) =< n(\boldsymbol{r}')n(\boldsymbol{r}' + \boldsymbol{r}) >$
$\Phi_n(\boldsymbol{k})$	Power spectrum of the refractive index fluctuations, in 1/m^{-3}, with $\boldsymbol{k}$, the three-dimensional spatial wave vector $\Gamma_n(\boldsymbol{r}) = \mathcal{F}_{\boldsymbol{k}}(\Phi_n(\boldsymbol{k}))$
$\varphi(\boldsymbol{\xi})$	Random phase in a plane $\boldsymbol{\xi}$
$\Gamma_\varphi(\boldsymbol{\xi})$	Correlation function of $V_\varphi(\boldsymbol{\xi})$ $\Gamma_\varphi(\boldsymbol{\xi}) =< V_\varphi(\boldsymbol{\xi}')V^*_\varphi(\boldsymbol{\xi}' + \boldsymbol{\xi}) >= \exp(-\frac{1}{2}D_\varphi(\boldsymbol{\xi}))$
$D_\varphi(\boldsymbol{\xi})$	Phase structure function, in rad^2 $D_\varphi(\boldsymbol{\xi}) =< \vert\varphi(\boldsymbol{\xi}') - \varphi(\boldsymbol{\xi}' + \boldsymbol{\xi})\vert^2 >$
$\Phi(\boldsymbol{\kappa})$	Kolmogorov spectrum, i.e. power spectrum of the phase fluctuations, in rad^2/m^{-2}, with $\boldsymbol{\kappa}$, the two-dimensional spatial frequency vector
$V_\varphi(\boldsymbol{\xi})$	Amplitude after propagation through turbulence $V_\varphi(\boldsymbol{\xi}) = \exp(\mathrm{i}\varphi(\boldsymbol{\xi}))$

Optical System

$A(\boldsymbol{\xi}), A(\boldsymbol{f}_\xi)$	Aperture function, D, respectively D/λ diameter of a circular aperture
$P(\boldsymbol{f}_\xi)$	Modulus of $A(\boldsymbol{f}_\xi)$
$\varphi(\boldsymbol{f}_\xi)$	Phase of $A(\boldsymbol{f}_\xi)$, optical aberration (or random phase due to atmospheric turbulence)
$V_{\mathrm{ap}}(\boldsymbol{\xi})$	Amplitude in the aperture of an optical system $V_{\mathrm{ap}}(\boldsymbol{\xi}) = V(\boldsymbol{\xi})A(\boldsymbol{\xi})$
$L(\boldsymbol{\xi}), L(\boldsymbol{f}_\xi)$	Lens function
$\mathrm{OTF}(\boldsymbol{R})$	Optical transfer function $\mathrm{OTF}(\boldsymbol{R}) := \frac{\lambda^2}{F^2} \int A(\boldsymbol{f}_\xi)A^*(\boldsymbol{f}_\xi - \boldsymbol{R})\,\mathrm{d}\boldsymbol{f}_\xi$
$\mathrm{PSF}(\boldsymbol{\alpha})$	Point spread function $\mathrm{PSF}(\boldsymbol{\alpha}) = \mathcal{F}_{\boldsymbol{R}}(\mathrm{OTF}(\boldsymbol{R}))$
$\mathrm{OTF}_{\mathrm{coh}}(\boldsymbol{f}_\xi)$	Coherent optical transfer function $\mathrm{OTF}_{\mathrm{coh}}(\boldsymbol{f}_\xi) := \frac{\lambda_0}{F}A(\boldsymbol{f}_\xi)$
$\mathrm{PSF}_{\mathrm{coh}}(\boldsymbol{\alpha})$	Coherent point spread function $\mathrm{PSF}_{\mathrm{coh}}(\boldsymbol{\alpha}) = \mathcal{F}_{\boldsymbol{f}_\xi}(\mathrm{OTF}_{\mathrm{coh}}(\boldsymbol{f}_\xi))$
$\hat{I}(\boldsymbol{R})$	Spatial frequency spectrum of the image intensity, dimensionless $\hat{I}(\boldsymbol{R}) = \mathcal{F}^{-1}_{\boldsymbol{\alpha}}(I(\boldsymbol{\alpha}))$
$O(\boldsymbol{\alpha}')$	Object intensity, same as $I(\boldsymbol{\alpha}')$ but strictly reserved for the object
$O(\boldsymbol{\alpha})$	Object intensity, projected into the image plane $\boldsymbol{\alpha}$
$\hat{O}(\boldsymbol{R})$	Spatial frequency spectrum of the object intensity, dimensionless $\hat{O}(\boldsymbol{R}) = \mathcal{F}^{-1}_{\boldsymbol{\alpha}}(O(\boldsymbol{\alpha}))$
$O(\boldsymbol{\alpha}', \nu)$	Object spectral intensity, in Hz^{-1} $O(\boldsymbol{\alpha}') = \int O(\boldsymbol{\alpha}', \nu)\mathrm{d}\nu$

$O_b(\boldsymbol{\alpha}')$	Object brightness distribution, same as $I_b(\boldsymbol{\alpha}')$ but strictly reserved for the object $O(\boldsymbol{\alpha}', \nu) = O_b(\boldsymbol{\alpha}')G(\nu)$
$O_b(\boldsymbol{\alpha})$	Object brightness distribution, projected into the image plane $\boldsymbol{\alpha}$
m_p	Demagnification factor of the telescope apertures D', with demagnified exit pupil diameter of $D = D'/m_p$ and magnified image coordinate, $\boldsymbol{\alpha} = m_p\boldsymbol{\alpha}'$
m_b	Demagnification factor of the baselines B' in the interferometer array, with demagnified exit pupil baseline of $B = B'/m_b$
γ_d	Pupil densification factor with $\gamma_d = m_b/m_p$

Constants, Units

c	Speed of light in vacuum
mas	Milliarcsec

Introduction

Less than 300 years after Galilei's first telescope observations of celestial objects, Hippolyte Fizeau suggested in 1868 to improve the measurement of stellar diameters by masking the telescope aperture leaving only two small sub-apertures [69]. Light passing through these sub-apertures would then interfere in the telescope focal plane. Following Fizeau's suggestion [129], the first demonstration of stellar interferometry was performed with masked telescopes by M. Stéphan in 1874 [219] and, independently, by A.A. Michelson in 1890 [156] measuring the diameter of the moons of Jupiter.

The first successful measurement of a stellar diameter with separate apertures was performed 30 years later by A.A. Michelson and F.G. Pease on Mt. Wilson, California, determining the diameter of α Orionis to 0.047 arcsec [158]. This was at a time when the smallest diameter that could be measured with a full aperture was about 1 arcsec, equivalent to the angular resolution of the telescope when observing through atmospheric turbulence.

Although the measurement of a stellar diameter is not the same as an image, the dramatic increase in angular resolution sparked enough interest in the new method so that it was soon understood how such contrast measurements with different pairs of sub-apertures – different in separation and orientation – can be combined to form a high resolution image not only of stars but of any type of celestial object.

However, due to insurmountable technical problems with the mechanical stability at larger separations of the sub-apertures, stellar interferometry in the optical[1] was abandoned in the late 1930s. It took until 1974, when A. Labeyrie combined the light from two independent telescopes at the Observatoire de la Côte d'Azur, France, demonstrating that stellar interferometry was feasible [121].

While the angular resolution increases linearly with the telescope diameter when eliminating atmospheric turbulence with adaptive optics, even today's largest telescopes cannot resolve features on the surface of individual stars. The diffraction limit is still so much larger than a star's disk that their images in the telescope focus

[1] Throughout the text, *optical* refers to both visible and infrared wavelengths as long as the light is manipulated by optics.

A. Glindemann, *Principles of Stellar Interferometry*, Astronomy and Astrophysics Library, DOI: 10.1007/978-3-642-15028-9_1,

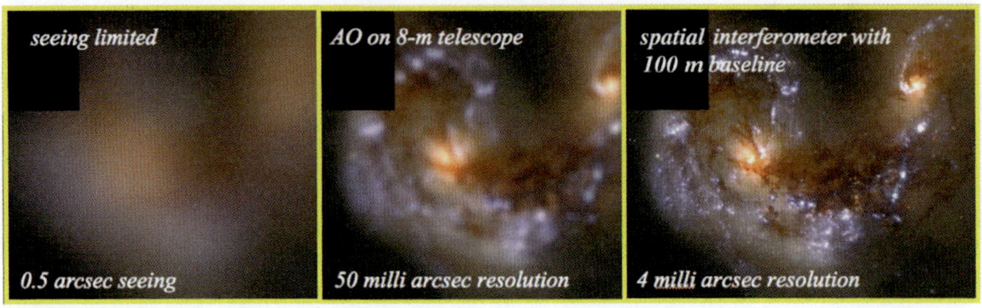

Fig. 1 Comparison of the image quality on an 8-m telescope at an observing wavelength of 2 μm. On the *left*, the seeing limited image is displayed for a seeing of 0.5 arcsec. In the *middle*, the diffraction limited image of an 8-m telescope with adaptive optics is shown. On the *right*, the reconstructed image of a stellar interferometer with a baseline of 100 m demonstrates the improvement over single telescope observations. These images are only simulations for comparison of the achievable angular resolution. Especially for stellar interferometers, an image of this size and this amount of detail has yet to be produced. (Courtesy R. Genzel)

are undistinguishable from point sources. For example, the angular resolution of 50 mas on an 8-m telescope is only just about the angular size of Betelgeuse.

Combining individual telescopes to form a *stellar interferometer*, the resolution is no longer determined by the individual telescope diameter but by the distance between the telescopes, called the *baseline, B*. Linking two 8-m telescopes that are separated by $B = 130$ m, like the telescopes of ESO's Very Large Telescope Interferometer (VLTI), improves the angular resolution by a factor of *baseline/telescope diameter* = 130/8 = 16 to about 3 mas in the near-infrared. Then, a large number of stars can be resolved revealing their shape that is not necessarily circular [55]. However, for a good image quality, one has to make many observations at different baselines, different in length and orientation (see Sect. 3.4), to obtain an image as in Fig. 1.

The major limitation for stellar interferometry – as for all observations from the ground – is set by the Earth's atmosphere. Just like for adaptive optics on individual telescopes, a bright reference star is needed for interferometry to reduce the effects of atmospheric turbulence so that faint objects can be observed. Since the reference star has to be typically within 1 arcmin of the faint object, the number of faint objects that can be observed is severely limited.

Thus, while the angular resolution as one parameter of the observing performance is improved drastically with stellar interferometers, the sensitivity as the second parameter suffers, unless a nearby bright reference star is available.

About this Book

Stellar interferometry is about observing interference patterns that are determined by the correlation of the amplitudes in the interferometer apertures. Therefore this is also called *amplitude interferometry*. The light of the star interfering with itself, we speak about *homodyne detection* to distinguish it from *heterodyne detection*, measuring the same quantity by having the light from each aperture interfere with a local oscillator with extremely stable frequency [109].

Apart from amplitude interferometry, one can also measure the correlation of intensities in each aperture, forming the basis of intensity interferometry that was developed by Hanbury Brown and Twiss [96]. They presented first results in the 1950s but owing to the very low sensitivity of the method, it has never been pursued rigorously.

Throughout the book, we will concentrate mainly on amplitude interferometry. Frequently, we will take a look beyond this border, discussing topics like intensity interferometry, speckle interferometry and adaptive optics in individual sections, highlighting what amplitude interferometry has in common with these topics and what distinguishes it from them.

One of the objectives of this book is to provide easy access to all topics. Short summaries at the end of each section, focussing on the main results, should enable the reader to step faster through the material up to the topic of interest.

Discussing stellar interferometers on the ground, we subdivide the field in the propagation of light through space and through optical systems, the imaging process in interferometers, the analysis of the statistics of atmospheric turbulence in free atmosphere and in telescopes, the principles of beam combination and of the layout of interferometric arrays, and the measurement of fringe patterns when observing through turbulence.

Accordingly, the book is organized as follows. In the second chapter, we will regard the basic properties of light propagating through space, discussing not only amplitude and intensity of the electro-magnetic wave but also its coherence function. Often, we will use Young's experiment to facilitate the interpretation of these quantities. Almost free of mathematical theory, interferometry will be explained in heuristic manner in Sect. 2.4.4.

In Chap. 3, we will use this formalism for the imaging process in interferometers, discussing the principles of the imaging process first for a single aperture and then for an interferometer with individual apertures. Introducing the uv-plane in Sect. 3.4, the impact of the number of baselines on the imaging process will be investigated and general principles of image reconstruction methods will be debated.

Chapter 4 is devoted to atmospheric turbulence. We will discuss the statistics of atmospheric turbulence, introducing basic parameters like the Fried parameter, and we will provide the statistical parameters of the perturbed wave after passage through apertures, with emphasis on spatial and temporal characteristics of fringe and image motion.

In Chap. 5, starting with the description of the Fizeau and Michelson configuration for two telescopes, we will expand the discussion to multiple apertures, and to multi-axial and co-axial beam combination. The layout of interferometric arrays and basic requirements for delay lines will be presented as well.

Finally, Chap. 6 combines the preceding results providing requirements and limitations for observing through turbulence. Beginning with visibility measurements in the presence of turbulence, we will then discuss how to beat the turbulence using fringe tracking and dual-feed systems. In the final section, the fundamental limits of adaptive optics systems will be discussed and compared to fringe tracking.

Focussing on the physical background, we left out technical developments and astronomical results that are in the process of constant advancement and are much better treated in review articles like those of Shao [211], Quirrenbach [188,189] and Monnier [161]. More information on these topics can also be found in the book of Labeyrie et al. [123].

Selected publications of interferometry, commencing with the original paper by Fizeau [69], were compiled by Lawson [134] who also edits a web page with optical long baseline interferometry news, http://olbin.jpl.nasa.gov, providing up-to-date information on scientific results, literature, meetings and links to interferometry projects.

Chapter 2
Propagation of Light

Propagating through space, light carries with it the information on the shape and the spectrum of the source. A straightforward way to extract this information is an optical system that forms an image, reproducing the source's spatial and spectral intensity distribution. However, there is a limit to the level of detail that can be resolved in the image of the source both due to fundamental physical boundaries – the diffraction at the aperture – and due to atmospheric turbulence.

While the increase in aperture size, improving the angular resolution, has its limitations, the combination of two or more apertures forming an interferometer has the maximum baseline as the limiting parameter. The imaging process in an interferometer requires to regard not only the amplitude of the light but also the coherence function, which is the correlation of the light amplitudes.

In this chapter, we will discuss the propagation of light, repeating briefly the scalar diffraction theory when the electromagnetic wave is represented by a scalar that is proportional to one component of the electric field vector. Using the Fresnel approximation we will introduce the thin lens as the simplest form of an optical system, and we will discuss the properties of the Airy disk.

The formalism for the propagation of the amplitude will then be expanded to the propagation of the coherence function, describing the coherence properties of the propagating wave. We will mainly restrict the discussion to incoherent sources and investigate partially coherent sources only as an exception. The main result is the van Cittert–Zernike theorem providing the link between the source intensity distribution and the coherence function in a plane far away from the source, e.g. in the aperture plane of our optical system.

Young's experiment with two pinholes will serve as an example for an interferometer, explaining the effect of the source shape on the coherence function by applying the theory developed in Sect. 2.3, and using a heuristic approach without coherence functions.

Finally, we will discuss the intensity interferometer involving higher order correlation functions of the amplitudes. We shall see that the intensity interferometer is less demanding in terms of opto-mechanical precision but in turn much less sensitive than an amplitude interferometer.

A. Glindemann, *Principles of Stellar Interferometry*, Astronomy and Astrophysics Library, DOI 10.1007/978-3-642-15028-9_2,

2.1 Preliminaries

2.1.1 Basic Properties of the Electromagnetic Wave

The electric field vector $\boldsymbol{E}$ of the electromagnetic wave is a function of space and time. Assuming a monochromatic plane wave in vacuum propagating in the z-direction (see Fig. 2.1), the x-component of $\boldsymbol{E}$ can be written as

$$E_x = E_{x0} \cos(\omega t - kz), \tag{2.1}$$

with $\omega = 2\pi\nu$, time t, and $k = 2\pi/\lambda$, where ν is the frequency and λ is the wavelength of the monochromatic wave. The phase velocity of the wave is $\omega/k = \nu\lambda$. The y-component can be described in the same way.

If the cosine functions of the x- and the y-component are not in phase, the resulting electric field vector $\boldsymbol{E}$ rotates around the propagation axis, describing a circle if the components are of equal amplitude with a phase difference of $\varphi = \pi$. The light is then called *circularly polarized*. If the phase difference is zero, the two components are in phase and the resulting field vector $\boldsymbol{E}$ oscillates in a plane. Then, the light is called *linearly polarized*. For all other phase differences, the light is *elliptically polarized*. When discussing polarization properties independent of a particular coordinate system one usually refers to the orthogonal components of the electric field vector as s and p polarization.

Since we will only describe the wave propagation in isotropic media the magnetic field $\boldsymbol{H}$ could be used just as well to describe the electromagnetic wave.

The intensity is the quantity that we usually measure with optical detectors. It is related to the energy flow density given by the *Poynting vector* $\boldsymbol{S} = \boldsymbol{E} \times \boldsymbol{H}$. The Poynting vector is perpendicular both to $\boldsymbol{E}$ and $\boldsymbol{H}$, and it points into the direction of propagation of the electromagnetic wave. If the electric field vector $\boldsymbol{E}$ is linearly polarized in the x-direction then the magnetic field vector $\boldsymbol{H}$ has a component only in the y-direction, and the wave propagates in the z-direction. Then, the Poynting

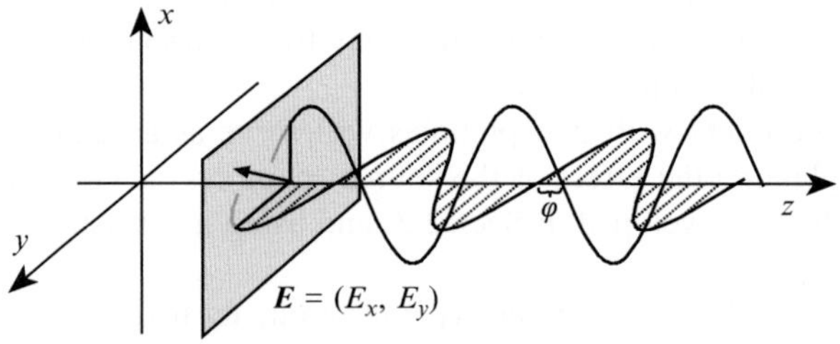

Fig. 2.1 The electric field vector $\boldsymbol{E}$, split into two orthogonal components E_x and E_y, propagating in the z-direction. If one of the two components is shifted by a phase difference $\varphi \neq 0$ the wave is polarized elliptically, i.e., the field vector – displayed as a *short arrow* – rotates around the z-axis describing an ellipse while propagating. If the phase difference φ is 0 the light is linearly polarized

vector has a component in the z-direction only, which can be written as

$$S_z = c\varepsilon_0 E_{x0}^2 \cos^2(\omega t - kz). \tag{2.2}$$

S_z has the dimension of W/m^2. c is the speed of light in vacuum with $c = 2.998 \times 10^8$ m/s, and ε_0 is the permittivity of vacuum with $\varepsilon_0 = 8.854 \times 10^{-12} \frac{\mathrm{A\,s}}{\mathrm{V\,m}}$. In a medium with refractive index n, the actual speed of light c/n has to be used, and ε_0 has to be replaced by the electric permittivity in the medium, usually expressed by $\varepsilon\varepsilon_0$, with ε the dielectric constant of the material.

The Poynting vector oscillates with twice the frequency ν of the electromagnetic wave, which is about 10^{15} Hz in the visible part of the spectrum. Since the temporal resolution of the available detectors is much lower than 10^{-15} s, one can only measure the time average of the Poynting vector defined as

$$< S_z > = \lim_{T\to\infty} \frac{1}{2T} c\varepsilon_0 \int_{-T}^{T} E_{x0}^2 \cos^2(\omega t - kz)\,\mathrm{d}t = \frac{c\varepsilon_0}{2} E_{x0}^2, \tag{2.3}$$

where $< . >$ denotes the time average as defined above.

In practice, it is sufficient if the integration interval T is much longer than any of the processes involved, thus $T \gg 1/\nu$. Even time intervals down to 10^{-12} s fulfil this requirement.

NB 1. *In some cases,* heterodyne detection *can be applied to detect amplitude and phase of the light wave by mixing two waves of very similar frequency – one of them precisely defined in amplitude and phase – and measuring the beat frequency. This was done the first time in 1955 by mixing Zeeman components of a visible spectral line [70]. In astronomy, this is applied in the mid-infrared around a wavelength of 10 μm, corresponding to a frequency of* 3×10^{13} *Hz. Here, light from a local oscillator, typically a* CO_2 *laser, is combined with the light to be detected producing a beat frequency signal in the GHz range [109]. This signal can be temporally resolved, and the amplitude and phase of the mid-infrared light can be determined. The limitation of heterodyne detection to the mid-infrared comes from the fact that at shorter wavelengths the sensitivity deteriorates very quickly [237], and that lasers with longer wavelengths are not available. However, if one aims at wavelengths beyond 100 μm, in micro-wave and radio interferometers, suitable tunable oscillators are available and heterodyne detection is the standard measurement method.*

The time average of the Poynting vector is called the *flux* (in astronomy) or the *irradiance* (in radiometry) of the electromagnetic wave in units of W m^{-2}. The measurable quantity in an optical detector is the integral of the flux over the area of the detector, i.e., the power in units of Watt.

Since, throughout this book, we are above all interested in the spatial flux distribution and not in absolute values we will work with dimensionless quantities. At first, we introduce a dimensionless scalar, the *optical disturbance* $v(\boldsymbol{r}, t)$ that is proportional to one component, e.g., E_x, of the electric field vector with $v = CE_x$ and

C a suitable constant to make v dimensionless. The y component of the electric field vector can be treated independently.

As long as optical systems like, e.g., astronomical telescopes interact linearly with the amplitude and the phase of the incoming light wave we can use product relations between the light wave and a so-called *transfer function* of the optical system. For the mathematical treatment, it is very convenient to extend the optical disturbance by an imaginary part so that v becomes a complex quantity. A linear optical system is then described by a complex transfer function, and the outgoing wave can be calculated as the product of two complex functions, the complex optical disturbance and the complex transfer function.

For a plane wave propagating along the z-axis, the extension of the optical disturbance by an imaginary part reads as

$$v(z,t) = v_0 \cos(\omega t - kz) - \mathrm{i}\, v_0 \sin(\omega t - kz) = v_0\, \mathrm{e}^{-\mathrm{i}(\omega t - kz)}. \tag{2.4}$$

It is important to keep in mind that the introduction of the complex optical disturbance is just a convenience and that only the real part has physical significance, since only the real part represents the electromagnetic wave.

The time average of the product $v v^*$ – the superscript * denoting the complex conjugate – can be used to define the *intensity* as

$$I(z) := \lim_{T\to\infty} \frac{1}{2T} \int_{-T}^{T} v(z,t) v^*(z,t)\, \mathrm{d}t = v_0^2. \tag{2.5}$$

The intensity is a dimensionless quantity that is proportional to the flux $< S_z >$ and, thus, proportional to the signal that is measured with optical detectors. Throughout this book, we will always use the intensity I instead of the flux.

For the propagation of light in space it is convenient to introduce the time independent dimensionless *amplitude* $V(\boldsymbol{r})$ at frequency ν so that the monochromatic optical disturbance can be written as

$$v(\boldsymbol{r},t) = V(\boldsymbol{r})\, \mathrm{e}^{-\mathrm{i}2\pi\nu t}. \tag{2.6}$$

With (2.5) one sees that the intensity is $I(\boldsymbol{r}) = |V(\boldsymbol{r})|^2$.

In the case of **polychromatic light**, the linear superposition of individual monochromatic waves forms the polychromatic optical disturbance. We introduce the time independent *spectral amplitude* $V(\boldsymbol{r},\nu)$ such that it is

$$v(\boldsymbol{r},t) = \int_0^{\infty} V(\boldsymbol{r},\nu)\, \mathrm{e}^{-\mathrm{i}2\pi\nu t}\, \mathrm{d}\nu. \tag{2.7}$$

$V(\boldsymbol{r},\nu)$ has the dimension of Hz^{-1}. The integration is restricted to the positive frequency arm since the spectral amplitude has to be zero for negative frequencies. Only then the physically relevant real part of the optical disturbance contains all the information, and the imaginary part can be ignored. Since this was pointed out by Gabor [76] the complex optical disturbance is also called *Gabor's analytic signal*. Formally, we can extend the integration to $-\infty$ writing the optical disturbance $v(\boldsymbol{r}, t)$ as the temporal Fourier transform of the spectral amplitude $V(\boldsymbol{r}, \nu)$ (see Sect. A.1).

The *spectral intensity* $I(\boldsymbol{r}, \nu)$[1] is defined such that it is

$$I(\boldsymbol{r}) = \int_0^\infty I(\boldsymbol{r}, \nu)\,\mathrm{d}\nu. \tag{2.8}$$

The dimension of $I(\boldsymbol{r}, \nu)$ is Hz^{-1}. The spectral intensity is proportional to the *flux density* in units of $\mathrm{W\,m^{-2}\,Hz^{-1}}$. In astronomy, a common unit for the flux density is 1 Jansky (Jy) $= 10^{-26}\,\mathrm{W\,m^{-2}\,Hz^{-1}}$.

The spectral intensity is linked to the spectral amplitude through an averaging process similar to (2.5). The exact definition will be given in Sect. 2.3.1 in the context of coherence functions. When discussing the scalar diffraction theory in the following we will simplify this relationship by setting $I(\boldsymbol{r}, \nu) = |V(\boldsymbol{r}, \nu)|^2$ without affecting the general conclusions of the discussions.

With the polychromatic intensity $I(\boldsymbol{r})$ being the integral of the spectral intensities $I(\boldsymbol{r}, \nu)$, the propagation of polychromatic light through space and through optical systems can be treated by first considering the monochromatic case and then adding up the spectral intensities at the very end of the propagation process.

Throughout the book, we will use the monochromatic amplitude $V(\boldsymbol{r})$ or the spectral amplitude $V(\boldsymbol{r}, \nu)$ to discuss diffraction effects and the propagation of light in space, and we will use (2.8) to calculate the intensity. When discussing the coherence function in Sect. 2.3 we will follow a similar principle. This facilitates the mathematical treatment enormously. But we should keep in mind that the simple relationship between monochromatic and polychromatic intensity (2.8) is valid only because optical detectors measure the intensity as a time average. Otherwise the measured signal would show high frequency oscillations and the relationship would be more complicated.

2.1.2 Young's Experiment

This is the classical diffraction experiment that we will discuss very often in the following sections. It is named after Thomas Young who conducted it in 1802, providing the experimental cornerstone for the demonstration of the wave nature of light [256]. Describing Young's experiment, we will introduce diffraction effects and the concept of coherence.

---

[1] In the literature, this is sometimes called the *power spectral density*, since it gives the power per Hz.

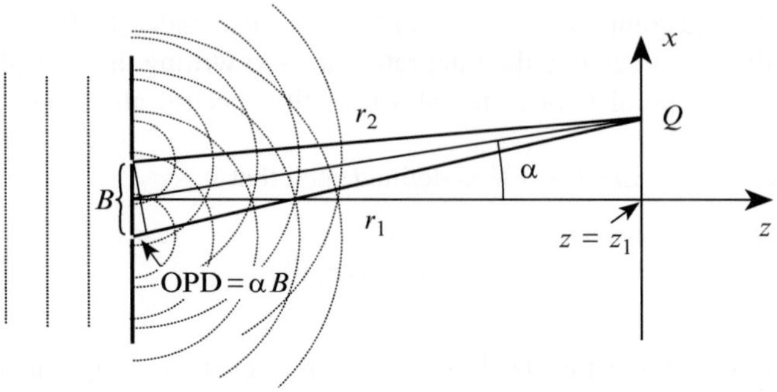

Fig. 2.2 The geometry of Young's experiment in a plane across the pinholes. A plane wave illuminates the two pinholes separated by a distance B. For small angles, α is approximately equal to x/z_1 with x the coordinate of Q, the point of observation. The difference $r_1 - r_2$ of the distances from the pinholes to Q is called the optical path difference (OPD), which equals αB for small angles α. The OPD is related to the difference in arrival time τ between the light from the two pinholes by $\tau = \mathrm{OPD}/c$

Young illuminated a screen with two pinholes with light from a single pinhole at a large distance. On passing through the pinholes the light was diffracted and the waves from the two pinholes interfered. On a second screen, the diffraction pattern could be observed showing the characteristic fringe pattern. In Fig. 2.2 the experiment is depicted schematically for an illuminating source at very large distance from the screen so that an approximately plane wave illuminates the two pinholes.

For the mathematical treatment we assume that the two pinholes are so small that they can be regarded as point sources of spherical waves. The spectral amplitude of a spherical wave at a distance $r = |\boldsymbol{r}|$ from its origin can be written as

$$V(\boldsymbol{r}, \nu) = \frac{V_0}{r}\,\mathrm{e}^{\mathrm{i}kr}. \tag{2.9}$$

The spectral intensity is the squared modulus of the amplitude, $I(\boldsymbol{r}, \nu) = V_0^2/r^2$, decreasing with the square of the distance from the source.

The assumption that spherical waves originate from each pinhole comes across quite naturally in this context. However, the underlying concept is the Huygens–Fresnel principle of elementary waves, forming the basis of scalar diffraction theory that will be treated more formally in Sect. 2.2.

The amplitude V at point $Q = (x, 0)$ in a plane at distance z_1 is then the sum of the two spherical waves originating from the pinholes,

$$\begin{aligned} V(x, \nu) &= \frac{V_0}{r_1}\,\mathrm{e}^{\mathrm{i}kr_1} + \frac{V_0}{r_2}\,\mathrm{e}^{\mathrm{i}kr_2} \\ &= \frac{V_0}{z_1}\,\mathrm{e}^{\mathrm{i}k(r_1+r_2)/2}\,2\cos\bigl(k(r_1 - r_2)/2\bigr). \end{aligned} \tag{2.10}$$

r_i is the distance between an individual pinhole and the point Q, with the approximation $r_1 = r_2 = z_1$ for the amplitudes V_0/r_i.

The spectral intensity is the squared modulus of the amplitude,

$$I(x,\nu) = |V(x,\nu)|^2 = \left(\frac{V_0}{z_1}\right)^2 2\big(1 + \cos(k(r_1 - r_2))\big). \tag{2.11}$$

For small diffraction angles α, it is $\alpha = x/z_1$, and $r_1 - r_2 = \alpha B$ is the *optical path difference* (OPD) between the optical paths from each pinhole to the point of observation. A difference in optical path length is at the same time a difference in arrival time called the time delay τ between the light from the two pinholes, with $\tau = \alpha B/c$. We will see later that the diffraction pattern for increasing diffraction angles α – corresponding to increasing time delays τ – is constrained by the temporal coherence of the incoming light.

We can now write the monochromatic intensity distribution of the diffraction pattern in its familiar form as a function of the diffraction angle α,

$$I(\alpha,\nu) \propto \big(1 + \cos(k\alpha B)\big). \tag{2.12}$$

This intensity distribution is also called the *fringe pattern* since the maxima of the cosine function appear as fringes on a two-dimensional screen. The first intensity minimum is at $\alpha_{\rm min} = \lambda/(2B)$ as stated by Young as a result of his experiment. The OPD at the first minimum is $r_1 - r_2 = \lambda/2$ and, consequently, the time delay is $\tau = (\lambda/2)/c = 1/(2\nu)$. Figure 2.3 shows the fringe pattern for a pinhole separation of 10 cm at wavelengths of 2.0, 2.2 and 2.4 μm.

The monochromatic fringe patterns have excellent contrast since the intensity oscillates between 0 and 1. This can be expressed more formally by defining the contrast called the *fringe visibility* as introduced by Michelson [156] as

$$\mathcal{V} = \frac{I_{\rm max} - I_{\rm min}}{I_{\rm max} + I_{\rm min}}. \tag{2.13}$$

With $I_{\rm min} = 0$ and $I_{\rm max} = 1$, the contrast of the fringe pattern is $\mathcal{V} = \infty$.

Stellar interferometry is about measuring the contrast of fringes. Hence, we should have a closer look at the result of Young's experiment. A contrast of 1 in a fringe pattern, i.e., perfect constructive and destructive interference in its maxima and minima, implies that the light waves from the two pinholes are perfectly coherent. In fact, the monochromatic plane wave illuminating the screen with the two pinholes and, thus, the light waves emerging from the pinholes are perfectly coherent.

The term *coherence* is linked to the existence of interference phenomena in diffraction experiments like Young's experiment. A fringe pattern with a good contrast requires a good coherence between the light waves from the two pinholes. Light of perfect coherence causes a fringe pattern with a contrast of 1 as stated above. If there is no coherence between the light from the pinholes, there is no fringe pattern

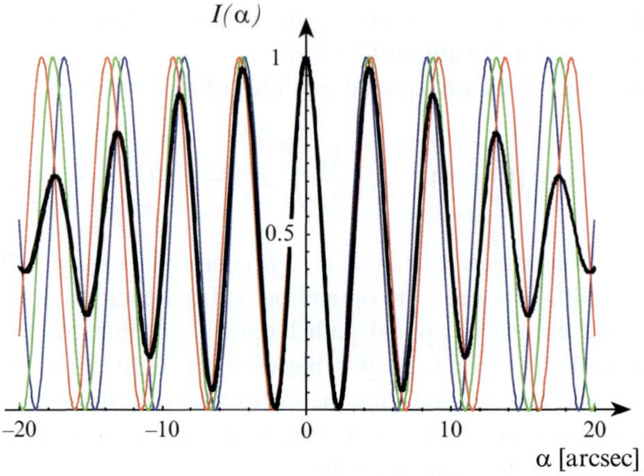

Fig. 2.3 Three individual monochromatic fringe patterns of Young's experiment, for $\lambda = 2.0, 2.2$ and $2.4\,\mu$m (*blue*, *green* and *red lines*), and the resulting polychromatic fringe pattern of the full K-band, $I(\alpha) = \int(I(\alpha, \nu)\mathrm{d}\nu$, displaying decreasing contrast for increasing diffraction angle α. The pinhole separation is 10 cm. The first minima of the individual monochromatic fringe patterns are at $\alpha_{\mathrm{min}} = \lambda/(2B) = 2.1, 2.3$ and 2.5 arcsec from the central maximum

but only a homogeneous illumination as a result of the diffraction of light at each individual aperture. Then, the light is called incoherent.

Implicitly, by discussing the behaviour of the amplitude only, we have used the fact that optical detectors perform a time average over a period much longer than $1/\nu$. If we had assumed that we have an ideal detector temporally resolving the oscillating electromagnetic wave we would then measure the instantaneous intensity of the light wave that is affected by the temporally varying parts $\exp(\mathrm{i}2\pi\nu t)$ of the optical disturbances in the two pinholes. The effect is that in any given moment, there is constructive or destructive interference at any point in the plane of observation regardless of the state of coherence. This diffraction pattern is oscillating typically with twice the light frequency. It settles to the familiar fringe pattern by the temporal averaging process when measuring the intensity with a real detector.

In Sect. 2.4, we will see in detail how the state of coherence of the light in the pinholes depends on the source properties. As an introduction, we now discuss the effect of a finite *spectral bandwidth* $\Delta\nu$ of the light source. The resulting fringe pattern is formed by adding up interference patterns like (2.12) at different frequencies to obtain the observed intensity distribution. This is displayed in Fig. 2.3 for the K-band,[2] with $\lambda = 2.2\pm0.2\,\mu$m, and $\Delta\lambda = 0.4\,\mu$m, and a pinhole separation of 10 cm. For zero OPD, at $\alpha = 0$, all wavelengths have an intensity maximum. This is why this fringe is called the *white-light fringe*. The position of the first

[2] The atmosphere transmits only certain bands in the infrared. One frequently used band in the near-infrared is the K-band at $2.2\pm0.2\,\mu$m, (see Sect. A.2). Most of the numerical examples will be given for this band.

minimum $\alpha_{\min} = \lambda/(2B)$ or OPD $= \lambda/2$ is then wavelength dependent, as well as the positions of the following maxima and minima.

This effect reduces the contrast of the resulting polychromatic fringe pattern for increasing diffraction angles α. Since α is related to the difference in arrival time, the time delay τ, through $\tau = \alpha B/c$, this effect can be reformulated by stating that the contrast of the resulting fringe pattern is reduced with increasing τ. The time delay that is related to the quasi loss of fringe contrast is called the *coherence time* τ_c, which is proportional to the reciprocal of the spectral bandwidth $\Delta\nu$. The exact relationship depends on the form of the spectral band.

This introduces the connection between temporal coherence, quantified by the coherence time τ_c, and the fringe contrast in a rather casual way. The spatial coherence of the incoming light, and its influence on the fringe contrast, is more difficult to come by. In Sect. 2.3 we will treat coherence as a statistical phenomena, and we will see that both temporal and spatial coherence are part of the more general coherence theory.

2.2 Scalar Diffraction Theory

The foundations of scalar diffraction theory have been discussed in most standard texts on optics [19]. Goodman [88] gave a very concise summary of the historical background and the mathematical deduction. We will briefly revisit the cornerstones of this theory in order to give an idea of how the propagation of a coupled vector field, i.e., of the electromagnetic wave can be reduced to the Fourier transform of a scalar.

2.2.1 The Rayleigh–Sommerfeld Diffraction Formula

Forming images in optical instruments, the propagation of light and the diffraction at physical boundaries needs to be dealt with. The fundamental equations treating the propagation of electromagnetic waves are *Maxwell's equations* [19]. They describe the propagation of both the electric and the magnetic field vectors $\boldsymbol{E}$ and $\boldsymbol{H}$, respectively, through coupled partial differential equations.

We will make a number of assumptions and approximations that are adequate for the imaging process. First, light is treated as a scalar quantity as introduced in Sect. 2.1, by defining a dimensionless scalar $v(\boldsymbol{r}, t)$, the *optical disturbance*, that is proportional to one component of, e.g., the electric field vector $\boldsymbol{E}$. Second, one assumes that the propagation of the two orthogonal vector components of $\boldsymbol{E}$ that are perpendicular to the direction of propagation can be treated independently.

This approximation is valid if (1) diffracting apertures are large compared to the wavelength, and if (2) the diffracted fields are not observed too close to the diffracting apertures. Both conditions are easily fulfilled in optical systems in general, and in astronomical telescopes in particular.

One solution of the time dependence of the differential equation describing the propagation of $v(\boldsymbol{r},t)$ is given by the monochromatic complex harmonic wave $V(\boldsymbol{r})\exp(-\mathrm{i}2\pi\nu t)$ in (2.6). Then, Maxwell's equation can be replaced by the *Helmholtz equation*, which is the time-independent form of the wave equation, acting on the amplitude $V(\boldsymbol{r})$:

$$(\nabla^2 + k^2)V(\boldsymbol{r}) = 0, \tag{2.14}$$

with ∇^2 being the Laplace operator $\frac{\partial^2}{\partial x^2} + \frac{\partial^2}{\partial y^2} + \frac{\partial^2}{\partial z^2}$, and $k = 2\pi/\lambda$. Solutions of this differential equation are a monochromatic plane wave, for example $V(\boldsymbol{r}) = V_0\exp(\mathrm{i}kz)$ for a wave propagating along the z-axis, or a monochromatic spherical wave $V(\boldsymbol{r}) = \frac{V_0}{r}\exp(\mathrm{i}kr)$ with $r = |\boldsymbol{r}|$.

In optical systems, one is interested in the amplitude V in the plane of observation, the image plane, as a function of the amplitude in the aperture plane, or as a function of the amplitude of the object. Any solution for V needs to satisfy the Helmholtz equation at all points $\boldsymbol{r}$.

A starting point for tackling this problem is Green's integral theorem assuming two functions V and G that are solutions of the Helmholtz equation. The choice of function G, also called Green's function, is now crucial. Kirchhoff's choice of a spherical wave $G = \exp(\mathrm{i}kr)/r$ leads to the *integral theorem of Huygens and Kirchhoff* obtaining the amplitude V at point P_0 as a function of the amplitude on a closed surface **A** surrounding P_0:

$$V(P_0) = \frac{1}{4\pi}\iint_{\mathbf{A}} (G\,\nabla V - V\,\nabla G)\mathrm{d}\boldsymbol{\xi}, \tag{2.15}$$

where $\nabla = (\frac{\partial}{\partial x}, \frac{\partial}{\partial y}, \frac{\partial}{\partial z})$ is the gradient or nabla operator.

Kirchhoff further developed this formula by performing the step from a closed surface to a plane surface limited by an aperture. Thus, he described a situation that is more similar to the setup in optical systems. This led to the Fresnel-Kirchhoff diffraction formula. Although it has been found experimentally to yield remarkably accurate results, the choice of Green's function required a number of assumptions known as the Kirchhoff boundary conditions implying, when applied rigorously, that the amplitude V needs to be identical to zero everywhere in space.

This inconsistency was removed by Sommerfeld by assuming two point sources emitting spherical waves that are placed symmetrically to the aperture on either side of it. Then, two new Green's functions G_- or G_+ can be defined as either the difference or the sum of the two spherical waves.

The final result known as the *Rayleigh–Sommerfeld diffraction formula* forms the basis of all our discussions on diffraction and propagation of light [88]. In the approximation of small diffraction angles, the propagation of the amplitude from the aperture plane $\boldsymbol{\xi}$ into the plane of observation $\boldsymbol{x}$, see Fig. 2.4, is computed by

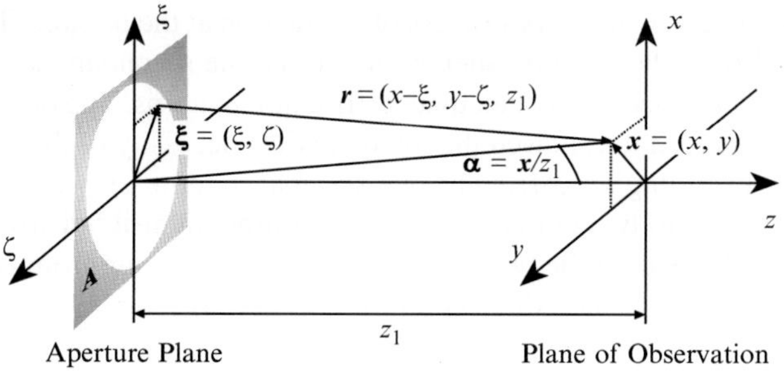

Fig. 2.4 The geometry for the diffraction at an aperture **A**. The aperture plane has coordinates (ξ, ζ) and the coordinate vector $\boldsymbol{\xi}$, the plane of observation is denoted by (x, y) with the vector $\boldsymbol{x}$. The distance between the planes is z_1 and the vector between two points (ξ, ζ) and (x, y) is $\boldsymbol{r} = (x-\xi, y-\zeta, z_1)$, with the notation $r = |\boldsymbol{r}|$. We call the z-axis of the coordinate system the *optical axis*. Usually, all elements of optical systems, i.e., apertures, lenses and mirrors, are centred on the optical axis

$$V(\boldsymbol{x}) = \frac{1}{\mathrm{i}\lambda} \iint_{\mathbf{A}} V(\boldsymbol{\xi}) \, \frac{\mathrm{e}^{\mathrm{i}kr}}{r} \, \mathrm{d}\boldsymbol{\xi}. \tag{2.16}$$

Let us assume for the moment that the incoming wave is a plane wave with $V(\boldsymbol{\xi}) = V_0$ in the aperture plane. Then the integration is performed over spherical waves $\frac{V_0}{r}\exp(\mathrm{i}kr)$ originating from each point in the aperture, and the superposition of these elementary waves forms the wave front[3] behind the aperture. This concept is called the *Huygens–Fresnel principle* of elementary waves. Huygens first expressed this principle out of intuition. In the context of seeking solutions for the integral theorem (2.15) the elementary waves come in as Green's functions.

In the extreme case of an infinitely large aperture **A** that is illuminated by a plane wave, a plane wave is expected to arrive in the plane of observation. In fact, the result of the integration (2.16) with infinite integration boundaries cancels with the factor $1/(\mathrm{i}\lambda)$ and a plane wave remains.

The other extreme is a single pinhole when a lone spherical wave originates from the pinhole. Since this is in conflict with the condition that apertures need to be large compared to the wavelength we will use the term *pinhole* if an aperture is small enough to describe the diffraction pattern for small diffraction angles (2.16) by a single spherical wave originating from the aperture. A pinhole diameter of a few wavelengths is sufficient to fulfil this condition.

One can now regard $1/(\mathrm{i}\lambda)$ as a weighting function of the spherical wave $\frac{V_0}{r}\exp(\mathrm{i}kr)$. Then, the amplitude of the spherical wave is reduced by $1/\lambda$ compared to that of the incident wave $V(\boldsymbol{\xi})$. This is intuitively comprehensible since long wavelengths are diffracted more strongly and the amplitude decreases over a given area in the plane of observation. The phase of $V(\boldsymbol{\xi})$ is shifted by $-\pi/2$ (since

[3] The term *wave front* describes the virtual surface of the same phase of a propagating wave.

$1/\mathrm{i} = \exp(-\mathrm{i}\pi/2)$) by the very process of diffraction at the pinhole. This is curious in a historical sense because Fresnel assumed that the secondary sources have this property in order to accurately calculate diffraction patterns. The deduction leading to the diffraction formula showed the mathematical necessity of this phase factor.

Returning to Young's experiment, we see that it represents the case when the aperture is reduced to two pinholes, and the resulting intensity distribution in plane $\boldsymbol{x}$ is calculated by adding the two spherical waves originating from the pinholes as a consequence of the general diffraction formula (2.16).

2.2.2 Fresnel Approximation

For small angles of the distance vector $\boldsymbol{r}$ with respect to the z-axis, $|\boldsymbol{x} - \boldsymbol{\xi}|/z_1$ (see Fig. 2.4), its length r can be approximated by z_1. However, for the phase kr in (2.16) the requirement is more stringent. Here, the residual of the approximation for r needs not only be relatively small but the approximation of the product kr has to be accurate within fractions of a radian. Developing r into a power series and neglecting orders higher than quadratic for $|\boldsymbol{x} - \boldsymbol{\xi}|^4/z_1^3 \ll \lambda$, yields the *quadratic Fresnel approximation*:

$$r = z_1 + \frac{|\boldsymbol{x} - \boldsymbol{\xi}|^2}{2z_1} - \ldots. \tag{2.17}$$

This approximation is of fundamental importance for Fourier optics.

Inserted in (2.16) one obtains

$$\begin{aligned} V(\boldsymbol{x}) &= \frac{1}{\mathrm{i}\lambda z_1} \quad \mathrm{e}^{\mathrm{i}kz_1} \iint_{\mathbf{A}} V(\boldsymbol{\xi})\,\mathrm{e}^{\mathrm{i}k|\boldsymbol{x}-\boldsymbol{\xi}|^2/(2z_1)}\mathrm{d}\boldsymbol{\xi} \\ &= \frac{1}{\mathrm{i}\lambda z_1} \quad \mathrm{e}^{\mathrm{i}kz_1}\,\mathrm{e}^{\mathrm{i}k|\boldsymbol{x}|^2/(2z_1)} \iint_{\mathbf{A}} V(\boldsymbol{\xi})\,\mathrm{e}^{\mathrm{i}k|\boldsymbol{\xi}|^2/(2z_1)}\,\mathrm{e}^{-\mathrm{i}k\boldsymbol{x}\cdot\boldsymbol{\xi}/z_1}\,\mathrm{d}\boldsymbol{\xi}. \end{aligned} \tag{2.18}$$

The two exponential functions before the integral describe a constant phase kz_1 due to the propagation between the planes, and a parabolic phase $k|\boldsymbol{x}|^2/(2z_1)$ due to the quadratic approximation for spherical waves originating from the aperture. These two functions disappear when calculating the intensity distribution $I(\boldsymbol{x}) = |V(\boldsymbol{x})|^2$ in a plane at z_1. We will omit them in the remainder of this section.

The shape of the aperture **A** is called $A(\boldsymbol{\xi})$, and it is incorporated under the integral through $V_{\mathrm{ap}}(\boldsymbol{\xi}) = A(\boldsymbol{\xi})V(\boldsymbol{\xi})$. Then, the amplitude for *Fresnel diffraction* [88] can be written as

$$V(\boldsymbol{x}) = \frac{1}{\mathrm{i}\lambda z_1} \iint_{-\infty}^{\infty} V_{\mathrm{ap}}(\boldsymbol{\xi})\mathrm{e}^{\mathrm{i}k|\boldsymbol{\xi}|^2/(2z_1)}\mathrm{e}^{-\mathrm{i}k\boldsymbol{x}\cdot\boldsymbol{\xi}/z_1}\,\mathrm{d}\boldsymbol{\xi}. \tag{2.19}$$

In the following, integration boundaries will only be given explicitly if they are finite.

At a very large distance from the aperture, i.e., for $|\boldsymbol{\xi}|^2/(2z_1) \ll \lambda$, the argument of the first exponential under the integral in (2.19) goes to zero. This approximation is called the *Fraunhofer approximation*. Then the amplitude $V(\boldsymbol{x})$ in the plane of observation and the amplitude $V_{\mathrm{ap}}(\boldsymbol{\xi})$ in the aperture are linked through a Fourier transform.

There is an apparent contradiction when considering the case of an infinitely large aperture. In the discussion following (2.16) we stated that an aperture of infinite extent lets a plane wave travel unaffected by any aperture boundaries into the plane of observation. This means that the diffraction pattern of the infinitely large aperture is a plane wave of infinite extent. This is in contrast to the diffraction pattern in Fraunhofer approximation that is computed by Fourier transforming the infinite aperture formally obtaining the Dirac δ-*function* as an intensity distribution (see Sect. A.1). However, the Fraunhofer approximation $|\boldsymbol{\xi}|^2/(2z_1) \ll \lambda$ requires not only a large distance z_1 but also a very small angular size $|\boldsymbol{\xi}|/z_1 \ll \sqrt{\lambda/z_1}$ of the aperture. If the aperture is of infinite extent the Fraunhofer approximation cannot be applied.

At the other extreme, we discussed in Sect. 2.2.1 that a very small aperture, a pinhole, cannot be infinitely small but needs a diameter of a few wavelengths. Formally such an aperture can be expressed by a δ-function since we investigate only small diffraction angles within the Fresnel approximation.

Returning to apertures of finite size, we regard an aperture with a diameter of 10 cm and an observing wavelength of 2.2 μm. Then, the Fresnel approximation is valid for distances $z_1 \gg 4$ m while in Fraunhofer approximation the distance has to be much larger than 2,500 m in order to properly compute the diffraction pattern with a Fourier transform (see Fig. 2.5).

Using a lens (or a parabolic mirror), Fraunhofer diffraction can be observed closer to the aperture. For our purpose, it is sufficient to regard the lens as a simple focusing element that converts an incoming plane wave into a spherical wave converging in the focus. The distance between the lens and the focus is called the focal length.

In quadratic approximation, the lens can be described by

$$L(\boldsymbol{\xi}) = \mathrm{e}^{-\mathrm{i}k|\boldsymbol{\xi}|^2/(2F)}, \tag{2.20}$$

with F the focal length [88]. The ideal lens has no absorption, hence $|L(\boldsymbol{\xi})| = 1$, it affects the phase of the incoming wave by adding a parabolic phase term $k|\boldsymbol{\xi}|^2/(2F)$, and it is infinitely thin. A lens in this definition is called a *thin lens*, and a mirror a *thin mirror*.

Inserting $L(\boldsymbol{\xi})$ in (2.19) one obtains

$$V(\boldsymbol{x}) = \frac{1}{\mathrm{i}\lambda z_1} \iint V_{\mathrm{ap}}(\boldsymbol{\xi}) \mathrm{e}^{-\mathrm{i}k|\boldsymbol{\xi}|^2/(2F)} \, \mathrm{e}^{\mathrm{i}k|\boldsymbol{\xi}|^2/(2z_1)} \, \mathrm{e}^{-\mathrm{i}k\boldsymbol{x}\cdot\boldsymbol{\xi}/z_1} \, \mathrm{d}\boldsymbol{\xi}. \tag{2.21}$$

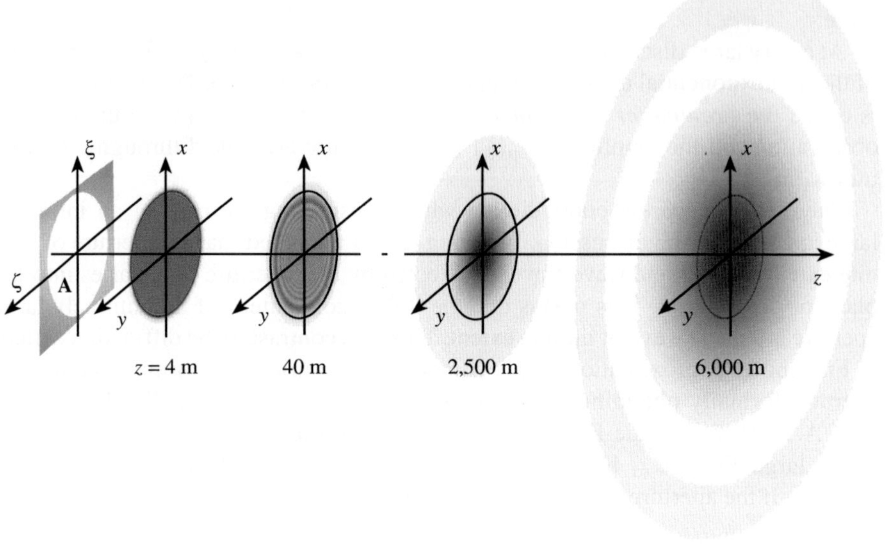

Fig. 2.5 Intensity distributions of the diffraction patterns of an aperture with 10 cm diameter and an observing wavelength of $\lambda = 2.2\,\mu$m at different distances. 4 m after the aperture, the intensity distribution is very similar to the geometrical shadow pattern of the aperture that is indicated as a *circle* in each diffraction pattern. At about 40 m the Fresnel approximation describes the situation properly showing typical ringing effects inside the aperture and a fast decrease to zero outside. At about 2,500 m the intensity distribution resembles the Fraunhofer diffraction pattern, which is fully developed at 6,000 m. The diameter of the first diffraction ring is 12 cm at 2,500 m and 30 cm at 6,000 m. Note that this diameter is never smaller than the diameter of the aperture. For an observing wavelength of 10 μm the Fresnel approximation is a proper approximation at 21 m (instead of 36 m), and the Fraunhofer diffraction pattern appears at 600 m (instead of 6,000 m). (Courtesy R. Wilhelm)

In the focal plane, with $z_1 = F$, the first two exponential functions under the integral cancel. Following common practice, we replace the spatial coordinate $\boldsymbol{x}$ in the focal plane by the angle coordinate $\boldsymbol{\alpha} = \boldsymbol{x}/F$. Then, the amplitude $V(\boldsymbol{\alpha})$ in the focal plane can be expressed as the Fourier transform of the amplitude $V_{\mathrm{ap}}(\boldsymbol{\xi})$ in the aperture [88]

$$V(\boldsymbol{\alpha}) = \frac{1}{\mathrm{i}\lambda F} \iint V_{\mathrm{ap}}(\boldsymbol{\xi})\, \mathrm{e}^{-\mathrm{i}k\boldsymbol{\alpha}\cdot\boldsymbol{\xi}}\, \mathrm{d}\boldsymbol{\xi}\,. \tag{2.22}$$

With this formula we can describe the situation in a telescope. The incoming plane wave stems from a point-like star approximately at infinity, i.e., the phase $\varphi(\boldsymbol{\xi})$ of the complex amplitude is zero, and $V(\boldsymbol{\xi}) = V_0 = \text{constant}$ so that $V_{\mathrm{ap}}(\boldsymbol{\xi}) = V_0 A(\boldsymbol{\xi})$. The light is diffracted at the aperture, and the telescope optics form the Fraunhofer diffraction pattern of the aperture in the telescope focal plane with the intensity distribution given by $I(\boldsymbol{\alpha}) = |V(\boldsymbol{\alpha})|^2$. Thus, this intensity distribution is the diffraction limited image of the point-like star, and the star is called *unresolved*.

In the theory of linear systems, the diffraction pattern represents the response of the optical system to an impulse, in this case the approximately point-like intensity distribution of an unresolved star. This response is called the *point-spread function* (PSF) of the optical system. The PSF is dimensionless and describes the spread of the intensity in the focal plane.

Aberrations of the telescope optics are incorporated in the phase $\varphi(\boldsymbol{\xi})$ of the aperture function $A(\boldsymbol{\xi})$. Then, the subsequent PSF describes no longer the diffraction limited but the aberrated image of the point source.

NB 2. *Apart from placing the lens directly in the aperture plane, we can also place the aperture at a distance z_0 in front of the lens. Then, the amplitude in the back focal plane of the lens can be written as [88]*

$$V(\boldsymbol{\alpha}) = \frac{1}{\mathrm{i}\lambda F} \, \mathrm{e}^{\mathrm{i}k\left(1-\frac{z_0}{F}\right)|\boldsymbol{\alpha}|^2 F/2} \iint V_{\mathrm{ap}}(\boldsymbol{\xi}) \, \mathrm{e}^{-\mathrm{i}k\boldsymbol{\alpha}\cdot\boldsymbol{\xi}} \, \mathrm{d}\boldsymbol{\xi} \,. \tag{2.23}$$

The amplitude in the back focal plane is again given as a Fourier transform of the amplitude $V_{\mathrm{ap}}(\boldsymbol{\xi})$ in the aperture, multiplied by a quadratic phase term. Following (2.18), we argued that this phase term needs not be considered since we are mainly interested in intensity distributions.

It is interesting, however, to discuss how this phase term is affected by the distance z_0 from the lens. We see that for $z_0 = F$, i.e., with the aperture in the front focal plane of the lens, the phase term disappears, and we have a Fourier connection between the amplitude in the front focal plane and in the back focal plane without any further approximation.

Thus, the position of the aperture with respect to the lenses in optical system does not affect the Fourier connection and, unless choosing extreme parameters for focal length and distance, the quadratic phase term can be disregarded.

2.2.3 The Airy Disk

As shown in the preceding section, the response of an optical system to a unresolved source is never point-like but enlarged by the diffraction at the telescope aperture. We now discuss the circular aperture as the most common case of a telescope aperture.

The circular aperture is described by the circ-function that is defined as $\mathrm{circ}(\frac{|\boldsymbol{\xi}|}{R}) = 1$ if $|\boldsymbol{\xi}| \leq R$ and 0 elsewhere. A circular aperture with diameter D is then given by $A(\boldsymbol{\xi}) = \mathrm{circ}\left(\frac{|\boldsymbol{\xi}|}{D/2}\right)$, with the area of the circular aperture called $A_0 = \pi(D/2)^2$. $A(\boldsymbol{\xi})$ is illuminated by a point source at infinity with $V(\boldsymbol{\xi}) = V_0 =$ const. in the aperture plane.

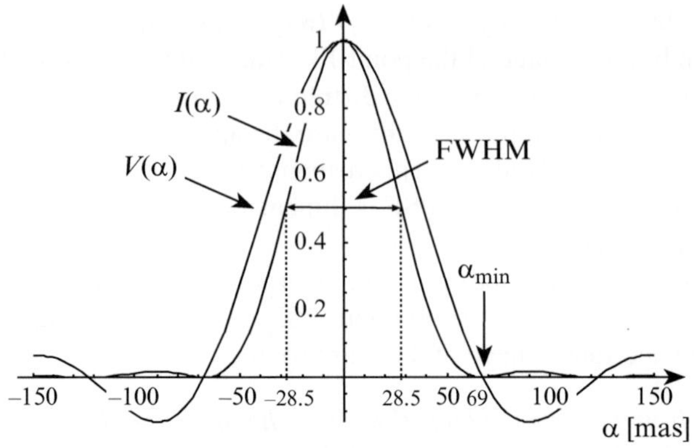

Fig. 2.6 Profiles of the amplitude $V(\boldsymbol{\alpha})$, normalised to unity, and of the subsequent intensity distribution $I(\boldsymbol{\alpha})$ of the point-spread function (PSF) for a circular aperture with a diameter of 8 m, and an observing wavelength of 2.2 μm. $V(\boldsymbol{\alpha})$ has the form of the Besinc-function. The first minimum of $I(\boldsymbol{\alpha})$ at $\alpha_{\min} = 1.22\,\lambda/D = 69$ mas (we use the notation mas for milliarcsec), and the full width at half maximum (FWHM) of $\alpha_{\text{FWHM}} = \lambda/D = 57$ mas are indicated. For the computation of the angle in mas, the formula $\alpha_{\text{FWHM}}/\text{mas} = 206\frac{\lambda/\mu\text{m}}{D/\text{m}}$ is useful, and for the FWHM as a length in the focal plane, we can use $x_{\text{FWHM}} = f_{\#}\lambda$, with $f_{\#}$ the *f-number* of the optical system defined as $f_{\#} = F/D$. Note that the diffraction rings in the intensity distribution at $\alpha \approx \pm 90$ mas reach only about 2% of the peak intensity

Using (2.22), the diffraction limited amplitude in the focal plane can be written as

$$\begin{aligned} V(\boldsymbol{\alpha}) &= \frac{V_0}{\mathrm{i}\lambda F} \iint \operatorname{circ}\left(\frac{|\boldsymbol{\xi}|}{D/2}\right) \mathrm{e}^{-\mathrm{i}k\boldsymbol{\alpha}\cdot\boldsymbol{\xi}} \mathrm{d}\boldsymbol{\xi} \\ &= \frac{V_0}{\mathrm{i}\lambda F} A_0 \operatorname{Besinc}(k\alpha D/2), \end{aligned} \tag{2.24}$$

with $\alpha = |\boldsymbol{\alpha}|$.

The result of the Fourier transform (2.24) of the circ-function is called the *Besinc-function* (see Fig. 2.6) defined as $\operatorname{Besinc}(x) := 2J_1(x)/x$, with $\mathrm{J}_1(x)$ the first order Bessel function. The Besinc-function is a real function that is point-symmetric with respect to $x = 0$. With A_0, the area of the circular aperture, and λF both having the dimension of *length*^{2}, the quotient is dimensionless and the amplitude $V(\boldsymbol{\alpha})$ has the same dimension as V_0, the amplitude of the incoming plane wave.

Splitting the amplitude $V(\boldsymbol{\alpha})$ in (2.24) into modulus and phase, one finds that the modulus is simply the absolute value of the Besinc-function, and the phase takes on distinct values of $-\pi/2$, with $1/\mathrm{i} = \exp(-\mathrm{i}\pi/2)$, when the Besinc-function is positive and $+\pi/2$ when it is negative. A phase function in the focal plane jumping between values of $+\pi/2$ and $-\pi/2$ means that the wave front in the focal plane is not curved but essentially flat with jumps of π.

This represents the transition from a converging spherical wave immediately after the optical system with the centre of the sphere in the focal plane, and a diverging spherical wave after the focal plane. However, despite the flat wave front the amplitude in (2.24) is not a plane wave because of the phase discontinuities at the phase jumps and the modulus following a Besinc-function. This causes the wave to expand into a diverging spherical wave in the course of the propagation. Further calculations of the phase just before and after the focal plane show that there is a phase jump from $-\pi$ to 0 when moving along the optical axis through the focal plane [19].

The intensity distribution of the diffraction limited PSF of telescopes with circular aperture is called the *Airy disk* [19]. The squared modulus of the amplitude $V(\boldsymbol{\alpha})$ yields the Airy disk as

$$I(\boldsymbol{\alpha}) = V(\boldsymbol{\alpha})V^*(\boldsymbol{\alpha}) = \frac{V_0^2}{(\lambda F)^2} A_0^2 \,\mathrm{Besinc}^2(k\alpha D/2). \tag{2.25}$$

The dimensionless PSF describing the shape of the intensity can now be defined as:

$$\mathrm{PSF}(\boldsymbol{\alpha}) := \frac{I(\boldsymbol{\alpha})}{V_0^2} = \frac{1}{(\lambda F)^2} A_0^2 \,\mathrm{Besinc}^2(k\alpha D/2). \tag{2.26}$$

As long as we discuss monochromatic light, the wavelength appears as a constant parameter in the formula, and we write the PSF as a function of $\boldsymbol{\alpha}$ only. Later, in Sect. 3.3.3, when we will investigate the imaging process in polychromatic illumination we will write $\mathrm{PSF}(\boldsymbol{\alpha}, \nu)$.

The first minimum of the PSF, the first *dark ring* is at $\alpha_{\mathrm{min}} = 1.22\,\lambda/D$. 84% of the total intensity are confined to the inside of the first dark ring. For a binary star with a separation α_{min}, the resulting image, which is the sum of two individual Airy disks, shows a local minimum between the peaks of the Airy disk (see Fig. 2.7). Therefore the two stars of the binary can be identified as individual objects in the

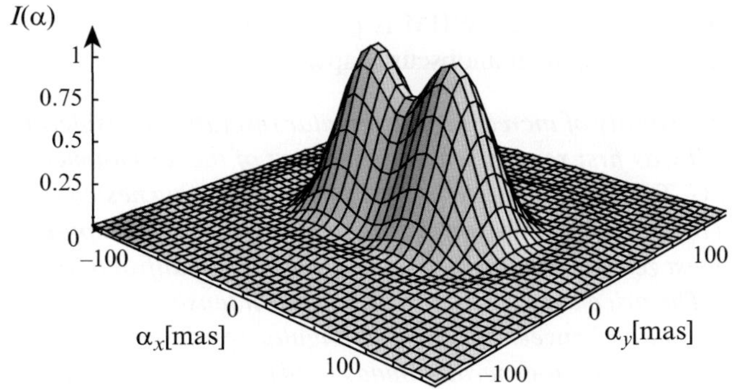

Fig. 2.7 The joint intensity distribution in the diffraction limited image of a binary star separated by the Rayleigh limit of resolution of 69 mas for an aperture diameter of 8 m and an observing wavelength of 2.2 μm. The local minimum between the two peaks is 19% lower than the intensity of the peaks

image. This criterion of angular resolution, when the smallest resolvable angle is α_{min}, is called the *Rayleigh criterion* of resolution of a telescope [19].

This definition is, however, somewhat arbitrary. Why should a local minimum with a larger value than 84% between the two peaks not be sufficient to resolve them? Therefore, another resolution criterion is often used. This criterion is related to the *full width at half maximum (FWHM)*, i.e., the diameter of the Airy disk at half its maximum intensity, which is in good approximation $\alpha_{\text{FWHM}} = \lambda/D$ (see Fig. 2.6). About 50% of the total intensity is within the FWHM of the Airy disk.

If two stars are separated by α_{FWHM} the intensity distribution between the two peaks of the Airy disks is almost flat so that the FWHM represents the limit to identify the peaks as separate. One could, however, argue that a much smaller separation could be spotted by precisely measuring the deviation of the resulting intensity distribution from the perfectly circular shape of an individual Airy disk. There is virtually no limit for the resolution unless the reality is taken into account with measurements that are never perfect [56, 87]. Implicitly, we have assumed a certain measurement quality in the discussion of Rayleigh criterion and FWHM that is sufficient for the identification of local intensity maxima, which can then be attributed to maxima of the object intensity.

Thus, the Rayleigh criterion has its virtue in providing a conservative estimate for the angular resolution that is linked to a prominent feature, the first minimum, of the Airy disk. We will see in Chap. 3 that the matter of angular resolution is more complex depending on many parameters of the imaging situation.

It should be noted that both quantities, α_{min} and α_{FWHM}, depend on the shape of the telescope aperture that usually has a central obscuration due to the telescope design with the secondary mirror centrally above the primary mirror. For comparison, Fig. 2.8 displays three PSF: without central obscuration (black curve), with a central obscuration of 40% of the telescope aperture (red) and with an annular aperture (blue). The latter represents the theoretical limit with close to zero transmission. One can see that for the practical cases between 0 and 40% obscuration the PSF varies only slightly. While the position of the first minimum is reduced from $1.22\,\lambda/D$ to about λ/D, the FWHM is practically unchanged. Therefore we will always discuss the case of an unobscured aperture.

NB 3. *The possibility of increasing the angular resolution considerably beyond the Rayleigh limit was first recognised in the context of the development of microwave antennas by G. Toraldo di Francia in 1952 [58]. Combining nested ring apertures he showed that the intensity in the first diffraction rings could be suppressed so that the PSF was almost zero around its core up to a very bright diffraction ring limiting the field of view. The price to pay was a reduced peak intensity of the central core. Soon after, the connection between increase of angular resolution and decrease of peak intensity was analysed on general grounds [144], and more recently a quantitative analysis was presented [205].*

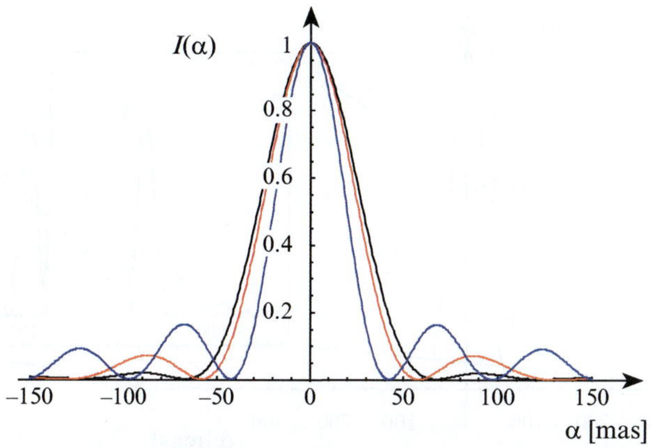

Fig. 2.8 Examples of PSF with different central obscuration and a circular 8-m aperture for an observing wavelength of 2.2 μm. In *black* the PSF without central obscuration is displayed, as in Fig. 2.6. The PSF with a circular central obscuration of 40% of the telescope aperture is shown in *red*, and the PSF of an infinitely thin annular aperture is shown in *blue*. The first minimum of the PSF at $1.22\,\lambda/D = 69$ mas (*black curve*) is reduced to about $\lambda/D = 57$ mas (*red curve*) and further to $0.76\,\lambda/D = 43$ mas (*blue curve*). For comparison, all PSF are normalised to unity

Polychromatic Airy Disk

So far, we have treated the case of monochromatic light. We now discuss the case of polychromatic illumination before moving on to the coherence function in the next section. The polychromatic intensity distribution $I(\boldsymbol{\alpha})$ of an Airy disk can be calculated as the integral of the monochromatic Airy disks (2.25) over the spectral band. Using the atmospheric K-band (2.2±0.2 μm), the result of the K-band Airy disk as well as the monochromatic Airy disk are displayed in Fig. 2.9. Around the peak the difference is barely noticeable. The ringing, however, is reduced in the minima – not reaching zero any more – as well as in the maxima – being lower than in the monochromatic case – both due to the wavelength dependence of the minima and maxima positions. Beyond the third sidelobe this effect wipes out any diffraction rings in the polychromatic Airy disk, albeit at extremely low light levels.

The influence of the wavelength on the diffraction pattern was much more prominent in Young's experiment (Fig. 2.3) when the contrast of the fringe pattern was reduced considerably. However, the physical process, namely the wavelength dependent diffraction, is the same in both cases.

This exercise of moving from monochromatic to polychromatic illumination by integrating over the spectral band will come back to us when treating the diffraction process in a stellar interferometer. There the situation is more complex since the properties of spatial coherence need also to be treated.

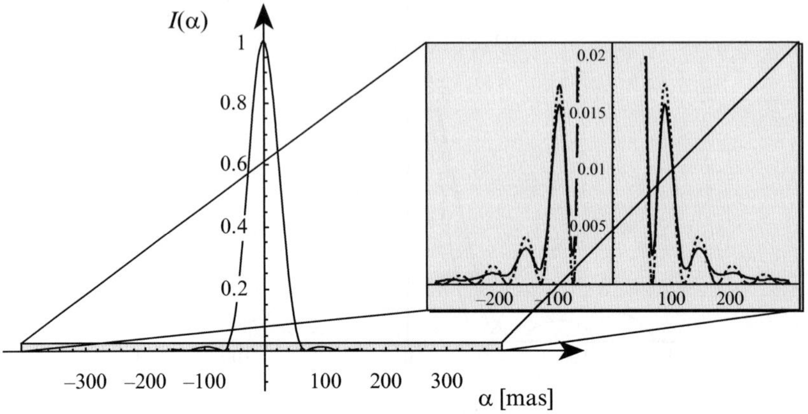

Fig. 2.9 The Airy disk of an 8-m telescope, observing the full K-band, 2.2±0.2 µm. In the *grey box*, the Airy disk is displayed with an enlarged intensity scale. Compared to the monochromatic Airy disk displayed by the *dashed line*, the sharp monochromatic minima disappear in the polychromatic Airy disk since their position is a function of wavelength

Scalar Diffraction Theory: Summary

The Rayleigh–Sommerfeld diffraction formula (2.16) describes the propagation of the amplitude $V(\boldsymbol{\xi})$ in space. Reducing the discussion to the propagation between two planes – the aperture plane and the plane of observation – in Fresnel approximation (2.17), the diffraction formula can be simplified considerably.

In addition, a lens with focal length F is introduced in the aperture plane. Then, the diffraction formula can be simplified further (2.21), and the Fraunhofer diffraction pattern of the aperture that usually appears at a very large distance can now be found in the back focal plane of the lens.

For Fraunhofer diffraction, the amplitude $V_{\text{ap}}(\boldsymbol{\xi})$ in the aperture and the amplitude $V(\boldsymbol{\alpha})$ in the focal plane are linked through a Fourier transform:

$$V(\boldsymbol{\alpha}) = \frac{1}{\mathrm{i}\lambda F} \iint V_{\text{ap}}(\boldsymbol{\xi})\, \mathrm{e}^{-ik\boldsymbol{\alpha}\cdot\boldsymbol{\xi}}\, \mathrm{d}\boldsymbol{\xi}. \tag{2.22}$$

$V_{\text{ap}}(\boldsymbol{\xi})$ is the product of the amplitude of the incoming wave $V(\boldsymbol{\xi})$ with the aperture function $A(\boldsymbol{\xi})$, the latter incorporating the shape of the aperture in its modulus and optical aberrations in its phase.

If the aperture is illuminated by a plane wave, $V(\boldsymbol{\xi}) = V_0 \exp(ikz)$, from a point source at a very large distance (e.g., an unresolved star), then the intensity distribution $I(\boldsymbol{\alpha}) = |V(\boldsymbol{\alpha})|^2$ in the focal plane is the diffraction limited image of the point source. In telescopes, the diffraction limited intensity distribution is called the Airy disk. For a circular aperture of diameter D the

Airy disk is given by

$$I(\boldsymbol{\alpha}) = |V(\boldsymbol{\alpha})|^2 = \frac{V_0^2}{(\lambda F)^2} A_0^2 \, \mathrm{Besinc}^2(k\alpha D/2). \tag{2.25}$$

Treating the optical system as a linear system, it is customary to introduce the dimensionless *point-spread function* (PSF) describing the form of the Airy disk. The PSF is the squared modulus of the Fourier transform of the telescope aperture $A(\boldsymbol{\xi})$ (2.24) linked to the intensity distribution $I(\boldsymbol{\alpha})$ in the focal plane through

$$\mathrm{PSF}(\boldsymbol{\alpha}) = \frac{I(\boldsymbol{\alpha})}{V_0^2} = \frac{1}{(\lambda F)^2} A_0^2 \, \mathrm{Besinc}^2(k\alpha D/2). \tag{2.26}$$

The Rayleigh criterion of angular resolution defines the resolution limit of a telescope as the angular distance between the peak of the Airy disk (or the PSF) and its first minimum, $\alpha_{\mathrm{min}} = 1.22\,\lambda/D$, with D the diameter of the telescope aperture. Stars that are separated by this distance can still be identified as individual objects, i.e., they can be resolved. Another resolution criteria is given by the full width at half maximum (FWHM) that is approximately given by $\alpha_{\mathrm{FWHM}} = \lambda/D$. The latter is easier to measure in real astronomical images, the former slightly more conservative measure has the historical merit of having introduced diffraction theory into observations with astronomical telescopes.

2.3 The Coherence Function

The nature of coherence was discussed in text books on optics [19], on statistical optics [87], and, very specialised, in books on coherence theory [147, 148]. The coherence function was introduced as a measure for the coherence of the light. If the light has good coherence, the value of the coherence function is high, and interference phenomena like the fringes in Young's experiment show a good contrast.

We will introduce the concept of the coherence function more formally in this section, we will present its different flavours, and we will provide the necessary detail for observing with stellar interferometers. By and large, we will follow the formalism laid out in [87] and [148].

Starting with this section, two-dimensional integrals will be denoted by a single integral sign. The vector as integration variable shall indicate the integration over two dimensions.

2.3.1 Varieties of the Coherence Function

Before defining the coherence function, the nature of the random process leading to coherence and to incoherence needs to be discussed. So far, the optical disturbance v has been regarded as a deterministic signal throughout the propagation and diffraction process, forming a plane or a spherical wave. Now, approaching large celestial bodies emitting (mostly) thermal radiation, their light cannot be regarded as monochromatic and only approximately as a plane wave. Very close to the surface of a star, it is almost impossible to define a wave front and a direction of propagation. At a very large distance from the star, a point is a good approximation for its shape, and a plane wave describes the situation rather well.

In any moment during the propagation process, the optical disturbance, fed by light from individual, independently radiating points on the star, takes on random values that fluctuate typically at timescales $1/\nu$, the reciprocal of the average frequency. If one could take a series of snapshots with femto second exposure time, the pictures would all look different. However, these fluctuations average out over time intervals longer than $1/\nu$. Thus, snapshots with longer exposure time would all look similar.

We will regard the individual wave fronts as possible realizations or members of the ensemble of the random process, and the optical disturbance $v(\boldsymbol{x}, t)$ as the random variable. Figure 2.10 illustrates the situation when the wave fronts, composed by values of the optical disturbance, are individual realisations of the random process.

We will make two assumptions on the random process that will make our life much easier:

First, we assume that the random process is statistically stationary in time. This means that the statistical properties are the same over the ensemble, i.e., that the average is independent of the absolute moment in time t when it is taken and that the correlation only depends on the time difference $t_1 - t_2$.

Second, we assume that the statistics over one particular wave front at a given moment is the same as the statistics at a given point waiting a "long" time. In other words, the statistics, e.g., the average, over the complete wave front as an individual realization of the random process can be replaced by the average over many different realizations that appear in temporal succession. A "long" time providing the average over a sufficient number of realisations is defined as $T \gg 1/\nu$, i.e., much longer than the oscillation of the electromagnetic wave.

If these two conditions are met, the process is called *ergodic* and the ensemble average can be replaced by the time average. Figure 2.10 illustrates the situation.

Incidentally, the measurement with an optical detector provides a time average $< v(\boldsymbol{x}, t)v^*(\boldsymbol{x}, t) >$ (see (2.5)) that is sufficiently long to replace the ensemble average of the intensity. In Sect. 2.4 we will show how optical detectors – although only measuring the intensity – can be used to measure the coherence function, which is the correlation function of the optical disturbance as we shall see in the following discussion.

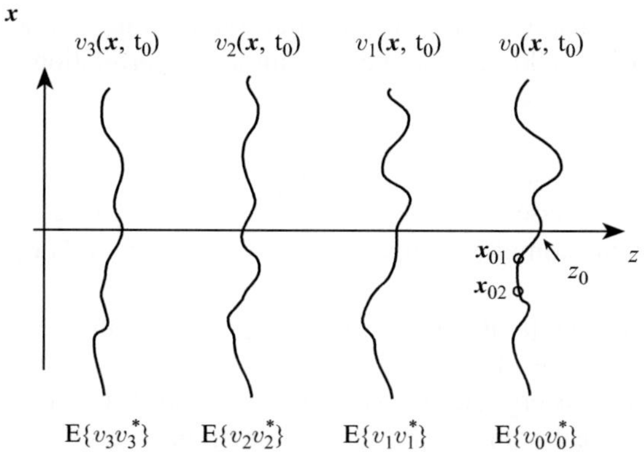

Intensity: $I(\boldsymbol{x}_{01}) = < v_{z_0}(\boldsymbol{x}_{01}, t)\, v^*_{z_0}(\boldsymbol{x}_{01}, t) > = E\{v_0(\boldsymbol{x}, t_0)\, v_0^*(\boldsymbol{x}, t_0)\}$

Mutual Coherence Function (MCF):
$\Gamma(\boldsymbol{x}_{01}, \boldsymbol{x}_{02}, \tau) = < v_{z_0}(\boldsymbol{x}_{01}, t+\tau)\, v^*_{z_0}(\boldsymbol{x}_{02}, t) > = E\{v_0(\boldsymbol{x}_1, t_0+\tau)\, v_0^*(\boldsymbol{x}_2, t_0)\}$

Fig. 2.10 Illustration of the propagation of light as an ergodic random process. The wave fronts of optical disturbances v_i are displayed at a given moment in time, $t = t_0$, and at different locations on the z-axis. The wave fronts were emitted at different times from an extended source at a large distance, having, thus, travelled different distances in the z-direction at $t = t_0$. Each wave front is an individual realisation of the random emission process. Due to the ergodicity of the random process the ensemble average $E\{v_0(\boldsymbol{x}, t_0)v_0^*(\boldsymbol{x}, t_0)\}$ "across" the wave front for instance at $z = z_0$ is identical to the time average $< v_{z_0}(\boldsymbol{x}_{01}, t)v_{z_0}^*(\boldsymbol{x}_{01}, t) >$ of different wave fronts passing through $z = z_0$ if the integration time is sufficiently long, i.e., $T \gg 1/\nu$. The correlation function of optical disturbances, the MCF $\Gamma(\boldsymbol{x}_{01}, \boldsymbol{x}_{02}, \tau)$, is computed as a time average $< v_{z_0}(\boldsymbol{x}_{01}, t + \tau)$ $v_{z_0}^*(\boldsymbol{x}_{02}, t) >$, (2.27), depending only on the time difference τ. The ensemble average over pairs of points $\boldsymbol{x}_1, \boldsymbol{x}_2$ across the wave front at $z = z_0$ would yield the same result

Unlike the time averaged intensity, the time average of the optical disturbance $< v(\boldsymbol{x}, t) >$ and, thus, its ensemble average $E\{v_i\}$ are zero since the optical disturbance fluctuates around zero.

The Mutual Coherence Function (MCF)

The correlation function,[4] also called the *second order moment*, of the optical disturbances v at positions $\boldsymbol{x}_1$ and $\boldsymbol{x}_2$ in a plane at times $t + \tau$ and t is called the *mutual coherence function* (MCF) [87, 148]:

$$\begin{aligned}\Gamma(\boldsymbol{x}_1, \boldsymbol{x}_2, \tau) &= \lim_{T\to\infty} \frac{1}{2T} \int_{-T}^{T} v(\boldsymbol{x}_1, t + \tau) v^*(\boldsymbol{x}_2, t)\, \mathrm{d}t \\ &= < v(\boldsymbol{x}_1, t + \tau) v^*(\boldsymbol{x}_2, t) > . \qquad (2.27)\end{aligned}$$

[4] Here, the correlation function $< v_1 v_2 >_{\text{cor}}$ is identical to the covariance function $< v_1 v_2 >_{\text{cov}} =$ $< v_1 v_2 >_{\text{cor}} - < v >^2$, as the optical disturbance has zero mean $< v >= 0$.

Remembering that the optical disturbance is proportional to one component of the electrical field vector, one could say that we determine the correlation of two electrical field vectors at two points in space (e.g., at two telescopes) and two moments in time.

The MCF is dimensionless. Depending on the optical disturbance v, the MCF can be complex with the phase denoted by $\phi_{\mathrm{MCF}}(\boldsymbol{x}_1, \boldsymbol{x}_2, \tau)$. Using (2.27), it can easily be shown that the complex conjugate of the MCF is $\Gamma^*(\boldsymbol{x}_1, \boldsymbol{x}_2, \tau) = \Gamma(\boldsymbol{x}_2, \boldsymbol{x}_1, -\tau)$.

The time average over the period T at two fixed points $\boldsymbol{x}_1$ and $\boldsymbol{x}_2$ replaces the ensemble average $\mathrm{E}\{v(\boldsymbol{x}_1, t + \tau)v^*(\boldsymbol{x}_2, t)\}$ "across" the wave front at a fixed time t since the random process is ergodic as discussed above (see Fig. 2.10).

The MCF at two identical points at the same moment provides the time average of the product of the optical disturbances, which is the intensity $I(\boldsymbol{x})$ as defined in (2.5):

$$I(\boldsymbol{x}) = \Gamma(\boldsymbol{x}, \boldsymbol{x}, 0) = \lim_{T\to\infty} \frac{1}{2T} \int_{-T}^{T} v(\boldsymbol{x}, t) v^*(\boldsymbol{x}, t)\, \mathrm{d}t. \tag{2.28}$$

The MCF inherently varies for different pairs of points if the intensity at these points varies. In order to obtain the pure correlation of optical disturbances, i.e., the probability to measure the same value, the MCF has to be calibrated by the intensities at the individual coordinates $\boldsymbol{x}_1$ and $\boldsymbol{x}_2$. We call this function the *degree of coherence* or the correlation coefficient of the optical disturbances:

$$\gamma(\boldsymbol{x}_1, \boldsymbol{x}_2, \tau) = \frac{\Gamma(\boldsymbol{x}_1, \boldsymbol{x}_2, \tau)}{\sqrt{\Gamma(\boldsymbol{x}_1, \boldsymbol{x}_1, 0)\, \Gamma(\boldsymbol{x}_2, \boldsymbol{x}_2, 0)}}, \tag{2.29}$$

which is in general a complex function with $0 \le |\gamma(\boldsymbol{x}_1, \boldsymbol{x}_2, \tau)| \le 1$.

The dependence of the MCF and of the degree of coherence on the time difference τ and on the spatial coordinates $\boldsymbol{x}_1$ and $\boldsymbol{x}_2$ characterises two different aspects, the temporal and the spatial coherence of the light. In Sect. 2.1.2, we gave an introduction to temporal coherence. In this section, we will treat both cases under the umbrella of a general coherence theory.

The Mutual Spectral Density Function (MSDF)

We introduced the spectral amplitude $V(\boldsymbol{x}, \nu)$ of the harmonic wave in Sect. 2.1.1 as a time independent function so that the Helmholtz equation (2.14) could be used to describe the propagation of light. The spectral amplitude $V(\boldsymbol{x}, \nu)$ is related to the optical disturbance $v(\boldsymbol{x}, t)$ through a temporal Fourier transform (2.7).

Here, in the domain of coherence functions, we follow the same path. We define the *mutual spectral density function* (MSDF) [148], also called the *cross spectral density function* [87], as the correlation function of the spectral amplitudes:

$$\hat{\Gamma}(\boldsymbol{x}_1, \boldsymbol{x}_2, \nu) := \lim_{T\to\infty} \frac{1}{2T} E\left\{V_T(\boldsymbol{x}_1, \nu) V_T^*(\boldsymbol{x}_2, \nu)\right\}, \tag{2.30}$$

where the subscript T indicates that the spectral amplitude for the interval $-T$ to T is taken. The dimension of the MSDF is Hz^{-1}.

Thus, the MSDF describes the spatial coherence at the frequency ν as a correlation of spectral amplitudes between two individual points $\boldsymbol{x}_1$ and $\boldsymbol{x}_2$.

The MCF is related to the MSDF through a Fourier transform,

$$\Gamma(\boldsymbol{x}_1, \boldsymbol{x}_2, \tau) = \int \hat{\Gamma}(\boldsymbol{x}_1, \boldsymbol{x}_2, \nu) \mathrm{e}^{-\mathrm{i}2\pi\nu\tau}\, \mathrm{d}\nu, \tag{2.31}$$

keeping in mind that the MSDF being deduced from the spectral amplitude is zero for negative frequencies.

In general, the MSDF is a complex quantity and its phase will be denoted by $\hat{\phi}(\boldsymbol{x}_1, \boldsymbol{x}_2, \nu)$. The complex conjugate of the MSDF is given by $\hat{\Gamma}^*(\boldsymbol{x}_1, \boldsymbol{x}_2, \nu) = \hat{\Gamma}(\boldsymbol{x}_2, \boldsymbol{x}_1, \nu)$.

We will use the MSDF to describe the propagation of the coherence properties in space and through optical systems. This is in analogy to the scalar diffraction theory when the propagation of the spectral amplitude V was discussed.

The MSDF at two identical points defines the spectral intensity:

$$I(\boldsymbol{x}, \nu) := \hat{\Gamma}(\boldsymbol{x}, \boldsymbol{x}, \nu) = \lim_{T\to\infty} \frac{1}{2T} E\left\{V(\boldsymbol{x}, \nu) V^*(\boldsymbol{x}, \nu)\right\}. \tag{2.32}$$

Using the connection between MCF and MSDF in (2.31) we find that the intensity $I(\boldsymbol{x})$ can be written as

$$I(\boldsymbol{x}) = \Gamma(\boldsymbol{x}, \boldsymbol{x}, 0) = \int \hat{\Gamma}(\boldsymbol{x}, \boldsymbol{x}, \nu)\, \mathrm{d}\nu = \int I(\boldsymbol{x}, \nu)\, \mathrm{d}\nu\,, \tag{2.33}$$

which is the integral of the spectral intensity $I(\boldsymbol{x}, \nu)$ over the frequency band. This expresses in the language of coherence functions what we found under very general assumptions in (2.8).

In the following, we will call the intensity $I(\boldsymbol{x})$ the *polychromatic* or *white-light intensity* in order to clearly distinguish it from the spectral intensity $I(\boldsymbol{x}, \nu)$.

The Self-Coherence Function

The MCF $\Gamma(\boldsymbol{x}, \boldsymbol{x}, \tau)$ at two identical points and arbitrary τ is called the *self-coherence function* [87, 148]. With (2.31) it can be written as the Fourier transform of the spectral intensity,

$$\Gamma(\boldsymbol{x},\boldsymbol{x},\tau) = \int \hat{\Gamma}(\boldsymbol{x},\boldsymbol{x},\nu)\mathrm{e}^{-\mathrm{i}2\pi\nu\tau}\,\mathrm{d}\nu = \int I(\boldsymbol{x},\nu)\mathrm{e}^{-\mathrm{i}2\pi\nu\tau}\,\mathrm{d}\nu. \tag{2.34}$$

The self-coherence function is a measure for the temporal coherence of the light source. If the spectral intensity distribution $I(\boldsymbol{x},\nu)$ is very narrow with respect to ν, i.e., if the spectral bandwidth $\Delta\nu$ is very small, the self-coherence function $\Gamma(\boldsymbol{x},\boldsymbol{x},\tau)$ is very wide with respect to the time difference τ. This can be quantified by defining the *coherence time* as that value τ_c when the self-coherence function is reduced significantly.

Describing, e.g., $I(\boldsymbol{x},\nu)$ by a rectangular function with a width of $\Delta\nu$, one finds that $\Gamma(\boldsymbol{x},\boldsymbol{x},\tau)$ as the Fourier transform of $I(\boldsymbol{x},\nu)$ has the form of a sinc-function, the sine function divided by its argument, with a width of about $1/\Delta\nu$. A suitable choice of the coherence time would be $\tau_c = 1/\Delta\nu$ when the self-coherence function has its first zero.

At the end of Sect. 2.1.2 we discussed the contrast of the fringe pattern in Young's experiment for different spectral bandwidths of the illuminating light. We saw that the contrast as a function of diffraction angle, which is proportional to the difference in arrival time τ, depends on the spectral bandwidth. The contrast goes down faster, i.e., at smaller τ, with increasing bandwidth $\Delta\nu$.

Here, in the context of coherence functions we deduced the mathematical relationship between spectral bandwidth and the self-coherence function. Both the contrast and the self-coherence function show the same behaviour with respect to the spectral bandwidth. This is intuitively comprehensible since the contrast in Young's experiment depends on the correlation of the interfering amplitudes. In Sect. 2.4 this will be discussed in full detail.

Very often the term *coherence length* l_c is used to express the permitted optical path difference before interference phenomena disappear. The coherence length can be defined as

$$\begin{aligned} l_c &= c\tau_c \\ &\approx c/\Delta\nu = \lambda\nu/\Delta\nu = \lambda^2/\Delta\lambda \end{aligned} \tag{2.35}$$

As a numerical example we regard again the K-band, $2.2\pm0.2\,\mu\mathrm{m}$, with $\Delta\lambda = 0.4\,\mu\mathrm{m}$ and a spectral bandwidth $\Delta\nu = 0.25\times10^{14}$ Hz. The coherence time is then approximately $\tau_c = 4\times10^{-14}$ s and the coherence length $l_c = 12\,\mu\mathrm{m}$.

Thus, the physical quantity governing the temporal coherence is the spectral width of the light source. In Sect. 2.3.4 we will see that the spatial coherence is determined by the angular size of the source, which is an entirely different physical property.

Special Cases: Coherence and Incoherence

Perfect coherence means perfect correlation of the optical disturbances throughout time and space. The requirement for coherence is that the degree of coherence (2.29) is unity, $|\gamma(\boldsymbol{x}_1,\boldsymbol{x}_2,\tau)| \equiv 1$, for all combinations of $\boldsymbol{x}_1$ and $\boldsymbol{x}_2$ in a domain D, and for arbitrary large time differences τ [87,148]. In terms of fringe contrast in Young's experiment (see Sect. 2.1.2) it is easy to comprehend that perfect coherence means a contrast of unity. We will see soon that the degree of coherence is a measure for the contrast.

The requirement for coherence is fulfilled for instance by a monochromatic plane wave with

$$v(\boldsymbol{x},t) = \int V(\boldsymbol{x},\nu)\delta(\nu-\nu_0)\exp(-\mathrm{i}2\pi\nu t)\mathrm{d}\nu = V(\boldsymbol{x})\exp(-\mathrm{i}2\pi\nu_0 t), \quad (2.36)$$

since using (2.27) we obtain

$$\begin{aligned}\Gamma(\boldsymbol{x}_1,\boldsymbol{x}_2,\tau) &= < v(\boldsymbol{x}_1,t+\tau)v^*(\boldsymbol{x}_2,t) > \\ &= V(\boldsymbol{x}_1)V^*(\boldsymbol{x}_2)\mathrm{e}^{-\mathrm{i}2\pi\nu_0\tau},\end{aligned} \quad (2.37)$$

and

$$\gamma(\boldsymbol{x}_1,\boldsymbol{x}_2,\tau) = \frac{\Gamma(\boldsymbol{x}_1,\boldsymbol{x}_2,\tau)}{\sqrt{\Gamma(\boldsymbol{x}_1,\boldsymbol{x}_1,0)\Gamma(\boldsymbol{x}_2,\boldsymbol{x}_2,0)}} = \exp(-\mathrm{i}2\pi\nu_0\tau), \quad (2.38)$$

with a modulus equal to one. In general, all solutions $V(\boldsymbol{x})$ of the Helmholtz equation (2.14) provide a perfectly coherent wave.

Strictly speaking, only monochromatic fields, having an infinitely narrow spectral line by definition, are perfectly coherent. This is not a very practical definition. The measurement of the MCF would have to last infinitely long in order to determine perfect coherence. For measurements of shorter duration a spectral line of finite width would deliver a result that is undistinguishable from perfect coherence. Laser light with a coherence length of many hundred metres is a typical example for light that is practically coherent.

We call light incoherent if the optical disturbances are perfectly uncorrelated even for infinitely small distances in space and time. Then, each point of the source radiates independently of its neighbour.

In terms of the coherence functions, the requirement for incoherence is $|\Gamma(\boldsymbol{x}_1,\boldsymbol{x}_2,\tau)| \equiv 0$ unless $\boldsymbol{x}_1 = \boldsymbol{x}_2$ and unless $\tau = 0$ [87, 148]. In incoherent illumination there would be no fringes in Young's experiment. Their contrast like the value of the MCF would be zero.

Although the MCF is infinitely narrow with respect to the coordinate difference $\boldsymbol{x}_1 - \boldsymbol{x}_2$, the spatial extension of the intensity distribution with respect to $\boldsymbol{x}$ can be very large like, e.g., the incoherently radiating surface of a star.

The Dirac δ-function seems to be ideally suited to express the MCF of an incoherent source, writing

$$\Gamma(\boldsymbol{x}_1,\boldsymbol{x}_2,\tau) = \chi I(\boldsymbol{x}_1)\delta(\boldsymbol{x}_1-\boldsymbol{x}_2)\delta(\tau), \quad (2.39)$$

where χ is a constant.

However, an incoherent light source that is defined as above cannot radiate because either the value of the MCF and, thus, the intensity is infinitely large, $\delta(\boldsymbol{x}_1 - \boldsymbol{x}_2) \to \infty$ for $\boldsymbol{x}_1 = \boldsymbol{x}_2$ and $\delta(\tau) \to \infty$ for $\tau = 0$, which is physically impossible. Or, if the values of the δ-functions are reduced to finite values, the integral over the infinitely narrow functions is zero, setting the emitted intensity to zero [13, 87, 148].

Therefore, we look for a more practical definition of incoherence. Instead of the MCF, we will discuss the MSDF $\hat{\Gamma}(\boldsymbol{x}_1, \boldsymbol{x}_2, \nu)$ that is related to the MCF through a Fourier transform (2.31). The MSDF as a function of ν contains the spectrum of the light source. If the spectrum were infinitely wide, the Fourier transform of the MSDF would be infinitely narrow with respect to τ. In practice, a very wide spectrum will be sufficient to establish incoherence if the absence of interference effects in any practical experimental configuration, e.g., zero contrast fringes in Young's experiment, is regarded as a sufficient criteria.

For the spatial characteristics, we replace $\delta(\boldsymbol{x}_1 - \boldsymbol{x}_2)$ in (2.39) by a narrow function that has non-zero values not only for $\boldsymbol{x}_1 = \boldsymbol{x}_2$ but for an interval of the order of the wavelength λ with $|\boldsymbol{x}_1 - \boldsymbol{x}_2| \leq \mathcal{O}(\lambda)$. For example, a Gaussian function with a width of $\lambda/\sqrt{2\pi}$ fulfils this requirement. The narrow Gaussian function has the same effect on the coherence function at some distance from the source as a δ-function if regarding only small angles.

A similar argument was used in Sect. 2.2.1 when discussing a diffracting pinhole with a diameter of the order of the wavelength λ. We found that a wave front passing through this pinhole can formally be replaced by a spherical wave emerging from the centre of the pinhole due to the restriction to small diffraction angles.

We will therefore keep the notation with the δ-function and choose $\chi = \lambda^2$ in order to obtain the same integral value for the δ-function as for the Gaussian function: $\int \chi\delta(\boldsymbol{x})\,\mathrm{d}\boldsymbol{x} = \chi = \int \exp(-|\boldsymbol{x}|^2/(\lambda^2/\pi))\,\mathrm{d}\boldsymbol{x}$.

The MSDF of an incoherent source [27] now reads as

$$\hat{\Gamma}(\boldsymbol{x}_1, \boldsymbol{x}_2, \nu) = \lambda^2 I(\boldsymbol{x}_1, \nu)\delta(\boldsymbol{x}_1 - \boldsymbol{x}_2). \tag{2.40}$$

This definition of incoherence allows us to describe light that is spatially incoherent but that has an arbitrary spectrum. We can thus model for instance a spatially incoherent but monochromatic source. This makes it much easier to separate the effects of spatial and temporal coherence.

2.3.2 Generalised van Cittert–Zernike Theorem

The generalised van Cittert–Zernike Theorem [248, 259] describes how the statistical properties, the coherence, of light change when it propagates. In analogy to the scalar diffraction theory (see Sect. 2.2) using the spectral amplitude we discuss the propagation of the MSDF. In this section, we provide a formalism for the computation of the MSDF of the propagating light field in different planes along the path of propagation.

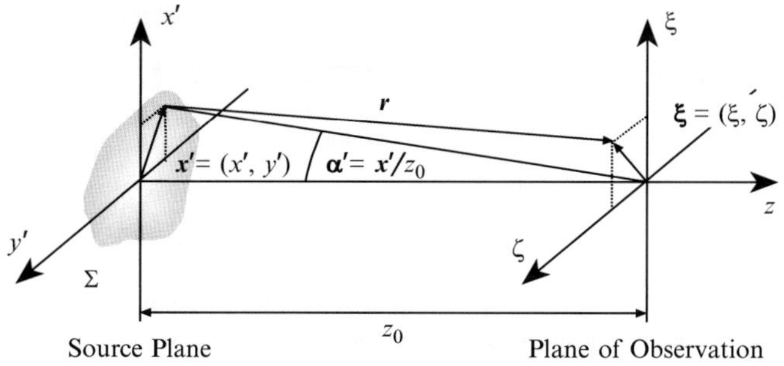

Fig. 2.11 The geometry for the propagation of the coherence function of an incoherent source in plane Σ. The coordinate vector in the source plane is denoted by $\boldsymbol{x}'$, the plane of observation has the coordinate vector $\boldsymbol{\xi}$. The size of the source and the area of interest in the plane of observation are much smaller than the distance z_0 allowing for small angle approximations, $\boldsymbol{\alpha}' = \boldsymbol{x}'/z_0$. In a stellar interferometer the source would be a star or another celestial body, and the plane of observation would be the aperture plane of an interferometric array

The propagation of the MSDF can be derived in the same way as the Rayleigh–Sommerfeld diffraction formula (2.16) yielding [87, 148] the MSDF in the plane of observation with coordinate vector $\boldsymbol{\xi}$ (see Fig. 2.11) as a function of the MSDF in the source plane Σ as

$$\hat{\Gamma}(\boldsymbol{\xi}_1, \boldsymbol{\xi}_2, \nu) = \frac{1}{(\lambda z_0)^2} \iint_{\Sigma} \hat{\Gamma}(\boldsymbol{x}'_1, \boldsymbol{x}'_2, \nu)\, \mathrm{e}^{\mathrm{i}k(r_1 - r_2)}\, \mathrm{d}\boldsymbol{x}'_1 \mathrm{d}\boldsymbol{x}'_2. \tag{2.41}$$

This is the basic equation for the propagation of the MSDF in space. It describes the propagation essentially as the product of two Rayleigh–Sommerfeld integrals (2.16) at individual points $\boldsymbol{\xi}_1$ and $\boldsymbol{\xi}_2$. Equation (2.41) is valid if the angles involved are small, i.e., if the diameter of the light source and the observed area in the plane of observation are much smaller than the distance between them.

The argument in the exponential can be simplified by using the Fresnel approximation (2.17):

$$r = z_0 + \frac{|\boldsymbol{x}' - \boldsymbol{\xi}|^2}{2z_0} - \ldots, \text{ and}$$

$$r_1 - r_2 = \frac{|\boldsymbol{\xi}_1|^2 - |\boldsymbol{\xi}_2|^2}{2z_0} - \frac{\boldsymbol{\xi}_1 \cdot \boldsymbol{x}'_1 - \boldsymbol{\xi}_2 \cdot \boldsymbol{x}'_2}{z_0} + \frac{|\boldsymbol{x}'_1|^2 - |\boldsymbol{x}'_2|^2}{2z_0} - \ldots.$$

Then, the propagation of the MSDF in Fresnel approximation is calculated by

$$\begin{aligned} \hat{\Gamma}(\boldsymbol{\xi}_1, \boldsymbol{\xi}_2, \nu) = {} & \frac{1}{(\lambda z_0)^2}\, \mathrm{e}^{\mathrm{i}k(|\boldsymbol{\xi}_1|^2 - |\boldsymbol{\xi}_2|^2)/(2z_0)} \\ & \times \iint \hat{\Gamma}(\boldsymbol{x}'_1, \boldsymbol{x}'_2, \nu) \mathrm{e}^{\mathrm{i}k(|\boldsymbol{x}'_1|^2 - |\boldsymbol{x}'_2|^2)/(2z_0)} \mathrm{e}^{-\mathrm{i}k(\boldsymbol{\xi}_1 \cdot \boldsymbol{x}'_1 - \boldsymbol{\xi}_2 \cdot \boldsymbol{x}'_2)/z_0}\, \mathrm{d}\boldsymbol{x}'_1 \mathrm{d}\boldsymbol{x}'_2. \end{aligned} \tag{2.42}$$

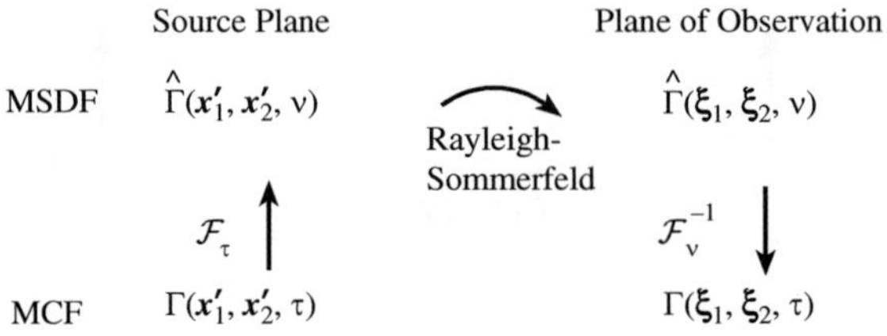

Fig. 2.12 The principle of the generalized van Cittert–Zernike theorem. The MCF $\Gamma(\boldsymbol{x}_1', \boldsymbol{x}_2', \tau)$ in the source plane is Fourier transformed yielding the MSDF $\hat{\Gamma}(\boldsymbol{x}_1', \boldsymbol{x}_2', \nu)$. Applying (2.41) yields the MSDF $\hat{\Gamma}(\boldsymbol{\xi}_1, \boldsymbol{\xi}_2, \nu)$ in the plane of observation and a Fourier back-transform provides the MCF $\Gamma(\boldsymbol{\xi}_1, \boldsymbol{\xi}_2, \tau)$ in the plane of observation

This formula describes the propagation between two planes that are separated by a sufficiently long distance z_0 to justify the Fresnel approximation.

Treating the general case of polychromatic light requires discussing the mutual coherence function, starting with the source MCF $\Gamma(\boldsymbol{x}_1', \boldsymbol{x}_2', \tau)$. Using the relationship between the MCF and the MSDF (2.31), the first step is to calculate the MSDF $\hat{\Gamma}(\boldsymbol{x}_1', \boldsymbol{x}_2', \nu)$ in the source plane, and then to apply (2.41) describing the propagation of the MSDF into the plane of observation. There, the MCF $\Gamma(\boldsymbol{\xi}_1, \boldsymbol{\xi}_2, \tau)$ can be computed by Fourier transforming the MSDF $\hat{\Gamma}(\boldsymbol{\xi}_1, \boldsymbol{\xi}_2, \nu)$ in (2.41). The MCF completely describes the coherence properties of the polychromatic light in the plane of observation. This process is visualised in Fig. 2.12. It is called the *generalized van Cittert–Zernike theorem* [87, 148].

In the following, we will see how the situation can be simplified by reducing the discussion to an incoherent source.

2.3.3 Incoherent Sources of Light: Stars

Observing celestial bodies, we will use $\boldsymbol{\alpha}' = \boldsymbol{x}'/z_0$, the angle of observation, as a source coordinate (see Fig. 2.11). We also reduce our discussion to spatially incoherent sources since stars as thermal sources are spatially incoherent. This means that every point on the surface of the star radiates independently of its neighbour point as discussed in Sect. 2.3.1. This applies to all celestial bodies. In its simplest form a star is shaped like a disk with a diameter independent of wavelength.

If the shape of the source is independent of wavelength over the observed spectrum, the spectral intensity $I(\boldsymbol{\alpha}', \nu)$ can be split into the product of *source brightness distribution*, $I_b(\boldsymbol{\alpha}')$, and *source spectrum*, $G(\nu)$:

$$I(\boldsymbol{\alpha}', \nu) = I_b(\boldsymbol{\alpha}')G(\nu). \tag{2.43}$$

$I_b(\boldsymbol{\alpha}')$ is dimensionless and describes the shape of the source intensity assumed to be independent of wavelength. $G(\nu)$ has the dimension Hz^{-1} and, unless stated otherwise, it is calibrated to unity $\int G(\nu)\mathrm{d}\nu = 1$.

For ground-based observations the width of the spectrum is limited either by the width of the atmospheric bands like the K-band or by spectral filters in the astronomical instruments.

Similar to (2.40), the MSDF of the star in the source plane in angular coordinates now can be written as

$$\hat{\Gamma}(\boldsymbol{\alpha}_1', \boldsymbol{\alpha}_2', \nu) = \frac{\lambda^2}{z_0^2} I_b(\boldsymbol{\alpha}_1')\delta(\boldsymbol{\alpha}_1' - \boldsymbol{\alpha}_2')G(\nu). \tag{2.44}$$

Due to the change of variables, the factor $1/z_0^2$ comes in, keeping in mind the discussion at the end of Sect. 2.3.1.

Whenever we speak about incoherent sources it is in this definition of a spatially incoherent source with arbitrary spectrum $G(\nu)$.

The expression for the MSDF of an incoherent source (2.44) is inserted into (2.42) replacing $\boldsymbol{x}_i'/z_0$ by $\boldsymbol{\alpha}_i'$ and $\mathrm{d}\boldsymbol{x}_i'$ by $z_0^2\mathrm{d}\boldsymbol{\alpha}_i'$. The quadratic exponential before the integral in (2.42) describes the planes of equal phase, similar to the wave front when describing the propagation of the spectral amplitude (2.18). In stellar interferometers, the source is usually a star at a fairly large distance z_0, and the plane of observation is on the surface of the Earth or, for space interferometers, in a near Earth orbit. Then the quadratic exponential can be approximated by unity, yielding the MSDF in the plane of observation

$$\begin{aligned}\hat{\Gamma}(\boldsymbol{\xi}_1, \boldsymbol{\xi}_2, \nu) &= G(\nu) \\ &\quad \times \iint I_b(\boldsymbol{\alpha}_1')\delta(\boldsymbol{\alpha}_1' - \boldsymbol{\alpha}_2')\mathrm{e}^{\mathrm{i}k(|\boldsymbol{\alpha}_1'|^2 - |\boldsymbol{\alpha}_2'|^2)z_0/2}\, \mathrm{e}^{-\mathrm{i}k(\boldsymbol{\xi}_1\cdot\boldsymbol{\alpha}_1' - \boldsymbol{\xi}_2\cdot\boldsymbol{\alpha}_2')}\mathrm{d}\boldsymbol{\alpha}_1'\mathrm{d}\boldsymbol{\alpha}_2' \\ &= G(\nu)\int I_b(\boldsymbol{\alpha}_1')\mathrm{e}^{-\mathrm{i}k(\boldsymbol{\xi}_1 - \boldsymbol{\xi}_2)\cdot\boldsymbol{\alpha}_1'}\mathrm{d}\boldsymbol{\alpha}_1',\end{aligned} \tag{2.45}$$

which is the product of the spectrum $G(\nu)$ with the Fourier transform of the source brightness distribution $I_b(\boldsymbol{\alpha}')$. Thus, the MSDF at frequency ν, i.e., the spatial coherence in the plane of observation, is determined by the Fourier transform of the source shape.

With the exponential kernel $\frac{2\pi}{\lambda}(\boldsymbol{\xi}_1 - \boldsymbol{\xi}_2)\cdot\boldsymbol{\alpha}_1'$, the Fourier transform is performed from $\boldsymbol{\alpha}'$ to $(\boldsymbol{\xi}_1 - \boldsymbol{\xi}_2)/\lambda$. This means firstly that the Fourier transform does not depend on the individual coordinates $\boldsymbol{\xi}_1$ and $\boldsymbol{\xi}_2$ but on their difference only, and secondly that the Fourier transform is a function of wavelength, $1/\lambda$.

The consequence of working with an incoherent source is, thus, that the MSDF is a function of coordinate difference only. Therefore, we will write the MSDF and the MCF in the plane of observation[5] as functions of coordinate difference in the following:

[5] More generally speaking, the coherence function depends on the coordinate difference in those planes that are illuminated by an incoherent source at large distance when the van Cittert–Zernike theorem applies.

$$\hat{\Gamma}(\xi_1 - \xi_2, \nu) = \hat{\Gamma}(\xi_1, \xi_2, \nu) \text{ and } \Gamma(\xi_1 - \xi_2, \tau) = \Gamma(\xi_1, \xi_2, \tau). \tag{2.46}$$

It is also important to note that the quadratic phase term $\exp(\mathrm{i}k(|\boldsymbol{\alpha}_1'|^2 - |\boldsymbol{\alpha}_2'|^2) z_0/2)$ in (2.45) that stems from the Fresnel approximation disappears only because the light of the celestial source is spatially incoherent. There is no further approximation required.

The spectral intensity at frequency ν is the MSDF for $\boldsymbol{\xi}_1 = \boldsymbol{\xi}_2$, i.e., $I(\boldsymbol{\xi}, \nu) = \hat{\Gamma}(\boldsymbol{\xi} - \boldsymbol{\xi}, \nu) = \hat{\Gamma}(0, \nu)$. Then, the Fourier transform in (2.45) is reduced to a simple integration $I(\boldsymbol{\xi}, \nu) = \hat{\Gamma}(0, \nu) = G(\nu) \int I_b(\boldsymbol{\alpha}')\mathrm{d}\boldsymbol{\alpha}' = G(\nu) I_0$ yielding the constant spectral intensity $G(\nu) I_0$ in the plane of observation.

In general, we are interested in **polychromatic light**. To determine the white-light intensity $I(\boldsymbol{\xi})$ in the plane of observation, we integrate the spectral intensity $I(\boldsymbol{\xi}, \nu) = G(\nu) I_0$ over the spectral band (see (2.33)) yielding the constant value I_0 since the integral over $G(\nu)$ is unity. Thus, not very surprising, the star sheds its light homogeneously over the surface of the Earth. Only the coherence functions reflect the properties – in particular the shape – of the source.

When it comes to coherence properties of polychromatic light we have to deal with the MCF in the plane of observation. The relationship between the MSDF and the MCF is defined by (2.31). However, this is not completely straightforward since a complicated integral with two entangled Fourier transforms needs to be resolved:

$$\begin{aligned} \Gamma(\xi_1 - \xi_2, \tau) &= \int \hat{\Gamma}(\xi_1 - \xi_2, \nu) \mathrm{e}^{-\mathrm{i}2\pi\nu\tau} \, \mathrm{d}\nu \\ &= \int G(\nu) \int I_b(\boldsymbol{\alpha}') \mathrm{e}^{-\mathrm{i}k(\xi_1 - \xi_2)\cdot\boldsymbol{\alpha}'} \, \mathrm{d}\boldsymbol{\alpha}' \, \mathrm{e}^{-\mathrm{i}2\pi\nu\tau} \, \mathrm{d}\nu. \end{aligned} \tag{2.47}$$

A common simplification of this double integral is based on the assumption that the source brightness distribution $I_b(\boldsymbol{\alpha}')$ is independent of wavelength. This is, however, not sufficient for splitting the formula into an integral over ν and an integral over $\boldsymbol{\alpha}'$ since – bearing in mind that the Fourier transform is done from $\boldsymbol{\alpha}'$ to $(\boldsymbol{\xi}_1 - \boldsymbol{\xi}_2)/\lambda$ – the Fourier transform of $I_b(\boldsymbol{\alpha}')$ is still a function of wavelength.

One has to assume additionally, that also the Fourier transform of $I_b(\boldsymbol{\alpha}')$ is invariant over the frequency range set by the spectrum. Then the MCF can be split into the product of the Fourier transform of the source brightness distribution $\int I_b(\boldsymbol{\alpha}') \exp(-\mathrm{i}k(\boldsymbol{\xi}_1 - \boldsymbol{\xi}_2) \cdot \boldsymbol{\alpha}')\mathrm{d}\boldsymbol{\alpha}'$ and the Fourier transform of the spectrum $\int G(\nu) \exp(-\mathrm{i}2\pi\nu\tau)\mathrm{d}\nu$. While the first Fourier transform can be regarded as the purely spatial part of the MCF, the second Fourier transform is the temporal part of the MCF like the self-coherence function.

If the source were approximately point-like, described by a δ-function, then its Fourier transform would be a constant, I_0, independent of the coordinate difference $\boldsymbol{\xi}_1 - \boldsymbol{\xi}_2$ and of λ. In this case, the MSDF in the plane of observation reads as

$$\hat{\Gamma}(\xi_1 - \xi_2, \nu) = G(\nu) I_0. \tag{2.48}$$

For a source of finite size, α_0, the Fourier transform would be approximately constant only either for very small coordinate differences $\boldsymbol{\xi}_1 - \boldsymbol{\xi}_2$ or over a very narrow spectrum. Writing the coordinate in Fourier space as $(\boldsymbol{\xi}_1 - \boldsymbol{\xi}_2)/\lambda$ we can also state that the Fourier transform is invariant over any coordinate range smaller than $1/\alpha_0$. This puts a constraint on the size of the source. For a larger source or for a wider spectrum the Fourier transform would vary over the observed spectrum, and the separation of the two integrals and the approximation no longer holds.

This approximation restricting the width of the spectrum and, implicitly, the size of the source is also known as the *quasi-monochromatic approximation*.

2.3.4 Quasi-Monochromatic Approximation

We simplify the double integral for the propagation of the MCF (2.47) to a manageable formula by assuming a narrow spectrum $G(\nu)$ with $\Delta\nu \ll \nu_0$ and by observing only time differences shorter than the coherence time, $\tau \ll 1/\Delta\nu$ [148]. The latter means that the corresponding optical path difference (OPD) must be much smaller than the coherence length l_c.

A narrow spectrum in connection with a small source means that, as argued above, the Fourier transform of $I_b(\boldsymbol{\alpha}')$ is invariant over the spectrum and the Fourier transform is represented by the value at the average frequency ν_0 (and at the average wave number k_0). The restriction to small time differences, $\tau \ll 1/\Delta\nu$, means that the second integral over ν is reduced to the function value $G(\nu_0)\exp(-\mathrm{i}2\pi\nu_0\tau)\Delta\nu = \exp(-\mathrm{i}2\pi\nu_0\tau)$ due to the calibration of $G(\nu)$.

Now the MCF in the plane of observation can be written in quasi-monochromatic approximation as

$$\begin{aligned}\Gamma_{\mathrm{qm}}(\boldsymbol{\xi}_1 - \boldsymbol{\xi}_2, \tau) &= \int I_b(\boldsymbol{\alpha}')\mathrm{e}^{-\mathrm{i}k_0(\boldsymbol{\xi}_1-\boldsymbol{\xi}_2)\cdot\boldsymbol{\alpha}'}\,\mathrm{d}\boldsymbol{\alpha}'\,\mathrm{e}^{-\mathrm{i}2\pi\nu_0\tau}\\ &= \Gamma_{\mathrm{qm}}(\boldsymbol{\xi}_1 - \boldsymbol{\xi}_2, 0)\,\mathrm{e}^{-\mathrm{i}2\pi\nu_0\tau}. \end{aligned} \tag{2.49}$$

Not very surprisingly, the MCF in quasi-monochromatic approximation (2.49) has some similarity to the MCF of coherent light (2.37). While in the coherent case the product of two deterministic amplitudes $V_{\nu_0}(\boldsymbol{\xi}_1)V^*_{\nu_0}(\boldsymbol{\xi}_2)\exp(-2\pi\nu_0\tau)$ formed the MCF for any time delay τ, the quasi-monochromatic MCF is valid for small time delays τ only.

However, unlike in the case of coherent light, there is not a product of deterministic amplitudes in quasi-monochromatic approximation but a correlation function $\Gamma_{\mathrm{qm}}(\boldsymbol{\xi}_1 - \boldsymbol{\xi}_2, 0)$ that is determined by the Fourier transform of the source shape.

With the MCF at $\tau = 0$ we can now write down the fundamental *van Cittert–Zernike theorem* determining the spatial coherence in the plane of observation as

$$\Gamma_{\mathrm{qm}}(\boldsymbol{\xi}_1 - \boldsymbol{\xi}_2, 0) = \int I_b(\boldsymbol{\alpha}')\mathrm{e}^{-\mathrm{i}k_0(\boldsymbol{\xi}_1-\boldsymbol{\xi}_2)\cdot\boldsymbol{\alpha}'}\,\mathrm{d}\boldsymbol{\alpha}'. \tag{2.50}$$

The brightness distribution $I_b(\boldsymbol{\alpha}')$ of an incoherent source is linked to the MCF $\Gamma_{\rm qm}(\boldsymbol{\xi}_1 - \boldsymbol{\xi}_2, 0)$ in the plane of observation at $\tau = 0$ through a Fourier transform. The spatial coherence depends on the shape of the intensity distribution of the source, and it is a function of the coordinate difference. The size of the source has to be sufficiently small so that its Fourier transform is invariant over the spectral band $\Delta\nu$ and can be replaced by its value at the average frequency ν_0 and k_0 respectively.

Dividing the MCF $\Gamma_{\rm qm}(\boldsymbol{\xi}_1 - \boldsymbol{\xi}_2, 0)$ by the geometric mean of the intensities $I(\boldsymbol{\xi}_i) = \Gamma_{\rm qm}(\boldsymbol{\xi}_i - \boldsymbol{\xi}_i, 0)$ at position $\boldsymbol{\xi}_1$ and $\boldsymbol{\xi}_2$, we obtain the normalised MCF, which is the degree of coherence, γ, (2.29) at $\tau = 0$. In quasi-monochromatic approximation, this is called the complex *visibility function* $\mu_{\nu_0}(\boldsymbol{\xi}_1 - \boldsymbol{\xi}_2)$.

Then, the van Cittert–Zernike theorem can be written in its most common form as:

$$\mu_{\nu_0}(\boldsymbol{\xi}_1 - \boldsymbol{\xi}_2) = \frac{\Gamma_{\rm qm}(\boldsymbol{\xi}_1 - \boldsymbol{\xi}_2, 0)}{\sqrt{I(\boldsymbol{\xi}_1)I(\boldsymbol{\xi}_2)}} = \frac{\int I_b(\boldsymbol{\alpha}')\mathrm{e}^{-\mathrm{i}k_0(\boldsymbol{\xi}_1-\boldsymbol{\xi}_2)\cdot\boldsymbol{\alpha}'}\,\mathrm{d}\boldsymbol{\alpha}'}{\int I_b(\boldsymbol{\alpha}')\,\mathrm{d}\boldsymbol{\alpha}'}. \tag{2.51}$$

so that $0 \le |\mu_{\nu_0}(\boldsymbol{\xi}_1 - \boldsymbol{\xi}_2))| \le 1$ and $\mu_{\nu_0}(0) = 1$. The normalisation by the geometric mean of the intensities $I(\boldsymbol{\xi}_i)$ is replaced by the integral over $I_b(\boldsymbol{\alpha}')$ that has the constant value $I_0 = \int I_b(\boldsymbol{\alpha}')\mathrm{d}\boldsymbol{\alpha}'$. However, observing with a real interferometer through atmospheric turbulence the intensity fluctuates randomly and the values at $\boldsymbol{\xi}_1$ and $\boldsymbol{\xi}_2$ are no longer constant and have to be accounted for individually. This calibration is a very important part of the interferometric data processing.

The phase of the visibility function will be denoted by $\phi_{\nu_0}(\boldsymbol{\xi}_1 - \boldsymbol{\xi}_2)$, which is identical to $\phi_{\rm MCF}(\boldsymbol{\xi}_1 - \boldsymbol{\xi}_2, 0)$, the phase of the MCF at $\tau = 0$.

The definition of the visibility function in (2.51) can be used to formally rewrite the MSDF in (2.45), as

$$\hat{\Gamma}(\boldsymbol{\xi}_1 - \boldsymbol{\xi}_2, \nu) = G(\nu) I_0 \mu_\nu(\boldsymbol{\xi}_1 - \boldsymbol{\xi}_2), \tag{2.52}$$

at any frequency ν. The phase, $\hat{\phi}(\boldsymbol{\xi}_1 - \boldsymbol{\xi}_2, \nu)$, of the MSDF is then identical to the phase, $\phi_\nu(\boldsymbol{\xi}_1 - \boldsymbol{\xi}_2)$, of the visibility function since I_0 and the spectrum $G(\nu)$ are real functions.

Thus, the MSDF is composed of the spectrum $G(\nu)$ determining the temporal coherence, and of the visibility function $\mu_\nu(\boldsymbol{\xi}_1 - \boldsymbol{\xi}_2)$ describing the spatial coherence. This facilitates the interpretation of the MSDF when discussing polychromatic illumination in the following sections. Because of the exponential kernel, $\frac{2\pi}{\lambda}(\boldsymbol{\xi}_1 - \boldsymbol{\xi}_2) \cdot \boldsymbol{\alpha}'$, of the Fourier transform, μ_ν varies with wavelength (indicated by subscript ν) although the source brightness distribution $I_b(\boldsymbol{\alpha}')$ is independent of wavelength. In quasi-monochromatic approximation, regarding only the average frequency ν_0, we use the subscript ν_0.

Similar to the coherence time that was defined in the context of the self-coherence function in Sect. 2.3.1 as the reciprocal of the spectral bandwidth (2.34), we can now define the *coherence width* as that coordinate difference $\boldsymbol{\xi}_1 - \boldsymbol{\xi}_2$ when the value of the visibility function is reduced significantly. Due to the Fourier connection between the visibility function and the source intensity distribution (2.51), the coherence width is inversely proportional to the angular size of the source.

The symmetry characteristics of the visibility function are driven by the fact that the source brightness distribution is a real and positive function by definition. Then, its Fourier transform is a Hermitian function (see Sect. A.1), meaning that the modulus of the visibility function is symmetric with respect to $|\boldsymbol{\xi}_1 - \boldsymbol{\xi}_2| = 0$ and that the phase ϕ_ν of the visibility function is anti-symmetric:

$$|\mu_\nu(\boldsymbol{\xi}_1 - \boldsymbol{\xi}_2)| = |\mu_\nu(\boldsymbol{\xi}_2 - \boldsymbol{\xi}_1)| \text{ and}$$
$$\phi_\nu(\boldsymbol{\xi}_1 - \boldsymbol{\xi}_2) = -\phi_\nu(\boldsymbol{\xi}_2 - \boldsymbol{\xi}_1).$$

How does the propagation of the coherence function described by the van Cittert–Zernike theorem (2.50) compare to the propagation of the spectral amplitude as discussed in Sect. 2.2.2? Both the Rayleigh–Sommerfeld diffraction integral in Fresnel approximation (2.22) and the van Cittert–Zernike theorem link physical properties in two separate planes through a Fourier transform. While the Rayleigh–Sommerfeld diffraction integral connects the spectral amplitude in the plane of observation with the spectral amplitude in an aperture – which are the same physical quantities – the van Cittert–Zernike theorem links the intensity distribution of an incoherent source with the visibility function in the plane of observation, which are seemingly different physical quantities. However, this should be well understood by now since the intensity is a special case of the coherence function, which, as a product of two optical disturbances, has the dimension of an intensity.

Both formulae are valid in Fresnel approximation and quadratic phase terms cancel in both cases albeit for different reasons: The diffraction integral is reduced to a Fourier transform by introducing a lens (2.21), and the van Cittert–Zernike theorem uses a Fourier transform because the light source is spatially incoherent.

The Visibility Function of Venus

An example illuminates the situation: we model Venus as a uniform disk with angular diameter α'_0 varying between about 15 and 45 arcsec depending on the mutual positions of Venus and Earth. The circular source brightness distribution is then represented by $I_b(\boldsymbol{\alpha}') = (\pi(\alpha'_0/2)^2)^{-1} \operatorname{circ}\left(\frac{|\alpha'|}{\alpha'_0/2}\right)$. The factor $(\pi(\alpha'_0/2)^2)^{-1}$ is a normalization factor so that the integral of the intensity over the source and, thus, the denominator in (2.51), equals one, and we compute the visibility function as the Fourier transform of $I_b(\boldsymbol{\alpha}')$ yielding

$$\begin{aligned}\mu_\nu(\boldsymbol{\xi}_1 - \boldsymbol{\xi}_2) &= \frac{1}{\pi(\alpha'_0/2)^2} \int \operatorname{circ}\left(\frac{|\boldsymbol{\alpha}'|}{\alpha'_0/2}\right) \mathrm{e}^{-ik(\boldsymbol{\xi}_1 - \boldsymbol{\xi}_2)\cdot\boldsymbol{\alpha}'}\, \mathrm{d}\boldsymbol{\alpha}' \\ &= \operatorname{Besinc}(k|\boldsymbol{\xi}_1 - \boldsymbol{\xi}_2|\alpha'_0/2). \qquad (2.53)\end{aligned}$$

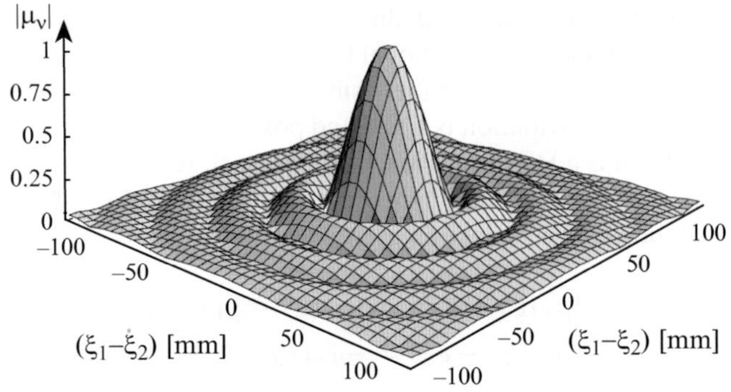

Fig. 2.13 The modulus of the visibility function $|\mu_\nu(\boldsymbol{\xi}_1 - \boldsymbol{\xi}_2)|$ of Venus in the plane of observation, e.g., on the surface of the Earth. The visibility function has the form of a Besinc-function since Venus is a light source at a very large distance, modelled as a uniform disk (2.53). If Venus has an angular diameter of $\alpha_0' = 15$ arcsec the visibility function has its first zero for a coordinate difference of $|\boldsymbol{\xi}_1 - \boldsymbol{\xi}_2|_0 = 37$ mm with $|\boldsymbol{\xi}_1 - \boldsymbol{\xi}_2|_0 = 1.22\lambda/\alpha_0'$ and for an observing wavelength of 2.2 μm (in the visible, it is $|\boldsymbol{\xi}_1 - \boldsymbol{\xi}_2|_0 = 8$ mm). The second zero of the visibility function is at $|\boldsymbol{\xi}_1 - \boldsymbol{\xi}_2|_0 = 2.233\lambda/\alpha_0' = 67.5$ mm. If Venus were closer to the Earth displaying an angular diameter of e.g. 45 arcsec, the coherence width would be reduced to 12 mm

Figure 2.13 displays the modulus of the visibility function for Venus' smallest angular diameter of $\alpha_0' = 15$ arcsec as a function of the coordinate difference in mm. A suitable measure for the coherence width is the first zero of the visibility function, which occurs at a coordinate difference of $|\boldsymbol{\xi}_1 - \boldsymbol{\xi}_2|_0 = 1.22\lambda/\alpha_0' = 37$ mm for an observing wavelength of $\lambda = 2.2$ μm. For larger coordinate differences the visibility function slowly oscillates between negative and positive values with decreasing amplitude.

Applying the quasi-monochromatic approximation we would assume that the visibility function at the average frequency ν_0 is a suitable approximation for the visibility function at all frequencies within the spectral band. If the spectral band is too wide or if the size of the source is too large this approximation no longer holds and we have to compute the integral over the spectral band in (2.47) by properly considering the visibility function as a function of wavelength.

The angular diameter of stars is typically in the milliarcsecond range, i.e., a factor of 1,000 smaller than Venus. If a star is modelled as a uniform disk the shape of the visibility function is again described by a Besinc-function but the coherence width has values of several 10 m and not millimetres. Separating the two points $\boldsymbol{\xi}_1$ and $\boldsymbol{\xi}_2$ by only a few metres, the light coming from the star still has a very high coherence indicated by values of the visibility function that are close to unity.

Another way of putting this is to say that the incoherent light emitted by the star formally has acquired coherence by the very process of propagation. While the coherence function on the surface of the incoherent star (2.44) is zero unless the coordinate difference is zero, the increased coherence width of the light at a large distance from the star allows for a coordinate difference of a few metres before the coherence function in the plane of observation drops substantially.

In contrast, we will see later that light from a coherent source does not change its state of coherence when propagating in free space. Coherent light remains coherent throughout space and time. Thus, the visibility function of a coherent source is always unity independent of the coordinate difference.

For these examples of uniform disks the visibility function as a Besinc-function has the same mathematical form as the spectral amplitude in the diffraction pattern of a circular aperture (2.24). While the diffraction pattern can be observed with the naked eye, the visibility function describing the correlation of optical disturbances at different points cannot be observed directly. It can only be observed through an interferometer experiment.

Revisiting Young's experiment in the following section we will see how the contrast of a fringe pattern as a measurable quantity can be related to the visibility function.

NB 4. *Expanding the scope of our discussion to sources of arbitrary spatial coherence we replace the δ-function in (2.44) by a function of finite extent describing the source's spatial coherence. Describing the spatial coherence by the visibility function $\mu(\boldsymbol{\alpha}'_1 - \boldsymbol{\alpha}'_2)$ – without the subscript ν since we are free to define the source's spatial coherence independent of wavelength – we write the MSDF of a partially coherent source as*

$$\hat{\Gamma}(\boldsymbol{\alpha}'_1, \boldsymbol{\alpha}'_2, \nu) = \frac{\lambda}{z_0^2} \sqrt{I_b(\boldsymbol{\alpha}'_1)} \sqrt{I_b(\boldsymbol{\alpha}'_2)}\, \mu(\boldsymbol{\alpha}'_1 - \boldsymbol{\alpha}'_2)\, G(\nu). \tag{2.54}$$

Then the coherence width of the source is no longer zero but it is determined by the shape of the visibility function μ.

In this notation, the visibility function is linked to the degree of coherence $\gamma(\boldsymbol{\alpha}'_1 - \boldsymbol{\alpha}'_2, \tau)$, using (2.29) and (2.31), through

$$\gamma(\boldsymbol{\alpha}'_1 - \boldsymbol{\alpha}'_2, \tau) = \int \mu(\boldsymbol{\alpha}'_1 - \boldsymbol{\alpha}'_2) G(\nu) \mathrm{e}^{-\mathrm{i}2\pi\nu\tau} \mathrm{d}\nu, \tag{2.55}$$

like the MCF to the MSDF. This is why the visibility function is sometimes called the spatial degree of coherence.

Choosing a Gaussian function for both the visibility function $\mu(\boldsymbol{\alpha}'_1 - \boldsymbol{\alpha}'_2)$ with a width of σ_μ and for the brightness distribution $I_b(\boldsymbol{\alpha}')$ with a width of $\sigma_{\alpha'}$, we introduce Gaussian Schell-model *(GSM) sources [147, Sect. 5.3 and 5.4] with an MSDF defined as*

$$\hat{\Gamma}(\boldsymbol{\alpha}'_1, \boldsymbol{\alpha}'_2, \nu) = \frac{\lambda}{z_0^2} \exp\left(-\frac{|\boldsymbol{\alpha}'_1 + \boldsymbol{\alpha}'_2|^2}{8\sigma_{\alpha'}^2}\right) \exp\left(-\frac{|\boldsymbol{\alpha}'_1 - \boldsymbol{\alpha}'_2|^2}{2\sigma_\mu^2}\right) G(\nu), \tag{2.56}$$

with the condition $\sigma_\mu \ll \sigma_{\alpha'}$, i.e., the coherence width is much smaller than the size of the source.

The propagation of light from the GSM source is described by the van Cittert–Zernike theorem replacing the δ-function in (2.45) by $\mu(\boldsymbol{\alpha}'_1 - \boldsymbol{\alpha}'_2)$. The spatial *coherence in the plane of observation in Fraunhofer approximation is again described by the Fourier transform of the intensity distribution in the source plane that – being a Gaussian function – is itself a Gaussian function of width* $1/(2\pi\sigma_{\alpha'})$. *The coherence width is thus proportional to the reciprocal of the angular source diameter,* $\sigma_{\alpha'}$, *just like for an incoherent source.*

It was shown by Carter and Wolf [27], that the intensity distribution $I(\boldsymbol{\xi})$ *in the plane of observation, which was constant in the case of an incoherent source, is the Fourier transform of the source's visibility function* $\mu(\boldsymbol{\alpha}'_1 - \boldsymbol{\alpha}'_2)$. *Then,* $I(\boldsymbol{\xi})$ *is a Gaussian function of width* $1/(2\pi\sigma_\mu)$. *In the incoherent limit when the coherence width in the source plane approaches zero,* $\sigma_\mu \to 0$, *the width of the intensity distribution steadily increases until it fills the plane of observation homogeneously.*

Thus, the source visibility function determines the intensity distribution in the plane of observation and, vice versa, the source intensity distribution determines the visibility function in the plane of observation. This reciprocity theorem was first pointed out by Walther [251].

In practice, GSM sources can be used to describe the propagation of multi-mode laser light [89, 217, 240]. Observing celestial bodies we rarely come across multi-mode laser light but we do encounter partially coherent light during the imaging process. In the preceding section, discussing the visibility function of Venus we found that the coherence width in the plane of observation, e.g., on the surface of the Earth, is some 10 mm. The light is then partially coherent. The aperture of a ground-based telescope is filled by this partially coherent light and, computing the propagation of light from the aperture plane into the image plane, we treat a case similar to the one above. In Sect. 3.2.3, discussing the coherence properties in the image plane, we will return to this subject.

The Coherence Function: Summary

The mutual coherence function is the critical quantity for understanding stellar interferometry. It is the second order correlation function of the optical disturbance as a function of time difference and spatial coordinates. The mutual coherence function (MCF) is defined as

$$\Gamma(\boldsymbol{x}_1, \boldsymbol{x}_2, \tau) =< v(\boldsymbol{x}_1, t+\tau) v^*(\boldsymbol{x}_2, t) >, \tag{2.27}$$

where $\boldsymbol{x}_i$ are the coordinate vectors and τ is the time difference.

For the propagation of the MCF in space it is convenient to introduce the mutual spectral density function (MSDF) that is the correlation function of spectral amplitudes, defined as

$$\hat{\Gamma}(\boldsymbol{x}_1, \boldsymbol{x}_2, \nu) := \lim_{T\to\infty} \frac{1}{2T} E\left\{V_T(\boldsymbol{x}_1, \nu) V_T^*(\boldsymbol{x}_2, \nu)\right\}. \tag{2.30}$$

The MSDF is related to the MCF through a Fourier transform,

$$\Gamma(\boldsymbol{x}_1, \boldsymbol{x}_2, \tau) = \int \hat{\Gamma}(\boldsymbol{x}_1, \boldsymbol{x}_2, \nu) \mathrm{e}^{-\mathrm{i}2\pi\nu\tau} \, \mathrm{d}\nu. \tag{2.31}$$

The polychromatic or white-light intensity is then

$$I(\boldsymbol{x}) = \Gamma(\boldsymbol{x}, \boldsymbol{x}, 0) = \int \hat{\Gamma}(\boldsymbol{x}, \boldsymbol{x}, \nu) \, \mathrm{d}\nu = \int I(\boldsymbol{x}, \nu) \, \mathrm{d}\nu, \tag{2.33}$$

with $I(\boldsymbol{x}, \nu) = \hat{\Gamma}(\boldsymbol{x}, \boldsymbol{x}, \nu)$, the spectral intensity.

The propagation of the coherence functions in space is described by applying the generalised van Cittert–Zernike theorem (2.41) that was derived from the Rayleigh–Sommerfeld diffraction formula (2.16) yielding the MSDF in the plane of observation with coordinate $\boldsymbol{\xi}$ as the Fourier transform of the source brightness distribution $I_b(\boldsymbol{\alpha}')$:

$$\hat{\Gamma}(\boldsymbol{\xi}_1, \boldsymbol{\xi}_2, \nu) \ = G(\nu) \int I_b(\boldsymbol{\alpha}') \mathrm{e}^{-ik(\boldsymbol{\xi}_1 - \boldsymbol{\xi}_2)\cdot\boldsymbol{\alpha}'} \mathrm{d}\boldsymbol{\alpha}', \tag{2.45}$$

with $\boldsymbol{\alpha}' \ = \ \boldsymbol{x}'/z_0$, and $G(\nu)$ the source spectrum. The source brightness distribution is assumed independent of wavelength.

For simplification, a number of assumptions and approximations are made, which are suitable for the situation in a stellar interferometer:

- Fresnel approximation (all involved angles are small)
- Spatially incoherent light sources (e.g., stars)
- The quasi-monochromatic approximation when
  (1) The spectral bandwidth $\Delta\nu$ is assumed to be much smaller than the average frequency ν_0
  (2) The time difference τ is much smaller than $1/\Delta\nu$ corresponding to an OPD much smaller than the coherence length l_c

With the assumption of an incoherent light source the coherence functions in the plane of observation can be written as functions of coordinate difference only

$$\hat{\Gamma}(\boldsymbol{\xi}_1 - \boldsymbol{\xi}_2, \nu) = \hat{\Gamma}(\boldsymbol{\xi}_1, \boldsymbol{\xi}_2, \nu) \text{ and } \Gamma(\boldsymbol{\xi}_1 - \boldsymbol{\xi}_2, \tau) = \Gamma(\boldsymbol{\xi}_1, \boldsymbol{\xi}_2, \tau).$$

The van Cittert–Zernike theorem (2.50) in quasi-monochromatic approximation describes the propagation of the mutual coherence function under these conditions stating that the MCF $\Gamma_{\mathrm{qm}}(\boldsymbol{\xi}_1 - \boldsymbol{\xi}_2, 0)$ in the plane of observation, i.e., the spatial coherence, is given by the Fourier transform of the source brightness distribution $I_b(\boldsymbol{\alpha}')$,

$$\Gamma_{\mathrm{qm}}(\boldsymbol{\xi}_1 - \boldsymbol{\xi}_2, 0) = \int I_b(\boldsymbol{\alpha}') \mathrm{e}^{-ik_0(\boldsymbol{\xi}_1 - \boldsymbol{\xi}_2)\cdot\boldsymbol{\alpha}'} \, \mathrm{d}\boldsymbol{\alpha}'. \tag{2.50}$$

The Fourier transform of $I_b(\boldsymbol{\alpha}')$ is taken at the average frequency ν_0 and k_0 respectively, under the assumption that the size of the source is sufficiently small so that its Fourier transform is invariant over the spectral band $\Delta\nu$.

The normalized MCF is called the *complex visibility function* μ_{ν_0}, defined as

$$\mu_{\nu_0}(\boldsymbol{\xi}_1 - \boldsymbol{\xi}_2) = \frac{\Gamma_{\mathrm{qm}}(\boldsymbol{\xi}_1 - \boldsymbol{\xi}_2, 0)}{\sqrt{I(\boldsymbol{\xi}_1)I(\boldsymbol{\xi}_2)}} = \frac{\int I_b(\boldsymbol{\alpha}')\mathrm{e}^{-\mathrm{i}k_0(\boldsymbol{\xi}_1 - \boldsymbol{\xi}_2)\cdot\boldsymbol{\alpha}'}\,\mathrm{d}\boldsymbol{\alpha}'}{\int I_b(\boldsymbol{\alpha}')\,\mathrm{d}\boldsymbol{\alpha}'}, \tag{2.51}$$

so that $0 \leq |\mu_{\nu_0}(\boldsymbol{\xi}_1 - \boldsymbol{\xi}_2))| \leq 1$ and $\mu_{\nu_0}(0) = 1$. We often denote the integral over $I_b(\boldsymbol{\alpha}')$ in the denominator by $I_0 = \int I_b(\boldsymbol{\alpha}')\,\mathrm{d}\boldsymbol{\alpha}'$, describing the homogeneous intensity in the plane of observation.

It is very important to note that the visibility function is a function of coordinate difference only. The absolute positions of the two points $\boldsymbol{\xi}_1$ and $\boldsymbol{\xi}_2$ are not relevant.

The visibility function $\mu_{\nu_0}(\boldsymbol{\xi}_1 - \boldsymbol{\xi}_2)$ is a complex function with phase $\phi_{\nu_0}(\boldsymbol{\xi}_1 - \boldsymbol{\xi}_2)$. Due to the source brightness distribution being real and positive by definition, the modulus $|\mu_{\nu_0}|$ is symmetric and the phase ϕ_{ν_0} is anti-symmetric:

$$|\mu_{\nu_0}(\boldsymbol{\xi}_1 - \boldsymbol{\xi}_2)| = |\mu_{\nu_0}(\boldsymbol{\xi}_2 - \boldsymbol{\xi}_1)| \text{ and}$$
$$\phi_{\nu_0}(\boldsymbol{\xi}_1 - \boldsymbol{\xi}_2) = -\phi_{\nu_0}(\boldsymbol{\xi}_2 - \boldsymbol{\xi}_1).$$

Using the definition of the visibility function, the MSDF in polychromatic illumination, (2.45), can be written as

$$\hat{\Gamma}(\boldsymbol{\xi}_1 - \boldsymbol{\xi}_2, \nu) = G(\nu)I_0\mu_\nu(\boldsymbol{\xi}_1 - \boldsymbol{\xi}_2), \tag{2.52}$$

splitting the MSDF in a term $G(\nu)$ determining the temporal coherence, and a term $\mu_\nu(\boldsymbol{\xi}_1 - \boldsymbol{\xi}_2)$ describing the spatial coherence.

This facilitates the interpretation of the MSDF. The phase, $\hat{\phi}(\boldsymbol{\xi}_1 - \boldsymbol{\xi}_2, \nu_0)$, of the MSDF is then identical to the phase, $\phi_\nu(\boldsymbol{\xi}_1 - \boldsymbol{\xi}_2)$, of the visibility function since I_0 and the spectrum $G(\nu)$ are real functions.

2.4 Young's Experiment Revisited

So far, we have discussed the properties of the coherence functions under various conditions and in different planes along the direction of propagation. But how do we relate these properties to a measurable quantity? In Sect. 2.1.2 it was noted that coherence effects have an influence on the interference pattern in Young's two-pinhole experiment. This will be discussed now in the context of the coherence functions.

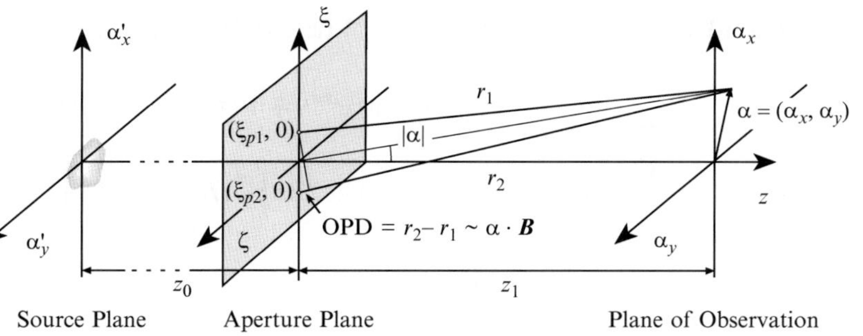

Fig. 2.14 Young's experiment with an incoherent source at large distance z_0. The two pinholes are at positions $\boldsymbol{\xi}_{p1} = (\xi_{p1}, 0)$ and $\boldsymbol{\xi}_{p2} = (\xi_{p2}, 0)$ on the ξ-axis. This plane is now called the aperture plane. The vector between the pinholes is called *baseline vector* with $\boldsymbol{B} = \boldsymbol{\xi}_{p1} - \boldsymbol{\xi}_{p2}$. The diffraction pattern is calculated as a function of observing angle $\boldsymbol{\alpha}$ in the plane of observation. The source could be a star and the aperture plane could be on the surface of the Earth

2.4.1 The Coherence Function in Young's Experiment

We modify the experiment in Sect. 2.1.2 by replacing the plane wave illumination of the two pinholes[6] by an illumination from an incoherent source at a large distance from the pinholes (see Fig. 2.14). Therefore, we replace the constant amplitudes V_0 in (2.10) by variable amplitudes $\lambda^2 V(\boldsymbol{\xi}_{p1}, \nu)$ and $\lambda^2 V(\boldsymbol{\xi}_{p2}, \nu)$, with λ^2, the square of the wavelength, accounting for the size of the pinholes.

Then, the spectral amplitude $V(\boldsymbol{\alpha}, \nu)$ in the plane of observation (see Fig. 2.14) is the sum of two elementary (spherical) waves originating from the pinholes, weighted by the spectral amplitudes $V(\boldsymbol{\xi}_{pi}, \nu)$ in the pinholes,

$$V(\boldsymbol{\alpha}, \nu) = \frac{\lambda^2 V(\boldsymbol{\xi}_{p1}, \nu)}{\mathrm{i}\lambda z_1} \mathrm{e}^{\mathrm{i}kr_1} + \frac{\lambda^2 V(\boldsymbol{\xi}_{p2}, \nu)}{\mathrm{i}\lambda z_1} \mathrm{e}^{\mathrm{i}kr_2}. \tag{2.57}$$

This result is identical to applying the Rayleigh–Sommerfeld diffraction formula (2.16) to an aperture with two pinholes of diameter λ.

We call the vector between the two pinholes the *baseline vector* $\boldsymbol{B} = \boldsymbol{\xi}_{p1} - \boldsymbol{\xi}_{p2}$, and the distance between the pinholes $B = |\boldsymbol{\xi}_{p1} - \boldsymbol{\xi}_{p2}|$. The optical path difference (OPD) between the light from the two pinholes is the scalar product of the baseline vector $\boldsymbol{B}$ with the coordinate vector $\boldsymbol{\alpha}$, OPD = $\boldsymbol{\alpha} \cdot \boldsymbol{B}$.

Then, the spectral intensity, i.e., the intensity distribution of the diffraction pattern at frequency ν, is

---

[6] As discussed at the end of Sect. 2.2.1 we define a pinhole as a small aperture with a diameter of a few wavelengths with the approximation that diffraction effects can be calculated using a single spherical wave emerging from the centre of the aperture.

$$\begin{aligned} I(\boldsymbol{\alpha}, \nu) = \hat{\Gamma}(\boldsymbol{\alpha}, \boldsymbol{\alpha}, \nu) &= \lim_{T\to\infty} \frac{1}{2T} E\{V(\boldsymbol{\alpha}, \nu)V^*(\boldsymbol{\alpha}, \nu)\} \\ &= \frac{\lambda^2}{z_1^2} \lim_{T\to\infty} \frac{1}{2T} E\{|V(\boldsymbol{\xi}_{p1}, \nu)\,\mathrm{e}^{\mathrm{i}kr_1} + V(\boldsymbol{\xi}_{p2}, \nu)\,\mathrm{e}^{\mathrm{i}kr_2}|^2\} \\ &= \frac{\lambda^2}{z_1^2} \left(I(\boldsymbol{\xi}_{p1}, \nu) + I(\boldsymbol{\xi}_{p2}, \nu) + 2\mathrm{Re}\left[\hat{\Gamma}(\boldsymbol{B}, \nu)\mathrm{e}^{-\mathrm{i}k\boldsymbol{\alpha}\cdot\boldsymbol{B}}\right]\right), \end{aligned} \tag{2.58}$$

using $\boldsymbol{\alpha} \cdot \boldsymbol{B} = r_2 - r_1$ (see Fig. 2.14), and

$$\begin{aligned} I(\boldsymbol{\xi}_{pi}, \nu) &= \lim_{T\to\infty} \frac{1}{2T} E\{V(\boldsymbol{\xi}_{pi}, \nu)V^*(\boldsymbol{\xi}_{pi}, \nu)\}, i = 1, 2, \text{ and} \\ \hat{\Gamma}(\boldsymbol{B}, \nu) &= \lim_{T\to\infty} \frac{1}{2T} E\{V(\boldsymbol{\xi}_{p1}, \nu)V^*(\boldsymbol{\xi}_{p2}, \nu)\}. \end{aligned}$$

In the discussion following (2.45), we stated that the intensity in a plane illuminated by an incoherent source is a constant independent of the position $\boldsymbol{\xi}$ in the aperture plane. Assuming that the two pinholes are illuminated by an incoherent source at a large distance we find indeed that the intensity in the aperture plane is constant, $I(\boldsymbol{\xi}_{p1}, \nu) = I(\boldsymbol{\xi}_{p2}, \nu) = G(\nu)I_0$.

Now, we replace the MSDF, $\hat{\Gamma}(\boldsymbol{B}, \nu)$, by the product of spectrum and visibility function, $G(\nu)I_0\mu_\nu(\boldsymbol{B})$, see (2.52), and we write the real part in (2.58) as the product of its modulus and the cosine of its phase. We obtain the spectral intensity distribution of the diffraction pattern $I(\boldsymbol{\alpha}, \nu)$ in the plane of observation as

$$\begin{aligned} I(\boldsymbol{\alpha}, \nu) &= 2\frac{\lambda_0^2}{z_1^2} \left(G(\nu)I_0 + |G(\nu)I_0\mu_\nu(\boldsymbol{B})| \cos(\phi_\nu(\boldsymbol{B}) - k\boldsymbol{\alpha}\cdot\boldsymbol{B})\right) \\ &= 2G(\nu)I_0' \left(1 + |\mu_\nu(\boldsymbol{B})| \cos(\phi_\nu(\boldsymbol{B}) - k\boldsymbol{\alpha}\cdot\boldsymbol{B})\right), \end{aligned} \tag{2.59}$$

with λ_0 the mean wavelength, and $I_0' = \frac{\lambda_0^2}{z_1^2} I_0$. The result is a fringe pattern essentially proportional to $1 + \cos(.)$ like in (2.12). The fringe spacing in the direction parallel to $\boldsymbol{B}$ is λ/B. The modulus of the visibility function, $|\mu_\nu(\boldsymbol{B})|$, taking values between 0 and 1, acts as a damping factor on the cosine function, determining the contrast of the fringe pattern, and the phase of the visibility function, $\phi_\nu(\boldsymbol{B})$, determines the position of the white-light fringe.

While the simplifying assumption of equal intensities in the two pinholes permits to simplify the formulae without restricting their general validity, we should be aware that in practice, the intensities in two apertures are rarely constant. Writing the intensities in the individual apertures as $I(\boldsymbol{\xi}_{pi}, \nu) = G(\nu)I_{pi}$, with $i = 1, 2$, we modify (2.59), yielding

$$I(\boldsymbol{\alpha}, \nu) = G(\nu) \left(I_{p1}' + I_{p2}' + 2\sqrt{I_{p1}'I_{p2}'}|\mu_\nu(\boldsymbol{B})| \cos(\phi_\nu(\boldsymbol{B}) - k\boldsymbol{\alpha}\cdot\boldsymbol{B})\right). \tag{2.60}$$

The contrast of the fringe pattern is now given by $2(I'_{p1}I'_{p2})^{1/2}/(I'_{p1} + I'_{p2}) \times |\mu_\nu(\boldsymbol{B})|$. We will use this formula when discussing observations through atmospheric turbulence in Sect. 6.1. In the absence of turbulence, we will apply the simplified approach in (2.59).

The fringe pattern of a point source on axis is centred at the position of zero OPD, $\alpha = 0$. In case of a non-zero phase $\phi_\nu(\boldsymbol{B})$, the fringe pattern would be shifted by α_0, with $k\boldsymbol{\alpha}_0 \cdot \boldsymbol{B} = \phi_\nu(\boldsymbol{B})$. A phase of π is equivalent to replacing the cos(.) by −cos(.), replacing a fringe maximum by a minimum and vice versa.

Before we discuss this in detail we will step from the spectral intensity $I(\boldsymbol{\alpha}, \nu)$ to the white-light intensity by integrating over the observed frequency band, $I(\boldsymbol{\alpha}) = \int I(\boldsymbol{\alpha}, \nu)\mathrm{d}\nu$. For a sufficiently narrow spectral band we replace the factor λ^2 before the bracket by the average wavelength λ_0^2 since we are primarily interested in the influence of the wavelength on the shape and not in its contribution to the absolute intensity.

For the discussion of the polychromatic case we start by assuming a **point source** on axis, with $\mu_\nu(\boldsymbol{B}) = \int \delta(\boldsymbol{\alpha}') \exp(-\mathrm{i}k\boldsymbol{B} \cdot \boldsymbol{\alpha}')\mathrm{d}\boldsymbol{\alpha}' = 1$ in the aperture plane.

Integrating monochromatic fringe patterns over the spectral band, the integration is applied individually to each term in (2.59). The first term yields the white-light intensity in the aperture plane $\int G(\nu)I_0\mathrm{d}\nu = I_0$. The second term is again written as the real part of a complex function as in (2.58). Integrating over the frequency ν, we replace $k\boldsymbol{\alpha} \cdot \boldsymbol{B}$ by $2\pi\nu\tau$, writing the polychromatic intensity distribution as a function of time delay τ as

$$I(\tau) = 2I_0'\left(1 + \mathrm{Re}\left[\int G(\nu)\mathrm{e}^{-\mathrm{i}2\pi\nu\tau}\mathrm{d}\nu\right]\right). \tag{2.61}$$

Replacing τ by $\boldsymbol{\alpha} \cdot \boldsymbol{B}/c$ we would obtain the diffraction pattern as a function of $\boldsymbol{\alpha}$ again.

The spectrum $G(\nu)$ has its centroid at the average frequency ν_0. Introducing the centred spectrum $G_c(\nu) = G(\nu + \nu_0)$, we write

$$\begin{aligned} I(\tau) &= 2I_0'\left(1 + \mathrm{Re}\left[\int G_c(\nu - \nu_0)\mathrm{e}^{-\mathrm{i}2\pi\nu\tau}\mathrm{d}\nu\right]\right) \\ &= 2I_0'\left(1 + \mathrm{Re}\left[\int G_c(\nu')\mathrm{e}^{-\mathrm{i}2\pi\nu'\tau}\mathrm{d}\nu'\,\mathrm{e}^{-\mathrm{i}2\pi\nu_0\tau}\right]\right), \end{aligned} \tag{2.62}$$

using the coordinate transform $\nu' = \nu - \nu_0$, see Sect. A.1. The Fourier transform of the centred spectrum is called $g(\tau)$ that is a real function if $G_c(\nu)$ is symmetric, for example rectangular or gaussian.

The polychromatic fringe pattern of a point source can now be written as

$$I(\tau) = 2I_0'\big(1 + g(\tau)\cos(2\pi\nu_0\tau)\big), \tag{2.63}$$

with $g(\tau) = \mathcal{F}_\nu\left(G_c(\nu)\right)$ and $g(0) = 1$. $\mathcal{F}_\nu$ denotes the Fourier transform with respect to coordinate ν.

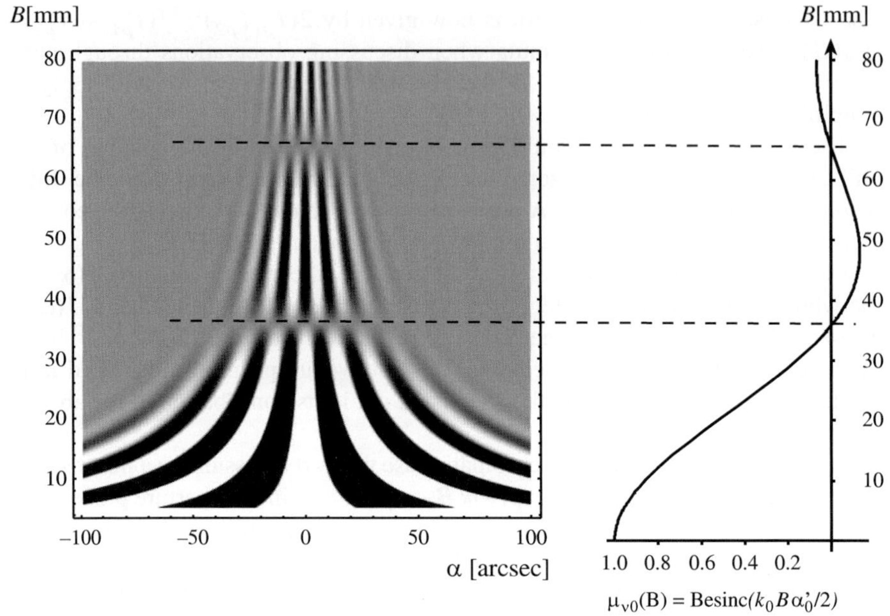

Fig. 2.15 Fringe patterns of Venus in Young's experiment. On the *left*, the intensity distributions along horizontal lines display the fringe pattern for pinhole separations B between 10 and 80 mm. The fringe spacing is λ_0/B. The light source illuminating the pinholes is Venus with a uniform disk diameter of 15 arcsec. The spectral band is the K-band (2.2±0.2 μm), i.e., $\lambda_0/\Delta\lambda = 5.5$. The finite spectral bandwidth makes the fringe visibility disappear for diffraction angles $|\boldsymbol{\alpha}|$ larger than about $5.5\lambda_0/B$, i.e., there are about 11 fringes in the fringe pattern (compare to (2.63)). The fringe spacing decreases with increasing pinhole separation, and the fringe visibility is reduced to zero at $B = 37$ mm. For 37 mm $\leq B \leq$ 67.5 mm, the fringe pattern inverts its sign displaying a *black* fringe at $\alpha = 0$. On the *right*, the visibility function $|\mu_{\nu_0}(\boldsymbol{B})|$ as a function of pinhole separation B is displayed following a Besinc-function (see also Fig. 2.13)

The variable part – the cosine function – of a fringe pattern with frequency ν_0 is multiplied by $g(\tau)$, the (real) Fourier transform of the (centred and symmetric) spectrum $G_c(\nu)$, that acts like an envelope on the fringe pattern. If the fringe pattern is observed in the K-band (2.2±0.2 μm) with $\Delta\lambda = 0.4$ μm and $\Delta\nu = 2.5 \times 10^{13}$ Hz, then $g(\tau)$ has its first zero for a time delay $\tau = \pm 1/\Delta\nu = \pm 4 \times 10^{-14}$ s (see Sect. A.1). This time delay corresponds to an optical path difference (OPD) equal to the coherence length $l_c = \tau c = \pm 12$ μm corresponding to $\pm 5.5\ \lambda_0$, i.e., the fringe pattern has about 11 perceivable fringes (see Fig. 2.15).

Returning to the general case of an **extended source** we write the integral of the spectral intensity distributions $I(\boldsymbol{\alpha}, \nu)$ (2.59) over the observed frequency band in the complex notation as

$$I(\boldsymbol{\alpha}) = \int I(\boldsymbol{\alpha}, \nu)\mathrm{d}\nu$$
$$= 2I_0'\left(1 + \int G(\nu)|\mu_\nu(\boldsymbol{B})|\cos\big(\phi_\nu(\boldsymbol{B}) - k\boldsymbol{\alpha}\cdot\boldsymbol{B}\big)\mathrm{d}\nu\right), \qquad (2.64)$$

or, with $k\boldsymbol{\alpha}\cdot\boldsymbol{B} = 2\pi\nu\tau$,

$$I(\tau) = 2I_0'\left(1 + \mathrm{Re}\left[\int G(\nu)\mu_\nu(\boldsymbol{B})\mathrm{e}^{-\mathrm{i}2\pi\nu\tau}\mathrm{d}\nu\right]\right). \qquad (2.65)$$

Disentangling the integral over the product of spectrum $G(\nu)$ and the visibility function $\mu_\nu(\boldsymbol{B})$ – both functions of ν – we can apply the same reasoning as after (2.47), bringing back the quasi-monochromatic approximation with the assumption of a narrow spectral bandwidth $\Delta\nu$, implicitly of a source of size $\boldsymbol{\alpha}_0$ that is larger than a point but still so small that its visibility function $\mu_\nu(\boldsymbol{B})$ does not vary over the spectral band.

We obtain the intensity distribution of the diffraction pattern of Young's experiment in quasi-monochromatic approximation as

$$I_{\mathrm{qm}}(\tau) = 2I_0'\left(1 + \mathrm{Re}\left[\mu_{\nu_0}(\boldsymbol{B})\int G(\nu)\mathrm{e}^{-\mathrm{i}2\pi\nu\tau}\mathrm{d}\nu\right]\right), \qquad (2.66)$$

with $\mu_{\nu_0}(\boldsymbol{B})$ the visibility function at ν_0, (2.51).

If the real part of the complex function is written as the modulus of $\mu_{\nu_0}(\boldsymbol{B})$ times the cosine of the phase we obtain the intensity distribution of the fringe pattern

$$I_{\mathrm{qm}}(\tau) = 2I_0'\big(1 + g(\tau)|\mu_{\nu_0}(\boldsymbol{B})|\cos(\phi_{\nu_0}(\boldsymbol{B}) - 2\pi\nu_0\tau)\big), \qquad (2.67)$$

or in the more familiar notation with diffraction angle $\boldsymbol{\alpha}$:

$$I_{\mathrm{qm}}(\boldsymbol{\alpha}) = 2I_0'\big(1 + g_B(\boldsymbol{\alpha})|\mu_{\nu_0}(\boldsymbol{B})|\cos(\phi_{\nu_0}(\boldsymbol{B}) - k_0\boldsymbol{\alpha}\cdot\boldsymbol{B})\big), \qquad (2.68)$$

with $k_0\boldsymbol{\alpha}\cdot\boldsymbol{B} = 2\pi\nu_0\tau$, and $g_B(\boldsymbol{\alpha}) = g(\boldsymbol{\alpha}\cdot\boldsymbol{B}/c) = g(\tau)$ using $\boldsymbol{\alpha}\cdot\boldsymbol{B} = \tau c$.

Like for the spectral intensity distribution in (2.59) the modulus of the visibility function determines the contrast of the fringe pattern around the white-light fringe, when $g_B(\boldsymbol{\alpha})$ is approximately constant, and the phase, $\phi_{\nu_0}(\boldsymbol{B})$, determines the position of the fringe pattern with respect to the position of zero OPD at $\alpha = 0$.

This result is identical to (2.12) when the two pinholes were illuminated by a monochromatic plane wave, so that $g_B(\boldsymbol{\alpha})$ is infinitely wide, since the modulus of the visibility function of a plane wave is $|\mu_{\nu_0}(\boldsymbol{B})| \equiv 1$ and the phase is $\phi_{\nu_0}(\boldsymbol{B}) \equiv 0$.

In Sect. 2.1.2, the contrast was called the visibility $\mathcal{V}$ of the fringe pattern and it was defined (2.13) as

$$\mathcal{V} = \frac{I_{\max} - I_{\min}}{I_{\max} + I_{\min}}. \qquad (2.69)$$

Comparing this definition with (2.68) it is easy to see that, in the ideal case, $|\mu_{\nu_0}(\boldsymbol{B})| = \mathcal{V}$ around the white-light fringe. Thus, we can measure the modulus of the visibility function by determining the fringe contrast in Young's experiment. However, if the fringe pattern is affected by atmospheric turbulence or other disturbances, we distinguish $\mathcal{V}$, the contrast of the measured fringe pattern, from the source visibility function μ_{ν_0}.

Very often, the visibility is interpreted as the quotient between the *correlated flux* and the *total flux* of the light since the visibility function is defined (2.51) as the quotient of the MCF, which is the correlation function of the optical disturbances, and the integrated intensity.

We can now understand the effect of a visibility function on the fringe pattern. Regarding for instance the visibility function of Venus, displayed in Fig. 2.13, that is shaped like a Besinc-function (2.53) we see in Fig. 2.15 that the visibility function shows up in the varying contrast of the fringe patterns. Each horizontal line shows a fringe pattern for an individual pinhole separation B, the contrast of which is determined by the modulus $|\mu_{\nu_0}(\boldsymbol{B})|$ of the visibility function.

The Besinc-function is a real function with zero phase. However, working with the modulus of the visibility function, negative values of the Besinc-function have to be accounted for by a phase of $\phi = \pi$, since $|\mu_{\nu_0}(\boldsymbol{B})|\mathrm{e}^{\mathrm{i}\pi} = -|\mu_{\nu_0}(\boldsymbol{B})|$. The fringe pattern is shifted by π, producing a black fringe at $\alpha = 0$ for pinhole separations between 37 mm and 67.5 mm when the Besinc-function has negative values (see Fig. 2.15).

The two pinholes can be regarded as an instrument to measure the coherence properties by probing the wave front with two pinholes, and determining the spatial coherence as the visibility of the fringe pattern. This result is interesting in two respects. First, we found how to relate a measurable quantity to the complex visibility function. Second, by doing so, we derived the visibility function in the aperture plane from characteristics of the intensity distribution (fringe contrast and centre fringe position) in the plane of observation.

2.4.2 *ABCD Method*

A method to derive both modulus and phase of the complex visibility from the fringe pattern was described by Shao [214], originally proposed by Wyant [254]. It is called the *ABCD method* since it relies on measuring the intensity at four different points around the white-light fringe of the fringe pattern $I_{\mathrm{qm}}(\boldsymbol{\alpha})$, (2.68), that are separated by 1/4 of the fringe spacing of λ_0/B. If we denote the intensities by I_A, I_B, I_C, I_D and their sum by I_{tot} we write

$$|\mu_{\nu_0}(\boldsymbol{B})| = \mathcal{V} = C\,\frac{\sqrt{(I_A - I_C)^2 + (I_B - I_D)^2}}{I_{\mathrm{tot}}}$$

$$\phi_{\nu_0}(\boldsymbol{B}) = \tan^{-1}\left(\frac{I_A - I_C}{I_B - I_D}\right). \tag{2.70}$$

The value of the constant C depends on the measurement method. If the four intensities are determined at individual points on the fringe pattern, it is $C = 2$. If the intensities are measured by pixels that are $\lambda/(4B)$ wide, integrating the intensity over $1/4$ of the fringe spacing and, thus, damping the fringe pattern, it is $C = \pi/\sqrt{2}$. However, this value is only of academic interest since the visibility always has to be calibrated for instance by observing a point source with a nominal visibility $\mu_{\nu_0} = 1$. This will be discussed in Sect. 6.1.

Thus, we can determine the visibility by applying the ABCD method to the white-light fringe. However, if it is difficult to identify the white-light fringe and by mistake another fringe in the field is used, its contrast is affected additionally by the temporal coherence, $g_B(\boldsymbol{\alpha})$, so that the estimate for $|\mu_{\nu_0}|$ is too small, while the phase estimate remains unaffected. Therefore, the ABCD method is particularly well suited to measure the fringe position.

As an alternative, the Fourier spectrum of the fringe pattern can be processed providing an estimator for the complex visibility function.

2.4.3 Power Spectrum of the Fringe Pattern

The fringe pattern can be regarded as a function of either the diffraction angle $\boldsymbol{\alpha}$ or of the time delay τ. The two quantities are linked through the optical path difference $\boldsymbol{\alpha} \cdot \boldsymbol{B} = \text{OPD} = \tau c$, with the baseline vector $\boldsymbol{B} = \xi_{p1} - \xi_{p2}$. In the following, we will choose the variable τ and we will perform the Fourier transform between time and frequency space. The properties of the Fourier spectrum will be discussed in detail before we introduce the power spectrum, which is the squared modulus of the Fourier spectrum.

We will first treat a monochromatic fringe pattern to develop a feeling for the process, and then we will discuss the general expression for the polychromatic diffraction pattern of Young's experiment (2.65).

The monochromatic fringe pattern is calculated in (2.65) by integrating the spectral intensity $I(\boldsymbol{\alpha}, \nu)$ (2.64) with a monochromatic spectrum $G(\nu) = \delta(\nu - \nu_0)$ as

$$I(\tau) = 2I_0' \left(1 + \text{Re}\left[\int G(\nu)\mu_\nu(\boldsymbol{B})\text{e}^{-\text{i}2\pi\nu\tau}\text{d}\nu\right]\right), \tag{2.71}$$

using $I_0' = \frac{\lambda_0^2}{z_1^2} I_0$. Then the Fourier transform $\hat{I}(\nu)$ of this fringe pattern reads as

$$\begin{aligned}\hat{I}(\nu) &= \int 2I_0' \left(1 + \text{Re}\left[\int \delta(\nu - \nu_0)\mu_\nu(\boldsymbol{B})\text{e}^{-\text{i}2\pi\nu\tau}\text{d}\nu\right]\right) \text{e}^{\text{i}2\pi\nu\tau}\,\text{d}\tau \\ &= I_0' \left(2\delta(\nu) + \delta(\nu - \nu_0)\mu_{\nu_0}(\boldsymbol{B}) + \delta(\nu + \nu_0)\mu_{\nu_0}^*(\boldsymbol{B})\right),\end{aligned} \tag{2.72}$$

Note that, depending on the properties of $I(\tau)$, $\hat{I}(\nu)$ can be a complex function.

Despite the syntax with time delay τ and frequency ν this is not a temporal Fourier transform since the replacement of the diffraction angle $\boldsymbol{\alpha}$ by the time difference τ does not make $I(\tau)$ a temporal signal but a signal of time difference. However, in analogy to (2.34) when introducing the self-coherence function, the coordinate ν is the frequency of the light, and we can draw conclusions on the shape of the spectrum $G(\nu)$ as will be discussed in the following.

The peak of the Fourier spectrum at $\nu = 0$, sometimes called the *photometric peak*, describes the constant component of the fringe pattern proportional to the constant intensity I_0 in the aperture plane. The δ-peaks at ν_0 and $-\nu_0$, sometimes called the *interferometric peak*, are weighted by the visibility function $\mu_{\nu_0}(\boldsymbol{B})$ at baseline $\boldsymbol{B}$ determining the amplitude of the cosine pattern.

The fringe pattern of a **polychromatic point source** was discussed in Sect. 2.4.1, and we found the intensity distribution of the fringe pattern (2.61) to be

$$I(\tau) = 2I_0'\left(1 + \mathrm{Re}\left[\int G(\nu)\mathrm{e}^{-\mathrm{i}2\pi\nu\tau}\mathrm{d}\nu\right]\right), \tag{2.73}$$

or, in its more familiar form (2.63), using $g(\tau) = \mathcal{F}_\nu\left(G(\nu+\nu_0)\right)$,

$$I(\tau) = 2I_0'\left(1 + g(\tau)\cos(2\pi\nu_0\tau)\right), \tag{2.74}$$

when the Fourier transform of the centred spectrum is the envelope of the cosine pattern reducing its visibility with increasing time delay τ, or diffraction angle $\boldsymbol{\alpha}$.

Using (2.61), it is straightforward to calculate the Fourier spectrum of the fringe pattern of a polychromatic point source as

$$\begin{aligned}\hat{I}(\nu) &= \int 2I_0'\left(1 + \mathrm{Re}\left[\int G(\nu)\mathrm{e}^{-\mathrm{i}2\pi\nu\tau}\,\mathrm{d}\nu\right]\right)\mathrm{e}^{\mathrm{i}2\pi\nu\tau}\,\mathrm{d}\tau \\ &= I_0'\left(2\delta(\nu) + G(\nu) + G(-\nu)\right),\end{aligned} \tag{2.75}$$

which is displayed in Fig. 2.16. If the width of the spectrum $G(\nu)$ increases, the width of its Fourier transform decreases and the fringe pattern loses contrast after fewer and fewer periods. Another way of putting this is to say that each individual frequency ν_i within the spectral band provides a cosine of frequency ν_i in the fringe pattern. Adding up these cosine functions results in a cosine pattern of the average frequency ν_0 that loses contrast depending on the spread of contributing frequencies, i.e., the width of the spectrum.

A similar exercise leads us to the Fourier spectrum of an **extended source**. We write the polychromatic fringe pattern of an extended source $I(\boldsymbol{\alpha})$ (2.65) as

$$I(\tau) = 2I_0'\left(1 + \mathrm{Re}\left[\int G(\nu)\mu_\nu(\boldsymbol{B})\mathrm{e}^{-\mathrm{i}2\pi\nu\tau}\,\mathrm{d}\nu\right]\right), \tag{2.76}$$

and perform the Fourier transform, yielding the Fourier spectrum as

$$\hat{I}(\nu) = I_0'\left(2\delta(\nu) + G(\nu)\mu_\nu(\boldsymbol{B}) + G(-\nu)\mu_\nu^*(\boldsymbol{B})\right). \tag{2.77}$$

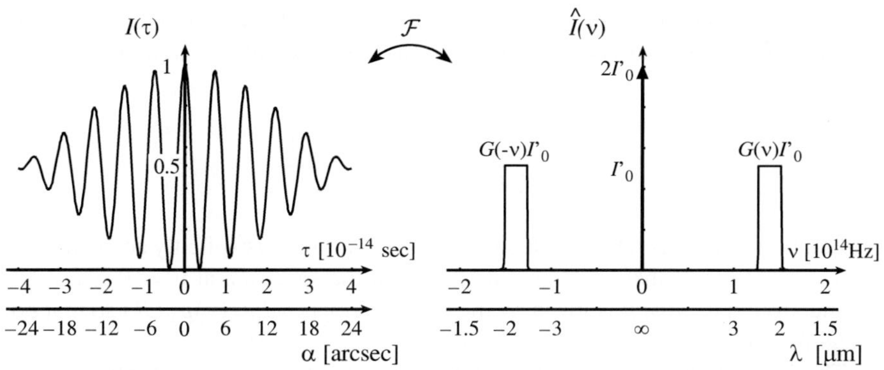

Fig. 2.16 On the *left*, the K-band fringe pattern of a point source for a pinhole separation of 10 cm is displayed both as a function of time difference τ and of diffraction angle $\boldsymbol{\alpha}$ (for $\boldsymbol{\alpha} \| \boldsymbol{B}$), linked through $\tau = \boldsymbol{\alpha} \cdot \boldsymbol{B}/c$. On the *right*, its Fourier transform is shown both as a function of frequency ν and of wavelength λ. The δ-peak is at $\nu = 0$, and the shape of the spectrum $G(\nu)$ is given by a rectangular function over the K-band, $\nu = 1.25 \times 10^{14} - 1.5 \times 10^{14}$ Hz (respectively $\lambda = 2.0 - 2.4\,\mu$m)

The peak of the Fourier spectrum at $\nu = 0$ describes again the constant component of the fringe pattern, proportional to the white-light intensity I_0 in the aperture plane. The width of the spectrum $G(\nu)$ and its mirror function at $-\nu$ is given by the width of the spectral band, and the values of the visibility function $\mu_\nu(\boldsymbol{B})$ over the spectral band are determined by the shape of the spectrum and by the spatial coherence at the given baseline $\boldsymbol{B}$.

Now it seems to be straightforward to determine the values of the spectrally resolved visibility function – in modulus and in phase – in the Fourier spectrum. However, in reality the situation is not quite that simple. Celestial bodies at a large distance are observed through the Earth's atmosphere so that the fringe pattern is temporally unstable and weak, and the white-light intensity I_0 in the aperture plane is neither temporally nor spatially constant. The implications will be discussed in detail in Chaps. 5 and 6.

Instead of exploiting the spectral information in the Fourier spectrum, one can do the opposite, averaging over the spectrum to improve the signal quality and directly measure the visibility function.

It was the idea of Roddier and Léna [196] to determine the average visibility function by calculating in the Fourier spectrum the ratio between the integral of the MSDF, the interferometric peak, over the spectral band and the integral of the δ-peak, the photometric peak, at $\nu = 0$.

Collapsing the spectrally resolved visibility function into a single value means to lose the spectral information. If the spectral band is sufficiently narrow and if the source is sufficiently small we can replace the visibility function by its value at the average frequency ν_0 yielding $\mu_{\nu_0}(\boldsymbol{B})$ in quasi-monochromatic approximation as given in the van Cittert–Zernike theorem (2.50).

Integrating the visibility function over the spectral band and dividing it by the integral of the spectrum at $\nu = 0$ yields the estimator for the visibility function $\mu_{\nu_0}(\boldsymbol{B})$ as

$$\frac{\int I'_0 G(\nu)\mu_\nu(\boldsymbol{B})\,\mathrm{d}\nu}{\int 2I'_0\delta(\nu)\mathrm{d}\nu} \approx \frac{\mu_{\nu_0}(\boldsymbol{B})}{2}. \tag{2.78}$$

Although for this theoretical deduction it is not required to explicitly account for I'_0, we must not forget that working with real data the spectrum at $\nu = 0$ is given by I'_0 and around $\nu = \nu_0$ it is given by the product $\hat{I}(\nu) = I'_0 G(\nu)\mu_\nu(\boldsymbol{B})$ so that both integrals in (2.78) have to be performed.

This conceptually simple computation provides a direct measure for the visibility of the fringe pattern. Due to the δ-peak at $\nu = 0$ being a real number we also measure the phase of the visibility function.

Discussing the quasi-monochromatic approximation in Sect. 2.4.1, we emphasised that $\mu_{\nu_0}(\boldsymbol{B})$ describes the fringe visibility only for small diffraction angles, i.e., around the white-light fringe, (2.68). It is the integral over the spectral band in (2.78) that reduces the information about the visibility over the full fringe pattern to the visibility around the white-light fringe as discussed before. The advantage of this method is that in case of noisy signals more values of the fringe pattern contribute to the measurement than just a few points around the white-light fringe.

The treatment of noise in the interferogram requires to calculate the power spectrum of the fringe pattern in order to have an unbiased measurement of the visibility function [224]. The power spectrum reads as

$$|\hat{I}(\nu)|^2 = I_0'^2\left(4\delta^2(\nu) + |G(\nu)\mu_\nu(\boldsymbol{B})|^2 + |G(-\nu)\mu^*_\nu(\boldsymbol{B})|^2\right), \tag{2.79}$$

and, using again the visibility value at the average frequency ν_0, we obtain the squared visibility function as

$$\frac{\int |I'_0 G(\nu)\mu_\nu(\boldsymbol{B})|^2\mathrm{d}\nu}{\int 4I_0'^2\delta^2(\nu)\mathrm{d}\nu} \approx \frac{|\mu_{\nu_0}(\boldsymbol{B})|^2}{4}. \tag{2.80}$$

While this quadratic estimator has advantages when measuring noisy signals, the phase $\phi_\nu(\boldsymbol{R})$ of the visibility function is lost (see Sect. 6.1.1). However, in practice it is very difficult to determine the position of the white-light fringe since atmospheric turbulence constantly moves the fringe pattern. Even when processing the fringe pattern directly instead of the power spectrum the phase cannot be determined unambiguously without further calibration methods (see Sect. 6.1).

The result of the Fourier transform of the fringe pattern is familiar from the *Michelson Fourier Spectrometer* measuring the source spectrum $G(\nu)$. There, the fringe pattern is created by splitting a light beam with a beam splitter and introducing an optical path length modulation in one of the two arms. The fringe pattern is then registered as a function of the OPD modulation after recombining the beams with a second beam combiner. Formally, the two spatial coordinates are then identical, $\boldsymbol{\xi}_{p1} = \boldsymbol{\xi}_{p2} = \boldsymbol{\xi}$ and $B = 0$, and we obtain the spectral intensity $I(\boldsymbol{\xi}, \nu) = G(\nu) \int I_b(\boldsymbol{\alpha}')\, \mathrm{d}\boldsymbol{\alpha}' = G(\nu) I_0$ with a homogeneous intensity distribution I_0.

In Young's experiment, we do not use beam splitters but two pinholes, and the spatial correlation between the two different points $\boldsymbol{\xi}_{p1}$ and $\boldsymbol{\xi}_{p2}$ across the wave front enters the result.

Thus, by Fourier transforming the polychromatic fringe pattern $I(\boldsymbol{\alpha})$ we receive information not only on the spectrum $G(\nu)$ but also on the spatial coherence properties of the light determined by the Fourier transform of the source brightness distribution. The term *double Fourier spatio-spectral interferometry* was phrased to emphasise this characteristic [150].

Example: A Uniform Disk

Stars can be modelled as a uniform disk, writing the source brightness distribution $I_b(\boldsymbol{\alpha}') = \frac{1}{\pi(\alpha_0'/2)^2} \mathrm{circ}\left(\frac{|\alpha'|}{\alpha_0'/2}\right)$ for a disk with diameter α_0'. This formula implies that the disk diameter is independent of the wavelength over the observed spectral band, which is a common assumption when attempting for instance to determine stellar diameters.

The visibility function is given by the spatial Fourier transform of the circular source brightness distribution (2.53) as

$$\mu_\nu(\boldsymbol{B}) = \mathrm{Besinc}\left(\frac{2\pi}{c}\, \nu B \frac{\alpha_0'}{2}\right), \tag{2.81}$$

with $\frac{2\pi}{c}\nu = \frac{2\pi}{\lambda}$.

So far, we have always discussed how the visibility function varies with pinhole separation $B = |\boldsymbol{\xi}_{p1} - \boldsymbol{\xi}_{p2}|$, and we have taken the wavelength as a fixed parameter. Here, we regard B as constant and discuss the Besinc as a function of ν, i.e., we regard the visibility function for a fixed pinhole separation at different frequencies. Although the disk diameter is wavelength independent, its Fourier transform, the Besinc-function, varies with wavelength since with the exponential kernel $\frac{2\pi}{\lambda} \boldsymbol{B} \cdot \boldsymbol{\alpha}'$ the Fourier transform is performed from $\boldsymbol{\alpha}'$ to $\boldsymbol{B}/\lambda = \boldsymbol{B}\nu/c$. If the stellar diameter varied with wavelength, the Besinc-function would be distorted reflecting the actual stellar shape at every frequency.

In Fig. 2.17 both the spectrum $G(\nu)$ and the Besinc-function are displayed as a function of ν. The consequences of the averaging process (2.78) over the spectral band are apparent in this figure. As long as the spectral band is sufficiently narrow for the visibility function to be linear within the band, the function value at the average frequency is a good estimate for the integral. It is very obvious that this approximation cannot be fulfilled with the same quality for the visibility function and for the squared visibility function.

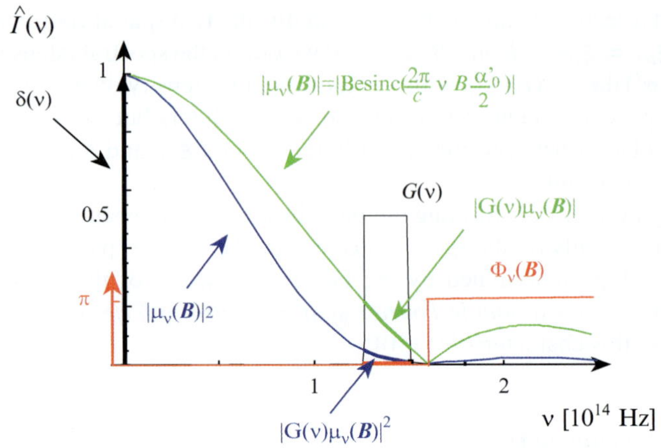

Fig. 2.17 The spectrum $\hat{I}(\nu)$ of a K-band fringe pattern of a uniform disk. The δ-peak at $\nu = 0$ and the visibility function $\mu_\nu(\boldsymbol{B})$ at a fixed baseline are displayed. The effective visibility function is the product of the spectrum $G(\nu)$ with the Besinc-function, which is the visibility function of a uniform disk. Since we discuss the visibility function as a complex function, the Besinc-function is split into its modulus (*green lines*) and its phase $\phi_\nu(\boldsymbol{B})$ (*red lines*). The phase is zero up to the first minimum of the Besinc-function and jumps to π beyond when the Besinc-function has negative values. For this example the pinhole separation is chosen as $B = 100\,\mathrm{m}$ and the stellar diameter as $\alpha'_0 = 4.7\,\mathrm{mas}$ yielding the first zero of the Besinc-function at $\nu_0 = 1.6 \times 10^{14}\,\mathrm{Hz}$, using $\nu_0/\mathrm{Hz} = 755/((B/\mathrm{m})(\alpha'_0/\mathrm{mas}))10^{14}$. The squared modulus of the Besinc-function (*blue lines*) is also displayed forming the power spectrum of the fringe pattern

2.4.4 Heuristic Approach

Since the connection between the coherence function and the intensity as a measurable quantity (2.68) is one of the fundamental pillars of stellar interferometry, we also look at it from a different perspective. Why is the fringe contrast affected by the shape of the source? We regard an individual point at angle position $\boldsymbol{\alpha}'_0$ on the surface of the source with intensity $I(\boldsymbol{\alpha}'_0)$. Then, the incoming plane wave is slightly tilted and there is a difference in arrival time at the pinholes that can be expressed as an OPD of $\boldsymbol{\alpha}'_0 \cdot \boldsymbol{B}$. The zero OPD position in the plane of observation is no longer on the optical axis but at position $\boldsymbol{\alpha} = -\boldsymbol{\alpha}'_0$ so that the incoming and outgoing OPDs, $\boldsymbol{\alpha}'_0 \cdot \boldsymbol{B}$ and $-\boldsymbol{\alpha} \cdot \boldsymbol{B}$ in Fig. 2.18, cancel.

We regard the fringe pattern in the plane of observation as given in (2.68), with $|\mu_\nu(\boldsymbol{B})| \equiv 1$ and $\phi_\nu(\boldsymbol{B}) \equiv 0$ for a point source on-axis in quasi-monochromatic approximation with a spectral width $\Delta\nu$ sufficiently narrow to allow for an OPD of several wavelengths without losing the fringe contrast. For an off-axis point source at angle position $\boldsymbol{\alpha}'_0$, the visibility function is $\mu_\nu(\boldsymbol{B}) = \exp(-ik\boldsymbol{\alpha}'_0 \cdot \boldsymbol{B})$ (using (2.51) with $\boldsymbol{\xi}_1 - \boldsymbol{\xi}_2 = \boldsymbol{B}$), and the intensity distribution reads as

$$I(\boldsymbol{\alpha}, \boldsymbol{\alpha}'_0) = 2\frac{\lambda^2}{z_1^2} I(\boldsymbol{\alpha}'_0)_\xi \Big(1 + \cos\big(-k(\boldsymbol{\alpha} + \boldsymbol{\alpha}'_0) \cdot \boldsymbol{B}\big)\Big).$$

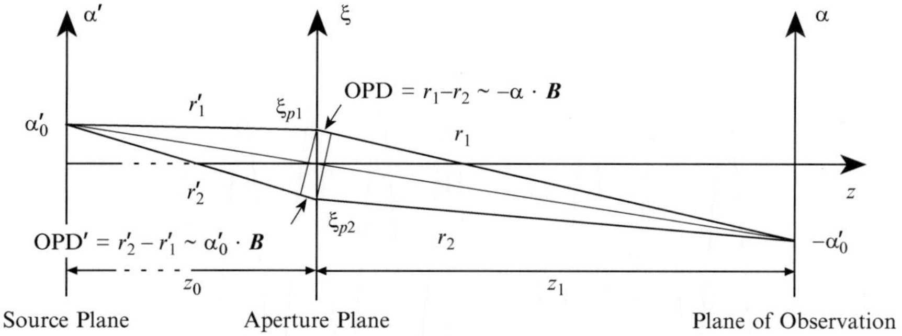

Fig. 2.18 Young's experiment with a point source at angle position $\boldsymbol{\alpha}'_0$. For the sake of simplicity, only one coordinate axis per plane is displayed. The optical path lengths between the source and the pinholes are slightly different with OPD' $= \boldsymbol{\alpha}'_0 \cdot \boldsymbol{B}$. The position of zero OPD in the plane of observation, i.e., the position where the total optical path lengths are equal, $r'_1 + r_1 = r'_2 + r_2$, is then at angle position $\boldsymbol{\alpha} = -\boldsymbol{\alpha}'_0$. This means that the centre of the fringe pattern, the white-light fringe, is shifted to $-\boldsymbol{\alpha}'_0$

The homogeneous intensity distribution in the aperture plane due to a source point at $\boldsymbol{\alpha}'_0$ is denoted by $I(\boldsymbol{\alpha}'_0)_\xi$.

The step from a point source to an extended source is done by summing up the fringe patterns of the individual points of the source, as in Fig. 2.19. This is correct since we assumed the light source to be spatially incoherent with every point radiating independently.

The subsequent integral over the fringe patterns of each point of an incoherent source with diameter $|\boldsymbol{\alpha}'_0|$ reads as

$$\begin{aligned} I(\boldsymbol{\alpha}) &= \int_{-|\alpha'_0|/2}^{|\alpha'_0|/2} I(\boldsymbol{\alpha}, \boldsymbol{\alpha}')\mathrm{d}\boldsymbol{\alpha}' \\ &= 2\frac{\lambda^2}{z_1^2}\left(I_0 + \int_{-|\alpha'_0|/2}^{|\alpha'_0|/2} I(\boldsymbol{\alpha}')_\xi \cos\bigl(-k(\boldsymbol{\alpha}+\boldsymbol{\alpha}')\cdot \boldsymbol{B}\bigr)\,\mathrm{d}\boldsymbol{\alpha}'\right), \end{aligned}$$

which is a convolution between the source intensity $I(\boldsymbol{\alpha}')$ and the fringe pattern of an on-axis point source.

The first term I_0 is a constant intensity as a result of the integration over the source $I(\boldsymbol{\alpha}')$. The second term is the real part of the complex Fourier transform $\int I(\boldsymbol{\alpha}') \exp(-\mathrm{i}k(\boldsymbol{\alpha}+\boldsymbol{\alpha}')\cdot \boldsymbol{B})\mathrm{d}\boldsymbol{\alpha}'$. We extract $\exp(-\mathrm{i}k\boldsymbol{\alpha}\cdot\boldsymbol{B})$ since this term does not depend on $\boldsymbol{\alpha}'$. The remaining expression $\int I(\boldsymbol{\alpha}')\exp(-\mathrm{i}k\boldsymbol{\alpha}'\cdot\boldsymbol{B})\mathrm{d}\boldsymbol{\alpha}' = \mu_\nu(\boldsymbol{B})I_0$ is the van Cittert–Zernike theorem as given by (2.51) with $\boldsymbol{\xi}_1 - \boldsymbol{\xi}_2 = \boldsymbol{B}$. Reducing the discussion to the real part again we obtain the final result as in (2.68),

$$I(\boldsymbol{\alpha}) = 2I'_0\bigl(1 + |\mu_\nu(\boldsymbol{B})|\cos(\phi(\boldsymbol{B}) - k\boldsymbol{\alpha}\cdot\boldsymbol{B})\bigr),$$

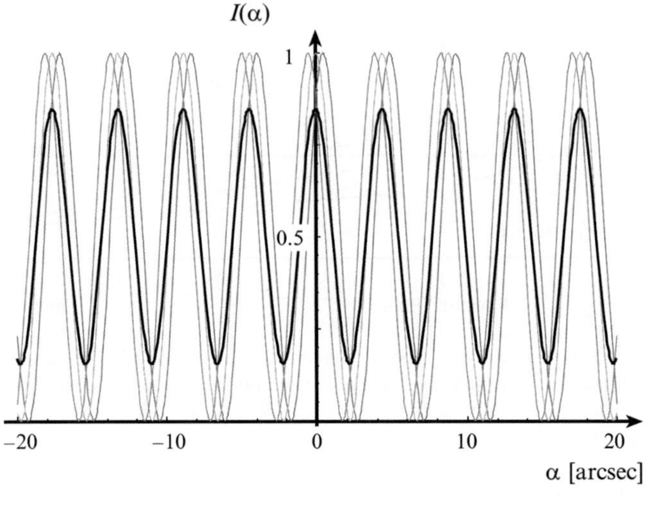

Fig. 2.19 Three individual monochromatic intensity distributions of the fringe patterns in Young's experiment for point source positions at $|\boldsymbol{\alpha}'| = -1$ arcsec, 0 and +1 arcsec (*grey lines*), and the resulting fringe pattern of an extended source with diameter 2 arcsec. $\boldsymbol{\alpha}'$ and $\boldsymbol{\alpha}$ are parallel to the baseline vector $\boldsymbol{B}$. The pinhole separation B is 10 cm, the observing wavelength $\lambda = 2.2\,\mu$m and the fringe spacing $\lambda/B = 4.5$ arcsec. The resulting fringe pattern is reduced in contrast since it is the sum of the individual fringes. Note that the contrast would be even smaller if either the pinhole separation were larger – and the fringe spacing smaller – or if the source were larger in diameter

describing a fringe pattern following in principle a 1+cos(.) function with a contrast of $|\mu_\nu(\boldsymbol{B})|$. It is $I_0' = \frac{\lambda_0^2}{z_1^2} I_0$.

The interpretation of the phase ϕ of the visibility function μ_ν is now rather simple. Returning to the process of adding up fringe patterns originating from individual points one can think of cases when the resulting fringe pattern does not have its white-light fringe at $\alpha = 0$. For example a source that is not extended between + and $-|\boldsymbol{\alpha}_0'/2|$ but between 0 and $|\boldsymbol{\alpha}_0'|$ would display the white-light fringe at $\boldsymbol{\alpha} = -\boldsymbol{\alpha}_0'/2$.

In the language of the coherence functions this is another way of saying that the phase of the visibility function is $-k(\boldsymbol{\alpha}_0'/2) \cdot \boldsymbol{B}$. If the source were not constant in intensity but, e.g., be brighter on one side, then fringe patterns with varying intensities have to be added causing the resulting fringe pattern to be shifted sideways. Thus, it is essential to measure not only the modulus but also the phase of the visibility function, i.e., the position of the white-light fringe, if one wants to determine the position of the star or to obtain its shape, which means its image.

The answer to the question at the beginning of this section, why the fringe contrast is affected by the source shape, is thus, that fringe patterns with slightly different white-light fringe positions are overlaid loosing contrast by this very process. The formal treatment of this integration reintroduces the coherence functions as discussed in Sect. 2.4.1.

Figure 2.20 illustrates the equivalence of regarding individual source points and regarding individual wavelengths. In both cases, the intensity distributions either of individual source points or of individual wavelengths are integrated. While the width of the spectrum $G(\nu)$, in this case the K-band, determines the temporal coherence, the width of the angular intensity distribution $I(\boldsymbol{\alpha}')$, in this case the diameter of the star, determines the spatial coherence. The temporal coherence affects the visibility as a function of diffraction angle $\boldsymbol{\alpha}$ that translates into a time delay τ (Fig. 2.20b), the spatial coherence affects the visibility as a function of baseline $\boldsymbol{B}$ (Fig. 2.20c). Thus, short baselines display a fringe pattern with higher visibility than long baselines.

In this context, a monochromatic source with infinitely narrow spectrum – or an emission line source – $G(\nu)$ is equivalent to a point source with infinitely narrow angular intensity distribution $I(\boldsymbol{\alpha}')$, since the fringe pattern of the monochromatic

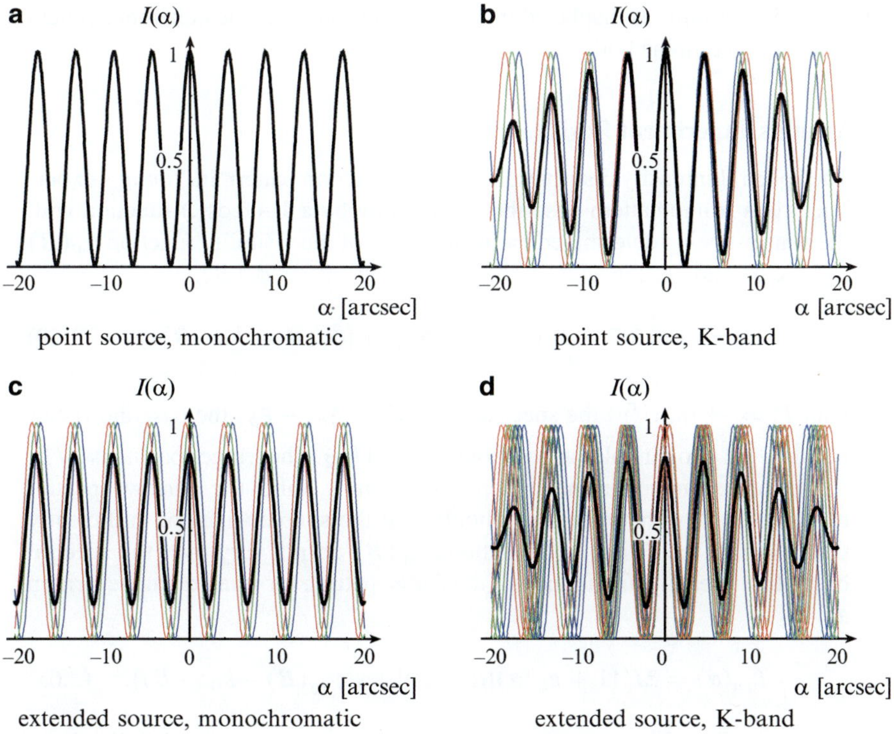

Fig. 2.20 Summary of the influence of source size and spectral bandwidth on the fringe pattern. The pinhole separation B is 10 cm in all figures, and we only observe at diffraction angles $\boldsymbol{\alpha}$ with $\boldsymbol{\alpha} \| \boldsymbol{B}$. In (**a**) an individual fringe pattern for an observing wavelength of $\lambda = 2.2\,\mu$m and a point source is displayed. In (**b**) the K-band fringe pattern is shown when observing a point source (as in Fig. 2.3). In (**c**) the monochromatic illumination of a source with diameter 2 arcsec produces a fringe pattern with reduced contrast (as in Fig. 2.19). In (**d**) the resulting fringe pattern in K-band illumination with a 2 arcsec source is displayed. The visibility is reduced around $\alpha = 0$ due to the source diameter and it is further reduced for increasing diffraction angles $\boldsymbol{\alpha}$ due to the finite spectral bandwidth

source has a constant visibility for any value of the diffraction angle $\boldsymbol{\alpha}$, and the fringe pattern of a point source displays a constant visibility for any baseline $\boldsymbol{B}$.

It is important to note the different starting point of the discussion in this section compared to the preceding sections. Before, we discussed the propagation of the coherence functions from the source to the aperture plane dealing with the coherence properties along the way. Here, on the contrary, we have regarded a monochromatic point source, for which we calculated the diffraction pattern, and we have drawn our conclusions from integrating the diffraction patterns of many point sources, thus introducing heuristically the concept of coherence at the very end of the process. This is a very convenient method to explain and understand the impact of coherence properties. We will see in Chap. 3 that by summarising integrals in different ways one can focus on different aspects of the imaging process.

In the real world, when stars are neither point sources nor monochromatic one could calculate two sets of fringe patterns, for each point and for each wavelength, and then do the sums as displayed in Fig. 2.20. Or one uses the coherence functions that elegantly combine both cases.

Young's Experiment Revisited: Summary

The spectral intensity distribution of the diffraction pattern in Young's experiment is a fringe pattern described essentially by a $1 + \cos(.)$ function with the contrast being determined by the modulus of the visibility function $\mu_\nu(\boldsymbol{B})$ and the position of the white-light fringe by its phase, $\phi_\nu(\boldsymbol{B})$,

$$I(\boldsymbol{\alpha}, \nu) = 2G(\nu)I_0' \left(1 + |\mu_\nu(\boldsymbol{B})| \cos\!\left(\phi_\nu(\boldsymbol{B}) - k\boldsymbol{\alpha} \cdot \boldsymbol{B}\right)\right), \qquad (2.59)$$

with $I_0' = \frac{\lambda_0^2}{z_1^2} I_0$, $G(\nu)$ the spectrum and $\boldsymbol{B} = \boldsymbol{\xi}_{p1} - \boldsymbol{\xi}_{p2}$ the baseline vector between the two pinholes at positions $\boldsymbol{\xi}_{p1}$ and $\boldsymbol{\xi}_{p2}$. The fringe spacing is λ/B.

The quasi-monochromatic approximation is valid for narrow spectral bands, $\Delta\nu \ll \nu$ that means implicitly that the sources must be sufficiently small so that their visibility functions $\mu_\nu(\boldsymbol{B})$ do not vary over the spectral band. The intensity distribution of the fringe pattern for small $\boldsymbol{\alpha}$ can be written as

$$I_{\mathrm{qm}}(\boldsymbol{\alpha}) = 2I_0'\left(1 + g_B(\boldsymbol{\alpha})|\mu_{\nu_0}(\boldsymbol{B})| \cos(\phi_{\nu_0}(\boldsymbol{B}) - k_0\boldsymbol{\alpha} \cdot \boldsymbol{B})\right), \qquad (2.68)$$

with $k_0\boldsymbol{\alpha} \cdot \boldsymbol{B} = 2\pi\nu_0\tau$, ν_0 the average frequency, $\mu_{\nu_0}(\boldsymbol{B})$ the visibility function at ν_0, (2.51), and $g_B(\boldsymbol{\alpha}) = g_B(\tau c/B) = g(\tau)$ the envelope function of the fringe pattern, which is the Fourier transform of the spectrum $G(\nu)$.

The visibility function $\mu_{\nu_0}(\boldsymbol{B})$ is a function of baseline vector $\boldsymbol{B}$ and not of the individual coordinates $\boldsymbol{\xi}_{p1}$ and $\boldsymbol{\xi}_{p2}$ (see Fig. 2.14). For very small diffraction angles around the white-light fringe, when $g_B(\boldsymbol{\alpha})$ is approximately constant, the visibility of the fringe pattern is given by $|\mu(\boldsymbol{B})|$. The phase $\phi_{\nu_0}(\boldsymbol{B})$ determines the position of the fringe pattern.

The *ABCD method* can be used to derive both modulus and phase of the complex visibility from the fringe pattern [214,254]. It relies on measuring the intensity around the white-light fringe at four different points that are spaced by $1/4$ of the fringe spacing λ/B. If we denote the intensities by I_A, I_B, I_C, I_D and their sum by I_{tot}, we write

$$|\mu_{\nu_0}(\boldsymbol{B})| = \mathcal{V} = C\frac{\sqrt{(I_A - I_C)^2 + (I_B - I_D)^2}}{I_{\mathrm{tot}}}$$
$$\phi_{\nu_0}(\boldsymbol{B}) = \tan^{-1}\left(\frac{I_A - I_C}{I_B - I_D}\right). \tag{2.70}$$

The value of the constant C is $\pi/\sqrt{2}$ if the intensities are measured with pixels that are $\lambda/(4B)$ wide. In practice, the visibility always has to be calibrated, for instance by observing a point source with a nominal visibility $\mathcal{V} = 1$.

Applying the ABCD method to any other fringe than the white-light fringe provides too small an estimate for $|\mu_{\nu_0}|$ since the fringe visibility $\mathcal{V}$ is reduced by the temporal coherence.

Returning to polychromatic illumination with spectrum $G(\nu)$, the Fourier transform (2.77) of the fringe pattern – expressed as a function of τ, using $\tau c = \boldsymbol{\alpha B}$ – provides information on the temporal and the spatial spectrum of the source (see Fig. 2.17) without the restriction of the quasi-monochromatic approximation to the white-light fringe:

$$\hat{I}(\nu) = I_0'\left(2\delta(\nu) + G(\nu)\mu_\nu(\boldsymbol{B}) + G(-\nu)\mu_\nu^*(\boldsymbol{B})\right), \tag{2.77}$$

displaying a peak at $\nu = 0$, and the visibility function $\mu_\nu(\boldsymbol{B})$ and its mirror function $\mu_\nu^*(\boldsymbol{B})$ (see Fig. 2.16). In the absence of spectral information of the source, one can only measure the integral of the visibility function over the spectral band.

Under the assumption that the visibility function is linear over the spectral band the value of the integral is approximately equal to the value of the visibility function at the average frequency ν_0 as requested in quasi-monochromatic approximation.

The ratio of the integral of the visibility function over the spectral band in the Fourier transform, the interferometric peak, (2.77) and the integral of the peak at $\nu = 0$, the photometric peak, is a measure for the average visibility function $\mu_{\nu_0}(\boldsymbol{B})$, with

$$\frac{\int I_0' G(\nu)\mu_\nu(\boldsymbol{B},\nu)\,\mathrm{d}\nu}{\int 2I_0'\delta(\nu)\mathrm{d}\nu} \approx \frac{\mu_{\nu_0}(\boldsymbol{B})}{2}. \tag{2.78}$$

If the signals are noisy the power spectrum of the fringe pattern provides an unbiased estimate of the visibility as

$$\frac{\int |I_0' G(\nu)\mu_\nu(\boldsymbol{B},\nu)|^2 \mathrm{d}\nu}{\int 4 I_0'^2 \delta^2(\nu)\mathrm{d}\nu} \approx \frac{|\mu_{\nu_0}(\boldsymbol{B})|^2}{4}. \tag{2.80}$$

Up to now, we have focused on determining the visibility function that we expressed as the Fourier transform of the source brightness distribution according to the van Cittert–Zernike theorem. This incorporates the possibility to reconstruct the source brightness distribution from the visibility function through a Fourier back transform, i.e., to form an image. For this it is required to measure the visibility function for many different vectors $\boldsymbol{B}$. However, without the phase of the visibility function the imaging capability is very limited.

We have, thus, introduced the topic of imaging while originally discussing Young's experiment to explain the basic principles of the measurement of the coherence function. We will see in Chap. 3 that the conceptual step towards an imaging system like a telescope and a stellar interferometer is relatively small.

2.5 Higher Order Correlation Functions: Intensity Interferometry

In 1949, R. Hanbury Brown developed the idea of correlating the intensities measured by individual telescopes rather than the amplitudes. The motivation was to determine stellar diameters using very long baselines avoiding – coming from radio interferometry – the use of local oscillators. First results at radio wavelengths were published only three years after the first idea [94]. Together with R.Q. Twiss, Hanbury Brown developed the theory, first for radio interferometers [95] and then for electromagnetic waves in general [97]. An intensity interferometer for the visible was eventually built by them near Narrabri in Australia in the early 1960s [96]. The history and theory of intensity interferometry are competently summarised in Hanbury Brown's book *The Intensity Interferometer* [93].

We will concentrate in the following on the concept of intensity interferometers in the general context of coherence functions, and we will discuss the example of a binary star to explain the concept.

We start by defining the instantaneous intensity

$$i(\boldsymbol{\xi},t) = v(\boldsymbol{\xi},t)v^*(\boldsymbol{\xi},t) \tag{2.82}$$

Table 2.1 This table summarizes the propagation of light in Young's experiment. Three different quantities, the MSDF, the spectral and the white-light intensity, are regarded in three planes (see Fig. 2.14): the source plane with an incoherent source, the aperture plane with two pinholes, and the plane of observation. In the source plane, the situation is completely described by the spectral intensity distribution $I(\boldsymbol{\alpha}', \nu)$ since the MSDF $\hat{\Gamma}(\boldsymbol{\alpha}'_1, \boldsymbol{\alpha}'_2, \nu)$ of an incoherent source only has non-zero values for $\boldsymbol{\alpha}'_1 = \boldsymbol{\alpha}'_2$, (2.44). The MSDF in the aperture plane is the product of the spectrum $G(\nu)$ and the Fourier transform of the source brightness distribution $I_b(\boldsymbol{\alpha}')$, (2.45). The latter is called the visibility function μ_ν, (2.52), which – taken at the average frequency ν_0 of a narrow spectral band – forms the core of the van Cittert–Zernike theorem in quasi-monochromatic approximation providing the visibility function $\mu_{\nu_0}(\boldsymbol{\xi}_1 - \boldsymbol{\xi}_2)$ in the aperture plane, (2.51). The MSDF in the plane of observation $\hat{\Gamma}(\boldsymbol{\alpha}_1, \boldsymbol{\alpha}_2, \nu)$ will be given in (3.48) in Sect. 3.3.1. This quantity is only of limited practical interest, and one would have to design another interferometer experiment to measure it. The white-light intensity $I(\boldsymbol{\alpha})$ is the quantity that is measured by an optical detector. In the source plane, the white-light intensity is given by the source brightness distribution. The aperture plane is illuminated homogeneously by the source, and the intensity takes the constant value I_0. In the image plane, the quasi-monochromatic approximation has to be applied to obtain a solution of the integral over the spectrum, (2.59), yielding a fringe pattern proportional to $1 + \cos(.)$, when the contrast of the fringes is determined by the modulus of the visibility $|\mu_{\nu_0}(\boldsymbol{B})|$, (2.68). The baseline vector is defined as $\boldsymbol{B} = \boldsymbol{\xi}_{p_1} - \boldsymbol{\xi}_{p_2}$

Source plane Incoherent star	Aperture plane with two Pinholes at $\boldsymbol{\xi}_{p1}$ and $\boldsymbol{\xi}_{p2}$	Plane of observation with $\boldsymbol{B} = \boldsymbol{\xi}_{p1} - \boldsymbol{\xi}_{p2}$
	MSDF	
$\hat{\Gamma}(\boldsymbol{\alpha}'_1, \boldsymbol{\alpha}'_2, \nu)$ $= \frac{\lambda^2}{z_0^2} G(\nu) I_b(\boldsymbol{\alpha}'_1) \delta(\boldsymbol{\alpha}'_1 - \boldsymbol{\alpha}'_2)$ (2.44) with $\boldsymbol{\alpha}' = \boldsymbol{x}/z_0$	$\hat{\Gamma}(\boldsymbol{\xi}_1 - \boldsymbol{\xi}_2, \nu)$ $= G(\nu) \int I_b(\boldsymbol{\alpha}') \mathrm{e}^{-\mathrm{i}k(\boldsymbol{\xi}_1 - \boldsymbol{\xi}_2)\cdot\boldsymbol{\alpha}'} \,\mathrm{d}\boldsymbol{\alpha}'$ (2.45) $= G(\nu) I_0 \mu_\nu(\boldsymbol{\xi}_1 - \boldsymbol{\xi}_2)$ (2.52) Quasi-monochromatic approx.: Visibility function $\mu_{\nu_0}(\boldsymbol{\xi}_1 - \boldsymbol{\xi}_2)$ $= \frac{\int I_b(\boldsymbol{\alpha}') \mathrm{e}^{-\mathrm{i}k_0(\boldsymbol{\xi}_1 - \boldsymbol{\xi}_2)\cdot\boldsymbol{\alpha}'} \,\mathrm{d}\boldsymbol{\alpha}'}{\int I_b(\boldsymbol{\alpha}') \,\mathrm{d}\boldsymbol{\alpha}'}$ (2.51)	$\hat{\Gamma}(\boldsymbol{\alpha}_1, \boldsymbol{\alpha}_2, \nu)$ $= 2\frac{\lambda^2}{z_1^2}\Big(G(\nu) I_0 \cos\big(2\pi \frac{\boldsymbol{R}_B}{2} \cdot (\boldsymbol{\alpha}_1 - \boldsymbol{\alpha}_2)\big)$ $+ \lvert\hat{\Gamma}(\boldsymbol{R}_B, \nu)\rvert \cos\big(\hat{\phi}(\boldsymbol{R}_B, \nu) - 2\pi \frac{\boldsymbol{R}_B}{2} \cdot (\boldsymbol{\alpha}_1 + \boldsymbol{\alpha}_2)\big)\Big)$ (3.48) in Sect. 3.3.1, with $\boldsymbol{R}_B = \boldsymbol{B}/\lambda$

(*continued*)

Table 2.1 (Continued)

Source plane Incoherent star	Aperture plane with two Pinholes at $\boldsymbol{\xi}_{p1}$ and $\boldsymbol{\xi}_{p2}$	Plane of observation with $\boldsymbol{B} = \boldsymbol{\xi}_{p1} - \boldsymbol{\xi}_{p2}$
	Spectral intensity	
$I(\boldsymbol{\alpha}', \nu)$ $= G(\nu) I_b(\boldsymbol{\alpha}')$	$I(\boldsymbol{\xi}, \nu) = \hat{\Gamma}(\boldsymbol{\xi} - \boldsymbol{\xi}, \nu)$ $= G(\nu) \int I_b(\boldsymbol{\alpha}')\, \mathrm{d}\boldsymbol{\alpha}'$ $= G(\nu) I_0$	$I(\boldsymbol{\alpha}, \nu) = \hat{\Gamma}(\boldsymbol{\alpha}, \boldsymbol{\alpha}, \nu)$ $= 2G(\nu) I_0'$ $\times \big(1 + \lvert\mu_\nu(\boldsymbol{B})\rvert \cos\left(\phi_\nu(\boldsymbol{B}) - k\boldsymbol{\alpha} \cdot \boldsymbol{B}\right)\big)$ (2.59)
	White-light intensity	
$I(\boldsymbol{\alpha}')$ $= \int I(\boldsymbol{\alpha}', \nu) \mathrm{d}\nu$ $= \int G(\nu) I_b(\boldsymbol{\alpha}') \mathrm{d}\nu = I_b(\boldsymbol{\alpha}')$	$I(\boldsymbol{\xi})$ $= \int I(\boldsymbol{\xi}, \nu) \mathrm{d}\nu$ $= \int G(\nu) I_0 \mathrm{d}\nu = I_0$	$I(\boldsymbol{\alpha}) = \int I(\boldsymbol{\alpha}, \nu) \mathrm{d}\nu = 2I_0'$ $\times \big(1 + \int G(\nu) \lvert\mu_\nu(\boldsymbol{B})\rvert \cos\left(\phi_\nu(\boldsymbol{B}, \nu) - k\boldsymbol{\alpha} \cdot \boldsymbol{B}\right) \mathrm{d}\nu\big)$ (2.64) Quasi-monochromatic approximation: $I_{\mathrm{qm}}(\boldsymbol{\alpha}) = 2I_0'$ $\times \big(1 + g_B(\boldsymbol{\alpha}) \lvert\mu_{\nu_0}(\boldsymbol{B})\rvert \cos\left(\phi_{\nu_0}(\boldsymbol{B}) - k_0 \boldsymbol{\alpha} \cdot \boldsymbol{B}\right)\big)$ (2.68)

as the product of optical disturbances. The instantaneous intensity like the optical disturbance is a fast oscillating signal that cannot be measured directly. Therefore, in Sect. 2.1, the intensity was defined as the time average (2.5) of this product, reading

$$I(\boldsymbol{\xi}) = < i(\boldsymbol{\xi}, t) > . \tag{2.83}$$

In the present context however, we will discuss the instantaneous intensity before calculating the time average. This is mathematically correct since the ergodicity of the random emission process of light allows us first to calculate statistical averages like the correlation and then to apply temporal averaging [87].

When introducing the coherence function as the second order moment of optical disturbances in Sect. 2.3.1 it was not necessary to make any assumptions on the form of the probability density function of the optical disturbances. Now, computing the correlation of instantaneous intensities and, thus, the *fourth order moment* of optical disturbances, we make use of the common assumption that optical disturbances $v(\boldsymbol{\xi}, t)$ follow a circular Gaussian random process [87] meaning that the real and imaginary part are independent, identically distributed zero-mean Gaussian random numbers. Then, higher order moments are reduced to second order moments [147], and the correlation of instantaneous intensities, measured at two different points $\boldsymbol{\xi}_1, \boldsymbol{\xi}_2$ in the aperture plane (e.g., at two telescopes) and at different times t_1, t_2, is written as

$$\begin{aligned} < i(\boldsymbol{\xi}_1, t_1) i(\boldsymbol{\xi}_2, t_2) > &= < v(\boldsymbol{\xi}_1, t_1) v^*(\boldsymbol{\xi}_1, t_1) v^*(\boldsymbol{\xi}_2, t_2) v(\boldsymbol{\xi}_2, t_2) > \\ &= < v(\boldsymbol{\xi}_1, t_1) v^*(\boldsymbol{\xi}_1, t_1) >< v^*(\boldsymbol{\xi}_2, t_2) v(\boldsymbol{\xi}_2, t_2) > \\ &\quad + < v(\boldsymbol{\xi}_1, t_1) v^*(\boldsymbol{\xi}_2, t_2) >< v^*(\boldsymbol{\xi}_1, t_1) v(\boldsymbol{\xi}_2, t_2) > \\ &= < i(\boldsymbol{\xi}_1, t_1) >< i(\boldsymbol{\xi}_2, t_2) > + | < v(\boldsymbol{\xi}_1, t_1) v^*(\boldsymbol{\xi}_2, t_2) > |^2 \\ &= I(\boldsymbol{\xi}_1) I(\boldsymbol{\xi}_2) + |\Gamma(\boldsymbol{\xi}_1 - \boldsymbol{\xi}_2, \tau)|^2, \end{aligned} \tag{2.84}$$

with τ the time difference $t_1 - t_2$. Being a stationary random process the correlation only depends on the time difference τ.

Since we are interested in the fluctuations of the intensity we write down the intensity covariance as

$$\begin{aligned} < i(\boldsymbol{\xi}_1, t + \tau) i(\boldsymbol{\xi}_2, t) >_{\mathrm{cov}} &= < i(\boldsymbol{\xi}_1, t + \tau) i(\boldsymbol{\xi}_2, t) > \\ &\quad - < i(\boldsymbol{\xi}_1, t + \tau) >< i(\boldsymbol{\xi}_2, t) > \\ &= | < v(\boldsymbol{\xi}_1, t + \tau) v^*(\boldsymbol{\xi}_2, t) > |^2 \\ &= |\Gamma(\boldsymbol{\xi}_1 - \boldsymbol{\xi}_2, \tau)|^2. \end{aligned} \tag{2.85}$$

Thus, the intensity covariance is the squared modulus of the MCF $\Gamma(\boldsymbol{\xi}_1 - \boldsymbol{\xi}_2, \tau)$, which is the correlation function of the optical disturbances (2.27).

We should keep in mind that in practice the intensities are measured in the focal plane of each telescope individually. There is no interference of the amplitudes in a common plane of observation as in Young's experiment. The coordinate $\boldsymbol{\xi}_i$ denotes the center of each telescope aperture, and the fact that the average of the intensity over the telescope aperture is actually measured does not affect the conclusion of this discussion.

One major shortcoming of the covariance of two intensities is that the phase of the MCF is lost when the squared modulus of the MCF is computed, and real images cannot be obtained. For binary stars this means that their separation can be measured with high resolution but one cannot determine which one of the two is the "brighter" star if they are of unequal intensity. More generally speaking, the symmetric content of the image can be determined but not the asymmetric content (see Sect. A.1).

Computing the **triple correlation** of intensities, the phase of the MCF can be recovered. We measure the intensities at three points $\boldsymbol{\xi}_1, \boldsymbol{\xi}_2, \boldsymbol{\xi}_3$ (three telescopes) and at three moments in time t_1, t_2 and t_3, writing the triple correlation as

$$
\begin{aligned}
< i_1(t_1) i_2(t_2) i_3(t_3) > &= \int i_1(t+\tau_1) i_2(t+\tau_2) i_3(t) \mathrm{d}t \\
&= < |v_1(t+\tau_1)|^2 |v_2(t+\tau_2)|^2 |v_3(t)|^2 >, \qquad (2.86)
\end{aligned}
$$

when the subscript i indicates the position $\boldsymbol{\xi}_i$ where the intensity is measured, and $t_{1/2}$ are replaced by $t + \tau_{1/2}$ and t_3 by t.

Under the same assumption as above, that the optical disturbances follow a circular Gaussian random process, we can reduce the sixth order moment of the disturbances to combinations of second order moments [147], yielding

$$
\begin{aligned}
< i_1(t+\tau_1) i_2(t+\tau_2) i_3(t) > = {} & I_1 I_2 I_3 \\
& + I_1 |\Gamma_{23}(\tau_2)|^2 + I_2 |\Gamma_{13}(\tau_1)|^2 + I_3 |\Gamma_{12}(\tau_1 - \tau_2)|^2 \\
& + |\Gamma_{12}(\tau_1 - \tau_2)| |\Gamma_{23}(\tau_2)| |\Gamma_{13}(\tau_1)| \\
& \times \cos(\phi_{12}(\tau_1 - \tau_2) + \phi_{23}(\tau_2) - \phi_{13}(\tau_1)).
\end{aligned}
\qquad (2.87)
$$

Similar to (2.84), we have the sum of the intensity products and of the squared MCF $|\Gamma_{ij}(\tau_i)|^2$ plus a term depending on the sum of MCF phases $\phi_{12}(\tau_1 - \tau_2) + \phi_{23}(\tau_2) - \phi_{13}(\tau_1)$. The latter is called the *closure phase* because it is the sum of phases around a closed loop of three telescopes. Repeating this measurement for a large number of configurations one can recover the phases ϕ_{ij} of the individual baselines from the closure phase [9, 84].

The intensity covariance in (2.85) represents the ideal result if the instantaneous intensities $i(\boldsymbol{\xi}, t)$ were measured with infinite temporal resolution. However, only the time averaged intensity is available. Applying a temporal averaging over a period T to the instantaneous intensities on the left hand side of (2.85) we have to calculate the time average of $|\Gamma(\boldsymbol{\xi}_1 - \boldsymbol{\xi}_2, \tau)|^2$, now with respect to the time difference τ:

$$
|\Gamma(\boldsymbol{\xi}_1 - \boldsymbol{\xi}_2, \tau)|_T^2 = \frac{1}{T} \int_{\tau - T/2}^{\tau + T/2} |\Gamma(\boldsymbol{\xi}_1 - \boldsymbol{\xi}_2, \tau')|^2 \mathrm{d}\tau'. \qquad (2.88)
$$

If T is much smaller than the width of the MCF, the time averaged signal is very similar to the original signal. Since the width of the MCF is determined by the coherence time τ_c the intensities would have to be measured with a temporal resolution better than the coherence time. However, even for a very narrow spectral bandwidth $\Delta\nu$ of 1/1,000 of the frequency, e.g., $\nu \approx 10^{15}$ Hz in the visible, the coherence time is approximately $\tau_c = 10^{-12}$ s, which is beyond the state of the art of today's detectors resolving signals only down to approximately $T = 10^{-9}$ s.

If T is much larger than the width of the MCF, the time average of $|\Gamma(\boldsymbol{\xi}_1 - \boldsymbol{\xi}_2, \tau)|^2$ is proportional to the coherence time τ_c [146]. The result is that the time average of the MCF and thus the covariance of time averaged intensities (2.85) are attenuated by a factor of $\tau_c/T = \Delta f/\Delta\nu$ compared to an ideal detector, reducing its sensitivity. $\Delta f = 1/T$ is the detector bandwidth. Diminishing the spectral bandwidth $\Delta\nu$ reduces the intensity so that the overall situation does not change.

Therefore, the signal-to-noise ratio (SNR) of this measurement is independent of the spectral bandwidth $\Delta\nu$ but it is proportional (1) to the spectral intensity of the source – strictly speaking to the number of detected photo electrons per unit optical bandwidth and per unit time –, (2) to $\sqrt{\Delta f}$, the square root of the detector bandwidth, and (3) to the square root of the integration time, limiting the sensitivity for a 5σ SNR to stars of approximately magnitude 5 for 10-m class telescopes and 10-min observations [97]. This does not compare very favourably to amplitude interferometry reaching stars of magnitude 10 in a few 10 msec.

The averaging process does not affect the behaviour of the MCF with respect to the spatial coordinates $\boldsymbol{\xi}_i$ that is determined by the shape of the object. Thus, the purpose of the intensity interferometer to measure the squared MCF as a function of baseline $\boldsymbol{B} = \boldsymbol{\xi}_1 - \boldsymbol{\xi}_2$ can be pursued.

Example: A Binary Star

The simple example of a binary star will give us an idea why the covariance of intensities turns out to be the square of the correlation of optical disturbances, i.e., the square of the MCF, and why only small frequency differences lead to a measurable signal. However, the reasoning in the following on the plausibility of the concept cannot replace a thorough mathematical analysis that can be found in [87, 93].

We place the individual stars at positions $+\alpha_0'/2$ and $-\alpha_0'/2$ and we assume that Fraunhofer conditions (see Sect. 2.2.2) apply for the setup displayed in Fig. 2.21. Then, the optical path length r_1 for the light from the star at $\alpha_0'/2$ can be approximated by $z_1 + \xi_1\alpha_0'/2$, and r_2 by $z_1 - \xi_1\alpha_0'/2$. We ignore the constant distance z_1, giving rise to a constant phase term, and we write the optical disturbance with frequency ν at position ξ_1 as the sum of the individual contributions

$$v(\xi_1, t) = V_0 \cos\left(k\xi_1 \frac{\alpha_0'}{2} + 2\pi\nu t + \varphi_1\right) + V_0 \cos\left(-k\xi_1 \frac{\alpha_0'}{2} + 2\pi\nu t + \varphi_2\right). \tag{2.89}$$

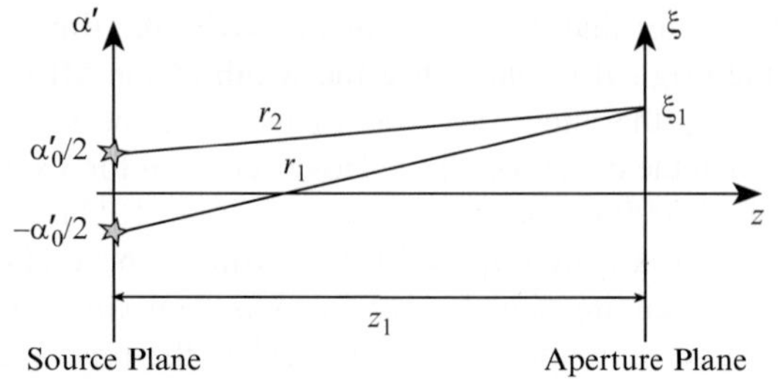

Fig. 2.21 A binary star separated by α'_0 at a distance z_1 from the aperture plane with coordinate ξ. The optical path lengths r_1 and r_2 between the individual stars and the position ξ_1 are displayed. The separation of the binary and the distance allows for the application of the Fraunhofer approximation with $r_{1/2} \approx z_1 \pm \xi_1\alpha'_0/2$

$\varphi_{1/2}$ are random phase terms considering the fact that the light from the two stars although formally of the same frequency will have a variable phase difference due to the random emission process in each star.

We can now write the instantaneous intensity as

$$\begin{aligned} i(\xi_1, t) &= v^2(\xi_1, t) \\ &= 4V_0^2 \cos^2\left(k\xi_1 \frac{\alpha'_0}{2} + \frac{\varphi_1 - \varphi_2}{2}\right) \cos^2\left(2\pi\nu t + \frac{\varphi_1 + \varphi_2}{2}\right). \end{aligned} \tag{2.90}$$

Since we cannot measure signals with a time resolution of the optical frequency we apply a moving time average over a period T longer than $1/\nu$ but shorter than the typical fluctuation of the random phases φ_1 and φ_2.

Then the second $\cos^2$ term that is a function of ν averages to $\frac{1}{2}$. We call the averaged signal $i_T(\xi_1, t)$, yielding

$$i_T(\xi_1, t) = \frac{1}{T}\int_{t-T}^{t} i(\xi_1, t')\mathrm{d}t' = 2V_0^2 \cos^2\left(k\xi_1 \frac{\alpha'_0}{2} + \frac{\Delta\varphi}{2}\right), \tag{2.91}$$

with $\Delta\varphi = \varphi_1 - \varphi_2$. This is a periodic signal, like a fringe pattern, as a function of the coordinate ξ_1 in the aperture plane. The period length is determined by the wavelength and by the separation of the binary. The time dependence of $i_T(\xi_1, t)$ is determined by the randomly varying values of $\Delta\varphi$. If the random phases were zero we formally would have a coherent binary in (2.89) – which is the same as a pair of pinholes illuminated by a plane wave – displaying a fringe pattern as in (2.59).

The random phases φ_1 and φ_2 were introduced because the phase of the wave with frequency ν changes randomly each time wave trains are emitted by individual atoms. If the integration time T in (2.91) is extended beyond the typical time constant of the fluctuations of the random phases, i.e., of their difference $\Delta\varphi$, the $\cos^2$ function is averaged to $\frac{1}{2}$, and one obtains the familiar expression of a constant intensity $i_T(\xi_1, t) = V_0^2$ in the aperture plane that is illuminated by two stars at a large distance.

What happens if light of two different frequencies interferes? We first have to add optical disturbances of two different frequencies ν_1 and ν_2 in the first and second term of (2.89), eventually obtaining the intensity time average over T as

$$i_T(\xi_1, t) = 2V_0^2 \cos^2\left(k\xi_1 \frac{\alpha_0'}{2} + \frac{\Delta\varphi}{2} + \pi(\nu_1 - \nu_2)t\right). \tag{2.92}$$

This means that as long as the frequency difference, the beat frequency, $\delta\nu = \nu_1 - \nu_2$ is smaller than $1/T$, the contribution of these two frequencies to the intensity is temporally resolved. However, we still cannot determine the separation α_0' of the binary because of the random phase $\Delta\varphi$ taking different values for each measurement of $i_T(\xi_1, t)$.

If the integration time T is both longer than $1/\delta\nu$ and longer than the time constant of the phase difference $\Delta\varphi$ then the fluctuations average out and the intensity is constant with $i_T(\xi_1, t) = 2V_0^2$.

Regarding the interference of optical disturbances with different frequencies in (2.92) seems to contradict our former statements that interference processes can be computed by first determining the spectral intensity, and then by integrating these intensities over the spectrum finding the polychromatic intensity (see e.g. Sect. 2.1). However, this approximation is limited to the cases when the integration time T is longer than the time scales of the fluctuations involved. Here in (2.92), we assume explicitly that the T is shorter than the fluctuations so that extra terms have to be considered. In the preceding sections, we were not interested in high frequency fluctuations of the intensity but we discussed the (long) time average intensity of the fringe pattern in order to determine the coherence function in the aperture plane.

The computation of the correlation between the two measured signals $i_T(\xi_1, t + \tau)$ and $i_T(\xi_2, t)$ with frequency ν – at positions ξ_1 and ξ_2 and at times $t + \tau$ and t – is now straightforward. With (2.91) and $V_0^2 = 1$ we obtain the product of the intensities as

$$\begin{aligned} i_T(\xi_1, t+\tau) i_T(\xi_2, t) &= 4\cos^2\left(k\xi_1 \frac{\alpha_0'}{2} + \frac{\Delta\varphi}{2}\right) \cos^2\left(k\xi_2 \frac{\alpha_0'}{2} + \frac{\Delta\varphi}{2}\right) \\ &= \left(1 + \cos(k\xi_1\alpha_0' + \Delta\varphi)\right)\left(1 + \cos(k\xi_2\alpha_0' + \Delta\varphi)\right) \\ &= 1 + \cos(k\xi_1\alpha_0' + \Delta\varphi) \\ &\quad + \frac{1}{2}\cos(k(\xi_1 - \xi_2)\alpha_0') + \frac{1}{2}\cos(k(\xi_1 + \xi_2)\alpha_0' + 2\Delta\varphi) \\ &\quad + \cos(k\xi_2\alpha_0' + \Delta\varphi). \end{aligned} \tag{2.93}$$

The correlation is the time average over a period that is long enough to average over all fluctuations of the random process, i.e., much longer than T (see Sect. 2.3.1). Then, all terms containing $\Delta\varphi$ in (2.93) disappear since their average is zero, yielding

$$< i_T(\xi_1, t+\tau) i_T(\xi_2, t) > = 1 + \frac{1}{2}\cos(k(\xi_1 - \xi_2)\alpha'_0)$$
$$= \frac{1}{2} + \cos^2\left(k(\xi_1 - \xi_2)\frac{\alpha'_0}{2}\right). \quad (2.94)$$

This is the correlation of the intensities at ξ_1 and ξ_2 when observing a binary star separated by α'_0 in monochromatic illumination at frequency ν. As in the general result in (2.84) we have a constant term and we found the squared modulus of the MCF of a binary (see (3.75)).

If we now look again at the signal with two different frequencies ν_1 and ν_2 with a frequency difference $\delta\nu$ we obtain the intensity correlation as

$$< i_T(\xi_1, t+\tau) i_T(\xi_2, t) >= \frac{1}{2} + \cos^2\left(k(\xi_1 - \xi_2)\frac{\alpha'_0}{2} + \pi\delta\nu\tau\right). \quad (2.95)$$

This is the result if the integration time T is shorter than $1/\delta\nu$. If the beat frequency $\delta\nu$ is larger than $1/T$ then the individual intensities $i_T(\xi_i, t)$ are constant and the intensity correlation is itself constant (see discussion after (2.92)) providing a constant background signal.

In our deduction, we assumed that one star radiates with frequency ν_1 and its neighbour with ν_2. If the spectrum were properly considered by having both stars emit light of two frequencies, ν_1 and ν_2, the $\cos^2$ function in (2.95) would be replaced by a product of two $\cos^2$ functions yielding

$$< i_T(\xi_1, t+\tau) i_T(\xi_2, t) > = 1 + \cos^2\left(k(\xi_1 - \xi_2)\frac{\alpha'_0}{2}\right)\cos^2(\pi\delta\nu\tau) \text{ and}$$
$$< i_T(\xi_1, t+\tau) i_T(\xi_2, t) >_{\text{cov}} = \cos^2\left(k(\xi_1 - \xi_2)\frac{\alpha'_0}{2}\right)\cos^2(\pi\delta\nu\tau). \quad (2.96)$$

Thus, the intensity covariance as a measure for the intensity fluctuations is exactly the squared modulus of the MCF of a binary observed at two frequencies. The MCF of a binary as the Fourier transform of its intensity distribution is $\cos(kB\alpha'_0/2)$, with $B = \xi_1 - \xi_2$, and $\cos(\pi\delta\nu\tau)$ is the Fourier transform of the spectrum consisting of two spectral lines at ν_1 and ν_2. For this result, we assumed that the integrals over ν and $\boldsymbol{\alpha}'$ in the van Cittert–Zernike theorem can be treated individually (see discussion following (2.47)).

As discussed above, only those frequencies contribute to the intensity fluctuations that are separated by less than $\delta\nu = 1/T$ and all those that are further apart contribute to the background signal. The temporal resolution of the available detectors limits T to about 10^{-9} s. Observing a finite spectrum that is usually much wider than 10^9 Hz, the measured signal of the intensity covariance is the squared MCF attenuated by $\delta\nu/\Delta\nu$, with $\Delta\nu$ the width of the observed spectrum [87]. Discussing the time average of the squared MCF (2.88) we came to the same conclusion stating that the fluctuations are reduced by a factor of τ_c/T, with $\tau_c = 1/\Delta\nu$ the coherence time.

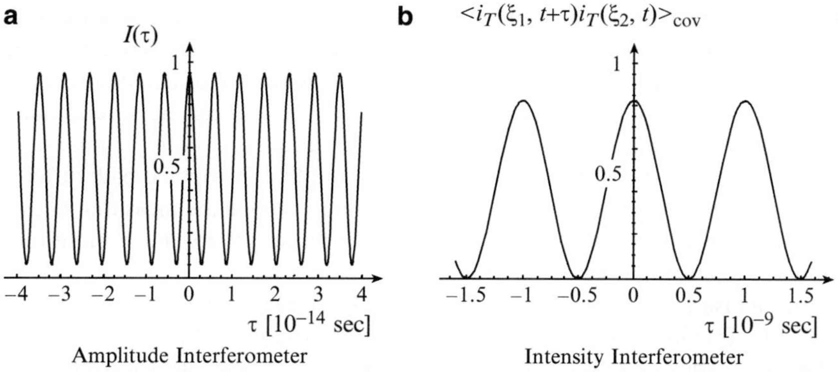

Fig. 2.22 The measurement of the MCF (**a**) with an amplitude interferometer, e.g., Young's experiment, and (**b**) with an intensity interferometer. The observed binary star is separated by $\alpha'_0 = 6$ mas and two spectral lines are accounted for, centred around $\lambda = 2.2\,\mu$m, i.e., $\nu = 1.37 \times 10^{14}$ Hz, and separated by $\delta\nu = 10^9$ Hz. The separation of the two pinholes in the aperture plane is the baseline $B = \xi_1 - \xi_2 = 10$ m. The fringe pattern in Young's experiment, (**a**), which is the intensity distribution in the plane of observation due to interference of the amplitudes, here as a function of time delay τ, displays a visibility of about 0.9 according to the binary's MCF $\cos(kB\alpha'_0/2)$ for the given parameters. With an intensity interferometer, (**b**), the measured signal is the covariance between the intensities when the two intensity detectors replace the pinholes in the aperture plane. Displayed is the covariance signal (2.96) given by $\cos^2(\pi\delta\nu\tau)$ attenuated by $\cos^2(kB\alpha'_0/2) = 0.9^2$, the latter being the squared modulus of the MCF

Figure 2.22 shows the comparison of interferometric measurements of a binary star, separated by $\alpha'_0 = 6$ mas, with an amplitude interferometer (Fig. 2.22a) and with an intensity interferometer (Fig. 2.22b). Out of the K-band, two spectral lines, separated by $\delta\nu = 10^9$ Hz, are used. The separation of the pinholes is $B = 10$ m. The amplitude interferometer as in Young's experiment shows a fringe pattern as a function of time delay τ with a fringe period of $1/\nu = 0.73\times10^{-14}$ s (see Fig. 2.22a) equivalent to $\lambda/B = 45$ mas fringe spacing. For these parameters the MCF of the binary, $\cos(kB\alpha'_0/2)$, has a value of 0.9 displayed by the reduced fringe contrast. The beat frequency of $\delta\nu = 10^9$ Hz is about 10^5 times smaller than the average frequency of $\nu = 1.37 \times 10^{14}$ Hz at $2.2\,\mu$m, and gives rise to a periodic envelope, $\cos(\pi\delta\nu\tau)$, of the fringe pattern with a period length of about 10^5 fringes, i.e., 2×10^{-8} s that cannot be displayed in Fig. 2.22a.

The square $\cos^2(\pi\delta\nu\tau)$ of this envelope with a period length of the reciprocal of the beat frequency of $1/\delta\nu = 10^{-9}$ s describes the intensity covariance (2.96) in an intensity interferometer as displayed in Fig. 2.22b. Here, the detectors are placed at the positions of the pinholes. The 6 mas separation of the binary reduces the maximum of the covariance to $\cos^2(kB\alpha'_0/2) = 0.9^2$. The latter, the spatial part of the MCF, is the quantity that we want to determine.

Writing these results as a function of OPD instead of time delay τ, using OPD $= \tau c$, we see that while the fringe spacing in the amplitude interferometer corresponds to an OPD of λ, i.e., $2.2\,\mu$m in the example above, the spacing of two maxima in

Fig. 2.22b corresponds to an OPD of $c/\delta\nu = 30\,\text{cm}$. This comparison illustrates the advantage of the intensity interferometer over the amplitude interferometer when it comes to real interferometers. If two telescopes are used and the detectors are placed at the end of an optical system, the optical path lengths can be several hundred metres long including a large number of optical surfaces. In both the intensity and the amplitude interferometer one has to make sure that the optical path difference, OPD, is about zero so that we know how to define $\tau = 0$ in the measurement setup. However, the requirements for an intensity interferometer are much more relaxed compared to an amplitude interferometer.

For precise measurements with an amplitude interferometer, the OPD has to be stable to a fraction of the wavelength, since the fringe pattern moves by one fringe if the OPD varies by λ. With an intensity interferometer, the relevant fringe spacing is about 30 cm (see Fig. 2.22) so that the required OPD accuracy is in the centimetre range.

By the same token, the random variations of the optical path lengths of several micrometers caused by atmospheric turbulence (see Chap. 4) do not disturb the measurement with an intensity interferometer since they are far below the relevant stability requirement of some centimetres.

At the time when Hanbury Brown came up with his idea this was an enormous advantage since the technology was not available to control the OPD with sub-μm accuracy as required for an amplitude interferometer.

In the meantime, the problems of OPD control have been resolved, and amplitude interferometers are in operation. Considering the severely limited sensitivity of intensity interferometry (see discussion following (2.88)) there have been no new attempts of building intensity interferometers. However, some new ideas based on using a large number of telescopes, that are relatively cheap due to the low requirement on optical quality, might lift the sensitivity to useful levels [173, 174].

Chapter 3
Imaging Process: Propagation Through Optical Systems

So far, we have discussed the propagation of light – of the amplitude and of the coherence function – through free space, from an incoherent light source at a very large distance to the plane of observation, and its diffraction at single and multiple apertures. We introduced Young's experiment as an instrument to measure the coherence function by determining contrast and position of the diffraction fringes. In Sect. 2.2, the diffraction of light at an aperture was investigated, and the case of an optical system was made by introducing the lens.

In this chapter, we will link these two formalism describing what is effectively the imaging process in optical systems.

This process involves multiple integrals that can be grouped in different ways in order to focus on different aspects of the propagation process. We will mainly restrict the discussion to incoherent light sources, and we will assume that Fresnel conditions apply and only the intensity is measured in the image plane. For the sake of completion we will take little excursions discussing coherent light sources and the coherence properties in the image plane.

In the first section, Fourier Optics as the "classical" imaging theory will be discussed, introducing the optical transfer function and optical aberrations. The second section will describe the propagation of the coherence function through optical systems. Regarding again the intensity distribution in the image plane, we will find a formalism that is identical to Fourier Optics. Two special cases, imaging of coherent sources and the coherence function in the image plane will be treated at the end.

In the third section, we will apply this formalism to interferometers, first discussing Young's Experiment as an imaging system and then interferometers with apertures of finite size. A brief section will be devoted to investigating the coherence properties of the light in the fringe pattern. We will look at the implications of the object size and the field of view connecting the findings in this chapter on imaging to the general concept of the quasi-monochromatic approximation in the first chapter. In the final section, the uv-plane coverage and image reconstruction methods will be discussed.

A. Glindemann, *Principles of Stellar Interferometry*, Astronomy and Astrophysics Library, DOI 10.1007/978-3-642-15028-9_3,

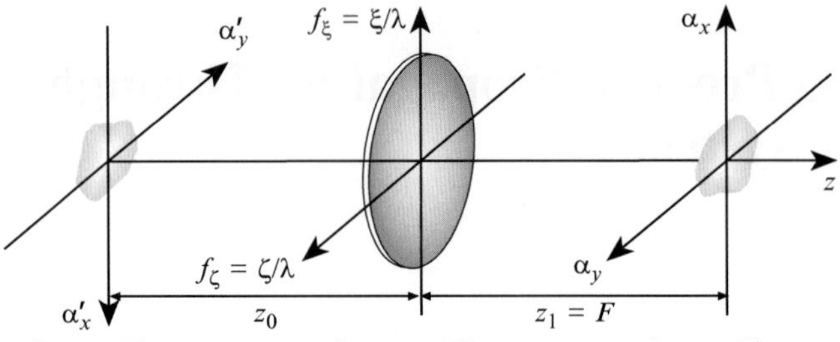

Fig. 3.1 The geometry of the imaging process with a source at a very large distance z_0. The aperture plane contains a thin lens of focal length F and diameter D, imaging the source into the image plane, in this case the focal plane, at distance F from the aperture plane. The coordinates in the source and in the image plane are angles $\boldsymbol{\alpha}'$ and $\boldsymbol{\alpha}$ respectively, and the z axis represents the optical axis. The axes in the source plane are deliberately inverted in order to consider the fact that the object is imaged upside down. Note that using a single lens for imaging, the angular magnification factor of the imaging process is always unity. In the aperture plane spatial frequencies $\boldsymbol{f}_\xi = (f_\xi, f_\zeta) = \boldsymbol{\xi}/\lambda$ are used

3.1 Fourier Optics

Assuming monochromatic illumination, we discuss the propagation of the amplitude through an optical system. This process can be separated into two steps. First, we calculate the propagation from the object into the aperture plane (see Fig. 3.1). Then, we calculate the propagation further into the image plane. For convenience, the optical system is represented by a single focussing element, a thin lens. Then, the entrance and the exit pupils[1] lie on top of each other in the aperture plane resulting in an angular *magnification factor* $m_p = \boldsymbol{\alpha}/\boldsymbol{\alpha}' = 1$. Optical systems with more than one optical element have the entrance and exit pupils at different positions and the angular magnification factor can be $m_p \neq 1$.

Spatial frequencies defined as $\boldsymbol{f}_\xi = \boldsymbol{\xi}/\lambda$ are introduced in the aperture plane. For the time being, we can regard $\boldsymbol{f}_\xi$ as a spatial coordinate calibrated by the wavelength, and we express the aperture function A as a function of $\boldsymbol{f}_\xi$. It should be noted that the diameter of a circular aperture, as a function of $\boldsymbol{f}_\xi$, is now wavelength dependent, $f_{\xi,D} = D/\lambda$. In the following section on Fourier optics, we will see why it is advantageous to define the aperture as a function of $\boldsymbol{f}_\xi$ and why the interpretation as a spatial frequency is adequate.

We first regard a single source point on-axis at $\alpha' = 0$ emitting a monochromatic spherical wave. Assuming that the source is at a very large distance z_0, an approximately plane wave arrives in the aperture plane. Introducing the thin lens, $L(\boldsymbol{f}_\xi) = \exp(-\mathrm{i}\ 2\pi|\boldsymbol{f}_\xi|^2\lambda/(2F))$, in the aperture plane, we write the monochromatic amplitude $V(\boldsymbol{\alpha})$ in the image plane, similar to (2.21), as

[1] The entrance and the exit pupils are images of the aperture that are produced by the optical elements before and after the real aperture respectively.

$$\begin{aligned} V(\boldsymbol{\alpha}) &= \frac{\lambda}{\mathrm{i}f} \int V_{\mathrm{ap}}(\boldsymbol{f}_\xi) L(\boldsymbol{f}_\xi) \mathrm{e}^{\mathrm{i}2\pi |\boldsymbol{f}_\xi|^2 \lambda/(2z_1)}\, \mathrm{e}^{-\mathrm{i}2\pi \boldsymbol{f}_\xi \cdot \boldsymbol{\alpha}}\, \mathrm{d}\boldsymbol{f}_\xi \\ &= \frac{\lambda}{\mathrm{i}f} \int V_{\mathrm{ap}}(\boldsymbol{f}_\xi) \mathrm{e}^{-\mathrm{i}2\pi \boldsymbol{f}_\xi \cdot \boldsymbol{\alpha}}\, \mathrm{d}\boldsymbol{f}_\xi\,, \end{aligned} \tag{3.1}$$

where the quadratic phase term $|\boldsymbol{f}_\xi|^2\lambda/(2z_1)$ due to Fresnel diffraction cancels with the quadratic phase term $-|\boldsymbol{f}_\xi|^2\lambda/(2F)$ that was introduced by the lens with focal length F, since we observe in the focal plane with $z_1 = F$, as discussed in Sect. 2.2.2. Replacing $\mathrm{d}\boldsymbol{\xi}$ in (2.22) by $\mathrm{d}\boldsymbol{f}_\xi\lambda^2$ yields the factor λ before the integral.

At a very small distance Δz_1 out of focus, at $z_1 = F + \Delta z_1$, a quadratic term $-|\boldsymbol{f}_\xi|^2\lambda\Delta z_1/(2F^2)$ remains. This phase term can be interpreted as an aberration function for defocus.

$V_{\mathrm{ap}}(\boldsymbol{f}_\xi)$ is the amplitude in the aperture plane, with $V_{\mathrm{ap}}(\boldsymbol{f}_\xi) = V(\boldsymbol{f}_\xi)A(\boldsymbol{f}_\xi)$, and $V(\boldsymbol{f}_\xi) = V_0 = \text{const}$ in the case of a single source point on-axis representing an approximately plane wave in the aperture plane.

Then, the shape of the amplitude $V(\boldsymbol{\alpha})$ in (3.1) is the Fourier transform of the aperture function $A(\boldsymbol{f}_\xi)$ (see the discussion following (2.22)), and the monochromatic intensity distribution $I(\boldsymbol{\alpha})$ in the image plane is given by

$$I(\boldsymbol{\alpha}) = |V(\boldsymbol{\alpha})|^2 = V_0^2 \frac{\lambda^2}{F^2} \left| \int A(\boldsymbol{f}_\xi) \mathrm{e}^{-\mathrm{i}2\pi \boldsymbol{f}_\xi \cdot \boldsymbol{\alpha}}\, \mathrm{d}\boldsymbol{f}_\xi \right|^2 . \tag{3.2}$$

This is the diffraction limited image of a point source at infinity. The shape of the amplitude is given by the Fraunhofer diffraction pattern of the aperture in the image plane (see Sect. 2.2.2).

The shape of the image intensity distribution is described by the dimensionless *point-spread function* (PSF) of the optical system denoted by

$$\mathrm{PSF}(\boldsymbol{\alpha}) = \frac{I(\boldsymbol{\alpha})}{V_0^2}. \tag{3.3}$$

Discussing monochromatic illumination in this section, we keep writing the PSF as a function of $\boldsymbol{\alpha}$ only, considering the wavelength as a constant parameter.

In astronomical telescopes with a circular aperture the diffraction limited intensity distribution is called the *Airy disk* (see Sect. 2.2.3) that has the form of a Besinc^2 function (2.26).

To find the intensity distribution in the image plane one has to sum up individual PSFs, each weighted according to the object intensity. This assumes that the PSF is shift-invariant over the field of view, i.e. that the so-called *isoplanatic angle* is larger than the size of the source. We will see in the context of imaging through atmospheric turbulence in Sect. 4.3 that the isoplanatic angle can be much smaller than the field of view.

3.1.1 The Optical Transfer Function

The linear superposition of shift-invariant PSFs is replaced by the integral over the source yielding the *convolution* of the object intensity distribution that, from now on, we will denote by $O(\boldsymbol{\alpha}')$, with the PSF:

$$\begin{aligned} I(\boldsymbol{\alpha}) &= \int O(\boldsymbol{\alpha}')\,\mathrm{PSF}(\boldsymbol{\alpha}-\boldsymbol{\alpha}')\,\mathrm{d}\boldsymbol{\alpha}' \\ &= O(\boldsymbol{\alpha}) * \mathrm{PSF}(\boldsymbol{\alpha}), \end{aligned} \tag{3.4}$$

with $*$ denoting the convolution. This formula is valid for an angular magnification factor $m_p = 1$. In general, one has to write $O(\boldsymbol{\alpha}/m_p) * \mathrm{PSF}(\boldsymbol{\alpha})$.

It is difficult to link the properties of the PSF through the convolution to the quality of the full image. In Sect. 2.2.3 we made an effort discussing the resolving power of an optical telescope by introducing the Rayleigh criterion defining the radius of the first minimum of the Airy disk as the limit of angular resolution. In general, one has to look at the transfer function of the optical system. The mathematical concept that we have for this purpose is called Fourier Optics.

The convolution of the object intensity with the PSF in image space can be replaced by a multiplication in Fourier space. This very general property, called the *convolution theorem*, can be used when Fourier transforming both sides of (3.4) turning the Fourier transform of the convolution into the product of the individual Fourier transforms of the object intensity $O(\boldsymbol{\alpha})$ and of $\mathrm{PSF}(\boldsymbol{\alpha})$ (see Appendix A.1).

Before doing so, we have, a closer look at the PSF. Rearranging the order of integration in (3.2) yields

$$\mathrm{PSF}(\boldsymbol{\alpha}) = \frac{\lambda^2}{F^2} \iint A(\boldsymbol{f}_{\xi,1})\, A^*(\boldsymbol{f}_{\xi,2}) \mathrm{e}^{-\mathrm{i}2\pi(\boldsymbol{f}_{\xi,1}-\boldsymbol{f}_{\xi,2})\cdot\boldsymbol{\alpha}}\,\mathrm{d}\boldsymbol{f}_{\xi,1}\mathrm{d}\boldsymbol{f}_{\xi,2}, \tag{3.5}$$

and introducing the difference of two *spatial frequency vectors* $\boldsymbol{R} = (u, v) = \boldsymbol{f}_{\xi,1} - \boldsymbol{f}_{\xi,2}$ as a new variable, the *optical transfer function* (OTF) is defined as

$$\mathrm{OTF}(\boldsymbol{R}) := \frac{\lambda^2}{F^2} \int A(\boldsymbol{f}_{\xi,1}) A^*(\boldsymbol{f}_{\xi,1} - \boldsymbol{R})\,\mathrm{d}\boldsymbol{f}_{\xi,1}\,, \tag{3.6}$$

which is the autocorrelation of the aperture function $A(\boldsymbol{f}_\xi)$.[2] Note that the shapes of the aperture function $A(\boldsymbol{f}_\xi)$ and of the OTF – being functions of spatial frequency – vary with wavelength. This is not particularly relevant for the discussion in this section treating only monochromatic light but it will have to be considered carefully when moving to polychromatic illumination.

[2] The correlation (3.6) differs from the convolution (3.4) by a change of sign in the coordinate of the second function and by using the conjugate complex of the second function.

In the presence of optical aberrations that are accounted for in the phase $\varphi(\boldsymbol{f}_\xi)$ of the aperture function $A(\boldsymbol{f}_\xi)$, the OTF is a complex quantity with the phase denoted by ϕ_{OTF}. A few cases of optical aberrations will be discussed in Sect. 3.1.2. In the case of chromatic aberrations, i.e. if the shape of the phase is wavelength dependent, $\varphi(\boldsymbol{f}_\xi)$ and the phase ϕ_{OTF} become functions of wavelength.

NB 5. *It is customary to define a normalised OTF with OTF(0) = 1. We would, thus, have to divide the OTF in (3.6) by* $(\lambda/F)^2 \int |A(\boldsymbol{f}_\xi)|^2 \mathrm{d}\boldsymbol{f}_\xi$. *This would simplify the treatment of the optical system in the context of linear systems theory. With a normalised OTF the relative influence on individual spatial frequencies is easy to judge. However, the normalization complicates the formal treatment that we apply. Therefore, we will maintain our definition but, for simplification, we will display the OTF always as a normalised function.*

It is important to note that the OTF is a function of the difference $\boldsymbol{R} = \boldsymbol{f}_{\xi,1} - \boldsymbol{f}_{\xi,2}$ of two spatial frequency vectors in the aperture plane and not a function of the spatial frequency vector $\boldsymbol{f}_\xi$ itself in the aperture plane.

This has to be kept in mind when linking the OTF to features of the aperture function. Discussing for example off-axis apertures one finds that the OTF will still be an even function centred at $R = 0$. When treating interferometric imaging with two and more telescopes we have to keep this in mind.

The OTF of a telescope without aberrations is displayed in Fig. 3.2 together with a graphical example of how to compute the value of the OTF at a given spatial frequency $\boldsymbol{R}_0$. Here, the OTF is a real function. Examples of complex OTF with optical aberrations will be discussed in the next section.

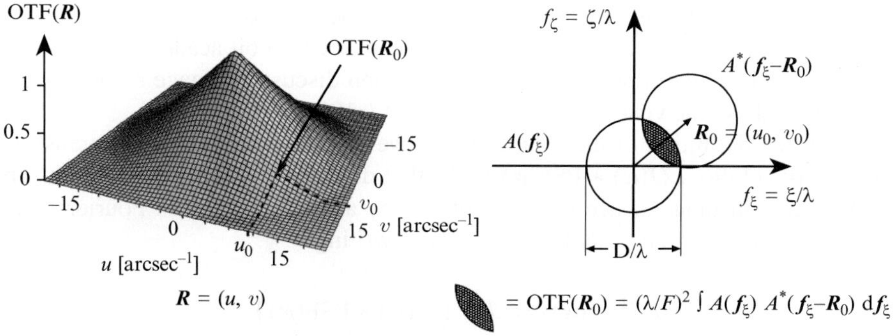

Fig. 3.2 The optical transfer function (OTF) of a circular aperture without aberrations. On the *left*, the real and positive OTF is displayed. It has a triangular shape reaching zero at $|\boldsymbol{R}| = D/\lambda = 17.6\,\mathrm{arcsec}^{-1}$ for an 8-m telescope in the K-band. Its Fourier transform, the PSF of a circular aperture, is the Airy disk. In general, if the telescope has aberrations, the OTF is a complex function. On the *right*, an example for the calculation of the autocorrelation of the aperture function $A(\boldsymbol{f}_\xi)$ is shown. The autocorrelation integral at the arbitrary value $\boldsymbol{R}_0$ is illustrated by the shaded overlapping area between the two apertures

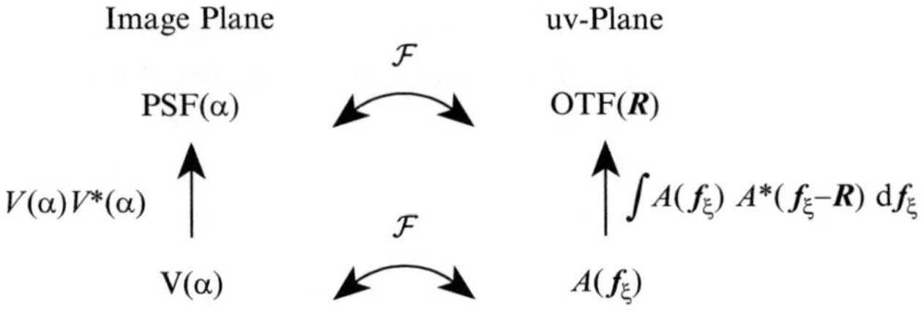

Fig. 3.3 Illustration of the autocorrelation theorem, called the Wiener–Khinchine theorem. The PSF being the modulus squared of the amplitude $V(\boldsymbol{\alpha})$ – and, thus, real and positive – implies that the OTF is the autocorrelation function of the aperture function $A(\boldsymbol{f}_\xi)$ of the optical system, which is the Fourier transform of the amplitude $V(\boldsymbol{\alpha})$

Inserting (3.6) in (3.5), we find that the PSF is the Fourier transform of the OTF,

$$\mathrm{PSF}(\boldsymbol{\alpha}) = \int \mathrm{OTF}(\boldsymbol{R})\mathrm{e}^{-\mathrm{i}2\pi \boldsymbol{R}\cdot\boldsymbol{\alpha}}\,\mathrm{d}\boldsymbol{R}\,. \tag{3.7}$$

With the PSF being a real function by definition, the OTF is Hermitian meaning that the modulus of the OTF is an even function, $|\mathrm{OTF}(\boldsymbol{R})| = |\mathrm{OTF}(-\boldsymbol{R})|$, and the phase is an odd function, $\Phi_{\mathrm{OTF}}(\boldsymbol{R}) = -\Phi_{\mathrm{OTF}}(-\boldsymbol{R})$ (see Appendix A.1).

The PSF is also a positive function by definition, given by the product of the amplitudes $V(\boldsymbol{\alpha})V^*(\boldsymbol{\alpha})$, see (3.2) and (3.3). Because of this, its Fourier transform, the OTF, is the autocorrelation of the aperture function $A(\boldsymbol{f}_\xi)$, (3.6), with $A(\boldsymbol{f}_\xi)$ being related to $V(\boldsymbol{\alpha})$ by a Fourier transform, (3.1). This is called the *autocorrelation theorem*, or the *Wiener–Khinchine theorem* that is illustrated in Fig. 3.3.

Generally speaking, any real and positive function, like the PSF, has a Fourier transform that is Hermitian, like the OTF, and can be expressed as an autocorrelation of another function, like $A(\boldsymbol{f}_\xi)$. While this point seems a bit academic in this context, we will see its importance in Sect. 3.4 when discussing image reconstruction from multi-baseline visibility measurements [17, 102, 119].

With the definition of the OTF at hand, we now discuss the Fourier transform of the convolution, $O(\boldsymbol{\alpha}) * \mathrm{PSF}(\boldsymbol{\alpha})$ (3.4), that describes the imaging process. The convolution in image space is then replaced by a multiplication in Fourier space. Formally, we use the Fourier back-transform, writing

$$\begin{aligned}\mathcal{F}_\alpha^{-1}\left(I(\boldsymbol{\alpha})\right) &= \mathcal{F}_\alpha^{-1}\Big(O(\boldsymbol{\alpha}) * \mathrm{PSF}(\boldsymbol{\alpha})\Big)\\ &= \mathcal{F}_\alpha^{-1}\big(O(\boldsymbol{\alpha})\big)\mathcal{F}_\alpha^{-1}\big(\mathrm{PSF}(\boldsymbol{\alpha})\big).\end{aligned} \tag{3.8}$$

Replacing the Fourier back-transform of the PSF by the OTF (see 3.7) and introducing the *spatial frequency spectra* $\hat{O}(\boldsymbol{R})$ and $\hat{I}(\boldsymbol{R})$ of the object and image intensity as

$$\hat{O}(\boldsymbol{R}) = \mathcal{F}_\alpha^{-1}\left(O(\boldsymbol{\alpha})\right) = \int O(\boldsymbol{\alpha})\mathrm{e}^{\mathrm{i}2\pi \boldsymbol{R}\cdot\boldsymbol{\alpha}}\,\mathrm{d}\boldsymbol{\alpha}, \tag{3.9}$$

and

$$\hat{I}(\boldsymbol{R}) = \mathcal{F}_{\alpha}^{-1}(I(\boldsymbol{\alpha})) = \int I(\boldsymbol{\alpha}) \mathrm{e}^{\mathrm{i}2\pi \boldsymbol{R}\cdot\boldsymbol{\alpha}} \,\mathrm{d}\boldsymbol{\alpha}, \tag{3.10}$$

we can write

$$\hat{I}(\boldsymbol{R}) = \hat{O}(\boldsymbol{R})\,\mathrm{OTF}(\boldsymbol{R}). \tag{3.11}$$

The image intensity spectrum $\hat{I}(\boldsymbol{R})$ is the product of OTF and $\hat{O}(\boldsymbol{R})$. This multiplication is equivalent to filtering the spatial frequency spectrum of the object intensity, with the optical transfer function acting as a filter.

In the context of this section, Fourier transforms of intensity distributions appear as rather abstract mathematical quantities that are very useful to describe the imaging process. However, when discussing the propagation of the coherence function in Sect. 3.2 we will be reminded that for monochromatic light, the spatial frequency spectrum as the Fourier transform of the object intensity distribution is identical to the visibility function in the aperture plane according to the van Cittert–Zernike theorem (2.45), putting the concept of Fourier optics in the context of stellar interferometry.

With the definition of the spatial frequency spectra $\hat{I}(\boldsymbol{R})$ and $\hat{O}(\boldsymbol{R})$ as Fourier transforms of the intensity distributions $I(\boldsymbol{\alpha})$ and $O(\boldsymbol{\alpha})$, the coordinate $\boldsymbol{R}$ can be interpreted with respect to the variables of the Fourier transform. Since the variable of the intensity distribution, $\boldsymbol{\alpha}$, is an angle (a spatial coordinate) the variable of its Fourier transform, $\boldsymbol{R}$, is a spatial frequency with the dimension 1/angle. The introduction of spatial frequency, $\boldsymbol{f}_{\xi} = \boldsymbol{\xi}/\lambda$, and $\boldsymbol{R} = \boldsymbol{f}_{\xi,1} - \boldsymbol{f}_{\xi,2}$, enables this interpretation since the exponential kernel of the Fourier transform, $\frac{2\pi}{\lambda}(\boldsymbol{\xi}_1 - \boldsymbol{\xi}_2)\cdot\boldsymbol{\alpha}$ in (2.45) reads as $2\pi\boldsymbol{R}\cdot\boldsymbol{\alpha}$ in (3.10), providing a Fourier transform between the coordinates $\boldsymbol{\alpha}$ and $\boldsymbol{R}$.

We will continue to discuss the imaging process in the notation with spatial frequency spectra in this section. Putting aside the aspects of the coherence properties we concentrate on the imaging process preparing the formalism that we will use later when discussing the situation in stellar interferometers.

The consequence of the definition of the spatial frequency is that the spectrum $\hat{O}(\boldsymbol{R})$ only depends on the shape of the object. An object with a wavelength independent circular diameter α_0 has a spatial frequency spectrum that is also independent of wavelength. $\hat{O}(\boldsymbol{R})$ has the form of a Besinc-function with its first minimum at $R_0 = 1.22/\alpha_0$ (see Fig. 3.4). Calculating the visibility function as a function of $\boldsymbol{\xi}_1 - \boldsymbol{\xi}_2$ the first minimum is at $|\boldsymbol{\xi}_1 - \boldsymbol{\xi}_2|_0 = 1.22\lambda/\alpha_0$ (see the example of Venus in Fig. 2.13).

The advantage of having wavelength independent spatial frequency spectra brings with it the disadvantage of a wavelength dependent OTF($\boldsymbol{R}$) being a function of spatial frequency. For a circular aperture with diameter D the wavelength dependent OTF radius is D/λ (see Fig. 3.2).

Apart from facilitating the formal treatment, the reasons for introducing spatial frequencies are partially historical. In the late 1930's, Fourier methods for the

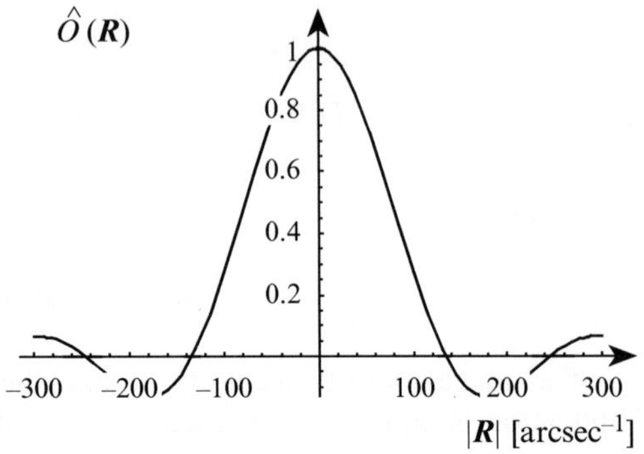

Fig. 3.4 Profile of the spatial frequency spectrum $\hat{O}(\boldsymbol{R})$ of a disk shaped object with a diameter of $\alpha_0 = 10$ mas as a function of $R = |\boldsymbol{R}|$. $\hat{O}(\boldsymbol{R})$ follows a Besinc-function with the first zero at $R_0 = 1.22/\alpha_0 = 122$ arcsec^{-1}

analysis of optical systems were described by P.M. Duffieux [64, 65]. One of the first applications was the optimization of television camera lenses, treating them as linear systems like in signal theory [208]. Then, the coordinate-pair *time-frequency* was replaced by the pair *spatial coordinate-spatial frequency* and the mathematical tools were applied to optical systems [177].

Since we are interested in computing the image intensity $I(\boldsymbol{\alpha})$, we rewrite (3.11) as

$$I(\boldsymbol{\alpha}) = \mathcal{F}_R\left(\hat{I}(\boldsymbol{R})\right) = \mathcal{F}_R\left(\hat{O}(\boldsymbol{R})\,\mathrm{OTF}(\boldsymbol{R})\right), \tag{3.12}$$

expressing the image intensity distribution as a Fourier transform of the product of the object's spatial frequency spectrum multiplied by the transfer function of the optical system.

The general concept of a spatial frequency spectrum calls for the interpretation of an object as a superposition of cosine functions with different spatial frequencies. Their amplitudes and phases form the Fourier spectrum.

For example, a source with the shape of a circular disk has a Fourier transform given by a Besinc-function displayed in Fig. 3.4. The spatial frequency spectrum in the form of a Besinc-function indicates that low spatial frequencies with rather high amplitude are combined with higher spatial frequencies with lower amplitudes of alternating signs since the Besinc-function has also negative values.

The product of a Besinc-function with an OTF like in Fig. 3.2 shows a sharper decrease than the Besinc-function alone, and eventually, beyond the cut-off frequency at D/λ, the product drops to zero. This has the effect that the sharp rectangular shape of the disk is softened and the edges are round rather than square. One could say that the "sharpness" of the image is lost if high frequencies are cut off.

Another example is that of a binary star that can be used to discuss criteria of **angular resolution** in the spatial frequency spectrum. A binary star is represented by the sum of two δ-functions:

$$O(\boldsymbol{\alpha}) = \frac{1}{2}\left(\delta\left(|\boldsymbol{\alpha} - \frac{\boldsymbol{\alpha}_0}{2}|\right) + \delta\left(|\boldsymbol{\alpha} + \frac{\boldsymbol{\alpha}_0}{2}|\right)\right), \tag{3.13}$$

with the components symmetrical to the optical axis at + and $-\boldsymbol{\alpha}_0/2$ and of equal intensity of 1/2. $\boldsymbol{\alpha}_0$ is called the separation vector.

The Fourier transform of the object intensity is then (see (A.4))

$$\hat{O}(\boldsymbol{R}) = \cos\left(2\pi \frac{\boldsymbol{\alpha}_0}{2} \cdot \boldsymbol{R}\right). \tag{3.14}$$

This cosine function – forming the spectrum of the binary – provides at each spatial frequency $\boldsymbol{R}$ the value of the amplitude, which in turn determines the weighting in the superposition of cosine functions when computing the object intensity as the Fourier back transform of the spectrum.

The spectrum $\hat{O}(\boldsymbol{R})$ is multiplied by the OTF in order to obtain the image intensity spectrum $\hat{I}(\boldsymbol{R})$, see (3.11) and Fig. 3.5.

Angular resolution criteria that were discussed in Sect. 2.2.3 by using properties of the Airy disk can now be discussed in frequency space. The binary star is called resolved if a full period of the cosine function (3.14) is transferred by the OTF. This is the case if the first zero of the OTF at $R = D/\lambda$ is equal to $1/\alpha_0$. The separation α_0 of the resolvable binary star is then $\alpha_0 = \lambda/D$. This resolution limit was given in Sect. 2.2.3 when it was linked to the full width at half maximum (FWHM) of the Airy disk.

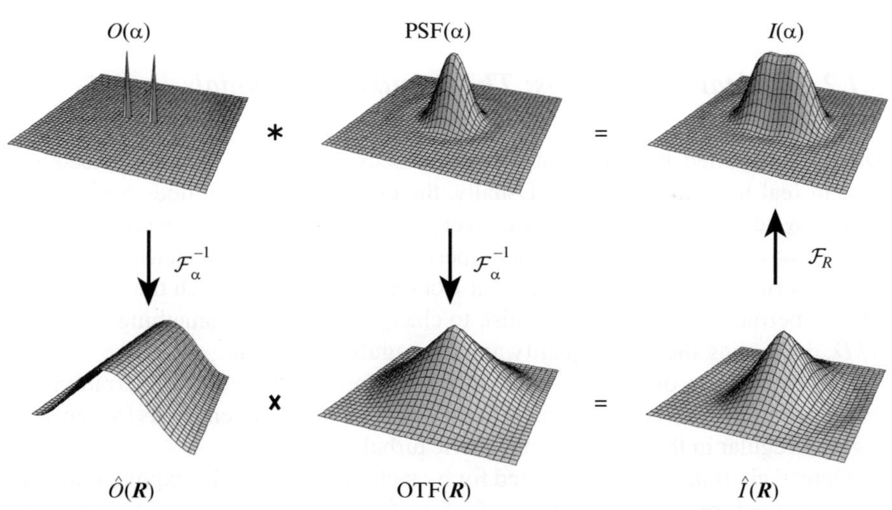

Fig. 3.5 The illustration of the imaging process at the limit of angular resolution with a circular aperture. The top row displays the convolution ($*$) of the object intensity $O(\boldsymbol{\alpha})$ – in this case a binary star separated by $\alpha_0 = \lambda/D$ – with the PSF, the Airy disk, yielding the image intensity $I(\boldsymbol{\alpha})$. The bottom row shows the situation in Fourier space: The OTF with radius D/λ is multiplied ($\times$) by the object intensity spectrum $\hat{O}(\boldsymbol{R})$ (3.14), a cosine function (one period is displayed) since we regard a binary star. If the separation α_0 of the binary is chosen as $\alpha_0 = \lambda/D$ then the OTF is wide enough to transfer one full period of the cosine function. The resulting image intensity spectrum $\hat{I}(\boldsymbol{R})$ is significantly narrower than the OTF along the separation vector $\boldsymbol{\alpha}_0$ and unaffected perpendicular to its direction. Its Fourier transform displays the image intensity $I(\boldsymbol{\alpha})$ that can just be resolved as a binary

Using $\alpha_0 = \lambda/D$ as a limit of resolution is reasonable since the system does not transfer spatial frequencies larger than D/λ. On the other hand, even a separation smaller than $\boldsymbol{\alpha}_0$, when one period of the cosine function is wider than the OTF, affects the resulting image spectrum still making it narrower in the direction of the separation vector, similar to Fig. 3.5. Equivalently, the resulting image intensity is wider than it would be for a single star, as discussed in Sect. 2.2.3. This deviation from the perfect circular symmetry can obviously not be measured for arbitrary small values since the measurement is never perfect. Thus, λ/D is a reasonable limit of resolution if the criterion is to resolve two objects as two local maxima. The actual detectability of two objects depends on the imaging situation, the quality of the measurement, the type of object and can be better than λ/D.

The Rayleigh criterion of resolution that was defined as the distance between the maximum of the PSF and the first minimum at 1.22 λ/D is more conservative requiring a larger separation $\alpha_0 = 1.22\,\lambda/D$ of a resolvable binary star. Then, the cosine in the spatial frequency spectrum in (3.14) is narrower and the OTF transfers slightly more than one period of the cosine function. Eventually, the more periods of the cosine functions are transferred, the more the image intensity distribution resembles a binary star.

Examples of complex OTF in the presence of aberrations will be discussed in the next section before we move on to the propagation of the coherence function.

3.1.2 Optical Aberrations: The Zernike Polynomials

Diffraction limited imaging that we have discussed so far is a rather exceptional case in real imaging situations. Usually, the optical wave front does not leave the optical system as a spherical wave converging in the focus, but there are aberrations from the spherical shape due to the imperfections of the optical elements or due to atmospheric turbulence when the light from a star travels through the atmosphere. These aberrations cause the Airy disk to change its shape to something wider than λ/D, decreasing the image quality and the angular resolution.

The exact form of the PSF depends on the form of the optical aberrations and can range from a well defined shape in the case of classical aberrations to something highly irregular in the case of atmospheric turbulence.

Optical aberrations are accounted for by a phase term φ in the exponential of the complex aperture function $A(\boldsymbol{f}_\xi) = P(\boldsymbol{f}_\xi)\mathrm{e}^{\mathrm{i}\varphi(\boldsymbol{f}_\xi)}$ as a function of spatial frequency $\boldsymbol{f}_\xi = \boldsymbol{\xi}/\lambda$, when $P(\boldsymbol{f}_\xi)$ denotes the shape of the telescope aperture.

The *Strehl ratio* is a simple metric that is often used when describing the image quality of adaptive optics systems (see Sect. 6.3). It is the defined as the ratio between the peak of the PSF and the peak of the Airy disk. The aberrations can be related to the Strehl ratio in a simple way, using the Maréchal approximation [19]. If the phase variance σ_φ^2 is smaller than about $\pi^2/4$, the Strehl ratio is approximated by

$$S = \mathrm{e}^{-\sigma_\varphi^2}. \tag{3.15}$$

In the following, we will discuss PSF and OTF for classical aberrations, and we will regard the symmetry properties. The case of atmospheric turbulence will be treated in Sect. 4.3.3.

Usually, *Zernike polynomials* are employed to describe aberrations. They were introduced in 1934 by F. Zernike who deduced them from the Jacobi polynomials and slightly modified them for the application in optics [258]. Zernike polynomials have the advantage that they are defined on the unit circle fitting the circular shape of most optical apertures and that the low order terms are related to the classical aberrations like astigmatism, coma and spherical aberration. The influence of the central obscuration due to the secondary mirror in astronomical telescopes is negligible for this discussion.

The Zernike polynomials form a set of orthogonal polynomials and it is convenient to write them as a function of polar coordinates ρ and θ, with $\rho = |f_\xi|/(f_{\xi,D}/2) \leq 1$, with $f_{\xi,D} = D/\lambda$ and D the diameter of the aperture. The normalization is chosen such that the variance of each polynomial over the circle is set equal to 1 rad^2, yielding

$$\begin{aligned} Z_{j_{\text{even}}} &= \sqrt{n+1}\, R_n^m(\rho)\sqrt{2}\, \cos(m\theta), \text{ for } m \neq 0, \\ Z_{j_{\text{odd}}} &= \sqrt{n+1}\, R_n^m(\rho)\sqrt{2}\, \sin(m\theta), \text{ for } m \neq 0, \\ Z_j &= \sqrt{n+1}\, R_n^0(\rho), \text{ for } m = 0, \end{aligned} \tag{3.16}$$

where

$$R_n^m(\rho) = \sum_{s=0}^{\frac{n-m}{2}} \frac{(-1)^s (n-s)!}{s!(\frac{n+m}{2}-s)!\,(\frac{n-m}{2}-s)!}\,\rho^{n-2s}. \tag{3.17}$$

Table 3.1 shows the low order Zernike polynomials when the columns m indicate the azimuthal orders and the rows n the radial orders.

Table 3.1 Zernike polynomials Z_j for $j = 1$ to 11. n is the radial order and m the azimuthal order. The modes are ordered such that even j correspond to the symmetric modes given by $\cos m\theta$ and odd j to the antisymmetric modes given by $\sin m\theta$

n	$m = 0$	1	2	3
0	$Z_1 = 1$ (piston)			
1		$Z_2 = 2\rho\cos\theta$ $Z_3 = 2\rho\sin\theta$ (tip and tilt)		
2	$Z_4 = \sqrt{3}\,(2\rho^2 - 1)$ (focus)		$Z_5 = \sqrt{6}\,\rho^2 \sin 2\theta$ $Z_6 = \sqrt{6}\,\rho^2 \cos 2\theta$ (astigmatism)	
3		$Z_7 = \sqrt{8}(3\rho^3 - 2\rho)\sin\theta$ $Z_8 = \sqrt{8}(3\rho^3 - 2\rho)\cos\theta$ (coma)		$Z_9 = \sqrt{8}\rho^3 \sin 3\theta$ $Z_{10} = \sqrt{8}\rho^3 \cos 3\theta$ (trefoil)
4	$Z_{11} = \sqrt{5}(6\rho^4 - 6\rho^2 + 1)$ (spherical aberration)			

If the optical aberrations are composed of a multitude of Zernike polynomials the weighted sum forms the aberration function

$$\varphi(\rho, \theta) = \sum_{i=1}^{\infty} a_i Z_i(\rho, \theta), \tag{3.18}$$

and the coefficients a_i, using the orthogonality, are given by

$$a_i = \int_{\text{aperture}} \phi(\rho, \theta) Z_i(\rho, \theta) \rho d\rho d\theta. \tag{3.19}$$

Here, we will discuss only two examples, the Zernike polynomials for the classical aberrations astigmatism, Z_8 with $m = 2$ and $n = 2$, and coma, Z_5 with $m = 1$ and $n = 3$, that are displayed in Fig. 3.6.

It is interesting to note that every Zernike polynomial is balanced in order to minimise the remaining aberration. For instance, if pure coma is considered to be $\propto \rho^3$, the Zernike coma (Z_7 and Z_8) has an additional linear term ρ that tilts the wave front so that the contribution of this aberration is minimised in the sense that the integral over the square of the aberration Z_7^2, the phase variance, is at a minimum.

The PSFs are computed by first Fourier transforming the aperture function $A(\boldsymbol{f}_\xi)$ obtaining the shape of the amplitude $V(\boldsymbol{\alpha})$ in the image plane (3.1). The PSF describes the shape of the squared modulus $|V(\boldsymbol{\alpha})|^2$, i.e. the intensity, using (3.2), (3.3) and $A(\boldsymbol{f}_\xi) = P(\boldsymbol{f}_\xi)\mathrm{e}^{\mathrm{i}\varphi(\boldsymbol{f}_\xi)}$:

$$\mathrm{PSF}(\boldsymbol{\alpha}) = \frac{\lambda^2}{F^2} \left| \int P(\boldsymbol{f}_\xi) \mathrm{e}^{-\mathrm{i}(2\pi \boldsymbol{f}_\xi \cdot \boldsymbol{\alpha} - \varphi(\boldsymbol{f}_\xi))} \, \mathrm{d}\boldsymbol{f}_\xi \right|^2. \tag{3.20}$$

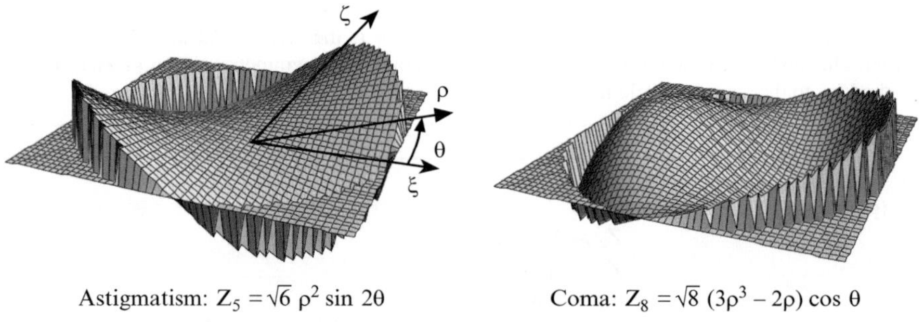

Fig. 3.6 Zernike polynomials for astigmatism $Z_5(\rho, \theta)$ and coma $Z_8(\rho, \theta)$ in the notation with polar coordinates $\boldsymbol{\xi} = (\rho, \theta)$. The aperture has the radius $\rho = 1$ since Zernike polynomials are defined on the unit circle. The polynomial for astigmatism is an even, or symmetric, function $Z_5(\boldsymbol{\xi}) = Z_5(-\boldsymbol{\xi})$ and the polynomial for coma is an odd, or anti-symmetric, function $Z_8(\boldsymbol{\xi}) = -Z_8(-\boldsymbol{\xi})$, both with respect to the centre of the aperture

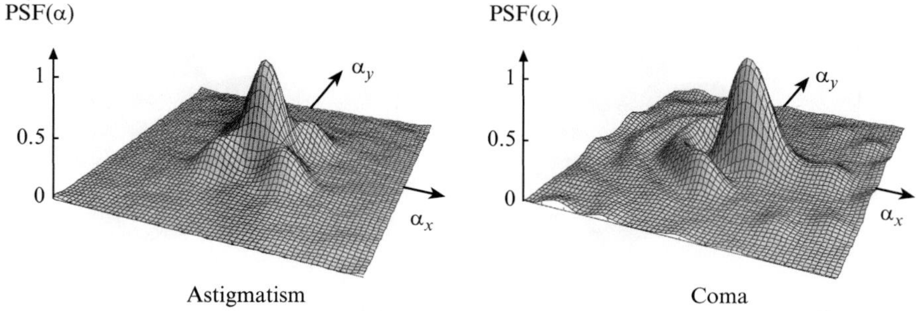

Fig. 3.7 Point-spread functions (PSFs) for astigmatism and coma. On the *left*, the PSF with astigmatism is an even function, point-symmetric with respect to the image centre, $\mathrm{PSF}(\boldsymbol{\alpha}) = \mathrm{PSF}(-\boldsymbol{\alpha})$. On the *right* the PSF with coma shows an axis-symmetry with respect to the α_x-axis

In the presence of astigmatism as an even function, the PSF is also an even function. In the case of coma as an odd function, $\varphi(\boldsymbol{f}_\xi) = -\varphi(-\boldsymbol{f}_\xi)$, the PSF shows an axis-symmetry with respect to the α_x-axis. Both PSFs are displayed in Fig. 3.7.

The influence of both astigmatism and coma on the image quality is difficult to interpret because of the convolution of the object intensity distribution with the PSF, (3.4). Therefore, we compute the OTFs by Fourier transforming the PSFs, (3.7). We should remember, that due to the PSF being real by definition, the OTF is Hermitian, with even modulus and odd phase function.

For astigmatism, with a real and even PSF, the OTF is also a real and even function with rather high values along the diagonals. Writing the OTF as modulus and phase, the negative values are denoted by an OTF phase of $\pm\pi$.

In the case of coma, the modulus of the OTF is not perfectly round but it is narrower along the direction of the large overall slope of the OTF phase. The latter is an odd function similar to the Zernike polynomial for coma. The axis-symmetry of the PSF is reflected by the axis-symmetry of the OTF phase with respect to the u-axis.

Compared to the diffraction limited OTF in Fig. 3.2, the moduli of the aberrated OTFs in Fig. 3.8 both show a narrow peak and a wide disk dropping to zero at the diffraction limit. This means that, while low spatial frequencies are transferred rather well, higher spatial frequencies are attenuated due to low values of the OTF modulus and may additionally suffer a phase shift due to the OTF phase. Compared to the diffraction limited case, the overall deterioration of the transfer of high spatial frequencies in the case of aberrations results in PSFs that are wider than in the diffraction limited case.

It is the goal to optimise the image quality by minimising optical aberrations when designing optical systems. We have seen here, that this is equivalent to maximising the OTF with the upper limit being the diffraction limited OTF as in Fig. 3.2. Especially if the optical quality is close to the diffraction limit it is important to optimise the OTF and not only the optical aberrations in the design process [60,105].

Since the role of the phase of the OTF and of the visibility function is important for the imaging process, we have a closer look.

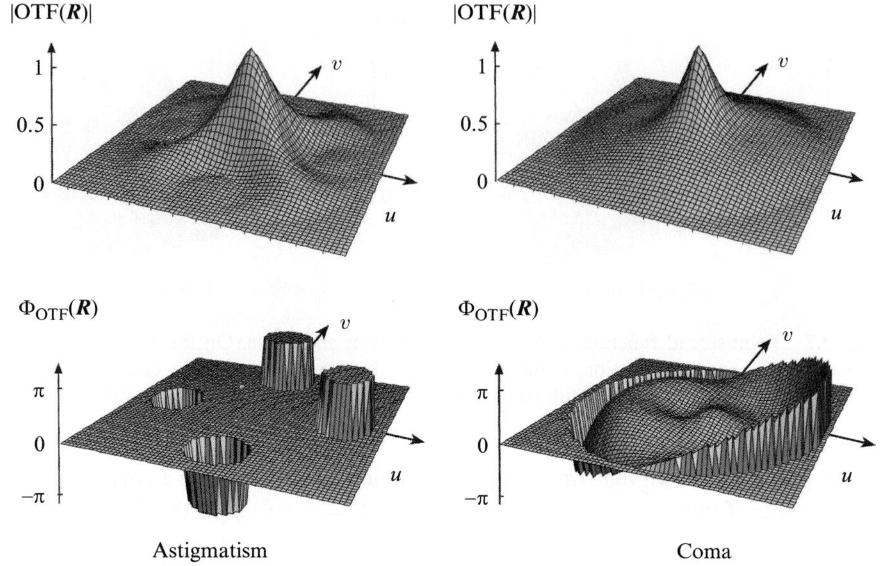

Fig. 3.8 Modulus (*top row*) and phase (*bottom row*) of optical transfer functions (OTF) with astigmatism (*left*) and coma (*right*) as a function of spatial frequency $\boldsymbol{R} = (u, v)$. The phase ϕ_{OTF} of the OTF with astigmatism (*left*) has only values of 0 and $\pm\pi$ since the OTF is a real function with negative values for $\phi_{\mathrm{OTF}} = \pm\pi$. While the OTF phase with astigmatism on the left lost all resemblance to the Zernike polynomial for astigmatism, the phase ϕ_{OTF} of the OTF with coma (*right*) is not dissimilar to the Zernike polynomial for coma (see Fig. 3.6)

What happens if we try to calculate a PSF with the modulus of the OTF only, ignoring its phase? In the case of astigmatism one would obtain a PSF that is very similar to the one in Fig. 3.7 since the non-zero phase values are $\pm\pi$ creating the negative parts of the OTF with very small values. However, in the case of coma, Fourier transforming only the OTF modulus gives a PSF that is very dissimilar to the correct PSF in Fig. 3.7. In fact, one would recover only the even parts of the PSF that look not at all like the PSF for coma in Fig. 3.7 (see Appendix A.1).

These examples show the importance of the phase of a Fourier transform. The characteristics of the Fourier pair *PSF–OTF* can be applied directly to the pair stellar *intensity distribution–coherence function*. If the phase of the coherence function is not determined, one assumes implicitly that the stellar intensity distribution must be an even function. Or one is content with just determining the even part of the stellar intensity distribution.

Fourier Optics: Summary

The propagation of light through optical systems is a two-step process, first from the source plane into the pupil plane, and then from the pupil plane into the image plane. Discussing the principle, we choose a thin lens as the simplest form of an optical system. Then, the aperture of the lens is the pupil of the system.

For the propagation of the amplitude we regard a single point at a large distance emitting a monochromatic spherical wave that arrives as an approximately plane wave in the aperture plane. The spatial coordinate $\boldsymbol{\xi}$ in the aperture plane is replaced by the spatial frequency $\boldsymbol{f}_\xi$ defined as $\boldsymbol{f}_\xi = \boldsymbol{\xi}/\lambda$. We write the amplitude of the plane wave as $V(\boldsymbol{f}_\xi) = V_0 = \text{const.}$

The amplitude $V(\boldsymbol{\alpha})$ in the image plane at a distance $z_1 = F$, i.e. in the focal plane of the thin lens, is then the Fourier transform of the aperture function, and the monochromatic intensity in the image plane can be written as

$$I(\boldsymbol{\alpha}) = |V(\boldsymbol{\alpha})|^2 = V_0^2 \frac{\lambda^2}{F^2} \left| \int A(\boldsymbol{f}_\xi) \mathrm{e}^{-\mathrm{i}2\pi \boldsymbol{f}_\xi \cdot \boldsymbol{\alpha}} \, \mathrm{d}\boldsymbol{f}_\xi \right|^2 , \tag{3.2}$$

with $A(\boldsymbol{f}_\xi)$ the aperture function, containing aberrations of the optical system in its phase and the transmission in its modulus.

The shape of the intensity distribution is called the *point-spread function* (PSF), with $\text{PSF}(\boldsymbol{\alpha}) = I(\boldsymbol{\alpha})/V_0^2$. In astronomical telescopes with a circular aperture, the intensity distribution is called the Airy disk and its shape has the form of a Besinc^2 function.

Rearranging the order of integration in (3.2), the *optical transfer function* (OTF) is defined as

$$\text{OTF}(\boldsymbol{R}) := \frac{\lambda^2}{F^2} \int A(\boldsymbol{f}_{\xi,1}) A^*(\boldsymbol{f}_{\xi,1} - \boldsymbol{R}) \, \mathrm{d}\boldsymbol{f}_{\xi,1} , \tag{3.6}$$

as a function of $\boldsymbol{R} = (u, v) = \boldsymbol{f}_{\xi,1} - \boldsymbol{f}_{\xi,2}$, and the PSF is the Fourier transform of the OTF

$$\text{PSF}(\boldsymbol{\alpha}) = \frac{I(\boldsymbol{\alpha})}{V_0^2} = \int \text{OTF}(\boldsymbol{R}) \mathrm{e}^{-\mathrm{i}2\pi \boldsymbol{R} \cdot \boldsymbol{\alpha}} \, \mathrm{d}\boldsymbol{R} . \tag{3.7}$$

Discussing instead of a point source an extended incoherent source emitting monochromatic light, the object intensity distribution $O(\boldsymbol{\alpha}')$ is written as a function of the image plane coordinate $\boldsymbol{\alpha}$, with $O(\boldsymbol{\alpha}) = O(\boldsymbol{\alpha}')$ since we discuss the case of a thin lens with an angular magnification factor of 1. $O(\boldsymbol{\alpha})$ is convolved with the PSF to obtain the intensity distribution in the image plane as

$$I(\boldsymbol{\alpha}) = O(\boldsymbol{\alpha}) * \text{PSF}(\boldsymbol{\alpha}), \tag{3.4}$$

with $O(\boldsymbol{\alpha})$ the source intensity distribution and $I(\boldsymbol{\alpha})$ the image intensity distribution.

It is a basic property of Fourier transforms that a convolution is equivalent to a multiplication in Fourier space. Therefore, we introduce the spatial frequency spectrum $\hat{O}(\boldsymbol{R})$, in general a complex function, as the Fourier transform of the source intensity $O(\boldsymbol{\alpha})$, writing the image intensity distribution as

$$I(\boldsymbol{\alpha}) = \mathcal{F}_R\left(\hat{O}(\boldsymbol{R})\,\mathrm{OTF}(\boldsymbol{R})\right). \tag{3.12}$$

In this notation, the imaging process is a linear process, with the OTF acting as a filter on the spatial frequency spectrum $\hat{O}(\boldsymbol{R})$ of the source intensity.

3.2 The Coherence Function

The propagation of the coherence function, i.e. the coherence properties of light propagating through an optical system, can be separated into two steps, from the star in the source plane into the aperture plane, and then further on into the image plane of the telescope, denoting planes and coordinates as in Fig. 3.1. We will apply the van Cittert–Zernike theorem describing the propagation of the MSDF (2.42) first for the propagation from the source plane into the aperture plane and a second time for the propagation into the image plane.

We will see that, observing incoherent objects, the source can be described by the intensity distribution instead of the MSDF, facilitating the step from the source plane into the aperture plane. In the image plane, using detectors that measure the intensity, we can also restrict the discussion to the intensity as a special case of the coherence function. Thus, we discuss the coherence function only in the aperture plane. Nevertheless the complete formalism is required to properly describe the imaging process.

For the sake of completion, the case of a coherent source and of the coherence properties in the image plane – i.e. the MSDF in the source and in the image plane – will also be discussed in Sects. 3.2.2 and 3.2.3.

3.2.1 Image Intensity Distribution

We start with the MSDF of the incoherent star in the source plane $\boldsymbol{\alpha}'$ (2.44) as discussed in Sect. 2.3.3:

$$\hat{\Gamma}(\boldsymbol{\alpha}'_1, \boldsymbol{\alpha}'_2, \nu) = \frac{\lambda^2}{z_0^2} O_b(\boldsymbol{\alpha}'_1)\,\delta(\boldsymbol{\alpha}'_1 - \boldsymbol{\alpha}'_2) G(\nu), \tag{3.21}$$

with $O_b(\boldsymbol{\alpha}'_1)$ the object brightness distribution and $G(\nu)$ the spectrum. Since this MSDF has non-zero values only for $\boldsymbol{\alpha}'_1 = \boldsymbol{\alpha}'_2$ we can replace it by the spectral

intensity distribution of the incoherent source is

$$O(\boldsymbol{\alpha}', \nu) = \hat{\Gamma}(\boldsymbol{\alpha}', \boldsymbol{\alpha}', \nu) = O_b(\boldsymbol{\alpha}')G(\nu). \tag{3.22}$$

We make the assumption that the object brightness distribution $O_b(\boldsymbol{\alpha}')$, i.e. the shape of the object is independent of wavelength over the observed spectral band.

Applying (2.45) in the notation with spatial frequencies $\boldsymbol{f}_\xi = \boldsymbol{\xi}/\lambda$, we write the MSDF in the aperture plane owing to an incoherent source at a large distance z_0 as

$$\begin{aligned}\hat{\Gamma}(\boldsymbol{f}_{\xi,1} - \boldsymbol{f}_{\xi,2}, \nu) &= \frac{z_0^2}{\lambda^2} \iint \hat{\Gamma}(\boldsymbol{\alpha}_1', \boldsymbol{\alpha}_2', \nu) \mathrm{e}^{\mathrm{i}k(|\alpha_1'|^2 - |\alpha_2'|^2)z_0/2} \\ &\quad \times \mathrm{e}^{\mathrm{i}2\pi(f_{\xi,1}\cdot\alpha_1' - f_{\xi,2}\cdot\alpha_2')}\, \mathrm{d}\boldsymbol{\alpha}_1' \mathrm{d}\boldsymbol{\alpha}_2' \\ &= G(\nu) \int O_b(\boldsymbol{\alpha}')\, \mathrm{e}^{\mathrm{i}2\pi(f_{\xi,1} - f_{\xi,2})\cdot\alpha'}\, \mathrm{d}\boldsymbol{\alpha}' \\ &= G(\nu) I_0 \mu(\boldsymbol{f}_{\xi,1} - \boldsymbol{f}_{\xi,2}).\end{aligned} \tag{3.23}$$

Note that the exponential has a positive sign since the axes in the source plane $\boldsymbol{\alpha}'$ are inverted with respect to the axes in the aperture plane.

While the spectrum $G(\nu)$ characterises the temporal coherence, the visibility function $\mu(\boldsymbol{f}_{\xi,1} - \boldsymbol{f}_{\xi,2})$ determines the spatial coherence of the light in the aperture plane.

The **visibility function** is defined in (2.51) as the Fourier transform of the source brightness distribution $O_b(\boldsymbol{\alpha}')$ divided by I_0. As a function of spatial frequencies $\boldsymbol{f}_\xi = \boldsymbol{\xi}/\lambda$, we now write the van Cittert–Zernike theorem as

$$\mu(\boldsymbol{f}_{\xi,1} - \boldsymbol{f}_{\xi,2}) = \frac{\int O_b(\boldsymbol{\alpha}')\mathrm{e}^{-\mathrm{i}2\pi(f_{\xi,1} - f_{\xi,2})\cdot\alpha'}\, \mathrm{d}\boldsymbol{\alpha}'}{\int O_b(\boldsymbol{\alpha}')\, \mathrm{d}\boldsymbol{\alpha}'}, \tag{3.24}$$

with $I_0 = \int O_b(\boldsymbol{\alpha}')\mathrm{d}\boldsymbol{\alpha}'$. It is $0 \leq |\mu(\boldsymbol{f}_{\xi,1} - \boldsymbol{f}_{\xi,2})| \leq 1$ and $\mu(0) = 1$.

The second step leads us from the aperture plane $\boldsymbol{f}_\xi$ into the image plane $\boldsymbol{\alpha}$. We modify (2.45) in order to describe the propagation of the MSDF from the aperture plane into the image plane at distance F, applying the aperture function $A(\boldsymbol{f}_\xi)$ and the lens function $L(\boldsymbol{f}_\xi) = \exp(-\mathrm{i}2\pi|\boldsymbol{f}_\xi|^2\lambda/(2F))$. Dealing only with the intensity in the image plane, we write the spectral image intensity distribution $I(\boldsymbol{\alpha}, \nu)$ in the notation with spatial frequencies as

$$\begin{aligned}I(\boldsymbol{\alpha}, \nu) = \hat{\Gamma}(\boldsymbol{\alpha}, \boldsymbol{\alpha}, \nu) &= \frac{\lambda^2}{F^2} \iint \hat{\Gamma}(\boldsymbol{f}_{\xi,1} - \boldsymbol{f}_{\xi,2}, \nu) A(\boldsymbol{f}_{\xi,1}) A^*(\boldsymbol{f}_{\xi,2}) L(\boldsymbol{f}_{\xi,1}) \\ &\quad \times L^*(\boldsymbol{f}_{\xi,2}) \mathrm{e}^{\mathrm{i}2\pi(|f_{\xi,1}|^2 - |f_{\xi,2}|^2)\lambda/(2F)} \mathrm{e}^{-\mathrm{i}2\pi(f_{\xi,1} - f_{\xi,2})\cdot\alpha}\, \mathrm{d}\boldsymbol{f}_{\xi,1} \mathrm{d}\boldsymbol{f}_{\xi,2} \\ &= \frac{\lambda^2}{z_1^2} G(\nu) I_0 \iint \mu(\boldsymbol{f}_{\xi,1} - \boldsymbol{f}_{\xi,2}) A(\boldsymbol{f}_{\xi,1}) A^*(\boldsymbol{f}_{\xi,2}) \\ &\quad \times \mathrm{e}^{-\mathrm{i}2\pi(f_{\xi,1} - f_{\xi,2})\cdot\alpha}\, \mathrm{d}\boldsymbol{f}_{\xi,1} \mathrm{d}\boldsymbol{f}_{\xi,2}\,.\end{aligned} \tag{3.25}$$

The quadratic phase terms $\exp(i2\pi|\boldsymbol{f}_{\xi,i}|^2\lambda/(2F))$ cancel with those of the lens $L(\boldsymbol{f}_{\xi,i})$ since we chose $z_1 = F$ and a star at a large distance z_0. This means that the image of the star appears in the back focal plane of our optical system.

NB 6. *It is interesting to discuss the case of a star (or any other incoherent source) at finite distance so that the image plane will be at a distance $z_1 \neq F$. In this case we have to keep the phase term $\exp(i2\pi(|\boldsymbol{f}_{\xi,1}|^2 - |\boldsymbol{f}_{\xi,2}|^2)\lambda/(2z_0))$ in front of the integral, as in (2.41), and we have to compute the product*

$$\exp\Big(i2\pi\lambda\big(|\boldsymbol{f}_{\xi,i}|^2/(2z_0) + |\boldsymbol{f}_{\xi,i}|^2/(2z_1) - |\boldsymbol{f}_{\xi,i}|^2/(2F)\big)\Big)$$
$$= \exp\Big(i2\pi\lambda|\boldsymbol{f}_{\xi,i}|^2\big(\frac{1}{z_0} + \frac{1}{z_1} - \frac{1}{F}\big)/2\Big).$$

The quadratic phase term disappears if $\frac{1}{z_0} + \frac{1}{z_1} = \frac{1}{F}$. This is the familiar lens law *from elementary geometrical optics where it is used to calculate the image position in the paraxial approximation [88]. Here, in the context of the propagation of the coherence function we obtain the same relationship to eliminate the quadratic phase terms in Fresnel approximation so that Fourier transforms can be used to describe the imaging process from the source plane through the aperture plane into the image plane.*

In the following, we will write the visibility function as a function of the spatial frequency vector $\boldsymbol{R} = (u, v) = \boldsymbol{f}_{\xi,1} - \boldsymbol{f}_{\xi,2}$, and we will call the spatial frequency plane the *uv-plane* of the interferometer. These are the most common notations for the visibility function and for the spatial frequency space.

With $\boldsymbol{R} = \boldsymbol{f}_{\xi,1} - \boldsymbol{f}_{\xi,2}$, the spectral intensity distribution in the image plane (3.25) reads as

$$I(\boldsymbol{\alpha}, \nu) = \frac{\lambda^2}{F^2} G(\nu) I_0 \iint \mu(\boldsymbol{R})\, A(\boldsymbol{f}_{\xi,1}) A^*(\boldsymbol{f}_{\xi,1} - \boldsymbol{R})\, e^{-i2\pi \boldsymbol{R}\cdot\boldsymbol{\alpha}} d\boldsymbol{f}_{\xi,1} d\boldsymbol{R}\,. \quad (3.26)$$

Performing the integration over $\boldsymbol{f}_{\xi,1}$ yields the OTF as defined in (3.6) and we obtain

$$I(\boldsymbol{\alpha}, \nu) = G(\nu) I_0 \int \mu(\boldsymbol{R})\, \mathrm{OTF}(\boldsymbol{R})\, e^{-i2\pi \boldsymbol{R}\cdot\boldsymbol{\alpha}} d\boldsymbol{R}$$
$$= G(\nu) I_0\, \mathcal{F}_R\big(\mu(\boldsymbol{R})\, \mathrm{OTF}(\boldsymbol{R})\big). \quad (3.27)$$

The advantage of this notation is that the visibility $\mu(\boldsymbol{R})$ only depends on the shape of the object, which is assumed independent of wavelength. The influence of the wavelength is considered only indirectly since moving in the uv-plane could mean either to change the coordinate difference $\boldsymbol{\xi}_1 - \boldsymbol{\xi}_2$ or the wavelength λ. It is their quotient, $\boldsymbol{R}$, that determines the function value of μ.

Now, the shape of the OTF depends not only on the aperture but also on the wavelength. For instance, the OTF of a telescope aperture of diameter D has a wavelength dependent radius of $R = D/\lambda$ in the uv-plane (see discussion in Sect. 3.1.1).

We will see in Sect. 3.4 that the formal treatment of the imaging process in an interferometer is simplified by using spatial frequencies.

Fourier Optics in monochromatic illumination was discussed in Sect. 3.1.1 when we introduced the spatial frequency spectrum $\hat{O}(\boldsymbol{R})$ as the Fourier transform of the object intensity distribution $O(\boldsymbol{\alpha}')$. Using the formalism of coherence functions, we find the following identities:

$$O(\boldsymbol{\alpha}') = \int O(\boldsymbol{\alpha}', \nu)\mathrm{d}\nu = \int \delta(\nu - \nu_0) O_b(\boldsymbol{\alpha}')\mathrm{d}\nu = O_b(\boldsymbol{\alpha}')$$

$$\hat{O}(\boldsymbol{R}) = I_0 \int G(\nu)\mu(\boldsymbol{R})\mathrm{d}\nu = I_0 \int \delta(\nu - \nu_0)\mu(\boldsymbol{R})\mathrm{d}\nu = I_0\mu(\boldsymbol{R}). \quad (3.28)$$

Thus, in the monochromatic case with $G(\nu) = \delta(\nu - \nu_0)$, the intensity distribution is identical to the brightness distribution, and the spatial frequency spectrum is identical to the visibility function. Consequently, the expression for the image intensity in Fourier Optics, describing it as the Fourier transform of the product of the spatial frequency spectrum with the OTF (3.12), is reproduced in (3.27) with the visibility function multiplied by the OTF.

The physical interpretation of the spatial frequency spectrum as the visibility function in the uv-plane explains immediately that $\hat{O}(\boldsymbol{R})$ is not an intensity distribution like $O(\boldsymbol{\alpha}')$ but a coherence function. And that $\boldsymbol{R}$ is a difference coordinate and not the coordinate vector in the aperture plane.

We should, however, keep in mind that the frequency spectrum $\hat{O}(\boldsymbol{R})$ was defined as an abstract mathematical quantity, following from the reasoning that the image intensity distribution of an incoherent object is the linear superposition of PSFs (3.4). In this section, discussing the propagation of the coherence function we have used a more elaborate definition of an incoherent source (2.44) taking physical properties into account.

Using (3.23), we simplify the formalism rewriting the imaging process as a convolution of individual Fourier transforms in (3.27) with $I_0\mathcal{F}_R(\mu(\boldsymbol{R})) = O_b(\boldsymbol{\alpha})$ – the object brightness distribution and the visibility function form a Fourier pair – and with the PSF as the Fourier transform of the OTF:

$$\begin{aligned} I(\boldsymbol{\alpha}, \nu) &= G(\nu) I_0 \mathcal{F}_R\big(\mu(\boldsymbol{R})\,\mathrm{OTF}(\boldsymbol{R})\big) \\ &= G(\nu) I_0 \Big(\mathcal{F}_R\big(\mu(\boldsymbol{R})\big) * \mathcal{F}_R\big(\mathrm{OTF}(\boldsymbol{R})\big)\Big) \\ &= G(\nu)\big(O_b(\boldsymbol{\alpha}) * \mathrm{PSF}(\boldsymbol{\alpha})\big). \end{aligned} \quad (3.29)$$

Note that we obtain the polychromatic intensity $I(\boldsymbol{\alpha})$ – as the quantity that we can measure – by integrating the spectral intensity $I(\boldsymbol{\alpha}, \nu)$ over the spectrum.

In the monochromatic case, with $G(\nu) = \delta(\nu - \nu_0)$, we obtain the image intensity very simply as

$$
\begin{aligned}
I(\boldsymbol{\alpha}) &= \int I(\boldsymbol{\alpha}, \nu)\mathrm{d}\nu = I_0 \mathcal{F}_R\big(\mu(\boldsymbol{R})\,\mathrm{OTF}(\boldsymbol{R})\big) \\
&= O_b(\boldsymbol{\alpha}) * \mathrm{PSF}(\boldsymbol{\alpha}),
\end{aligned}
\tag{3.30}
$$

with $\int G(\nu)\mathrm{d}\nu = 1$.

The formalism of the propagation of the coherence function through an optical system has led us to exactly the same result as in Fourier optics. This is due to the restriction of our discussion to an incoherent source, instead of a source with arbitrary coherence function, and because the intensity distribution is discussed instead of the coherence function in the image plane.

In the following two subsections, we expand the discussion by investigating coherent sources and the coherence properties in the image plane. We will return to the image intensity in Sect. 3.3 regarding the imaging process in interferometers.

3.2.2 *Coherent Imaging*

A coherent source in the source plane $\boldsymbol{\alpha}'$ is given by a plane wave travelling along the z-axis with the monochromatic amplitude $V(\boldsymbol{\alpha}')$. Plane waves, like all other solutions of the Helmholtz equation, form a coherent field (see Sect. 2.3.1), and the MSDF of a coherent field is simply the product of the amplitudes, using (2.37) and (2.31).

The shape of the object, its transmittance distribution is given by $V_o(\boldsymbol{\alpha}')$, and we set $V(\boldsymbol{\alpha}') = V_o(\boldsymbol{\alpha}')$ immediately after the object. Slide projectors and microscopes are examples for optical systems illuminating the object with a dedicated illumination system.

Coherent imaging is described by the propagation of the MSDF of the coherent field, that as the product of the amplitudes is given by $\hat{\Gamma}(\boldsymbol{\alpha}'_1, \boldsymbol{\alpha}'_2, \nu_0) = V_o(\boldsymbol{\alpha}'_1)V_o^*(\boldsymbol{\alpha}'_2)$. Introducing this in (3.23) yields the MSDF in the aperture plane as

$$
\begin{aligned}
\hat{\Gamma}(\boldsymbol{f}_{\xi,1}, \boldsymbol{f}_{\xi,2}, \nu_0) &= \frac{z_0^2}{\lambda_0^2} \iint V_o(\boldsymbol{\alpha}'_1)V_o^*(\boldsymbol{\alpha}'_2)\, \mathrm{e}^{\mathrm{i}2\pi(|\boldsymbol{\alpha}'_1|^2 - |\boldsymbol{\alpha}'_2|^2)z_0/(2\lambda_0)} \\
&\quad \times \mathrm{e}^{\mathrm{i}2\pi(\boldsymbol{f}_{\xi,1}\cdot\boldsymbol{\alpha}'_1 - \boldsymbol{f}_{\xi,2}\cdot\boldsymbol{\alpha}'_2)}\, \mathrm{d}\boldsymbol{\alpha}'_1 \mathrm{d}\boldsymbol{\alpha}'_2 \\
&= \frac{z_0^2}{\lambda_0^2} \int V_o(\boldsymbol{\alpha}'_1)\mathrm{e}^{\mathrm{i}2\pi|\boldsymbol{\alpha}'_1|^2 z_0/(2\lambda_0)}\, \mathrm{e}^{\mathrm{i}2\pi \boldsymbol{f}_{\xi,1}\cdot\boldsymbol{\alpha}'_1}\, \mathrm{d}\boldsymbol{\alpha}'_1 \\
&\quad \times \int V_o^*(\boldsymbol{\alpha}'_2)\mathrm{e}^{-\mathrm{i}2\pi|\boldsymbol{\alpha}'_2|^2 z_0/(2\lambda_0)}\, \mathrm{e}^{-\mathrm{i}2\pi \boldsymbol{f}_{\xi,2}\cdot\boldsymbol{\alpha}'_2}\, \mathrm{d}\boldsymbol{\alpha}'_2 \\
&= V_o(\boldsymbol{f}_{\xi,1})\, V_o^*(\boldsymbol{f}_{\xi,2}).
\end{aligned}
\tag{3.31}
$$

Thus, after propagation from the source into the aperture plane the light is still perfectly coherent and the MSDF remains the product of the amplitudes. Here, $\hat{\Gamma}(\boldsymbol{f}_{\xi,1}, \boldsymbol{f}_{\xi,2}, \nu_0)$ can no longer be written as a function of coordinate difference as in the case of an incoherent source but the individual coordinates $\boldsymbol{f}_{\xi,1}$ and $\boldsymbol{f}_{\xi,2}$ have to be considered.

The quadratic phase terms $\exp(\mathrm{i}2\pi|\boldsymbol{\alpha}_i'|^2 z_0/(2\lambda_0))$ in (3.31) that disappeared in (3.23) due to the incoherence of the source, do not disappear here unless the object is small. They indicate that the surface of equal phase of the amplitude is not plane but parabolic. Rigorous calculation without the Fresnel approximation shows that the surface of equal phase is a sphere with its centre in the aperture plane [148].

Most of our objects have a very small angular size $\boldsymbol{\alpha}'$ and we set the exponential to unity. The necessity to deal with small objects means that the isoplanatic angle is very small in coherent imaging. With this approximation, the amplitude $V_o(\boldsymbol{f}_\xi)$ in the aperture plane is the Fourier back transform of the amplitude in the source plane:

$$V_o(\boldsymbol{f}_\xi) = \frac{z_0}{\lambda_0} \int V_o(\boldsymbol{\alpha}')\mathrm{e}^{\mathrm{i}2\pi \boldsymbol{f}_\xi \cdot \boldsymbol{\alpha}'}\,\mathrm{d}\boldsymbol{\alpha}' . \tag{3.32}$$

$V_o(\boldsymbol{f}_\xi)$ is the spatial frequency spectrum of $V_o(\boldsymbol{\alpha}')$, i.e. of the object transmittance in the source plane.

The final step leads us into the image plane, by putting the result of the MSDF (3.31) into (3.25). With $I(\boldsymbol{\alpha}) = \int \hat{\Gamma}(\boldsymbol{\alpha}, \boldsymbol{\alpha}, \nu_0)\delta(\nu - \nu_0)\mathrm{d}\nu$ we obtain the intensity in the image plane as

$$\begin{aligned} I(\boldsymbol{\alpha}) &= \frac{\lambda_0^2}{F^2} \iint V_o(\boldsymbol{f}_{\xi,1})V_o^*(\boldsymbol{f}_{\xi,2})A(\boldsymbol{f}_{\xi,1})A^*(\boldsymbol{f}_{\xi,2})\mathrm{e}^{-\mathrm{i}2\pi(\boldsymbol{f}_{\xi,1}-\boldsymbol{f}_{\xi,2})\cdot\boldsymbol{\alpha}}\mathrm{d}\boldsymbol{f}_{\xi,1}\mathrm{d}\boldsymbol{f}_{\xi,2} \\ &= \frac{\lambda_0^2}{F^2} \int V_o(\boldsymbol{f}_{\xi,1})\, A(\boldsymbol{f}_{\xi,1})\mathrm{e}^{-\mathrm{i}2\pi \boldsymbol{f}_{\xi,1}\cdot\boldsymbol{\alpha}}\,\mathrm{d}\boldsymbol{f}_{\xi,1} \\ &\quad \times \int V_o^*(\boldsymbol{f}_{\xi,2})\, A^*(\boldsymbol{f}_{\xi,2})\mathrm{e}^{\mathrm{i}2\pi \boldsymbol{f}_{\xi,2}\cdot\boldsymbol{\alpha}}\,\mathrm{d}\boldsymbol{f}_{\xi,2} \\ &= \frac{\lambda_0^2}{F^2} \left|\mathcal{F}_s\big(V_o(\boldsymbol{f}_\xi)A(\boldsymbol{f}_\xi)\big)\right|^2 . \end{aligned} \tag{3.33}$$

The image intensity is the squared modulus of the amplitude $V_{\mathrm{im}}(\boldsymbol{\alpha})$ in the image plane, with $V_{\mathrm{im}}(\boldsymbol{\alpha}) = \frac{\lambda_0}{\mathrm{i}F}\mathcal{F}_{f\xi}(V_o(\boldsymbol{f}_\xi)A(\boldsymbol{f}_\xi))$,[3] which is the Fourier transform of the product of the spatial frequency spectrum $V_o(\boldsymbol{f}_\xi)$ of the amplitude with the aperture function $A(\boldsymbol{f}_\xi)$, defining the coherent optical transfer function as $\mathrm{OTF}_{\mathrm{coh}}(\boldsymbol{f}_\xi) := \frac{\lambda_0}{F}A(\boldsymbol{f}_\xi)$.

We now define the coherent point-spread function

$$\mathrm{PSF}_{\mathrm{coh}}(\boldsymbol{\alpha}) := \mathcal{F}_s\left(\mathrm{OTF}_{\mathrm{coh}}(\boldsymbol{f}_\xi)\right), \tag{3.34}$$

which is the amplitude (and not the intensity) of the diffraction pattern of a circular aperture (3.1) displayed in Fig. 3.9. In the image plane, the Fourier transform of the product in (3.33) is rewritten as a convolution of the amplitude in the source plane

[3] The quadratic phase factor $\exp(\mathrm{i}2\pi|\boldsymbol{\alpha}|^2 F/\lambda_0)$ (2.42) has been omitted since we only regard the intensity $I(\boldsymbol{\alpha})$ in the image plane.

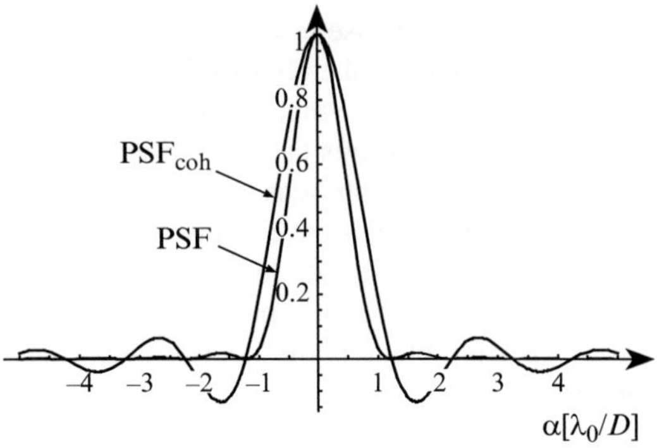

Fig. 3.9 Coherent and incoherent PSF as function of λ_0/D. The first minimum of the PSF (and the first zero of PSF_{coh}) is at $1.22\lambda_0/D$

$V_o(\boldsymbol{\alpha})$ with PSF_{coh},

$$I(\boldsymbol{\alpha}) = |V_o(\boldsymbol{\alpha}) * \text{PSF}_{\text{coh}}(\boldsymbol{\alpha})|^2 . \tag{3.35}$$

We have to keep in mind that this convolution is only applicable if the object is rather small, i.e. in coherent imaging the isoplanatic angle is small. In the incoherent case (3.4), the isoplanatic angle is unlimited within the Fresnel approximation.

What is the Difference Between Coherent and Incoherent Imaging?

Comparing the image intensity distribution of a point source in the incoherent case with that in the coherent case by using a δ-function to describe the object in (3.4) and in (3.35) we see that the intensity distributions are $\text{PSF}(\boldsymbol{\alpha})$ respectively $|\text{PSF}_{\text{coh}}(\boldsymbol{\alpha})|^2$ with, taking the discussion in Sect. 2.2.3 into account, $\text{PSF}(\boldsymbol{\alpha}) = |\text{PSF}_{\text{coh}}(\boldsymbol{\alpha})|^2$. This is not really surprising since the point source, although treated by the formalism of the incoherent imaging process, is a coherent source by definition.

The difference between the two processes lies in the treatment of extended objects. In the coherent case (3.35), the superposition of coherent PSFs forms the amplitude in the image, which is equivalent to convolving PSF_{coh} with the amplitude of the object. In the incoherent case (3.4), when the intensities add, the convolution is between the PSF and the intensity distribution of the object.

The difference between these two cases is particularly obvious when looking at the image of an edge, displayed in Fig. 3.10. The coherent image shows a prominent fringe structure with an intensity of 0.25 at the nominal edge position ($\alpha = 0$) while the incoherent image is rather smooth with $I(0) = 0.5$. Both curves have about the same gradient around the nominal edge position but the position of the edge, defined, e.g., as the position of 50% intensity, is different in coherent and incoherent illumination. This is important for instance in microlithography when the exact edge position in the image defines the width of the conductor strip on the integrated circuit.

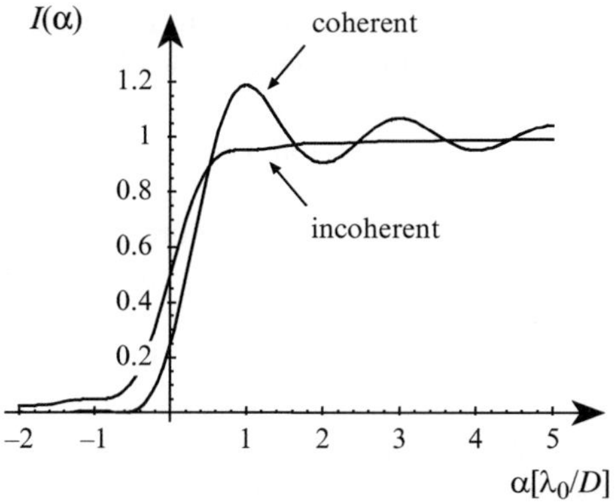

Fig. 3.10 The image intensity distribution of a straight edge in coherent and incoherent illumination. The fringe structure shows up in coherent illumination as a consequence of the convolution with the coherent point-spread function $\mathrm{PSF}_{\mathrm{coh}}$ while the curve is rather smooth in incoherent illumination. For an infinitely large aperture with $D \rightarrow \infty$ the fringe spacing goes to zero and the two curves are identical

Imaging the Telescope Aperture, Entrance and Exit Pupil

In astronomical telescopes, entrance and exit pupils are defined as the images (real or virtual) of the physical telescope aperture. The entrance pupil is then formed by all optical elements before the aperture and the exit pupil by those behind the aperture. If the only light source of this imaging process is a point-like star providing spatially coherent illumination, one has to apply the formalism of coherent imaging.

Thus, we will discuss an imaging process when the telescope aperture is the object. It is not completely straightforward to define the limiting optical aperture determining the point-spread function in the image of the telescope aperture. In a first attempt we assume that the involved optical elements have a very large diameter and that the limiting aperture is in the Fourier plane of the telescope aperture, which is the image plane of the telescope. We have then swapped the roles of image and aperture plane.

If there is only one focus in the optical path of the telescope, for instance the Cassegrain or the Nasmyth focus, the illumination of the telescope aperture is by the stars in the sky. However, very often there are one or more intermediate foci in the optical path for which the optical train for the Coudé focus is a typical example, as are most interferometers with lights paths of several hundred metres. Then the image of the telescope aperture is also formed several times along the optical train – alternating with intermediate foci – with the exit pupil being the last image before the detector.

Let us take the example of an optical train of an interferometer when the focal plane just before the science camera is limited by a field stop. Very often this field stop is as small as one Airy disk as we will see later. First, we set the size of the field stop to a few arc seconds. The camera optics will form the final image on the detector. The camera optics will also form an image, the exit pupil, of the telescope aperture somewhere in the camera. If the observation is done at infrared wavelengths, the camera and its optics are usually placed in a dewar ensuring the required low temperatures for the detector. The exit pupil is also inside the dewar and a physical aperture, the *cold stop* is placed at its position to make sure that only the light passing through the telescope aperture and no stray light from mirror mounts arrives on the detector. The cold stop is carefully dimensioned in order to match the size of the reimaged telescope aperture.

The intensity distribution in the exit pupil, i.e. the image of the telescope aperture, is determined by a point-spread function that is the Fourier transform of the field stop in the focal plane before the science camera. For a numerical example we assume an 8-m aperture and a field stop limiting the diameter of the focal plane to 2 arcsec. Then, in the K-band we will find an intensity distribution in the exit pupil as displayed in Fig. 3.11a when an individual star provides spatially coherent illumination for the aperture. There is a fringe structure similar to the fringes when imaging a straight edge (see Fig. 3.10) in coherent illumination. If the field stop is reduced to the diameter of the Airy disk, clipping its diffraction rings, the intensity distribution in the exit pupil in Fig. 3.11b is very smooth without fringes and without a sharp edge limiting the exit pupil. Inspecting the image formed on the detector of the camera, it is obvious that the point-spread function forming the image of the stellar object in case of the small field stop – having clipped the Airy rings in the preceding focus – no longer has the shape of an Airy disk but, as the Fourier transform of the exit pupil, is a smooth function similar to a Gaussian function.

These considerations illustrate the interplay between field stops and aperture stops along the optical train and their influence on the intensity distributions.

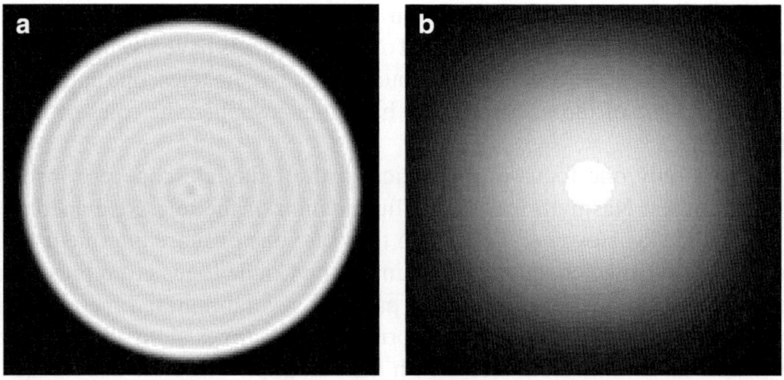

Fig. 3.11 The intensity distribution in the exit pupil of an 8-m telescope, illuminated by a single non resolved star in the K-band. If the field of view is limited by a physical field stop to 2 arcsec, the intensity distribution in the exit pupil (**a**) displays a fringe structure similar to a coherent edge image (see Fig. 3.10). A field stop of the size of the Airy disk, clipping its diffraction rings, has an intensity distribution (**b**) similar to a Gaussian

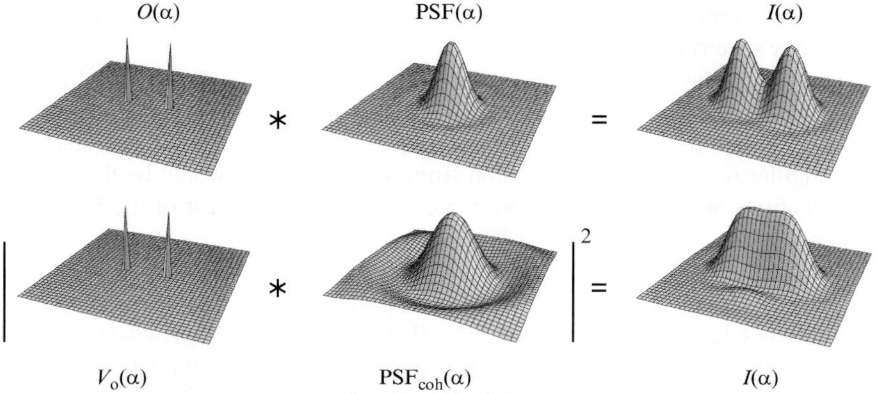

Fig. 3.12 Comparison of the incoherent (*top row*) and the coherent (*bottom row*) imaging process. A binary star with a separation $\alpha_0 = 1.5\lambda/D$ is displayed in the left column. Both the object intensity distribution $O(\boldsymbol{\alpha})$ and the object transmittance $V_o(\boldsymbol{\alpha})$ are represented by the sum of two δ-functions. The incoherent point-spread function, PSF($\boldsymbol{\alpha}$), in the top row and the coherent one, $\text{PSF}_{\text{coh}}(\boldsymbol{\alpha})$, in the bottom row are linked through $\text{PSF}(\boldsymbol{\alpha}) = |\text{PSF}_{\text{coh}}(\boldsymbol{\alpha})|^2$. With an aperture diameter of D the FWHM of PSF($\boldsymbol{\alpha}$) is about λ/D. In the incoherent case, the convolution of $O(\boldsymbol{\alpha})$ with PSF($\boldsymbol{\alpha}$) yields directly the image intensity distribution $I(\boldsymbol{\alpha})$. In the coherent case, the modulus squared of the convolution product $V_o(\boldsymbol{\alpha}) * \text{PSF}_{\text{coh}}(\boldsymbol{\alpha})$ provides the intensity. It is interesting to note the difference in angular resolution due to the different FWHM of the relevant point-spread functions. While the binary is clearly resolved in the incoherent case, using the PSF with a FWHM of λ/D, it is only barely resolved in the coherent case, when PSF_{coh} has a width of about $1.38\lambda/D$

A Binary Star

We use again the example of a binary star to illustrate the different behaviour of coherent and incoherent imaging keeping in mind that in reality a coherent binary star emitting radiation with a constant phase of the light of the two stars is an unrealistic situation. In Fig. 3.12, the two convolution products are displayed. It is interesting to note that the angular resolution of an optical system in coherent illumination is much lower than in incoherent illumination. This is due to the fact that the full width at half maximum (FWHM) of PSF_{coh} is about $1.38\lambda/D$ while it is λ/D for the PSF that is used in the case of incoherent illumination. Therefore, the binary star with a separation of $\alpha_0 = 1.5\lambda/D$ is clearly separated in the incoherent case by an optical system with a circular aperture of diameter D, but only barely resolved if it is illuminated coherently.

The advantage of incoherent over coherent illumination can also be explained regarding spatial frequency spectra and transfer functions in (3.12) and (3.33). The OTF as the autocorrelation function of the aperture $A(\boldsymbol{f}_\xi)$ has twice the width of the aperture function as illustrated in Fig. 3.2, and spatial frequencies up to D/λ (the radius of the OTF) are transferred through the system. In coherent imaging when the aperture function $\frac{\lambda}{F}A(\boldsymbol{f}_\xi)$ serves as the transfer function, the frequency cutoff is at $s_0 = \frac{1}{2}D/\lambda$. Due to the approximately triangular shape of the OTF – in contrast to the rectangular shape of the aperture function – the angular resolution in incoherent illumination is not twice that of coherent illumination since the spatial frequency spectrum is damped.

Discussing the image quality of a straight edge in Fig. 3.10, we found that the steepness of the intensity distribution around $\alpha = 0$ is similar in coherent and incoherent illumination. Using this steepness as a criterion for image sharpness and angular resolution the conclusion from the edge image would be that the two illuminations provide similar resolution contradicting the result of the binary star image.

However, the type of object is pivotal [3]. Individual points have to be treated like the binary above, and an extended source with a sharp edge behaves more like the straight edge in Fig. 3.10. In addition, there might be situations when the fringe structure at the edge ruins the image quality while for instance in micro lithography this might not be important if the detector has a suitable sensitivity curve.

In any case, in astronomy we will not observe objects emitting spatially coherent illumination. There is some speculation that pulsars are spatial coherent sources [220] but since there angular size is too small to be resolved they are indistinguishable from a point source and the imaging process can be described like that of an incoherent source. Thus, we will stay with incoherent sources for the remainder of this book.

Summary of Coherent and Incoherent Imaging

An incoherent source requires treating the transfer of the MSDF through the optical system. The Fourier transform of the spectral intensity distribution $O(\boldsymbol{\alpha}', \nu) = G(\nu)O_b(\boldsymbol{\alpha}')$ is the MSDF in the uv-plane $\hat{\Gamma}(\boldsymbol{R}, \nu) = G(\nu)I_0\mu(\boldsymbol{R})$ (see 3.23), which, in monochromatic illumination, is formally identical to the spatial frequency spectrum $\hat{O}(\boldsymbol{R}) = \mathcal{F}'_{\alpha}(O(\boldsymbol{\alpha}'))$, the Fourier transform of the object intensity distribution.

- The transfer function is the OTF, which is the autocorrelation of the aperture function $A(\boldsymbol{f}_\xi)$, acting on the visibility function $\mu(\boldsymbol{R})$.
- The Fourier transform of the product OTF and visibility function provides the intensity distribution in the image plane, see (3.29).
- The optical system is linear with respect to the transfer of the visibility function and spatially stationary since the isoplanatic angle is unlimited within the Fresnel approximation.

A coherent source allows for the transfer of the amplitude $V(\boldsymbol{\alpha}')$ through the optical system. The Fourier transform of the coherently illuminated object transmittance distribution $V_o(\boldsymbol{\alpha}')$ is the amplitude $V_o(\boldsymbol{f}_\xi)$ in the aperture plane, with $V_o(\boldsymbol{f}_\xi)$ containing the spatial frequency spectrum of $V_o(\boldsymbol{\alpha}')$.

- Here, the transfer function is the aperture function $A(\boldsymbol{f}_\xi)$ acting on the spatial frequency spectrum $V_o(\boldsymbol{f}_\xi)$ of the amplitude, see (3.33).
- The Fourier transform of this product provides the amplitude in the image plane, and its squared modulus is the image intensity distribution.

- The optical system is linear with respect to the transfer of the spatial frequency spectrum of the amplitude. However, the isoplanatic angle is limited.

In optical instruments, intermediate states of coherence, *partial coherence*, do also exist. Then, the state of partial coherence in the source plane has to be determined and the propagation of the coherence function through the optical system is calculated by multiple applications of the van Cittert–Zernike Theorem [80].

The important point to keep in mind is the fact that a perfectly coherent wave does not lose coherence by the process of propagation. The state of coherence remains the same from the source plane through the optical system into the image plane, even if we are observing through atmospheric turbulence, as we will discuss in Sect. 4.3.

This is in contrast to the wave field of an incoherent source that acquires coherence by the very process of propagation as described in Sect. 2.3.4 when discussing the coherence function of Venus.

3.2.3 Coherence Properties in the Image Plane

Although most of the time we measure the image intensity, it is interesting to investigate how the coherence properties in the image plane are affected by the imaging process. For an incoherent source we write the MSDF in the aperture again as a function of coordinate difference, $\hat{\Gamma}(\boldsymbol{f}_{\xi,1} - \boldsymbol{f}_{\xi,2}, \nu) = G(\nu) I_0 \mu(\boldsymbol{R})$. We obtain the MSDF in the image plane by applying the van Cittert–Zernike theorem. This was done in (3.25) for the spectral intensity in the image $I(\boldsymbol{\alpha}, \nu) = \hat{\Gamma}(\boldsymbol{\alpha}, \boldsymbol{\alpha}, \nu)$ by reducing the coordinates to $\boldsymbol{\alpha} = \boldsymbol{\alpha}_1 = \boldsymbol{\alpha}_2$. Allowing for different values of the image coordinates, we modify (3.26) to write the MSDF in the image plane

$$\begin{aligned}\hat{\Gamma}(\boldsymbol{\alpha}_1, \boldsymbol{\alpha}_2, \nu) &= \frac{\lambda^2}{F^2} G(\nu) I_0 \\ &\times \iint \mu(\boldsymbol{f}_{\xi,1} - \boldsymbol{f}_{\xi,2}) A(\boldsymbol{f}_{\xi,1}) A^*(\boldsymbol{f}_{\xi,2})\, \mathrm{e}^{-\mathrm{i}2\pi(\boldsymbol{f}_{\xi,1}\cdot\boldsymbol{\alpha}_1 - \boldsymbol{f}_{\xi,2}\cdot\boldsymbol{\alpha}_2)}\, \mathrm{d}\boldsymbol{f}_{\xi,1} \mathrm{d}\boldsymbol{f}_{\xi,2}.\end{aligned} \tag{3.36}$$

We replace the visibility function by the Fourier transform of the object brightness distribution assuming an incoherent light source (3.23).

The MSDF in the image plane can now be written as

$$\begin{aligned}\hat{\Gamma}(\boldsymbol{\alpha}_1, \boldsymbol{\alpha}_2, \nu) &= \frac{\lambda^2}{F^2} G(\nu) \\ &\times \iiint O_b(\boldsymbol{\alpha}') A(\boldsymbol{f}_{\xi,1}) A^*(\boldsymbol{f}_{\xi,2})\, \mathrm{e}^{-\mathrm{i}2\pi(\boldsymbol{f}_{\xi,1}\cdot(\boldsymbol{\alpha}_1 - \boldsymbol{\alpha}') - \boldsymbol{f}_{\xi,2}\cdot(\boldsymbol{\alpha}_2 - \boldsymbol{\alpha}'))}\, \mathrm{d}\boldsymbol{f}_{\xi,1} \mathrm{d}\boldsymbol{f}_{\xi,2} \mathrm{d}\boldsymbol{\alpha}' \\ &= G(\nu) \int O_b(\boldsymbol{\alpha}') \mathrm{PSF}_{\mathrm{coh}}(\boldsymbol{\alpha}_1 - \boldsymbol{\alpha}')\, \mathrm{PSF}^*_{\mathrm{coh}}(\boldsymbol{\alpha}_2 - \boldsymbol{\alpha}') \mathrm{d}\boldsymbol{\alpha}' .\end{aligned} \tag{3.37}$$

The MSDF is determined by a two-fold convolution of the object brightness distribution with the coherent point-spread function $\mathrm{PSF}_{\mathrm{coh}}$, which is the Fourier transform of the aperture $A(\boldsymbol{f}_\xi)$.

If we reduce the MSDF again to the spectral intensity $I(\boldsymbol{\alpha}, \nu) = \hat{\Gamma}(\boldsymbol{\alpha}, \boldsymbol{\alpha}, \nu)$, we obtain the familiar convolution between the object brightness $O_b(\boldsymbol{\alpha}')$ and the point-spread function $\mathrm{PSF}(\boldsymbol{\alpha} - \boldsymbol{\alpha}') = |\mathrm{PSF_{coh}}(\boldsymbol{\alpha} - \boldsymbol{\alpha}')|^2$ as described in (3.4).

Regarding the coherence properties, the coherence width – the permissible coordinate difference $\boldsymbol{\alpha}_1 - \boldsymbol{\alpha}_2$ before the value of the MSDF goes to zero – is a good measure for the size of the area with good coherence in the image plane.

In order to have an idea about the form of the MSDF we assume for the moment that the object brightness distribution is well resolved by the optical system, i.e. that $O_b(\boldsymbol{\alpha}')$ does not vary too much over the area of the coherent point-spread function $\mathrm{PSF_{coh}}(\boldsymbol{\alpha})$. We also apply a variable transform, $\boldsymbol{\alpha}' = \boldsymbol{\alpha}_1 - \boldsymbol{\alpha}$ to (3.37) yielding

$$\begin{aligned}\hat{\Gamma}(\boldsymbol{\alpha}_1, \boldsymbol{\alpha}_2, \nu) &= G(\nu) \int O_b(\boldsymbol{\alpha}_1 - \boldsymbol{\alpha}) \mathrm{PSF_{coh}}(\boldsymbol{\alpha})\, \mathrm{PSF^*_{coh}}(\boldsymbol{\alpha} - (\boldsymbol{\alpha}_1 - \boldsymbol{\alpha}_2)) \mathrm{d}\boldsymbol{\alpha} \\ &\approx G(\nu) O_b(\boldsymbol{\alpha}_1) \int \mathrm{PSF_{coh}}(\boldsymbol{\alpha})\, \mathrm{PSF^*_{coh}}(\boldsymbol{\alpha} - (\boldsymbol{\alpha}_1 - \boldsymbol{\alpha}_2)) \mathrm{d}\boldsymbol{\alpha} \\ &= G(\nu) O_b(\boldsymbol{\alpha}_1) \mathrm{PSF_{coh}}(\boldsymbol{\alpha}_1 - \boldsymbol{\alpha}_2). \end{aligned} \tag{3.38}$$

In this approximation, the correlation at two positions $\boldsymbol{\alpha}_1$ and $\boldsymbol{\alpha}_2$ is determined by the autocorrelation of $\mathrm{PSF_{coh}}$, which is $\mathrm{PSF_{coh}}$ itself. It drops to zero, indicating the coherence width, at $1.22\lambda/D$. Compared to the MSDF in the source plane, with $\delta(\boldsymbol{\alpha}'_1 - \boldsymbol{\alpha}'_2)$ characterizing the spatial incoherence with zero coherence width, we find that the coherence has improved in the image plane due to the propagation through an optical system with finite aperture diameter.

In the theoretical case of an infinitely large aperture and an infinitely narrow PSF the coherence width goes to zero and the light in the image plane is incoherent as it was in the source plane.

The Illuminated Aperture as a Light Source

This discussion has stayed very close to the topic of this section, which is the imaging process. Taking a different point of view, we ignore everything before the aperture plane and regard the light in the aperture plane as our source that propagates into the image plane. The light in the aperture plane is partially coherent, indicated by the visibility function $\mu(\boldsymbol{f}_{\xi,1} - \boldsymbol{f}_{\xi,2})$ as a function of finite extent, and the intensity distribution of our source in the aperture plane is determined by the shape of the aperture $A(\boldsymbol{f}_\xi)$.

We then have a situation similar to the discussion about partially coherent Gaussian Schell-model (GSM) sources at the end of Sect. 2.3.4. There we found the reciprocity relation stating that the coherence function of the source determines the intensity distribution in the plane of observation and, vice versa, that the intensity distribution of the source determines the coherence function in the plane of observation.

Here, we found the same relationship. The Fourier transform of the intensity distribution, its shape given by the aperture $A(\boldsymbol{f}_\xi)$, is $\mathrm{PSF_{coh}}(\boldsymbol{\alpha}_1 - \boldsymbol{\alpha}_2)$ determining the coherence function in the image plane as discussed above (3.38). And the Fourier transform of the visibility function $\mu(\boldsymbol{f}_{\xi,1} - \boldsymbol{f}_{\xi,2})$ in the aperture plane (3.23) provides the intensity distribution $O_b(\boldsymbol{\alpha})$ in the image plane, also as above. The approximation above, that the intensity distribution $O_b(\boldsymbol{\alpha})$ does not vary over the area of $\mathrm{PSF}_{\mathrm{coh}}(\boldsymbol{\alpha})$, i.e. that the object is well resolved by the telescope, can now be translated into the aperture plane stating that the coherence width in the aperture plane needs to be smaller than the diameter of the aperture. Similar approximations were used for GSM sources, see Sect. 5.4.2 in [147].

While at first sight a partially coherent light source seems to be a very academic problem we have shown here that the light in the aperture plane of a telescope can be regarded as a partially coherent source, thus, a very common situation. Being able to relate the coherence properties in the aperture plane to an incoherent source at a large distance through the van Cittert–Zernike theorem should be helpful for the interpretation of the characteristics of partially coherent sources in general and the propagation of its light.

The Coherence Function of Venus

A numerical example illuminates the situation. In Sect. 2.3.4 we looked at Venus and its coherence function at a very large distance, e.g., on Earth or in the aperture plane of a telescope on Earth. With Venus' angular diameter varying between 15 and 45 arcsec depending on the mutual position of Earth and Venus, we found that for 15 arcsec diameter the coherence width in the aperture plane is about 37 mm at a wavelength of 2.2 μm. It should be noted that the intensity distribution in the aperture plane is homogeneous.

If we choose an 8-m telescope producing a point-spread function with a FWHM of 57 mas 2,2 μm, the image of Venus is well resolved, about 260 times wider than the PSF. The coherence width in the image plane is determined by the width of $\mathrm{PSF}_{\mathrm{coh}}$ – the Fourier transform of the 8-m aperture – and is, thus, about 69 mas. This angular width is converted into a spatial coordinate through multiplication with the focal length of the telescope (see Sect. 2.2.2). The focal length is chosen usually such that there are about two detector pixels per PSF in order to ensure proper sampling of the image. With a typical pixel size of about 10 μm, the diameter of the PSF and, thus, the coherence width is about 24 μm.

Although the scale is rather arbitrary, it gives an idea about the basic properties of the propagation of the coherence function. While Venus causes homogeneous illumination in the telescope aperture with a coherence width of 37 mm, the subsequent imaging process can reduce the coherence width in the image plane to values that are substantially smaller than in the aperture plane. Multiple application of the reciprocity relation, from the source into the aperture plane and from the aperture plane into the image plane, can alter the degree of coherence from incoherent in the source, to partially coherent in the aperture plane (coherence width 37 mm), to almost incoherent (coherence width 24 μm) in the image plane.

The Coherence Function: Summary

For the propagation of the coherence function through an optical systems, a two-step process with the van Cittert–Zernike theorem has to be employed. We regard a spatially incoherent source as discussed in Sect. 2.3.3 with the MSDF as

$$\hat{\Gamma}(\boldsymbol{\alpha}'_1, \boldsymbol{\alpha}'_2, \nu) = \frac{\lambda^2}{z_0^2} O_b(\boldsymbol{\alpha}'_1)\, \delta(\boldsymbol{\alpha}'_1 - \boldsymbol{\alpha}'_2) G(\nu), \tag{3.21}$$

with $O_b(\boldsymbol{\alpha}'_1)$ the object brightness distribution and $G(\nu)$ the spectrum. We assume that the object brightness distribution, i.e. the shape of the object is independent of wavelength over the observed spectral band.

The MSDF, having non-zero values only for $\boldsymbol{\alpha}'_1 = \boldsymbol{\alpha}'_2$, is replaced by the spectral intensity distribution of the incoherent source:

$$O(\boldsymbol{\alpha}', \nu) = \hat{\Gamma}(\boldsymbol{\alpha}', \boldsymbol{\alpha}', \nu) = O_b(\boldsymbol{\alpha}') G(\nu). \tag{3.22}$$

Applying the van Cittert–Zernike theorem, the MSDF $\hat{\Gamma}(\boldsymbol{R}, \nu)$ in the uv-plane reads as

$$\begin{aligned} \hat{\Gamma}(\boldsymbol{R}, \nu) &= G(\nu) \int O_b(\boldsymbol{\alpha}')\, \mathrm{e}^{\mathrm{i}2\pi \boldsymbol{R}\cdot\boldsymbol{\alpha}'}\, \mathrm{d}\boldsymbol{\alpha}' \\ &= G(\nu) I_0 \mu(\boldsymbol{R}), \end{aligned} \tag{3.23}$$

when the uv-plane, with $\boldsymbol{R} = (u, v) = \boldsymbol{f}_{\xi,1} - \boldsymbol{f}_{\xi,2}$, is related to the aperture plane with coordinate vector $\boldsymbol{\xi}$ through $\boldsymbol{R} = (\boldsymbol{\xi}_1 - \boldsymbol{\xi}_2)/\lambda$.

The MSDF is the product of the spectrum $G(\nu)$, of the mean intensity $I_0 = \int O_b(\boldsymbol{\alpha}')\mathrm{d}\boldsymbol{\alpha}'$, and of the visibility function $\mu(\boldsymbol{R})$, which is the Fourier transform of the source brightness distribution $O_b(\boldsymbol{\alpha}')$ divided by I_0, so that it is $0 \leq |\mu(\boldsymbol{R}))| \leq 1$ and $\mu(0) = 1$.

Both $\hat{\Gamma}(\boldsymbol{R}, \nu)$ and $\mu(\boldsymbol{R})$ are complex functions. Under the assumption of an object brightness distribution that is independent of wavelength, its Fourier transform, the visibility function, only indirectly depends on wavelength through $\boldsymbol{R} = (\boldsymbol{\xi}_1 - \boldsymbol{\xi}_2)/\lambda$.

Using the van Cittert–Zernike theorem again, we propagate the MSDF from the uv-plane into the image plane. We restricted the discussion to the intensity in the image plane since optical detectors can only measure intensity. With the optical transfer function (OTF) as defined in (3.6) the spectral intensity in the image plane is

$$I(\boldsymbol{\alpha}, \nu) = G(\nu) I_0 \mathcal{F}_R\big(\mu(\boldsymbol{R})\, \mathrm{OTF}(\boldsymbol{R})\big), \tag{3.27}$$

yielding the product of the visibility function and the OTF, as in Fourier optics (3.12) between spatial frequency spectrum of the object intensity $\hat{O}(\boldsymbol{R})$ and the OTF.

Thus, the propagation of the coherence function from the source plane through the aperture plane into the image plane, using the van Cittert–Zernike theorem, is substantially simplified by observing incoherent sources and by determining the image intensity only. The coherence function, i.e. the visibility function, needs to be dealt with only in the aperture plane.

The spectral intensity in the image plane can also be expressed as a convolution of the individual Fourier transforms in (3.27). With $\mathcal{F}_R(\mu(\boldsymbol{R})) = O_b(\boldsymbol{\alpha})$ and the PSF as the Fourier transform of the OTF, we obtain the spectral intensity distribution as

$$I(\boldsymbol{\alpha}, \nu) = G(\nu)\big(O_b(\boldsymbol{\alpha}) * \mathrm{PSF}(\boldsymbol{\alpha})\big). \tag{3.29}$$

Note that we obtain the polychromatic intensity $I(\boldsymbol{\alpha})$ as the quantity that we can measure by integrating the spectral intensity $I(\boldsymbol{\alpha}, \nu)$ over the spectrum.

In the monochromatic case, with $G(\nu) = \delta(\nu - \nu_0)$, the image intensity reads as

$$\begin{aligned} I(\boldsymbol{\alpha}) &= \int I(\boldsymbol{\alpha}, \nu)\mathrm{d}\nu = I_0 \mathcal{F}_R\big(\mu(\boldsymbol{R})\,\mathrm{OTF}(\boldsymbol{R})\big) \\ &= O_b(\boldsymbol{\alpha}) * \mathrm{PSF}(\boldsymbol{\alpha}). \end{aligned} \tag{3.30}$$

Using the fact that the spatial frequency spectrum $\hat{O}(\boldsymbol{R})$ is identical to the visibility function, the complete formalism of Fourier Optics including the treatment of aberrations can be applied to the propagation of the MSDF through optical systems when observing the intensity in the image plane.

3.3 Propagation Through Interferometers

So far, we have described the imaging process in optical systems with the transfer function and the visibility function. We have seen how modulus and phase of both functions affect the properties of the image. The explicit measurement of the coherence function that we discussed in Sect. 2.4 when linking the contrast of a fringe pattern to the coherence function in the aperture plane has not been considered in this context. It is the nature of the imaging process to effectively multiply the OTF by the visibility function at all spatial frequencies inside the telescope aperture, and then to compose the Fourier transform of the product by propagating through the optical system to form the image. In this respect, the optical system acts as an analogue computer.

In this section, we discuss the propagation of light through an interferometer in the context of the formalism developed in Sect. 3.2, basically applying Fourier optics to the propagation of the MSDF.

In the case of an interferometer with separate apertures, or of masks in the aperture of a single telescope, we do not obtain an image in the classical sense but a fringe pattern that does not bear much resemblance to the object, although we can detect fine detail limited by the fringe spacing.

If we want to have an image we need to measure coherence functions in the aperture plane for many different baselines $\boldsymbol{B}$ and then do the Fourier transform into the image plane with a digital computer. The process of filling this virtual aperture that has a diameter given by the longest baseline is called *aperture synthesis*. This is a well established technique in radio interferometry obtaining images with a spatial resolution determined by the longest interferometer baseline and not by the telescope diameter [180, 203]. This will be discussed in Sect. 3.4.

Combining many apertures simultaneously, directly forming an image, avoids the measurement of the coherence function but requires an enormous technological effort since the OPDs between all apertures need to be controlled and minimised to the sub-wavelength level at all times. In Sect. 5.3, we will investigate different concepts of direct imaging.

3.3.1 Young's Experiment with a Lens

After discussing single aperture imaging, we will now do the step to an interferometer, first by introducing an aperture consisting of two sub-apertures the size of pinholes at position $\boldsymbol{\xi}_{p1}$ and $\boldsymbol{\xi}_{p2}$.

The vector between the two sub-apertures is called the *baseline vector* $\boldsymbol{B}$ of the interferometer that is defined as $\boldsymbol{B} = \boldsymbol{\xi}_{p1} - \boldsymbol{\xi}_{p2}$. The separation between the sub-apertures is $B = |\boldsymbol{B}|$.

In the notation with spatial frequencies $\boldsymbol{f}_\xi = \boldsymbol{\xi}/\lambda$, we write the aperture as

$$A(\boldsymbol{f}_\xi) = \delta(\boldsymbol{f}_\xi - \boldsymbol{f}_{\xi,p1}) + \delta(\boldsymbol{f}_\xi - \boldsymbol{f}_{\xi,p2}). \quad (3.39)$$

The pinhole size of the sub-apertures is accounted for by using δ-functions.

The baseline vector $\boldsymbol{R}_B$ in the uv-plane is given by $\boldsymbol{R}_B = \boldsymbol{f}_{\xi,p1} - \boldsymbol{f}_{\xi,p2} = \boldsymbol{B}/\lambda$ (see Fig. 3.13). The baseline vector in the uv-plane has units of spatial frequency.

The OTF is the autocorrelation of $A(\boldsymbol{f}_\xi)$ (see (3.6)) yielding three δ-peaks, one at the origin at $R = 0$, and two at $\boldsymbol{R} = \pm\boldsymbol{R}_B$:

$$\mathrm{OTF}(\boldsymbol{R}) = \frac{\lambda^2}{F^2}\big(\delta(\boldsymbol{R} - \boldsymbol{R}_B) + 2\delta(\boldsymbol{R}) + \delta(\boldsymbol{R} + \boldsymbol{R}_B)\big). \quad (3.40)$$

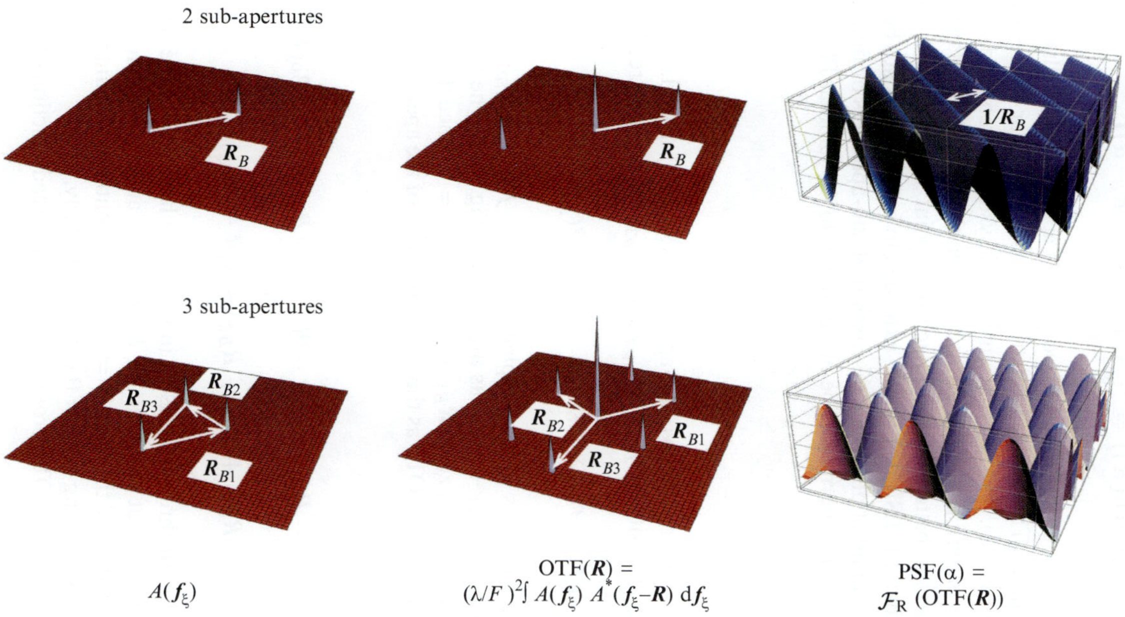

Fig. 3.13 Comparison of Young's Experiment with two (*top row*) and three (*bottom row*) sub-apertures displayed as δ-peaks. The aperture $A(f_\xi)$ and the baseline vector $\boldsymbol{R}_B = \boldsymbol{B}/\lambda$ are shown. In the case of three sub-apertures the three vectors, $\boldsymbol{R}_{B1}$, $\boldsymbol{R}_{B2}$ and $\boldsymbol{R}_{B3}$ are shown. The OTF is displayed with its central peak at $\boldsymbol{R} = 0$ and the peaks at $\boldsymbol{R} = \pm\boldsymbol{R}_B$ and $\boldsymbol{R} = \pm\boldsymbol{R}_{Bi}$ respectively. While the fringe pattern PSF($\boldsymbol{\alpha}$) of two pinholes is familiar, the fringe pattern of three pinholes, as the sum of three cosine functions (3.44), shows a regular bumpy pattern. Observing an extended source, contrast and phase of the fringe pattern are determined by the visibility at the given baseline. While this is straightforward for a single baseline, the bumpy pattern for three sub-apertures usually makes it more difficult to discern the visibilities of the individual baselines. This can be done in the uv-plane after Fourier transforming the fringe pattern

Note that the δ-peak at $R = 0$ is twice as large as the other two δ-peaks due to the autocorrelation process.

Then, the PSF as the Fourier transform of the OTF reads as

$$\mathrm{PSF}(\boldsymbol{\alpha}) = \mathcal{F}_R\big(\mathrm{OTF}(\boldsymbol{R})\big) = 2\frac{\lambda^2}{F^2}\big(1 + \cos(2\pi \boldsymbol{R}_B \cdot \boldsymbol{\alpha})\big). \tag{3.41}$$

This is the shape of the image intensity of a two-pinhole aperture that is illuminated by a point source at infinity. This intensity distribution is the familiar fringe pattern of Young's experiment (2.12). The lens in our optical system is required only to provide the diffraction pattern at a finite distance, i.e. in the back focal plane. The interpretation of the fringe pattern as the image of a point source should not be confused by the fact that this PSF is formally extended to infinity, thus not bearing much resemblance to a point at all. In the discussion of sub-apertures of finite size in Sect. 3.3.2 the diffraction patterns will be more similar to the familiar PSF.

Three Sub-Apertures

The step to apertures with more than two sub-apertures is straightforward. Assuming we have three pinholes at positions $\boldsymbol{f}_{\xi,p1}$, $\boldsymbol{f}_{\xi,p2}$ and $\boldsymbol{f}_{\xi,p3}$, the aperture reads as

$$A_3(\boldsymbol{f}_\xi) = \delta(\boldsymbol{f}_\xi - \boldsymbol{f}_{\xi,p1}) + \delta(\boldsymbol{f}_\xi - \boldsymbol{f}_{\xi,p2}) + \delta(\boldsymbol{f}_\xi - \boldsymbol{f}_{\xi,p3}). \tag{3.42}$$

We then have three baselines in the uv-plane, $\boldsymbol{R}_{B1} = \boldsymbol{f}_{\xi,p1} - \boldsymbol{f}_{\xi,p2}$, $\boldsymbol{R}_{B2} = \boldsymbol{f}_{\xi,p2} - \boldsymbol{f}_{\xi,p3}$ and $\boldsymbol{R}_{B3} = \boldsymbol{f}_{\xi,p3} - \boldsymbol{f}_{\xi,p1}$ as displayed in Fig. 3.13. The sum of the three baseline vectors amounts to zero, $\boldsymbol{R}_{B1} + \boldsymbol{R}_{B2} + \boldsymbol{R}_{B3} = 0$, since they connect the three points of a triangle. The OTF as the autocorrelation of the aperture function $A_3(\boldsymbol{f}_\xi)$ becomes

$$\begin{aligned}\mathrm{OTF}_3(\boldsymbol{R}) = \frac{\lambda^2}{F^2}\big(&\delta(\boldsymbol{R} - \boldsymbol{R}_{B1}) + \delta(\boldsymbol{R} - \boldsymbol{R}_{B2}) + \delta(\boldsymbol{R} - \boldsymbol{R}_{B3}) + 3\delta(\boldsymbol{R}) \\ &+ \delta(\boldsymbol{R} + \boldsymbol{R}_{B1}) + \delta(\boldsymbol{R} + \boldsymbol{R}_{B2}) + \delta(\boldsymbol{R} + \boldsymbol{R}_{B3})\big).\end{aligned} \tag{3.43}$$

Instead of a combination of three δ-peaks as in (3.40) we have now three pairs of δ-peaks at $\pm\boldsymbol{R}_{Bi}$ and the one at the origin at $R = 0$. The autocorrelation of three sub-apertures yields a factor of three at $R = 0$. Subsequently, combining four or more sub-apertures, the weighting at $R = 0$ increases linearly.

The Fourier transform of this OTF provides the point-spread function of a three-pinhole aperture as

$$\begin{aligned}\mathrm{PSF}_3(\boldsymbol{\alpha}) &= \mathcal{F}_R\big(\mathrm{OTF}_3(\boldsymbol{R})\big) \qquad (3.44)\\ &= \frac{\lambda^2}{F^2}\big(3 + 2\cos(2\pi \boldsymbol{R}_{B1} \cdot \boldsymbol{\alpha}) + 2\cos(2\pi \boldsymbol{R}_{B2} \cdot \boldsymbol{\alpha}) + 2\cos(2\pi \boldsymbol{R}_{B3} \cdot \boldsymbol{\alpha})\big).\end{aligned}$$

This PSF is always positive since the sum of the three baseline vectors $\boldsymbol{R}_{Bi}$ amounts to zero. This in turn is due to the fact that the OTF is the autocorrelation function of three sub-apertures so that its Fourier transform, the PSF, is real and positive by definition.

Observing real objects, the fringe pattern loses contrast and is shifted depending on modulus and phase of the visibility at the given baseline. With a single baselines these quantities can be detected in the fringe pattern. However, using three or more sub-apertures, it is very difficult to discern the visibilities of the individual baselines in the bumpy fringe pattern. Then, the Fourier transform of the fringe pattern can be analysed as discussed in Sects. 2.4.3 and 6.1.

Polychromatic Light

The spectral intensity distribution in the image of an extended incoherent source can be calculated, (3.29), by convolving the source brightness distribution $O_b(\boldsymbol{\alpha}')$ with the interferometer PSF,

$$I(\boldsymbol{\alpha}, \nu) = G(\nu)\,(O_b(\boldsymbol{\alpha}) * \mathrm{PSF}(\boldsymbol{\alpha}))\,. \tag{3.45}$$

This is equivalent to the linear superposition of individual fringe patterns, PSF $(\boldsymbol{\alpha} - \boldsymbol{\alpha}')$, created by individual source points at $\boldsymbol{\alpha}'$ (see 3.4).

The Fourier transform of this equation takes us into the uv-plane, forming the product of the OTF with the visibility function, $\mu(\boldsymbol{R})$, (3.27) as

$$\begin{aligned} I(\boldsymbol{\alpha}, \nu) &= G(\nu) I_0 \int \mu(\boldsymbol{R})\,\mathrm{OTF}(\boldsymbol{R})\,\mathrm{e}^{-\mathrm{i}2\pi \boldsymbol{R}\cdot\boldsymbol{\alpha}} \mathrm{d}\boldsymbol{R} \\ &= \frac{\lambda^2}{F^2} G(\nu) I_0 \int \mu(\boldsymbol{R})\left(\delta(\boldsymbol{R} - \boldsymbol{R}_B) + 2\delta(\boldsymbol{R}) + \delta(\boldsymbol{R} + \boldsymbol{R}_B)\right)\mathrm{e}^{-\mathrm{i}2\pi \boldsymbol{R}\cdot\boldsymbol{\alpha}} \mathrm{d}\boldsymbol{R} \\ &= \frac{\lambda^2}{F^2} G(\nu) I_0 \left(\mu(\boldsymbol{R}_B)\mathrm{e}^{-\mathrm{i}2\pi \boldsymbol{R}_B\cdot\boldsymbol{\alpha}} + 2\mu(0) + \mu(-\boldsymbol{R}_B)\mathrm{e}^{\mathrm{i}2\pi \boldsymbol{R}_B\cdot\boldsymbol{\alpha}}\right) \\ &= 2\frac{\lambda^2}{F^2} G(\nu) I_0 \Big(1 + |\mu(\boldsymbol{R}_B)| \cos\big(\phi(\boldsymbol{R}_B) - 2\pi \boldsymbol{R}_B \cdot \boldsymbol{\alpha}\big)\Big), \end{aligned} \tag{3.46}$$

with $\mu(0) = 1$ and $\mu(-\boldsymbol{R}_B) = \mu^*(\boldsymbol{R}_B)$.

In the case of three sub-apertures, the terms at $\pm\boldsymbol{R}_{B2}$ and $\pm\boldsymbol{R}_{B3}$ have to be added as in (3.43) and (3.44).

This is exactly the same result as in Young's experiment, (see (2.59) in Sect. 2.4), in the notation with spatial frequency $\boldsymbol{R}_B = \boldsymbol{B}/\lambda$.

The fringe pattern is proportional to $1 + \cos(.)$ with a visibility proportional to the modulus of the visibility function, $|\mu(\boldsymbol{R}_B)|$, and with a fringe position determined by the phase $\phi(\boldsymbol{R}_B)$.

Before discussing the intensity distribution of polychromatic light with a finite spectral band $\Delta\nu$, we will use the formalism developed in Sect. 3.1 to discuss the more realistic case of a stellar interferometer with finite apertures in monochromatic illumination in the next section.

Finishing this section, we will now have a look at the coherence properties – instead of the intensity distribution – of the fringe pattern.

Coherence Properties of the Fringe Pattern

For the sake of completion, we investigate the coherence properties of the fringe pattern in Young's Experiment. We start with the formula for the MSDF in the image plane of an optical system (3.37) writing the aperture function as $A(\boldsymbol{f}_\xi) = \delta(\boldsymbol{f}_\xi - \boldsymbol{f}_{\xi,p1}) + \delta(\boldsymbol{f}_\xi - \boldsymbol{f}_{\xi,p2})$, for two pinholes at position $\boldsymbol{f}_{\xi,p1}$ and $\boldsymbol{f}_{\xi,p2}$. Then, with $\boldsymbol{f}_{\xi,p1} = \boldsymbol{R}_B/2$ and $\boldsymbol{f}_{\xi,p2} = -\boldsymbol{R}_B/2$, its Fourier transform reads as

$$\mathrm{PSF_{coh}}(\boldsymbol{\alpha}) = 2\cos\left(2\pi\frac{\boldsymbol{R}_B}{2}\cdot\boldsymbol{\alpha}\right). \tag{3.47}$$

Inserting this function into (3.37) yields the MSDF in the image plane as

$$\begin{aligned}
&\hat{\Gamma}(\boldsymbol{\alpha}_1,\boldsymbol{\alpha}_2,\nu)\\
&= 4\frac{\lambda^2}{F^2}G(\nu)\int O_b(\boldsymbol{\alpha}')\cos\left(2\pi\frac{\boldsymbol{R}_B}{2}\cdot(\boldsymbol{\alpha}_1-\boldsymbol{\alpha}')\right)\cos\left(2\pi\frac{\boldsymbol{R}_B}{2}\cdot(\boldsymbol{\alpha}_2-\boldsymbol{\alpha}')\right)\mathrm{d}\boldsymbol{\alpha}'\\
&= \frac{\lambda^2}{F^2}G(\nu)\Bigg[\int O_b(\boldsymbol{\alpha}')\mathrm{d}\boldsymbol{\alpha}'\cos\left(2\pi\frac{\boldsymbol{R}_B}{2}\cdot(\boldsymbol{\alpha}_1-\boldsymbol{\alpha}_2)\right)\\
&\qquad + \int O_b(\boldsymbol{\alpha}')\cos\left(2\pi\frac{\boldsymbol{R}_B}{2}\cdot(\boldsymbol{\alpha}_1+\boldsymbol{\alpha}_2-2\boldsymbol{\alpha}')\right)\mathrm{d}\boldsymbol{\alpha}'\Bigg]\\
&= 2\frac{\lambda^2}{F^2}G(\nu)I_0\Bigg[\cos\left(2\pi\frac{\boldsymbol{R}_B}{2}\cdot(\boldsymbol{\alpha}_1-\boldsymbol{\alpha}_2)\right)\\
&\qquad + |\mu(\boldsymbol{R}_B)|\cos\left(\phi(\boldsymbol{R}_B)-2\pi\frac{\boldsymbol{R}_B}{2}\cdot(\boldsymbol{\alpha}_1+\boldsymbol{\alpha}_2)\right)\Bigg].
\end{aligned} \tag{3.48}$$

The MSDF is composed of the two terms inside the square brackets, the first being a cosine function of the coordinate difference, the second a cosine of the sum of the coordinates, $\boldsymbol{\alpha}_1 + \boldsymbol{\alpha}_2$. Setting $\boldsymbol{\alpha}_1 = \boldsymbol{\alpha}_2$, we obtain $\hat{\Gamma}(\boldsymbol{\alpha},\boldsymbol{\alpha},\nu) = I(\boldsymbol{\alpha},\nu)$, the spectral intensity as in (3.46).

The special case of $\boldsymbol{\alpha}_1 = \boldsymbol{\alpha}_0$ and $\boldsymbol{\alpha}_2 = -\boldsymbol{\alpha}_0$ gives an idea of the impact of the source size on the coherence properties of the fringe pattern. This choice of coordinates means that we compute the MSDF for two points that are symmetrical to the optical axis. The first term inside the square brackets of (3.48) now becomes $\cos(2\pi\boldsymbol{R}_B\cdot\boldsymbol{\alpha}_0)$, and the second one, $|\mu(\boldsymbol{R}_B)|\cos(\phi(\boldsymbol{R}_B))$, is a constant, independent of the sum of our coordinates, but depending on the size of the object $O_b(\boldsymbol{\alpha}')$. For a given baseline, $|\mu(\boldsymbol{R}_B)|$ is wider if the object is smaller.

A point source provides spatially coherent illumination, and it is $|\mu(\boldsymbol{R}_B)| = 1$ for any baseline so that the MSDF of a point source is

$$\hat{\Gamma}(\boldsymbol{\alpha}_0, -\boldsymbol{\alpha}_0, \nu) = 2\frac{\lambda^2}{F^2} G(\nu) I_0 \left(\cos(2\pi \boldsymbol{R}_B \cdot \boldsymbol{\alpha}_0) + 1\right). \tag{3.49}$$

As discussed in Sect. 3.2.2, coherent light remains coherent during the propagation process. Illuminating the two pinholes coherently, i.e. with a plane wave, we expect the light in the image plane to be coherent, too. The result for the MSDF in the image plane using (3.48) is $\cos(2\pi \boldsymbol{R}_B \cdot \boldsymbol{\alpha}_0) + 1$, i.e. varying periodically with $\boldsymbol{\alpha}_0$. Does this mean that the correlation varies with the position of the two points?

To answer this question we have to return to the MCF, which in the case of monochromatic illumination is simply $\Gamma(\boldsymbol{\alpha}_2, \boldsymbol{\alpha}_2, \tau) = \hat{\Gamma}(\boldsymbol{\alpha}_1, \boldsymbol{\alpha}_2, \nu_0) \exp(\mathrm{i}2\pi \nu_0 \tau)$, and $|\Gamma(\boldsymbol{\alpha}_2, \boldsymbol{\alpha}_2, \tau)| = |\hat{\Gamma}(\boldsymbol{\alpha}_1, \boldsymbol{\alpha}_2, \nu_0)|$. We have to calculate the degree of coherence (2.29) as the MCF calibrated by the intensities at positions $\boldsymbol{\alpha}_1$ and $\boldsymbol{\alpha}_2$. Then, we can check if the criterion for coherence, $|\gamma(\boldsymbol{\alpha}_1, \boldsymbol{\alpha}_2, \tau)| \equiv 1$, is fulfilled that was given in Sect. 2.3.1.

With $I(\boldsymbol{\alpha}, \nu) \propto (1 + \cos(2\pi \boldsymbol{R}_B \cdot \boldsymbol{\alpha}))$ and $I(\boldsymbol{\alpha}) = I(\boldsymbol{\alpha}, \nu_0)$ for monochromatic illumination, it is straightforward to see that

$$\begin{aligned} |\gamma(\boldsymbol{\alpha}_0, -\boldsymbol{\alpha}_0, \tau)| &= \frac{|\Gamma(\boldsymbol{\alpha}_0, -\boldsymbol{\alpha}_0, \tau)|}{\sqrt{I(\boldsymbol{\alpha}_0) I(-\boldsymbol{\alpha}_0)}} = \frac{|\hat{\Gamma}(\boldsymbol{\alpha}_0, -\boldsymbol{\alpha}_0, \nu_0)|}{\sqrt{I(\boldsymbol{\alpha}_0, \nu_0) I(-\boldsymbol{\alpha}_0, \nu_0)}} \\ &= \frac{\cos(2\pi \boldsymbol{R}_B \cdot \boldsymbol{\alpha}_0) + 1}{1 + \cos(2\pi \boldsymbol{R}_B \cdot \boldsymbol{\alpha}_0)} = 1, \end{aligned} \tag{3.50}$$

as requested for coherent light. Although this computation was done for two specific coordinates – two points at opposite sides from the optical axis – the result applies to any combination of coordinates.

This example shows very clearly, that MCF and MSDF do not provide directly the correlation as a probability value for measuring the same values of the amplitudes. The calibration with the individual intensities leading to the degree of coherence is required to obtain this probability value and to make a statement on coherence.

After the discussion of illumination by a point source, the other extreme, that of a very large source, is straightforward. A very large source makes the light in the aperture plane practically incoherent, and the contrast of the fringes goes to zero. Then, the visibility function in the aperture plane for a baseline $\boldsymbol{R}_B$, $\mu(\boldsymbol{R}_B)$, determining the contrast of the fringe pattern, goes to zero, and the MSDF in the image plane, (3.48), reads as $\hat{\Gamma}(\boldsymbol{\alpha}_1, \boldsymbol{\alpha}_2, \nu) = G(\nu) I_0 \cos(2\pi \boldsymbol{R}_B/2 \cdot (\boldsymbol{\alpha}_1 - \boldsymbol{\alpha}_2))$. Without any specific assumptions on the coordinates, we find the MSDF as a function of coordinate difference only.

If there are no fringes, the intensity distribution in the image plane is constant, and the calibration of the MSDF does not change the functional dependence on the coordinates. Thus, we can state that the correlation, like the MSDF, varies periodically with $\boldsymbol{\alpha}_1 - \boldsymbol{\alpha}_2$. If we forget for the moment about the interferometer, we can describe this setup as two incoherent point sources – the two pinholes that are illuminated incoherently – illuminating homogeneously the plane of observation – the image plane at a "large" distance due to the lens in the aperture plane – so that we can calculate the MSDF in the plane of observation through a Fourier transform of the intensity distribution of the two point sources, applying the van Cittert–Zernike theorem as in Sect. 2.3.3. According to the van Cittert–Zernike theorem we find that the MSDF depends on the coordinate difference $\boldsymbol{\alpha}_1 - \boldsymbol{\alpha}_2$ only.

The situation changes again if the source has a finite size. We then will find intermediate states of coherence in the fringe pattern that can be computed with (3.48).

3.3.2 Apertures of Finite Size

The aperture function of an interferometer can be expressed by a convolution of the individual telescope aperture $a(\boldsymbol{f}_\xi)$ with δ-functions setting the centre aperture positions at spatial frequencies $\boldsymbol{f}_{\xi,p1}$ and $\boldsymbol{f}_{\xi,p2}$:

$$A(\boldsymbol{f}_\xi) = a(\boldsymbol{f}_\xi) * \big(\delta(\boldsymbol{f}_\xi - \boldsymbol{f}_{\xi,p1}) + \delta(\boldsymbol{f}_\xi - \boldsymbol{f}_{\xi,p2})\big). \tag{3.51}$$

Formally, this describes the masking of a large aperture with two small sub-apertures $a(\boldsymbol{f}_\xi)$, each with diameter D as displayed in Fig. 3.14. Note that in the notation with spatial frequencies the shape of $A(\boldsymbol{f}_\xi)$ is wavelength dependent.

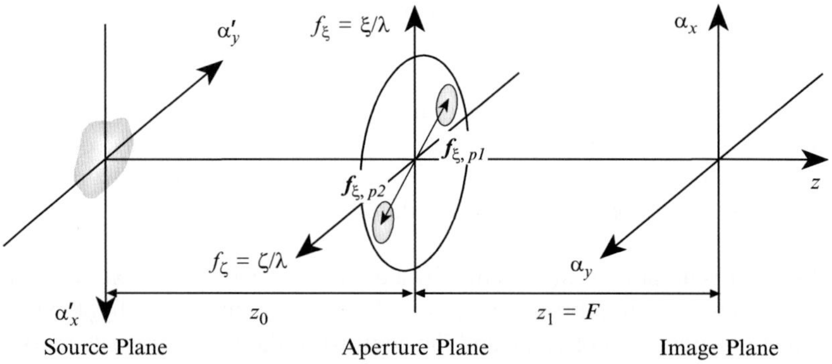

Fig. 3.14 The geometry of a stellar interferometer simplified as a masked aperture with two small sub-apertures $\boldsymbol{f}_{\xi,p1}$ and $\boldsymbol{f}_{\xi,p2}$, the so-called Fizeau configuration. Each sub-aperture has the diameter D. The distance between the apertures is related to the baseline vector $\boldsymbol{B}$ through $\boldsymbol{f}_{\xi,p1} - \boldsymbol{f}_{\xi,p2} = \boldsymbol{R}_B = \boldsymbol{B}/\lambda$. The source is at a very large distance z_0. The aperture plane contains a thin lens of focal length F, creating the diffraction pattern of the aperture in the back focal plane, in this case the image plane. With the two sub-apertures at the same distance $|\boldsymbol{f}_{\xi,p1}| = |\boldsymbol{f}_{\xi,p2}|$ from the optical axis z, the optical path lengths between each sub-aperture and the centre of the image plane at $\boldsymbol{\alpha} = 0$ are equal and the optical path difference, OPD, is zero

Doing the step from a masked aperture to the combination of individual telescopes, the optical paths between each telescope and the image plane are passing through optical elements and can be several hundred metres long. Controlling the optical path lengths and determining the OPD is a major issue in stellar interferometry. Due to the physical separation of the telescopes, the reimaging of the telescopes' exit pupils into the beam combining instrument also requires attention as will be discussed in Sect. 5.1. If the reimaged exit pupils are arranged such that the ratio of pupil diameter to pupil distance is the same as that of telescope aperture to baseline, called *homothetic mapping* of the telescope apertures, the interferometer can still be treated like a masked aperture. Then, the interferometric field of view, i.e. the area of the image with fringes, can be very large, only limited by the optical design of the interferometer. This configuration is named *Fizeau configuration*.

If the telescope apertures are not mapped homothetically but if the distance of the exit pupils in the beam combining instrument is chosen independent of the baseline, this is called the *Michelson configuration*. While this configuration puts fewer constraints on the optical design than the Fizeau configuration, the interferometric field of view is very small, basically confined to the Airy disk, i.e. to the diffraction limit of the individual apertures. In the extreme case, in *co-axial combination*, the exit pupils are projected on top of each other with a beam splitter and, unlike all configurations with separated exit pupils, we do not find fringes in the image plane. By modulating the optical path length in one of the two beams, constructive and destructive interference can then be produced, "switching" the Airy disk on and off. All these aspects will be discussed in more detail in Sect. 5.1.3.

We restrict the following discussions to two sub-apertures bearing in mind that the same formalism applies to three and more sub-apertures. Aspects of combining multiple apertures in different configurations will be investigated in Sect. 5.2.

In the Fizeau configuration, the interferometer OTF is the convolution of the individual aperture OTF_a with the OTF of two pinholes, (3.40), yielding the sum of three OTF_a at positions $\boldsymbol{R_B}, 0$ and $-\boldsymbol{R_B}$ (see Fig. 3.16), as

$$\mathrm{OTF}(\boldsymbol{R}) = \mathrm{OTF}_a(\boldsymbol{R}) * \big(\delta(\boldsymbol{R} - \boldsymbol{R_B}) + 2\delta(\boldsymbol{R}) + \delta(\boldsymbol{R} + \boldsymbol{R_B})\big). \tag{3.52}$$

While observations with a single telescope are limited to spatial frequencies $|\boldsymbol{R}| \leq D/\lambda$ inside OTF_a, the interferometric combination adds a range of higher spatial frequencies within a radius D/λ around $\boldsymbol{R_B}$ and $-\boldsymbol{R_B}$ to the imaging process. This range of baselines around the interferometer baseline $\boldsymbol{R_B} = \boldsymbol{B}/\lambda$ can also be explained as those baselines connecting individual points in the two apertures. It is exactly the information in the higher spatial frequencies that we are looking for and that provides increased angular resolution with an interferometer.

The interferometer PSF is the product of the individual Fourier transforms of the functions forming the convolution in (3.52). The Fourier transform of OTF_a is the point-spread function PSF_a of an individual aperture, and the Fourier transform of the sum of the δ-functions is the fringe pattern of two pinholes, $1 + \cos(.)$:

$$\mathrm{PSF}(\boldsymbol{\alpha}) = \mathcal{F}_R\left(\mathrm{OTF}(\boldsymbol{R})\right) = 2\mathrm{PSF}_a(\boldsymbol{\alpha})\big(1 + \cos(2\pi \boldsymbol{R_B} \cdot \boldsymbol{\alpha})\big). \tag{3.53}$$

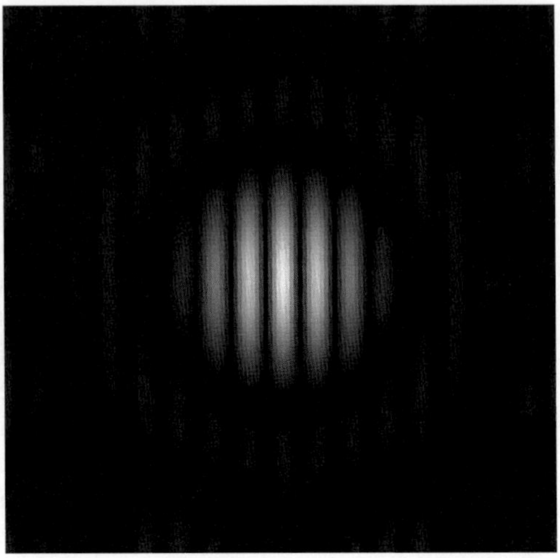

Fig. 3.15 The PSF of a stellar interferometer (3.53) with two apertures as in Fig. 3.14 in monochromatic illumination. The baseline B is three times larger than the aperture diameter D, resulting in a fringe spacing, λ/B, of 1/3 of the FWHM, λ/D, of the Airy disk. The fringes in the diffraction ring are also visible

If the individual apertures have a circular shape, then PSF_a has the form of an Airy disk with a FWHM of λ/D, called the *primary beam* in radio interferometry, and the complete PSF of this stellar interferometer is an Airy disk with fringes, with fringe spacing $1/|\boldsymbol{R}_B| = \lambda/B$, as displayed in Fig. 3.15.

Now, we have the interferometer with individual apertures as an optical system producing a PSF, an Airy disk with fringes, much better resembling a point source than the infinitely extended PSF in Young's experiment that we discussed in Sect. 3.3.1. However, it is still a long way from this high-resolution PSF, which allows to resolve detail limited by the fringe spacing and not by the first minimum of the Airy disk, to a high-resolution image. It involves the combination of observations with many different baselines so that the fringe structure in the reconstructed PSF disappears at the same rate as the "holes" in the uv-plane are filled. This will be discussed in Sect. 3.4.

Extended Objects

The next step expands the discussion to an incoherent source with angular brightness distribution $O_b(\boldsymbol{\alpha})$ that we assume independent of wavelength.

In monochromatic illumination with $G(\nu) = \delta(\nu - \nu_0)$, we obtain the image intensity distribution, $I(\boldsymbol{\alpha})$, by a convolution (3.30) of $O_b(\boldsymbol{\alpha})$ with the interferometer PSF (3.53), writing

$$\begin{aligned} I(\boldsymbol{\alpha}) &= O_b(\boldsymbol{\alpha}) * \text{PSF}(\boldsymbol{\alpha}) \\ &= 2\, O_b(\boldsymbol{\alpha}) * (\text{PSF}_a(\boldsymbol{\alpha})\, (1 + \cos(2\pi \boldsymbol{R}_B \cdot \boldsymbol{\alpha}))) \,. \end{aligned} \tag{3.54}$$

Equivalently one can discuss the imaging process in the uv-plane with the image intensity spectrum being the product of $\mu(\boldsymbol{R})$ and OTF (3.30), reading

$$
\begin{aligned}
\hat{I}(\boldsymbol{R}) &= \int I(\boldsymbol{\alpha})\, \mathrm{e}^{\mathrm{i}2\pi \boldsymbol{R}\cdot\boldsymbol{\alpha}} \mathrm{d}\boldsymbol{\alpha} \\
&= I_0 \mu(\boldsymbol{R})\, \mathrm{OTF}(\boldsymbol{R}) \qquad (3.55) \\
&= I_0 \mu(\boldsymbol{R}) \big(\mathrm{OTF}_a(\boldsymbol{R}) * (\delta(\boldsymbol{R} - \boldsymbol{R_B}) + 2\delta(\boldsymbol{R}) + \delta(\boldsymbol{R} + \boldsymbol{R_B})) \big) ,
\end{aligned}
$$

when the interferometer OTF itself is a convolution of the OTF of an individual aperture, OTF_a, with the OTF of a two-pinhole interferometer (3.52).

We can now discuss the imaging process in an interferometer either in the image plane $\boldsymbol{\alpha}$, (3.54), or in the uv-plane, $\boldsymbol{R} = (u, v)$, (3.55).

In the image plane, the convolution (3.54) between the object brightness and the interferometer PSF (see Fig. 3.15), means adding up interferometer PSFs of individual points of the object. Each point of the object gives rise to a PSF that is shifted and weighted according to the object point coordinate and brightness. As long as the object is smaller than the Airy disk the fringe patterns overlap, and the fringes lose contrast. This is what we found in Young's Experiment, and this is what we expect here: a loss of contrast if the object is extended.

However, in Young's Experiment, we did not discuss the convolution of the object intensity with the PSF since the fringe visibility directly provides the visibility function at the point $\boldsymbol{R_B}$ in the uv-plane (3.46). We obtain the same result here if we use the optical transfer function, $\mathrm{OTF}_a(\boldsymbol{R}) = \delta(\boldsymbol{R})$, of a pinhole so that $\mu(\boldsymbol{R})\mathrm{OTF}_a(\boldsymbol{R} \pm \boldsymbol{R_B})$ can be replaced by $\mu(\pm\boldsymbol{R_B})$.

With finite apertures instead of pinholes we do not find the same simple relationship between the visibility of the fringe pattern and the visibility function because we have to multiply $\mu(\boldsymbol{R})$ by OTF_a, which is shifted to $\boldsymbol{R} = \pm\boldsymbol{R_B}$, (3.55), and which is a function of roughly triangular shape and diameter $2D/\lambda$ (see Fig. 3.16). It is the Fourier transform of $\mu(\boldsymbol{R})$ weighted by OTF_a at $\boldsymbol{R} = 0$ and at $\pm\boldsymbol{R_B}$ that determines the intensity distribution of the fringe pattern. If the visibility function were a smooth function that does not vary over the width of OTF_a we could still replace it by single values. We will see in the following that the smoothness of $\mu(\boldsymbol{R})$ depends on the object size.

One aspect that has been left aside so far is that of an object larger than the Airy disk. The image would then be composed of many PSF that do not overlap for all parts of the object so that the visibility of the fringe pattern would vary over the image. Thus, there is no longer *the* visibility of *the* fringe pattern that is linked unambiguously to the shape of the object. We will come back to this point at the end of Sect. 3.3.4 discussing a wide binary star. Here, linking the results of finite apertures to Young's experiment we will discuss small objects.

Small Monochromatic Objects

For monochromatic objects that are much smaller than the Airy disk, PSF_a, the intensity distribution in the image plane of an individual telescope is about the same as that of the Airy disk, i.e. indistinguishable from a point source. The object cannot be resolved. However, if the source size is still larger than the fringe spacing its

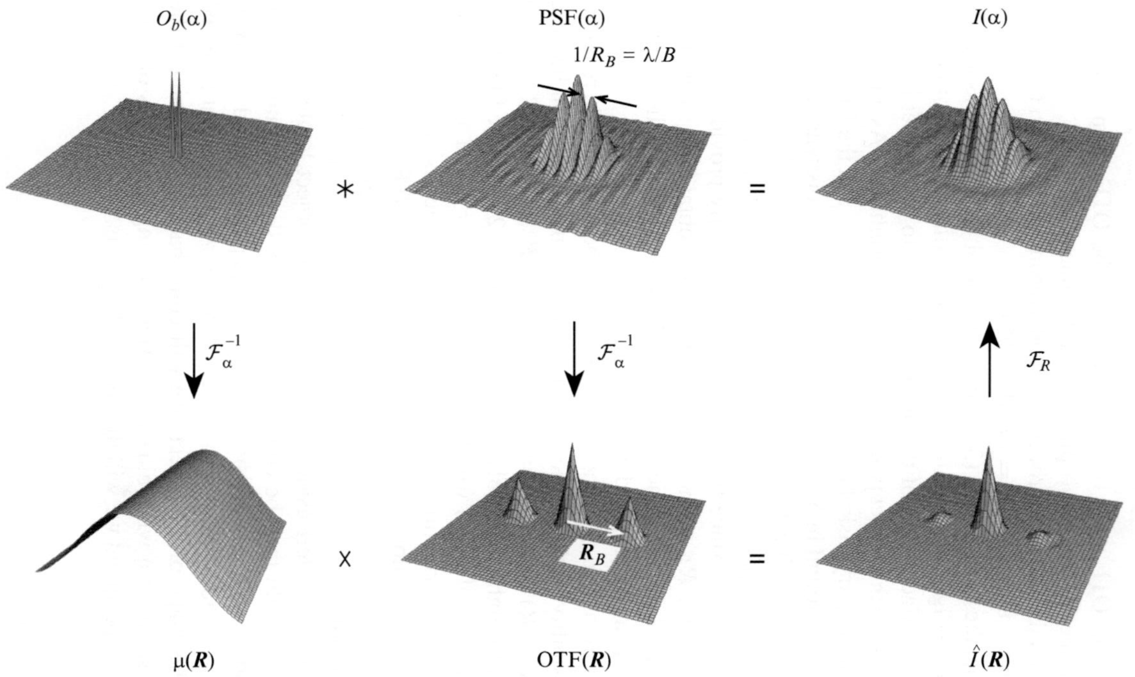

Fig. 3.16 Illustration of the imaging process of a narrow binary star in an interferometer in monochromatic illumination. The *top row* displays the convolution of the object brightness distribution $O_b(\boldsymbol{\alpha})$ with the PSF of the interferometer yielding the image intensity distribution $I(\boldsymbol{\alpha})$ like in (3.54). The contrast of the fringes is reduced due to the overlap of two identical but slightly shifted PSFs. The *bottom row* shows the situation in the uv-plane: The visibility function $\mu(\boldsymbol{R})$ of the binary star – a cosine function of which one period is displayed – is multiplied by the interferometer OTF consisting of three transfer functions of the individual apertures, OTF_a, centred at $|\boldsymbol{R}| = 0$ and at $\boldsymbol{R} = \pm\boldsymbol{R}_B$, see (3.55). The peak of OTF_a at $\boldsymbol{R} = \pm\boldsymbol{R}_B$ is exactly half of the peak at $|\boldsymbol{R}| = 0$, and the fringes of the PSF have maximum contrast. The resulting product has much reduced values around $\boldsymbol{R} = \pm\boldsymbol{R}_B$ and the contrast of the fringes in the image is reduced accordingly

contrast is reduced as a function of the baseline. This situation is not unusual when observing with stellar interferometers, and measuring stellar diameters is a typical example of an observation of an object that is smaller than the Airy disk.

For small objects, the convolution in (3.54) can be simplified by setting $\mathrm{PSF}_a(\boldsymbol{\alpha})$ as constant for all points of the object, $O_b(\boldsymbol{\alpha})$, and the image intensity distribution is written as

$$I(\boldsymbol{\alpha}) = 2\,\mathrm{PSF}_a(\boldsymbol{\alpha})\Big(\,O_b(\boldsymbol{\alpha}) * \big(1 + \cos(2\pi\,\boldsymbol{R_B}\cdot\boldsymbol{\alpha})\big)\Big). \tag{3.56}$$

This means that the structure of the small object affects the contrast of the fringes by adding up individual fringe patterns of individual object points – the convolution of $O_b(\boldsymbol{\alpha})$ with 1+cos(.) – without affecting the Airy disk as the envelope of the fringe pattern.

In the uv-plane, the Fourier transform of the object brightness distribution, the visibility function $\mu(\boldsymbol{R})$, is not only much wider than the Fourier transform of the Airy disk, $\mathrm{OTF}_a(\boldsymbol{R})$, but it is also a very smooth function that does not vary over distances given by the width of OTF_a. This is the consequence of the approximation above, assuming the object to be much smaller than PSF_a.

To explain this, we swap the roles of image and Fourier space for the moment. If we regard the brightness distribution $O_b(\boldsymbol{\alpha})$ in image space as a function containing the spatial frequency spectrum of $\mu(\boldsymbol{R})$ we can state that $O_b(\boldsymbol{\alpha})$ as the spatial frequency spectrum is limited to a maximum "frequency" that is determined by the size of the object, $\boldsymbol{\alpha}_0$. Thus, the structure of $\mu(\boldsymbol{R})$ does not contain detail finer than $1/\boldsymbol{\alpha}_0$. Together with our assumption that the object is much smaller than the Airy disk, its Fourier transform, OTF_a, is much narrower than $1/\boldsymbol{\alpha}_0$. Then, $\mu(\boldsymbol{R})$, containing only structure coarser than $1/\boldsymbol{\alpha}_0$, is completely represented by samples spaced by the width $2D/\lambda$ of OTF_a.

If we return to computing the intensity distribution by Fourier transforming the product of the visibility function $\mu(\boldsymbol{R})$ and the interferometer OTF, (3.55), we can replace the visibility function by the value of $\mu(\boldsymbol{R})$ at the centre of the respective OTF_a since $\mu(\boldsymbol{R})$ contains no significant information over the width of OTF_a.

We now rewrite the image intensity spectrum (3.55) and the monochromatic intensity distribution of a fringe pattern (3.56) in case of a small object as

$$\begin{aligned} I(\boldsymbol{\alpha}) &= I_0 \int \mathrm{OTF}_a(\boldsymbol{R}) \\ &\quad * \big(\mu(\boldsymbol{R_B})\delta(\boldsymbol{R}-\boldsymbol{R_B}) + 2\mu(0)\delta(\boldsymbol{R}) + \mu(-\boldsymbol{R_B})\delta(\boldsymbol{R}+\boldsymbol{R_B})\big)\mathrm{e}^{-\mathrm{i}2\pi\boldsymbol{R}\cdot\boldsymbol{\alpha}}\mathrm{d}\boldsymbol{R} \\ &= 2I_0\,\mathrm{PSF}_a(\boldsymbol{\alpha})\Big(1 + |\mu(\boldsymbol{R_B})|\cos\big(\phi(\boldsymbol{R_B}) - 2\pi\,\boldsymbol{R_B}\cdot\boldsymbol{\alpha}\big)\Big), \end{aligned} \tag{3.57}$$

with $\mu(-\boldsymbol{R_B}) = \mu^*(\boldsymbol{R_B})$, which is very similar to the fringe pattern of Young's Experiment (3.46). In contrast to the fringes in Young's Experiment, here the fringe pattern is limited in diameter, enveloped by the point-spread function of a single telescope, PSF_a.

Figure 3.16 visualises the imaging process in the case of a narrow binary star. The object brightness distribution $O_b(\boldsymbol{\alpha})$ is given by the sum of two δ-functions with a cosine as visibility function $\mu(\boldsymbol{R})$. Note, that the separation vector $\boldsymbol{\alpha}_0$ of the binary is parallel to the interferometer baseline $\boldsymbol{R_B}$. The top row displays the convolution in the image plane, (3.54), and the bottom row shows the multiplication in the uv-plane, (3.55), with the interferometer OTF composed of three individual OTF_a at $R = 0$ and at $\boldsymbol{R} = \pm\boldsymbol{R_B}$. Here, the value $\mu(\boldsymbol{R_B})$ can be used in good approximation to describe the visibility of the fringe pattern, (3.57).

Thus, for objects that are smaller than the Airy disk of an individual aperture the measurement of the complex visibility of the fringe pattern as a single parameter sufficiently characterises the fringe pattern, yielding the single parameter $\mu(\boldsymbol{R_B})$.

3.3.3 Spectra of Finite Width

The step from monochromatic illumination to a spectrum of finite width is done by integrating the spectral intensity distribution of the fringe pattern (3.54) over the spectrum. We now write the PSF explicitly as a function of $\boldsymbol{\alpha}$ and ν, properly considering the functional dependency in the following integrals over the spectral band.

We start by observing a point source with $O_b(\boldsymbol{\alpha}) = I_0\delta(\boldsymbol{\alpha})$. Then the convolution in (3.29) is reduced to a product and the spectral intensity distribution in the image plane is

$$\begin{aligned} I(\boldsymbol{\alpha},\nu) &= I_0 G(\nu)\delta(\boldsymbol{\alpha}) * \mathrm{PSF}(\boldsymbol{\alpha},\nu) \\ &= I_0 G(\nu)\mathrm{PSF}(\boldsymbol{\alpha},\nu). \end{aligned} \tag{3.58}$$

In addition to a wavelength independent object shape we also assume that the PSF of an individual aperture, PSF_a, is independent of wavelength. Being the envelope of the fringe pattern its effect on the polychromatic image is relatively benign.

The interferometer PSF (3.53) is now written as

$$\mathrm{PSF}(\boldsymbol{\alpha},\nu) = 2\mathrm{PSF}_a(\boldsymbol{\alpha})\big(1 + \cos(2\pi\boldsymbol{R_B}\cdot\boldsymbol{\alpha})\big), \tag{3.59}$$

with $\boldsymbol{R_B}\cdot\boldsymbol{\alpha} = \frac{\boldsymbol{B}}{\lambda}\cdot\boldsymbol{\alpha}$. The wavelength dependent part of the PSF is the fringe pattern with fringe spacing $1/\boldsymbol{R_B}$. The PSF of an individual telescope that we consider approximately independent of wavelength is the envelope of the fringe pattern. PSF_a is used at the mean wavelength λ_0.

Then, the polychromatic interferometer PSF is calculated as

$$\begin{aligned} \mathrm{PSF}_{pc}(\boldsymbol{\alpha}) &= \frac{I(\boldsymbol{\alpha})}{I_0} = \int G(\nu)\mathrm{PSF}(\boldsymbol{\alpha},\nu)\mathrm{d}\nu \\ &= 2\mathrm{PSF}_a(\boldsymbol{\alpha})\int G(\nu)\big(1 + \cos(2\pi\boldsymbol{R_B}\cdot\boldsymbol{\alpha})\big)\mathrm{d}\nu\,. \end{aligned} \tag{3.60}$$

The spectrum $G(\nu)$ is normalised, $\int G(\nu)\mathrm{d}\nu = 1$, so that the first part of the integral is equal to one. To solve the integral over $G(\nu)\cos(2\pi \boldsymbol{R}_B \cdot \boldsymbol{\alpha})$ we shift the spectrum $G(\nu)$ that is centred around ν_0, the average frequency, to $\nu = 0$ writing $G(\nu + \nu_0)$, like in Sect. 2.4.1 discussing Young's Experiment. The Fourier transform of $G(\nu + \nu_0)$ is a function of time difference τ, and we write $g(\tau) = \mathcal{F}_\nu(G(\nu + \nu_0))$, with $g(0) = 1$. The image coordinate $\boldsymbol{\alpha}$ is related to τ through $\tau = \boldsymbol{\alpha} \cdot \boldsymbol{R}_B/\nu = \boldsymbol{\alpha} \cdot \boldsymbol{B}/c$, and we define $g_B(\boldsymbol{\alpha}) = g(\boldsymbol{\alpha} \cdot \boldsymbol{B}/c) = g(\tau)$. While the width of $g(\tau)$ is given by $1/\Delta\nu$, the reciprocal of the spectral bandwidth, the width of $g_B(\boldsymbol{\alpha})$ – in the direction parallel to $\boldsymbol{B}$ – is given by l_c/B, with l_c the coherence length.

Then it is $\int G(\nu)\cos(2\pi \boldsymbol{R}_B \cdot \boldsymbol{\alpha})\mathrm{d}\nu = g_B(\boldsymbol{\alpha})\cos(2\pi \boldsymbol{R}_{0,B} \cdot \boldsymbol{\alpha})$, and the polychromatic PSF of a stellar interferometer becomes

$$\mathrm{PSF}_{pc}(\boldsymbol{\alpha}) = 2\mathrm{PSF}_a(\boldsymbol{\alpha})\big(1 + g_B(\boldsymbol{\alpha})\cos(2\pi \boldsymbol{R}_{0,B} \cdot \boldsymbol{\alpha})\big), \tag{3.61}$$

with $\boldsymbol{R}_{0,B} = \boldsymbol{B}/\lambda_0$ the baseline vector at the mean wavelength λ_0. The sum inside the brackets describes a fringe pattern with a spacing of $1/R_{0,B} = \lambda_0/B$ multiplied by $g_B(\boldsymbol{\alpha})$.

A very similar expression, (2.63), describes the polychromatic fringe pattern of the pinholes in Young's Experiment. Here, with real apertures, the fringe pattern has the point-spread function of the individual aperture, PSF_a, as an additional envelope as displayed in Fig. 3.17.

The consequence of multiplying the cosine by $g_B(\boldsymbol{\alpha})$ – a function with a width of about l_c/B in the direction parallel to $\boldsymbol{B}$ – is that the fringe contrast is reduced after a few periods depending on the coherence length l_c as discussed in Sect. 2.4.1. Thus, without additional assumptions, the polychromatic fringe pattern cannot be characterised by a single visibility value.

The polychromatic interferometer PSF is shown in Fig. 3.17 for an aperture diameter of 8 m and a baseline of 100 m observing in the K-band. Due to the width of the K-band, with $\lambda_0/\Delta\lambda = 5.5$, there are about 5 fringes visible on each side of the white-light fringe – corresponding to $5.5\lambda_0/B = 25$ mas – before the contrast goes to zero (compare to Fig. 2.15). The half-width of the fringe package is $(\lambda_0/\Delta\lambda)/R_{0,B} = (\lambda_0/\Delta\lambda)\lambda_0/B = l_c/B$.

In the following, we will mainly discuss cases when the fringe package is narrower than the Airy disk, $l_c/B < \lambda/D$, as in Fig. 3.17.

In the uv-plane, the situation is described by integrating OTF at individual wavelengths over the spectral band, keeping in mind that the OTF as a function of the spatial frequency $\boldsymbol{R}$ is a function of wavelength. For instance the physical baseline $\boldsymbol{B}$ is converted to the wavelength dependent spatial frequency $\boldsymbol{R}_B = \boldsymbol{B}/\lambda$. Integrating over the spectral band we therefore will have to consider OTF($\boldsymbol{R}$) as a function of wavelength.

A general expression for the fringe pattern as Fourier transform of the product of OTF and $\mu(\boldsymbol{R})$ was given in (3.29). For a point source, the visibility function is constant, $\mu(\boldsymbol{R}) = 1$. Integrating over the spectral band yields the intensity distribution

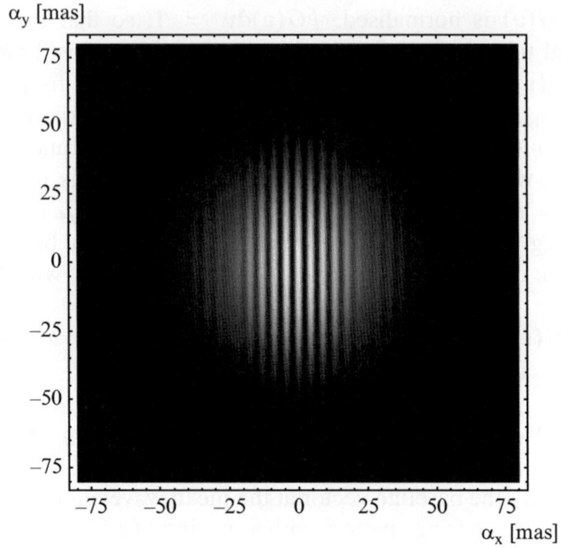

Fig. 3.17 The polychromatic PSF of a stellar interferometer (3.61) with two apertures as in Fig. 3.14, with telescope apertures of $D = 8$ m and a baseline B of 100 m in the K-band, with $\lambda = 2.2\pm0.2\,\mu$m. The FWHM of the Airy disk – only the central core is displayed – is $\lambda_0/D =$ 57 mas and the fringe spacing $\lambda_0/B = 4.5$ mas. The consequence of observing a spectral band instead of a single spectral line is a complete loss of contrast after $\lambda_0/\Delta\lambda = 2.2/0.4 = 5.5$ periods, i.e. after about ±25 mas from the white-light fringe at the centre. There is no single visibility value describing the polychromatic PSF. Setting the Airy disk independent of wavelength, the FWHM variation over the K-band, which is 57 ± 5 mas, is ignored and the average value is used for the whole K-band. The effect on the fringe package is negligible

of the polychromatic PSF as

$$\begin{aligned} I(\boldsymbol{\alpha}) &= I_0 \iint G(\nu)\mathrm{OTF}(\boldsymbol{R})\mathrm{e}^{-\mathrm{i}2\pi \boldsymbol{R}\cdot\boldsymbol{\alpha}}\mathrm{d}\boldsymbol{R}\,\mathrm{d}\nu \\ \mathrm{PSF}_{pc}(\boldsymbol{\alpha}) &= \frac{I(\boldsymbol{\alpha})}{I_0} = \int \left(\int G(\nu)\mathrm{OTF}(\boldsymbol{R})\mathrm{d}\nu\right)\mathrm{e}^{-\mathrm{i}2\pi \boldsymbol{R}\cdot\boldsymbol{\alpha}}\mathrm{d}\boldsymbol{R}\,. \end{aligned} \tag{3.62}$$

We move our discussion completely into the uv-plane, formally by regarding the Fourier transform of PSF_{pc} that is called the polychromatic optical transfer function, OTF_{pc}, with

$$\begin{aligned} \mathrm{OTF}_{pc}(\boldsymbol{R}) &= \int G(\nu)\mathrm{OTF}(\boldsymbol{R})\mathrm{d}\nu \\ &= \int G(\nu)\mathrm{OTF}_a(\boldsymbol{R}) * \big(\delta(\boldsymbol{R}-\boldsymbol{R_B}) + 2\delta(\boldsymbol{R}) + \delta(\boldsymbol{R}+\boldsymbol{R_B})\big)\mathrm{d}\nu\,, \end{aligned} \tag{3.63}$$

In Fig. 3.18, OTF at three different wavelengths are displayed, illustrating the impact of the wavelength on form and position of the OTF. The integral over

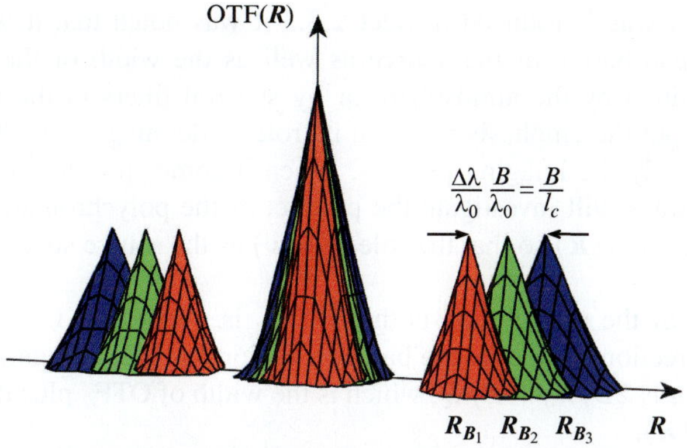

Fig. 3.18 The OTF at three different wavelengths as in (3.64). For a given physical baseline $\boldsymbol{B}$ there are three different baseline vectors $\boldsymbol{R}_{B_i} = \boldsymbol{B}/\lambda_i$, $i = 1, 2, 3$, in the uv-plane. The colours of the curves indicate their wavelengths. For longer wavelengths, towards the red end of the spectrum, $\boldsymbol{R}_B$ is shorter and the individual OTF_a are narrower. For a spectral bandwidth of $\Delta\nu$ the monochromatic OTF_a are distributed over a range with a width of $(\Delta\lambda/\lambda_0)B/\lambda_0 = B/l_c$, with l_c the coherence length. The sum of the monochromatic OTF forms the polychromatic OTF that is elongated by B/l_c

the three distinct areas – around $|\boldsymbol{R}| = 0$ and around $\boldsymbol{R} = \pm\boldsymbol{R}_{B_i}$ – forms the polychromatic OTF.

Our assumption that PSF_a is approximately independent of wavelength means that also OTF_a is not a function of ν. We use OTF_a at the mean frequency ν_0, writing the polychromatic OTF as

$$\begin{aligned}\mathrm{OTF}_{pc}(\boldsymbol{R}) &= \mathrm{OTF}_a(\boldsymbol{R}) * \int G(\nu)\big(\delta(\boldsymbol{R}-\boldsymbol{R}_B) + 2\delta(\boldsymbol{R}) + \delta(\boldsymbol{R}+\boldsymbol{R}_B)\big)\mathrm{d}\nu \\ &= \mathrm{OTF}_a(\boldsymbol{R}) * \big(G_B(\boldsymbol{R}) + 2\delta(\boldsymbol{R}) + G_B(-\boldsymbol{R})\big), \qquad (3.64)\end{aligned}$$

with $G_B(\boldsymbol{R})$ the spectrum as a function of spatial frequency $\boldsymbol{R}$ for the baseline $\boldsymbol{B}$, since $G_B(\boldsymbol{R}) := G_B(\frac{\boldsymbol{B}}{\lambda}) = G_B(\frac{\boldsymbol{B}}{c}\nu)$. $G_B(\boldsymbol{R})$ is dimensionless, with $\int G_B(\boldsymbol{R})\mathrm{d}\boldsymbol{R} = 1$. It forms a Fourier pair with $g_B(\boldsymbol{\alpha})$, like the spectrum $G(\nu)$ with $g(\tau)$.

If $G(\nu)$ is a rectangular function of width $\Delta\nu$ centred at the average frequency ν_0, the width of $g(\nu)$ is $1/\Delta\nu$. And if $G_B(\boldsymbol{R})$ is a rectangular function of width B/l_c, centred at the spatial frequency $\boldsymbol{R}_{0,B} = \boldsymbol{B}/\lambda_0$, $g_B(\boldsymbol{\alpha})$ has the width l_c/B.

The convolution of OTF_a with the three δ-peaks and the subsequent integration over the spectral band in (3.63) is replaced by the convolution of OTF_a with the spectra $G_B(\pm\boldsymbol{R})$ and the δ-function in (3.64). Both computations yield the same result, displayed in Fig. 3.18.

We have thus attached the source spectrum $G(\nu)$ to the OTF that characterises the optical system. We can do this since we assume that the object shape is independent of wavelength over the observed spectral band, so that the object is described by its brightness distribution $O_b(\boldsymbol{\alpha})$.

When $G(\nu)$ was introduced in Sect. 2.3.3 it was noted that it would describe the spectral distribution of the source as well as the width of the spectral band that is transmitted by the atmosphere or by spectral filters in the interferometer. In (3.64), we put the emphasis more on its role as defining the width of the spectral band to study the imaging process. When it comes to interpreting the fringe pattern we always will investigate the product of the polychromatic OTF and the visibility function $\mu(\boldsymbol{R})$ so that the role of $G(\nu)$ as the source spectrum is properly considered.

The result of the convolution is that OTF_{pc} is wider than the monochromatic OTF in the direction parallel to the baseline vector $\boldsymbol{R_B}$. The elongation of the areas around $\pm\boldsymbol{R}_{0,B}$ is $2D/\lambda_0 + B/l_c$, which is the width of OTF_a plus the width of the spectrum $G_B(\boldsymbol{R})$.

In Sect. 2.4.3, we discussed the power spectrum of the fringe pattern in Young's experiment, investigating the Fourier transform of the fringe pattern without any consideration of optical transfer functions. For that discussion, we expressed the fringe pattern as a function of time delay τ using $\tau = \boldsymbol{\alpha} \cdot \boldsymbol{R_B}/\nu$. The Fourier transform was then performed from τ to ν. We found that the Fourier transform of the fringe pattern consists of a δ-peak at $\nu = 0$ and $G(\nu)$ around $\pm\nu_0$ (2.75), see Fig. 2.16.

Equation (3.64) can easily be modified to describe the situation in Young's Experiment. For pinhole sized apertures the optical transfer function is given by a δ-function, $\mathrm{OTF}_a(\boldsymbol{R}) = \delta(\boldsymbol{R})$, and the convolution in (3.64) is reduced to the sum of three terms:

$$\mathrm{OTF}_{pc}(\boldsymbol{R}) = G_B(\boldsymbol{R}) + 2\delta(\boldsymbol{R}) + G_B(-\boldsymbol{R}), \tag{3.65}$$

when $G_B(\boldsymbol{R})$ is the spectrum as a function of spatial frequency centred around $\boldsymbol{B}/\lambda_0$. This is the same result as in (2.75) when the spectra were written as a function of ν.

Summary: The polychromatic interferometer PSF displays a fringe package within the Airy disk of an individual telescope when the fringes lose contrast after about $\lambda_0/\Delta\lambda$ periods and, thus, cannot be characterised by a single value. The fringe spacing is λ_0/B, and the FWHM of the Airy disk is λ_0/D, with λ_0 the mean wavelength of the spectral band, B the baseline length and D the telescope diameter (see Fig. 3.17). The polychromatic OTF consists of OTF of individual telescopes, OTF_a, with diameter $2D/\lambda_0$, centred at $R = 0$ and at $\pm\boldsymbol{R}_{0,B}$ (see Fig. 3.18). The distribution of individual OTF_a over $(\Delta\lambda/\lambda_0)\boldsymbol{B}/\lambda_0 = \boldsymbol{B}/l_c$ around $\pm\boldsymbol{R}_{0,B}$ makes OTF_{pc} wider than the monochromatic OTF. This is the equivalent of the decrease of fringe contrast after $\lambda_0/\Delta\lambda$ periods, making the area with fringes narrower than in the monochromatic PSF.

Small Objects with Wavelength Dependent Shape

Observing objects that are larger than a point but much smaller than the fringe package allows for a different interpretation of the polychromatic fringe pattern. Throughout this chapter we have used the approximation of a wavelength independent object shape. What happens if we observe such a small object with varying shape over the observed bandwidth? We no longer describe the imaging process by the product of the polychromatic OTF with the visibility function $\mu(\boldsymbol{R})$ but we have to multiply $\mu(\boldsymbol{R})$ at each wavelength by the monochromatic OTF, integrating over the spectrum at the end.

The polychromatic fringe pattern forms by adding up the resulting monochromatic fringe patterns in the image. It should be noted that this is the normal procedure when doing the step from monochromatic to polychromatic illumination. Using the polychromatic PSF and the polychromatic OTF is just a convenient approximation in case of a wavelength independent object shape.

Regarding the Fourier transform of the polychromatic fringe package we would still find an elongated area around $\boldsymbol{R}_{0,\boldsymbol{B}}$. While this could be represented by a single value in the case of wavelength independent object shape, we can now attribute the shape of the visibility function over this area to values at different wavelengths, remembering that the elongated areas are the integral over monochromatic OTF, see Fig. 3.18. However, the spectral resolution is limited since at a given spatial frequency $\boldsymbol{R}$, there is a mix of contributions from a wavelength λ_1 and a baseline $\boldsymbol{B}_1$ with those at another wavelength λ_2 and a baseline $\boldsymbol{B}_2$, as long as $\boldsymbol{R} = \boldsymbol{B}_1/\lambda_1 = \boldsymbol{B}_2/\lambda_2$. The baseline vectors $\boldsymbol{B}_1$ and $\boldsymbol{B}_2$ connect individual points in the telescope apertures that are separated by the interferometer baseline, the centre-to-centre distance of the telescopes, $\boldsymbol{B}$. This mixing reduces the spectral resolution to the width $2D/\lambda$ of OTF_a, which is equivalent to a resolvable spectral band of $\Delta\lambda = \lambda_0 2D/B$.

For a baseline of 100 m and a telescope diameter of 1.8 m the resolvable spectral bands in the K-band would be about $\Delta\lambda = \lambda_0/28 = 0.08\,\mu\mathrm{m}$ wide. Given that the width of the K-band is $0.4\,\mu\mathrm{m}$, one can distinguish about five spectral channels for this combination of baseline and telescope diameter.

If the baseline were reduced to $B = 20$ m the spectral resolution would be only $\Delta\lambda = 0.4\,\mu\mathrm{m}$, equivalent to the full K-band. Then, there would be no additional spectral information across the elongated OTF.

These values of the spectrally resolved visibility function can be used to reconstruct a polychromatic image. The values have to be processed according to the spectral channel, first reconstructing a monochromatic image and then adding up the monochromatic images at the end, copying the physical process that takes place in the optical system.

The variation of spectral resolution with baselines leads to some ambiguity when filling the uv-plane with values at different baselines. Therefore, one has to carefully consider if the lowest spectral resolution would be applied to all baselines or if one is only interested in the high spatial frequencies using the highest spectral resolution.

If the size of an object with wavelength dependent shape is unknown there is no choice but reducing the measured size to the half-width of the fringe package by integrating over OTF_{pc} around $\boldsymbol{R}_{0,B}$. Otherwise, the spatial and spectral information would be mixed in an indiscernible way.

Objects that are larger than the half-width of the fringe package cannot be resolved spectrally and require the approximation of a wavelength independent shape. This can also be achieved by reducing the width of the spectral bands by physically splitting the light in the interferometric instruments into narrow spectral channels like a spectrometer so that each channel creates a narrow-band fringe pattern. The spectral resolution of the instruments has to be such that the object shape is approximately independent of wavelength in each spectral channel.

3.3.4 Objects of Finite Size

After the discussion on the fringe pattern of very small sources, it is straightforward to move to larger objects. The results in this section investigating apertures, spectra and objects of realistic dimensions form the basis for the description of the imaging process in stellar interferometers.

We use again (3.29), writing the spectral intensity distribution as the convolution of the spectral intensity $O(\boldsymbol{\alpha}, \nu)$ with the interferometer PSF. Assuming that the object brightness distribution $O_b(\boldsymbol{\alpha})$ is independent of wavelength over the observed spectrum $G(\nu)$, we write

$$\begin{aligned} I(\boldsymbol{\alpha}, \nu) &= G(\nu) O_b(\boldsymbol{\alpha}) * \mathrm{PSF}(\boldsymbol{\alpha}, \nu) \\ &= O_b(\boldsymbol{\alpha}) * \big(G(\nu)\mathrm{PSF}(\boldsymbol{\alpha}, \nu)\big). \end{aligned} \tag{3.66}$$

The integral over the spectral band to obtain the polychromatic intensity distribution can be reduced to the integral of the product $G(\nu)\mathrm{PSF}(\boldsymbol{\alpha}, \nu)$, which is the polychromatic PSF as shown in (3.60).

Thus, the convolution of the object brightness distribution with the polychromatic PSF (3.61) forms the image intensity distribution of an extended object as

$$\begin{aligned} I(\boldsymbol{\alpha}) &= O_b(\boldsymbol{\alpha}) * \int G(\nu)\mathrm{PSF}(\boldsymbol{\alpha}, \nu)\mathrm{d}\nu \\ &= O_b(\boldsymbol{\alpha}) * \mathrm{PSF}_{pc}(\boldsymbol{\alpha}) \\ &= 2\, O_b(\boldsymbol{\alpha}) * \big(\mathrm{PSF}_a(\boldsymbol{\alpha})\,(1 + g_B(\boldsymbol{\alpha}) \cos(2\pi \boldsymbol{R}_{0,B} \cdot \boldsymbol{\alpha}))\big), \end{aligned} \tag{3.67}$$

with $g_B(\boldsymbol{\alpha})$ the Fourier transform of the spectrum $G_B(\boldsymbol{R})$, and with the approximation that the PSF of an individual aperture, $\mathrm{PSF}_a(\boldsymbol{\alpha})$, is independent of wavelength.

An example for a polychromatic PSF is displayed in Fig. 3.17. The image intensity of an extended object is thus a repetition of PSF_{pc} – each displaying a fringe pattern – all over the image plane, weighted and positioned according to the object brightness distribution $O_b(\boldsymbol{\alpha})$.

The polychromatic PSF displays a white-light fringe at its centre. A white-light fringe identifies the position of zero OPD for the optical paths from the source point through the apertures into the image plane. Thus, the distribution of point-spread functions in the image plane means that the light from each point of the object has zero OPD at its image position where its PSF forms. In fact, it is the nature of the imaging process that the image of each point forms at exactly that position where the optical paths through the optical system are equal and the OPD is zero.

We have to keep in mind that we are discussing the Fizeau configuration that is modelled by masking a large aperture with two small sub-apertures. As mentioned in Sect. 3.3.2, stellar interferometers with individual telescopes, requiring the reimaging of the telescope apertures, are often operated in Michelson configuration, when there is no homothetic mapping of the telescope apertures, i.e., when the reimaged apertures in the exit pupil are not a downscaled image of the interferometer array. Then, the imaging process cannot be modelled as a masked aperture. The OPD is not necessarily zero at the centre of each PSF and we do not see fringes in each PSF but only in those close to the image centre. This will be discussed in Sect. 5.1.2.

If the object is larger than PSF_{pc} there is no single fringe pattern since not all point-spread functions overlap in the image. The visibility of the fringes varies over the image both due to the object shape and to the spectral bandwidth. There is no single visibility value for the fringes in the image. However, this does not affect the ability to produce a high resolution image as we shall see in the following.

The image intensity can be expressed by the Fourier transform of the product of $\mu(\boldsymbol{R})$ with the OTF, (3.55). The spectral intensity in the image is then

$$I(\boldsymbol{\alpha}, \nu) = I_0 \int G(\nu)\mu(\boldsymbol{R})\mathrm{OTF}(\boldsymbol{R})\mathrm{e}^{-\mathrm{i}2\pi \boldsymbol{R}\cdot\boldsymbol{\alpha}}\mathrm{d}\boldsymbol{R}\,, \tag{3.68}$$

and the polychromatic intensity is

$$\begin{aligned} I(\boldsymbol{\alpha}) &= I_0 \iint G(\nu)\mu(\boldsymbol{R})\mathrm{OTF}(\boldsymbol{R})\mathrm{e}^{-\mathrm{i}2\pi \boldsymbol{R}\cdot\boldsymbol{\alpha}}\mathrm{d}\boldsymbol{R}\mathrm{d}\nu \\ &= I_0 \int \left(\mu(\boldsymbol{R}) \int G(\nu)\mathrm{OTF}(\boldsymbol{R})\mathrm{d}\nu\right)\mathrm{e}^{-\mathrm{i}2\pi \boldsymbol{R}\cdot\boldsymbol{\alpha}}\mathrm{d}\boldsymbol{R}\,, \end{aligned} \tag{3.69}$$

with $\int G(\nu)\mathrm{OTF}(\boldsymbol{R})\mathrm{d}\nu$ being the polychromatic optical transfer function, OTF_{pc}, (3.63). We should note again that the shape of $\mathrm{OTF}(\boldsymbol{R})$ varies with wavelength (see Fig. 3.18) so that it stays under the integral over the spectral band while the visibility function $\mu(\boldsymbol{R})$ as the Fourier transform of the object brightness is wavelength independent since the object brightness is also independent of wavelength over the observed spectral band.

What would happen if the object brightness distribution were a function of wavelength? The object brightness distribution in (3.67) and the visibility function in (3.69) would have to stay under the integral over ν and the complete imaging process would first be treated like in the monochromatic case integrating the spectral intensities over the spectral band at the end. The advantage of a wavelength independent object brightness, describing the imaging process as a linear process with a polychromatic optical transfer function, will be discussed in the following.

We move completely into the uv-plane by regarding the image intensity spectrum, the Fourier transform of the image intensity distribution in (3.69) (compare to 3.55), which is

$$\begin{aligned}\hat{I}(\boldsymbol{R}) &= \int I(\boldsymbol{\alpha})\mathrm{e}^{\mathrm{i}2\pi \boldsymbol{R}\cdot\boldsymbol{\alpha}}\mathrm{d}\boldsymbol{\alpha} \\ &= I_0\mu(\boldsymbol{R})\,\mathrm{OTF}_{pc}(\boldsymbol{R}) \\ &= I_0\mu(\boldsymbol{R})\big(\mathrm{OTF}_a(\boldsymbol{R}) * (G_B(\boldsymbol{R}) + 2\delta(\boldsymbol{R}) + G_B(-\boldsymbol{R}))\big). \end{aligned} \quad (3.70)$$

The spatial frequency spectrum $\hat{I}(\boldsymbol{R})$ of the image intensity is the product of the visibility function $\mu(\boldsymbol{R})$ of the object brightness distribution multiplied by the polychromatic OTF (3.64) that is displayed in Fig. 3.18. Like the PSF of an individual aperture its optical transfer function, OTF_a, is independent of wavelength using the function at the mean wavelength λ_0.

Thus, we have the expressions for the polychromatic image intensity $I(\boldsymbol{\alpha})$ as a convolution, (3.67), and as a Fourier transform of its visibility function, (3.70). For these formulae, we assumed that $O_b(\boldsymbol{\alpha})$ and the PSF_a, and therefore OTF_a, are independent of wavelength.

We will now discuss the two extreme cases of an object that is much smaller than the fringe package and one that is larger than an Airy disk.

Small Objects, the Quasi-Monochromatic Approximation

The notation can be further simplified by regarding small objects, as discussed in quasi-monochromatic approximation in Sect. 2.3.4. Indirectly, we have used this approximation when discussing the conditions under which the fringe pattern can be characterised by a single visibility value. Here, we will put this discussion in the general context of the quasi-monochromatic approximation.

We make the assumption that the width of the spectrum $G(\nu)$ is smaller than the mean frequency ν_0, or, more precisely, the width of the spectrum together with the object size is such that the visibility function $\mu(\boldsymbol{R})$ is invariant over the elongated area in the polychromatic OTF, which is about $B/l_c + 2D/\lambda_0$. If the object size α_0 is smaller than PSF_a and additionally smaller than the half-width l_c/B of the fringe package in PSF_{pc}, this condition is met.

We can then simplify the spatial frequency spectrum in (3.70), similar to (3.57), writing

$$\begin{aligned}\hat{I}(\boldsymbol{R}) = I_0\Big(&\mu(\boldsymbol{R}_{0,B})\mathrm{OTF}_a(\boldsymbol{R}) * G_B(\boldsymbol{R}) \\ &+ 2\mu(0)\mathrm{OTF}_a(\boldsymbol{R}) * \delta(\boldsymbol{R}) + \mu(-\boldsymbol{R}_{0,B})\mathrm{OTF}_a(\boldsymbol{R}) * G_B(-\boldsymbol{R})\Big),\end{aligned} \quad (3.71)$$

replacing the visibility function by its individual values at $\pm \boldsymbol{R}_{0,B}$, the spatial frequency at the mean wavelength λ_0, and at $|\boldsymbol{R}| = 0$ since $\mu(\boldsymbol{R})$ is invariant over the elongated OTF due a source that is smaller than the half-width of the fringe package.

Computing the image intensity distribution $I(\boldsymbol{\alpha})$ by Fourier transforming $\hat{I}(\boldsymbol{R})$ yields a fringe package with a width of $g_B(\boldsymbol{\alpha})$ that has a maximum contrast of $|\mu(\boldsymbol{R}_{0,B})|$ around the white-light fringe at $|\boldsymbol{\alpha}| = 0$, writing

$$\begin{aligned} I_{\mathrm{qm}}(\boldsymbol{\alpha}) &= 2 I_0 \, \mathrm{PSF}_a(\boldsymbol{\alpha}) \\ &\quad \times \left(1 + g_B(\boldsymbol{\alpha}) |\mu(\boldsymbol{R}_{0,B})| \cos\left(\phi(\boldsymbol{R}_{0,B}) - 2\pi \boldsymbol{R}_{0,B} \cdot \boldsymbol{\alpha}\right)\right), \end{aligned} \tag{3.72}$$

with $\phi(\boldsymbol{R}_{0,B})$ the phase of the visibility function and using $\int G_B(\boldsymbol{R}) \mathrm{d}\boldsymbol{R} = 1$.

We obtained the same expression in Sect. 2.4.1, describing the fringe pattern in Young's experiment in quasi-monochromatic approximation by (2.68).

Considering a real observing situation with two telescopes, the complex visibility can be determined directly, e.g., with the ABCD method as contrast and position (phase) of the white-light fringe, as discussed in Sect. 2.4.2, when only a small fraction of the intensity – that in the white-light fringe – would contribute to the measurement. If more than two telescopes are combined simultaneously, this direct measurement becomes increasingly difficult, as illustrated by Fig. 3.13. Alternatively, the visibility at each baseline can be determined individually in the uv-plane after Fourier transforming the complete fringe pattern, as discussed in Sect. 2.4.3 for Young's experiment.

Starting from (2.78) but now with real apertures instead of pinholes, we integrate the image intensity spectrum (3.71) over the elongated area of OTF_{pc}, given by $\mathrm{OTF}_a(\boldsymbol{R}) * G_B(\boldsymbol{R})$, the interferometric peak, (see Fig. 3.18) and divide it by the integral over OTF_a around $\boldsymbol{R} = 0$, the photometric peak, yielding the estimator for the complex visibility function $\mu(\boldsymbol{R}_{0,B})$ in quasi-monochromatic approximation as

$$\frac{\mu(\boldsymbol{R}_{0,B}) \int \mathrm{OTF}_a(\boldsymbol{R}) * G_B(\boldsymbol{R}) \, \mathrm{d}\boldsymbol{R}}{2\mu(0) \int \mathrm{OTF}_a(\boldsymbol{R}) \, \mathrm{d}\boldsymbol{R}} = \frac{\mu(\boldsymbol{R}_{0,B})}{2}, \tag{3.73}$$

using $\mu(0) = 1$ and $\int G_B(\boldsymbol{R}) \mathrm{d}\boldsymbol{R} = 1$.

Calculating the power spectrum $|\hat{I}(\boldsymbol{R})|^2$ of the fringe pattern, we obtain an estimator for the squared modulus of the visibility function $|\mu(\boldsymbol{R}_{0,B})|^2$ as

$$\frac{|\mu(\boldsymbol{R}_{0,B}) \int \mathrm{OTF}_a(\boldsymbol{R}) * G_B(\boldsymbol{R}) \, \mathrm{d}\boldsymbol{R}|^2}{|2\mu(0) \int \mathrm{OTF}_a(\boldsymbol{R}) \, \mathrm{d}\boldsymbol{R}|^2} = \frac{|\mu(\boldsymbol{R}_{0,B})|^2}{4}. \tag{3.74}$$

The power spectrum provides an unbiased measure for the visibility function in case of noisy measurements [224], at the price of losing the phase $\phi(\boldsymbol{R}_{0,B})$ of the visibility function (see Sect. 6.1.1).

A Wide Binary Star

Applying the convolution of the object brightness distribution with the polychromatic PSF (3.67) to an extended object, it is obvious that with distinct Airy disks the fringe patterns within the Airy disks will not overlap. Discussing the case of a wide binary in the following subsection we will see in detail what this means.

We call a binary star a *wide binary* if the separation is much larger than the Airy disk of an individual telescope, PSF_a. Its brightness distribution and visibility function are

$$O_b(\boldsymbol{\alpha}) = \frac{I_0}{2}\left(\delta\left(|\boldsymbol{\alpha} - \frac{\boldsymbol{\alpha}_0}{2}|\right) + \delta\left(|\boldsymbol{\alpha} + \frac{\boldsymbol{\alpha}_0}{2}|\right)\right) \quad \text{and}$$

$$\mu(\boldsymbol{R}) = \frac{\mathcal{F}_{\alpha}^{-1}(O_b(\boldsymbol{\alpha}))}{I_0} = \cos\left(2\pi \boldsymbol{R} \cdot \frac{\boldsymbol{\alpha}_0}{2}\right). \tag{3.75}$$

The separation of the binary is $\alpha_0 = |\boldsymbol{\alpha}_0|$ and the period length of the cosine is $2/\alpha_0$. Figure 3.19 illustrates the imaging process for the binary's orientation parallel to the interferometer baseline $\boldsymbol{\alpha}_0 \| \boldsymbol{B}$.

The Fourier transform of the two δ-functions yields a cosine function for the visibility function $\mu(\boldsymbol{R})$. This must not be confused with the intensity distribution of the fringe pattern in the image plane, which is also described by a cosine function.

The polychromatic image intensity distribution is calculated by inserting the binary's brightness distribution into the convolution in (3.67), writing

$$\begin{aligned} I(\boldsymbol{\alpha}) &= \frac{I_0}{2}\mathrm{PSF}_a\left(\boldsymbol{\alpha} - \frac{\boldsymbol{\alpha}_0}{2}\right)\left(1 + g_B(\boldsymbol{\alpha})\cos\left(2\pi \boldsymbol{R}_{0,B} \cdot \left(\boldsymbol{\alpha} - \frac{\boldsymbol{\alpha}_0}{2}\right)\right)\right) \\ &+ \frac{I_0}{2}\mathrm{PSF}_a\left(\boldsymbol{\alpha} + \frac{\boldsymbol{\alpha}_0}{2}\right)\left(1 + g_B(\boldsymbol{\alpha})\cos\left(2\pi \boldsymbol{R}_{0,B} \cdot \left(\boldsymbol{\alpha} + \frac{\boldsymbol{\alpha}_0}{2}\right)\right)\right). \end{aligned} \tag{3.76}$$

This is the sum of two Airy disks, one with its fringe pattern centred at $\boldsymbol{\alpha}_0/2$ and the second one at $-\boldsymbol{\alpha}_0/2$. The Fourier transform $g_B(\boldsymbol{\alpha})$ of the spectrum acts as envelope of each fringe pattern.

With PSF_a much narrower than the separation of the binary, we have two distinct Airy disks, each like in Fig. 3.17, that do not overlap at all. The fringe patterns in each Airy disk have a visibility of 1 like that of an individual point source.

NB 7. *Rather than discussing the visibility of the fringe pattern we could also determine directly the binary's separation α_0 by measuring the distance between the fringe packages in the image, see Fig. 3.19. This type of measurement determining the angular distance between celestial bodies is called* astrometry. *The precise measurement of star and planet positions is one of the oldest disciplines in astronomy. When this is done with single telescopes the accuracy is limited, amongst other factors, by the width of the Airy disk. The position of the Airy disk can be determined to a fraction of its width achieving an accuracy of about 1/100 of the diameter of the Airy disk.*

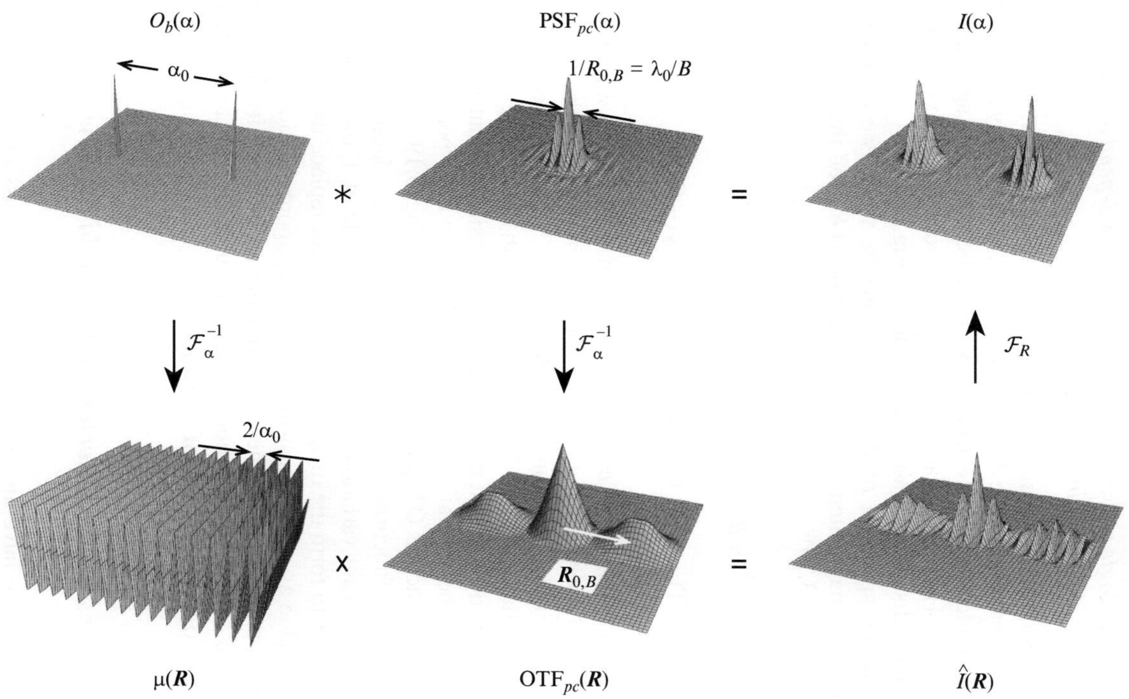

Fig. 3.19 Illustration of the imaging process of a wide binary star in an interferometer in polychromatic illumination with $\boldsymbol{\alpha}_0 \| \boldsymbol{B}$. The *top row* displays the convolution of the object brightness distribution $O_b(\boldsymbol{\alpha})$ with the polychromatic PSF of the interferometer yielding the image intensity $I(\boldsymbol{\alpha})$, (3.67). The separation of the binary is larger than the Airy disk so that two Airy disks form the image.
The *bottom row* shows the situation in the uv-plane: The visibility function $\mu(\boldsymbol{R})$ of the binary star is a cosine function, with period length $R = 2/\alpha_0$, that has many periods over the OTF. This can be seen in the spatial frequency spectrum $\hat{I}(\boldsymbol{R})$, which is the product of $\mu(\boldsymbol{R})$ with the polychromatic OTF. $\hat{I}(\boldsymbol{R})$ is related to the image intensity through a Fourier transform (3.69). The OTF is slightly elongated around $\boldsymbol{R} = \pm \boldsymbol{R}_{0,B}$ due to the polychromatic illumination

Using an interferometer for astrometry, the angular distance – or rather the differential OPD which is related to the angular distance by $OPD = \boldsymbol{\alpha}_0 \cdot \boldsymbol{B}$ *– between the white-light fringes provides the projection of the separation vector onto the baseline vector. The accuracy of this measurement, depending now on the width of the fringe instead of the Airy disk, is improved by a factor of about* B/D *compared to the observation through a single telescope. The absolute values depend on a number of parameters that will be discussed in Sect. 6.2.2.*

Having distinct Airy disks, we cannot determine a single visibility value that is related to the binary's separation as we did for small objects. If one is only interested in its separation one should measure the distance between the fringe packages in the image, i.e. do astrometry. Here, we regard the binary as an example for an extended object, and we move the discussion into the uv-plane (see Fig. 3.19).

The spatial frequency spectrum of the image intensity, which we obtain by Fourier transforming the image intensity, is the product of $\mu(\boldsymbol{R})$ and the polychromatic interferometer OTF, (3.70):

$$\hat{I}(\boldsymbol{R}) = I_0 \cos\left(2\pi \boldsymbol{R} \cdot \frac{\boldsymbol{\alpha}_0}{2}\right) \times \left(\mathrm{OTF}_a(\boldsymbol{R}) * \left(G_B(\boldsymbol{R}) + 2\delta(\boldsymbol{R}) + G_B(-\boldsymbol{R})\right)\right). \tag{3.77}$$

For a wide binary with a separation α_0 much larger than the FWHM, λ_0/D, of the PSF of an individual aperture (and much larger than the fringe package), the cosine function of its visibility function $\cos(2\pi \boldsymbol{R} \cdot \boldsymbol{\alpha}_0/2)$ (3.75) has a periodicity $R = 2/\alpha_0$ that is much smaller than $B/l_c + 2D/\lambda_0$, the width of the elongated parts of the optical transfer function, so that many periods of the cosine are transferred by the OTF. If, for instance, the separation is $\alpha_0 = 5\lambda_0/D$ there are 5 periods of the visibility's cosine function across OTF_a. This is displayed in Fig. 3.19.

In general, the visibility function of a large object of size α_0 – large compared to the PSF – shows significant variation over OTF_a on a scale of $1/\alpha_0$ that needs to be considered in order to form the image so that we cannot replace the visibility function by a single value as in the quasi-monochromatic approximation. For our example of a binary separation of α_0, the cosine function of its visibility function has a periodicity of $2/\alpha_0$, so that a resolution of $1/\alpha_0$ is required in order to properly sample the cosine and to form an image of size α_0.

3.3.5 Considerations on the Interferometric Field of View

Throughout this section, we have discussed the imaging process in interferometers in Fizeau configuration with, in principle, unlimited field of view. Here, we will investigate some limitations on the field of view that are related to the level of detail in the visibility function. Restrictions on the field of view that are determined by the interferometer configuration, for instance by the Michelson configuration, will be discussed in Sect. 5.1.2.

The question of maximum object size or field of view of an interferometer is linked to the level of detail in the visibility function. If the visibility function is invariant over the elongated optical transfer function, $\mathrm{OTF}_a(\boldsymbol{R}) * G_B(\boldsymbol{R})$, the object is smaller than the fringe package (and much smaller than the Airy disk), the visibility is represented by a single value at $\boldsymbol{R}_{0,B}$. If the visibility function varies over the width of the elongated parts of the optical transfer function, $\mathrm{OTF}_a(\boldsymbol{R}) * G_B(\boldsymbol{R})$, the object is larger. Or, turning this argument around, the form of the visibility function inside the OTF needs to be measured in order to determine an object or a field of view that is much larger than the Airy disk. If the object is smaller than an Airy disk, one value of the visibility function is sufficient to characterise the object at the given baseline.

One should note that even a very large object can have visibility values at very long baselines if the object has fine structure with a spatial frequency according to that baseline. Thus, the size of an object does not determine the extension of the visibility function in the uv-plane but its structure, i.e. the object's spatial frequency content does. If the object shape is so smooth that its visibility function is zero beyond the limit given by the optical transfer function of an individual telescope, this means simply that the object details are fully resolved with an individual telescope. But then observations with an interferometer are not required.

Step by step we have moved the discussion from determining *the* visibility of *the* fringe pattern to regarding the visibility, and the visibility function, as a quantity that is transferred by an optical system. It is not even required that a single fringe pattern forms in the image plane in order to determine the object shape with an interferometer. Thus, image formation in an interferometer is completely described by the characteristics of the optical transfer function in the uv-plane.

Having point-spread functions with fringes, the image might not resemble the object too much if the uv-plane is not well covered with values. A wide spectral band helps increasing the uv-plane coverage and, thus, reducing the number of fringes in the PSF. However, a wide spectrum puts a constraint on the object size and on the assumption that the object brightness distribution is invariable with wavelength.

One has to find a compromise between the width of the spectral band and the characteristics of the source to be studied. While for the measurement of a stellar diameter the assumption of a wavelength independent shape is usually fulfilled over a wide spectral band, other measurements require a spectral resolution that limits the width of the spectral band considerably. In fact, the most interesting astronomical results have been obtained with a medium to high spectral resolution, i.e. with rather narrow spectral bands.

Thus, the arbitrary increase of the spectral bandwidth is not an option to improve the OTF coverage of spatial frequencies, and we have to combine observations with different baselines $\boldsymbol{B}$ – different in length and orientation – to have a good coverage of the uv-plane and therefore a point-spread function without fringes, resembling a point again.

It is exactly this last point, filling the uv-plane to synthesize a large aperture, that is one of the key issues of stellar interferometry. Before we will discuss this in detail in Sect. 3.4, we have a look at the Large Binocular Telescope.

The optical design of the **Large Binocular Telescope (LBT)** on Mt. Graham in Arizona [101] is very close to the layout with masked apertures, the Fizeau configuration, that we have used so far (see Fig. 3.14). The LBT design is unique in having two 8.4-m primary mirrors mounted on the same telescope structure. The centre-to-centre distance of the two mirrors, i.e. the baseline is $B = 14.4$ m so that the OTF is similar to the one in Fig. 3.19. The result of the computation of the image intensity distribution is displayed in Fig. 3.20, illustrating the large field of view in Fizeau configuration by using a crowded field with stars of varying intensities instead of a binary. This large field has a visibility function that requires many data points across OTF_a, similar to the visibility function of the wide binary in Fig. 3.19. All the data points in the image spectrum $\hat{I}(\boldsymbol{R})$ across OTF_a need to be considered in modulus and phase, i.e. in fringe contrast and position, in order to reconstruct an image of this size.

Most interferometers have a baseline B that is much larger than the diameter D of the individual telescopes, and the telescopes are not mounted on the same mechanical structure. These interferometers are usually operated in Michelson configuration and not in Fizeau configuration, i.e. they cannot be modelled by masked apertures. The consequence is that the stars in a large field of view like in Fig. 3.20 would only

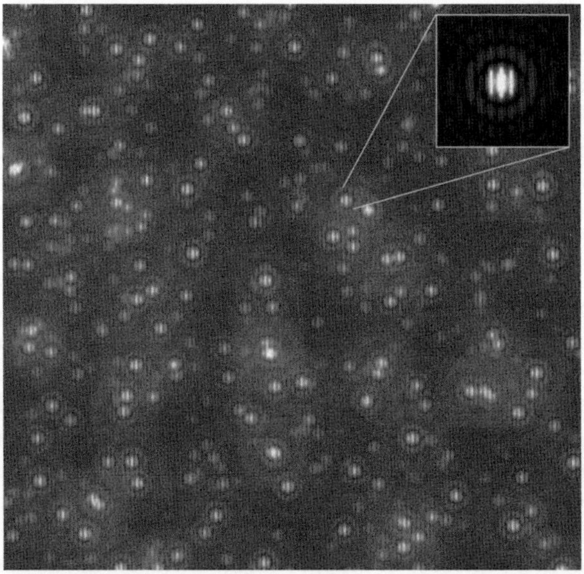

Fig. 3.20 Simulated intensity distribution in the image of the Large Binocular Telescope (LBT) on Mt. Graham in Arizona. The parameters are: telescope aperture $D = 8.4$ m, baseline $B = 14.4$ m, $\lambda = 2.2 \pm 0.2\,\mu$m (K-band). The first minimum of the PSF is at $1.22\lambda_0/D = 66$ mas and the fringe spacing is 32 mas. Each star in this crowded field shows an Airy disk with fringes. The inset displays an individual PSF. Due to the combination of values for D and B, there are only about three visible fringes across the Airy disk, and the loss of contrast due to the width of the spectral band is barely visible. (Courtesy T. Herbst)

show fringes if they are very close to a line that is parallel to the fringes and that is passing through the image centre. The usable field of view as the permitted distance from this line could be as small as the central core of an Airy disk of an individual telescope. While the impact of the beam combination on the usable field of view will be investigated in Sects. 5.1 and 5.2, we will discuss the interferometric field of view under very general assumptions in the following.

Content of Information

The interferometric field of view can also be discussed regarding the visibility function as a signal that is scanned at individual points in the uv-plane, leaving out all considerations of wavelength bands and aperture size. Assuming that we scan a regular pattern of $N \times N$ uv-plane values at multiples of the shortest baseline $R_{B_{\rm min}} = B_{\rm min}/\lambda$, we would find that the interferometric field[4] $\alpha_{\rm max}$ is limited by $1/R_B$, the reciprocal of the shortest baseline [123]. A larger object would not only have visibility values at baselines shorter than $R_{B_{\rm min}}$ that we cannot detect, but it would also create a variation of the visibility function on scales smaller than $\Delta R_{B_{\rm min}}$ elsewhere in the uv-plane, to which we are blind when scanning only a regular pattern spaced by $R_{B_{\rm min}}$.

It should be emphasised that it is not so much the smallest baseline that imposes this blindness but it is the fact that nowhere in the uv-plane there is a spacing smaller than $R_{B_{min}}$ between two measurements. For instance, regarding again a binary with a cosine visibility function, we cannot distinguish if we scan the first, second etc maximum of the cosine – for a binary separation smaller than $1/R_{B_{\rm min}}$ – or if it is the second, fourth etc maximum – for twice the separation. This effect is called *aliasing*.

With the angular resolution given by $1/R_{B_{\rm max}} = \lambda/B_{\rm max}$ and $B_{\rm max} = N B_{\rm min}$, the ratio (field of view/angular resolution) is given by $R_{B_{\rm max}}/R_{B_{\rm min}} = N$. Thus, the number of resolution elements to the field of view in the image plane is N^2. This is sometimes called the *crowding limit* since it corresponds to the number of sources that can be distinguished unambiguously within the field of view. This also corresponds to the Shannon theorem stating that the number of measurements, i.e. the content of information, and the number of independent points in the image are identical [116].

However, observing with real interferometers, the situation is not quite that simple. First, the spacing of the baselines in the uv-plane, after observations of several hours, is never regular, and we scan the uv-plane also in finer steps than given by the smallest baseline. In Sect. 3.4.2 (see Fig. 3.24), we shall discuss a typical example. Thus, visibility variations on a scale corresponding to the smallest distance between two measured values in the uv-plane affect the observations. Second, as illustrated in Fig. 3.19, the small scale variations of the visibility function over an individual

[4] In Sect. 5.4 discussing the situation for a variety of interferometer designs we will introduce the term *clean field of view* for this quantity.

aperture can be detected since the interferometer OTF does not only scan individual points in the uv-plane.

This does not mean that we create a content of information that is larger than the number of measured values, but it means that we cannot simply assume that there is a square (or circular) field of view that is composed of independent diffraction limited resolution elements. The resolution elements in the larger field of view, larger than $1/R_B$, are not independent but they are linked through missing spatial frequencies that are not measured in the uv-plane. This reduces the content of information in the image in a more subtle way than simply assuming that every pixel within the field of view is an independent piece of information. Thus, the answer to the question about the interferometric field of view is highly dependent on the object morphology.

3.3.6 *Masked Field of View*

While the Fizeau configuration with its large field of view is very desirable, most of today's interferometers are operated with a very limited field of view almost exclusively in Michelson configuration. For a number of reasons related to the opto-mechanical combination of individual telescopes with a long baseline (see Sect. 5.1), the field of view is reduced to substantially less than the Airy disk of an individual telescope and the quasi-monochromatic approximation applies.

Observing through turbulence, it is then advantageous to mask the image intensity distribution, a process called *spatial filtering*. This will be discussed in Sect. 6.1. The size of the mask is about the diameter of an Airy disk. In the following, we investigate the effect on the imaging process.

Usually, the images of the individual telescopes are masked *before* the beam combination. For the time being we assume perfect pointing of the telescopes so that the individually masked fields of view on a potentially large object match perfectly.

The masks are either pinholes with the diameter of an Airy disk [186], or so-called monomode optical fibres with a Gaussian transmittance function that is matched to the diameter of the Airy disk [75, 210]. Formally the masking is described by multiplying the image intensity distribution $I(\boldsymbol{\alpha})$ by a mask $M(\boldsymbol{\alpha})$ with rectangular or Gaussian shape and with the size of the Airy disk, about λ/D, writing

$$I_m(\boldsymbol{\alpha}) = M(\boldsymbol{\alpha})I(\boldsymbol{\alpha}). \tag{3.78}$$

Moving into the uv-plane, we introduce the spatial frequency spectrum of the mask as $\hat{M}(\boldsymbol{R})$ and of the masked intensity as $\hat{I}_m(\boldsymbol{R})$. The Fourier transform of the product $M(\boldsymbol{\alpha})I(\boldsymbol{\alpha})$ is then the convolution of the individual Fourier transforms, yielding

$$\hat{I}_m(\boldsymbol{R}) = \hat{M}(\boldsymbol{R}) * \hat{I}(\boldsymbol{R}). \tag{3.79}$$

Choosing a mask with a diameter of an Airy disk means that its spatial frequency spectrum $\hat{M}(\boldsymbol{R})$ has dimensions similar to the OTF of an individual telescope, OTF_a, which has a width of $2D/\lambda$. The convolution with the intensity spectrum $\hat{I}(\boldsymbol{R})$ then averages out the details of the spectrum on a scale smaller than about $2D/\lambda$.

What is the field of view with a mask the size of an Airy disk? If the mask were put directly on the object, it is straightforward that the field of view would be directly determined by the mask size. The image intensity distribution after diffraction at the telescope aperture would then be slightly wider than the mask. Applying the mask in the image plane, the image intensity distribution is limited by the mask size, and the unobscured field of view is smaller than the mask since Airy disks of all object points that are not exactly on-axis are truncated by the mask.

In the uv-plane, replacing the intensity spectrum $\hat{I}(\boldsymbol{R})$ by the product of visibility function $\mu(\boldsymbol{R})$ and polychromatic OTF, (3.70), we write

$$\hat{I}_m(\boldsymbol{R}) = I_0 \hat{M}(\boldsymbol{R}) * \mu(\boldsymbol{R}) \mathrm{OTF}_{pc}(\boldsymbol{R}). \tag{3.80}$$

Although the convolution of $\hat{M}(\boldsymbol{R})$ has to be done with the product of $\mu(\boldsymbol{R})$ with $\mathrm{OTF}_{pc}(\boldsymbol{R})$, we restrict the convolution to the visibility function defining the masked visibility function as:

$$\mu_m(\boldsymbol{R}) := \hat{M}(\boldsymbol{R}) * \mu(\boldsymbol{R}), \tag{3.81}$$

which is equivalent to masking the object instead of the image. This facilitates the description of the imaging process in the following.

With the polychromatic OTF in (3.70) we write the masked image intensity spectrum as

$$\begin{aligned} \hat{I}_m(\boldsymbol{R}) &= I_0 \mu_m(\boldsymbol{R}) \mathrm{OTF}_{pc}(\boldsymbol{R}) \\ &= I_0 \mu_m(\boldsymbol{R}) \Big(\mathrm{OTF}_a(\boldsymbol{R}) * \big(G_B(\boldsymbol{R}) + 2\delta(\boldsymbol{R}) + G_B(-\boldsymbol{R}) \big) \Big). \end{aligned} \tag{3.82}$$

Regarding first a **monochromatic** spectrum with $G_B(\pm\boldsymbol{R}) = \delta(\boldsymbol{R} \mp \boldsymbol{R}_B)$ the convolution is reduced to OTF_a centred at the positions of the δ-functions yielding

$$\begin{aligned} \hat{I}_m(\boldsymbol{R}) = I_0 \mu_m(\boldsymbol{R}) \\ \times \big(\mathrm{OTF}_a(\boldsymbol{R} - \boldsymbol{R}_B) + 2\mathrm{OTF}_a(\boldsymbol{R}) + \mathrm{OTF}_a(\boldsymbol{R} + \boldsymbol{R}_B) \big). \end{aligned} \tag{3.83}$$

We argued above that the masked visibility function would not vary over the width of OTF_a. Therefore, we can replace $\mu_m(\boldsymbol{R})$ by its value at the centre $\boldsymbol{R}_B$ of the respective OTF_a, writing the masked image intensity distribution as the Fourier transform of the product of the masked visibility function and the interferometer OTF (3.69) as

$$\begin{aligned} I_m(\boldsymbol{\alpha}) &= I_0 \int \big(\mu_m(\boldsymbol{R_B})\mathrm{OTF}_a(\boldsymbol{R}-\boldsymbol{R_B}) + 2\mu_m(0)\mathrm{OTF}_a(\boldsymbol{R}) \\ &\quad + \mu_m(-\boldsymbol{R_B})\mathrm{OTF}_a(\boldsymbol{R}+\boldsymbol{R_B})\big)\mathrm{e}^{-\mathrm{i}2\pi \boldsymbol{R}\cdot\boldsymbol{\alpha}}\mathrm{d}\boldsymbol{R} \\ &= 2I_0\mathrm{PSF}_a(\boldsymbol{\alpha})\Big(1 + |\mu_m(\boldsymbol{R_B})|\cos\big(\phi(\boldsymbol{R_B}) - 2\pi\boldsymbol{R_B}\cdot\boldsymbol{\alpha}\big)\Big), \end{aligned} \quad (3.84)$$

with $\mu_m(-\boldsymbol{R_B}) = \mu_m^*(\boldsymbol{R_B})$. Thus, the fringe pattern of a masked image has a constant visibility given by $|\mu_m(\boldsymbol{R_B})|$ enveloped by an Airy disk. In Sect. 3.3.2, discussing small objects we found exactly the same result, with the value of the visibility function at $\boldsymbol{R_B}$ determining the visibility of the fringe pattern, (3.57).

This is not very surprising since for small objects we took the visibility function as approximately invariant over the width of OTF_a, as we did here as a consequence of masking the image.

However, there is one significant difference between a small object and a masked large object. While the visibility of the small object is determined by its shape, the visibility of the masked field of view of the large object is biased, depending on the object shape inside the intersection of the masked images of the individual telescopes. The term *modal visibility* is sometimes used for the biased visibility [153]. The masked images can differ in the individual telescopes due to their pointing errors. This is illustrated by Fig. 3.21. In the extreme case, completely different parts of the object are observed and the modal visibility is zero although the visibility of the object is not zero. One has to carefully determine the average overlap following from the pointing error and calibrate the visibility.

In the **polychromatic** case the fringe pattern in the PSF cannot be described by a single visibility value since the visibility of the fringes varies over the Airy disk. The visibility is reduced from its maximum value at the white-light fringe to zero with the total number of fringes in PSF_{pc} being determined by the value of $\lambda_0/\Delta\lambda$ (see Fig. 3.17). Depending on the width of the fringe package with respect to the masked field of view, the object points within the field of view can produce partially overlapping or even distinct fringe packages. The variation of the visibility values is then due to a mix of spectral bandwidth and object shape.

The difference to the monochromatic case can be discussed in the uv-plane, regarding the polychromatic OTF, (3.82), with a spectral band given by $G_B(\boldsymbol{R})$. Individual OTF_a at different wavelengths overlap to form OTF_{pc} (see Fig. 3.18). The elongated areas around $\boldsymbol{R}_{0,B}$ are wider than the area $2D/\lambda_0$ over which the masked object brightness spectrum is invariant. Thus, we have to determine the visibility function with a spacing of $2D/\lambda_0$ and we cannot replace $\mu_m(\boldsymbol{R})$ by a single value like in the monochromatic case. Unless we assume that the object is smaller than the half-width of the fringe package, which is l_c/B. In this case, the visibility function is invariant over the elongated areas of OTF_{pc} and a single value of the

Fig. 3.21 Illustration of the effect of masking different parts of a densely crowded object. The *circles* indicate the mask diameters of Airy disk size. The visibility should be determined by the intensity distribution in the intersection of the two circles. However, the intensity in the non-intersecting parts of the circles contributes to the background intensity in the fringe pattern, increasing the total flux and, thus, reducing the measured visibility. Depending on the pointing stability of the individual telescopes these circles move with respect to each other, further complicating any attempt to calibrate this effect

visibility function at $\boldsymbol{R}_{0,B}$ is sufficient to characterise the visibility function. This will be discussed in the following section.

If the object is larger than the fringe package in PSF_{pc}, the averaging over OTF_{pc} cannot be applied. There, a large number of values across OTF_{pc} has to be considered in order to determine the intensity distribution of the object. Replacing the visibility function nonetheless by a single value the reconstructed image will be distorted. This effect is known as *bandwidth smearing* and has been studied extensively in radio astronomy [21,47].

Propagation Through Interferometers: Summary

We regard a stellar interferometer as an optical system consisting of two or more telescopes forming a joint image of an object after recombination of the light from the individual telescopes. Our general assumption is that the object brightness distribution $O_b(\boldsymbol{\alpha})$, the shape of the object, is independent of wavelength over the observed spectral band.

Observing a point source at a large distance, the fringe pattern of a stellar interferometer with two telescopes is like that in Young's experiment but with the Airy disk of a single telescope as an envelope. We call this fringe pattern the interferometer point-spread function (PSF) since we observe it in the focal plane of the interferometer, which is the image plane for an object at a very large distance (Sect. 3.3.2).

The monochromatic interferometer PSF reads like

$$\mathrm{PSF}(\boldsymbol{\alpha}) = 2\mathrm{PSF}_a(\boldsymbol{\alpha})\big(1 + \cos(2\pi \boldsymbol{R_B} \cdot \boldsymbol{\alpha})\big), \tag{3.53}$$

with PSF_a the Airy disk of an individual telescope, called the primary beam in radio interferometry. The FWHM of the Airy disk is λ/D, with D the aperture diameter, and the fringe spacing is $1/|\boldsymbol{R_B}| = \lambda/|\boldsymbol{B}|$, with $\boldsymbol{B}$ the baseline vector. The PSF is displayed in Fig. 3.15.

We assume that the interferometer is operated in the Fizeau configuration so that the combination of the individual telescopes can be treated like a large aperture masked by two small sub-apertures.

The optical transfer function (OTF) of the interferometer aperture is given by

$$\mathrm{OTF}(\boldsymbol{R}) = \mathrm{OTF}_a(\boldsymbol{R}) * \big(\delta(\boldsymbol{R} - \boldsymbol{R_B}) + 2\delta(\boldsymbol{R}) + \delta(\boldsymbol{R} + \boldsymbol{R_B})\big), \tag{3.52}$$

with $\boldsymbol{R} = (u, v)$ the coordinate vector in the uv-plane. The interferometer OTF is the sum of optical transfer functions of the sub-aperture, OTF_a, at positions $\boldsymbol{R_B}$, 0 and $-\boldsymbol{R_B}$, with $\boldsymbol{R_B} = \boldsymbol{B}/\lambda$.

The PSF and the OTF are linked by a Fourier transform. Therefore, the multiplication of PSF_a with the fringe pattern 1+cos(.) in (3.53) is replaced by the convolution of OTF_a with the δ-functions in (3.52).

Observing an object with brightness distribution $O_b(\boldsymbol{\alpha})$, the intensity distribution in the focal plane of the interferometer is the convolution of its brightness distribution with the interferometer PSF:

$$I(\boldsymbol{\alpha}) = 2\, O_b(\boldsymbol{\alpha}) * \big(\mathrm{PSF}_a(\boldsymbol{\alpha})\big(1 + \cos(2\pi \boldsymbol{R_B} \cdot \boldsymbol{\alpha})\big)\big). \tag{3.54}$$

If the object is smaller than the Airy disk, PSF_a, the observation with a single telescope does not reveal any object detail. Observing with an interferometer, the contrast of the fringes is affected by detail as fine as the fringe spacing. Eventually, after measuring the fringe contrast for many different baselines, a real image can be reconstructed when the angular resolution is determined by the fringe spacing. This illustrates the advantage in angular resolution of the interferometer over a single telescope.

This effect is best discussed by regarding the situation in the uv-plane, with $\boldsymbol{R} = (u, v)$, when the convolution (3.54) transforms into the product of $\mu(\boldsymbol{R})$, which is the visibility function of the object brightness distribution, and the interferometer OTF reading

$$\hat{I}(\boldsymbol{R}) = \int I(\boldsymbol{\alpha})\, \mathrm{e}^{\mathrm{i}2\pi \boldsymbol{R}\cdot\boldsymbol{\alpha}} \mathrm{d}\boldsymbol{\alpha} \tag{3.55}$$
$$= I_0\mu(\boldsymbol{R})\big(\mathrm{OTF}_a(\boldsymbol{R}) * (\delta(\boldsymbol{R} - \boldsymbol{R}_B) + 2\delta(\boldsymbol{R}) + \delta(\boldsymbol{R} + \boldsymbol{R}_B))\big)\,.$$

An object that is smaller than the Airy disk PSF_a has a visibility function $\mu(\boldsymbol{R})$, which is basically invariant over the width of the sub-aperture transfer function OTF_a. Then, the visibility function can be replaced by the value of $\mu(\boldsymbol{R})$ at the centre of the respective OTF_a since $\mu(\boldsymbol{R})$ contains no significant information over the width of OTF_a, and the convolution in (3.54) is simplified to

$$I(\boldsymbol{\alpha}) = 2\mathrm{PSF}_a(\boldsymbol{\alpha})\,(1 + |\mu(\boldsymbol{R}_B)|\cos(\phi(\boldsymbol{R}_B) - 2\pi \boldsymbol{R}_B\cdot\boldsymbol{\alpha}))\,. \tag{3.57}$$

The visibility of the fringe pattern is now given by a single function value of the visibility function μ at $\boldsymbol{R}_B$, with the contrast being determined by its modulus, and the fringe position by its phase. This is very similar to the result of Young's experiment (3.46).

Moving on to polychromatic objects (Sect. 3.3.3) – still with wavelength independent shape, for instance a uniform disk – we obtain the polychromatic point-spread function, PSF_{pc}, as

$$\mathrm{PSF}_{pc}(\boldsymbol{\alpha}) = 2\mathrm{PSF}_a(\boldsymbol{\alpha})\big(1 + g_B(\boldsymbol{\alpha})\cos(2\pi \boldsymbol{R}_{0,B}\cdot\boldsymbol{\alpha})\big), \tag{3.61}$$

with $\boldsymbol{R}_{0,B} = \boldsymbol{B}/\lambda_0$ the baseline vector at the mean wavelength λ_0, and with $g_B(\boldsymbol{\alpha})$ the Fourier transform of the spectrum $G_B(\boldsymbol{R})$, in the dimensionless notation $G_B(\boldsymbol{R}) = G_B(\boldsymbol{B}/\lambda) = G_B(\frac{\boldsymbol{B}}{c}\nu)$.

Thus, the fringe pattern within PSF_a has the Fourier transform of the spectrum as an envelope, reducing the contrast of the fringes after $\nu_0/\Delta\nu = \lambda_0/\Delta\lambda$ fringe periods, with $\Delta\nu$ the spectral bandwidth, as displayed in Fig. 3.17.

The polychromatic OTF incorporates the spectrum $G_B(\boldsymbol{R})$, yielding

$$\mathrm{OTF}_{pc}(\boldsymbol{R}) = \mathrm{OTF}_a(\boldsymbol{R}) * \big(G_B(\boldsymbol{R}) + 2\delta(\boldsymbol{R}) + G_B(-\boldsymbol{R})\big). \tag{3.64}$$

As a result of this convolution, the OTF is elongated around $\boldsymbol{R}_{0,B} = \boldsymbol{B}/\lambda_0$ where $G_B(\boldsymbol{R})$ is centred, see Fig. 3.18. We can attach the object spectrum to the OTF since we assume that the object brightness distribution $O_b(\boldsymbol{\alpha})$ and, thus, its spatial frequency spectrum $\hat{O}(\boldsymbol{R})$ are independent of wavelength over the observed spectral band $\Delta\nu$.

Then, the convolution of the object brightness distribution with the polychromatic PSF (3.61) forms the image intensity distribution of an extended object as

$$\begin{aligned} I(\boldsymbol{\alpha}) &= O_b(\boldsymbol{\alpha}) * \mathrm{PSF}_{pc}(\boldsymbol{\alpha}) \\ &= 2\, O_b(\boldsymbol{\alpha}) * \big(\mathrm{PSF}_a(\boldsymbol{\alpha})\,(1 + g_B(\boldsymbol{\alpha}) \cos(2\pi \boldsymbol{R}_{0,B} \cdot \boldsymbol{\alpha}))\big). \end{aligned} \tag{3.67}$$

If the object is larger than the PSF, for example a stellar cluster with many individual stars, the convolution results in individual fringe patterns all over the image plane as shown in Fig. 3.20 when observing in Fizeau configuration. In Michelson configuration, only those objects that are close to the image centre display fringes. In any case, one cannot attribute a single visibility value reflecting the object morphology to this extended intensity distribution.

In general, an object of size α_0 shows a variation in its visibility function $\mu(\boldsymbol{R})$ on a scale of $1/\alpha_0$ that needs to be accounted for in the uv-plane (Sect. 3.3.4). According to the number of individual fringe patterns of the extended object one has to process an equivalent number of values in the image intensity spectrum $\hat{I}(\boldsymbol{R})$.

Masking an object with a mask of Airy-disk size (Sect. 3.3.6) – a process called *spatial filtering* – smoothes the spatial frequency spectrum on a scale of about $2D/\lambda$, the diameter of the telescope OTF. This is the same effect as observing a small object. However, masking small parts of a large object with different telescopes, one has to ensure that the masked patches of the object match perfectly, see Fig. 3.21. Otherwise the visibility will be reduced until – in the extreme case of pointing at completely separate parts of the object – it goes to zero.

Restricting the observation to objects smaller than PSF_a and additionally smaller than the half-width of the fringe package, i.e. smaller than $(\lambda_0/\Delta\lambda)/R_{0,B} = (\lambda_0/\Delta\lambda)\lambda_0/B = l_c/B$, with l_c the coherence length, brings back the quasi-monochromatic approximation as in Sect. 2.3.4.

In this approximation, the visibility function in (3.55) is replaced by its individual values at $\pm\boldsymbol{R}_{0,B}$, the spatial frequency at the mean wavelength λ_0, and at $|\boldsymbol{R}| = 0$ since $\mu(\boldsymbol{R})$ is invariant over the elongated OTF due the source that is smaller than the half-width of the fringe package, yielding

$$\begin{aligned} \hat{I}(\boldsymbol{R}) = I_0\Big(&\mu(\boldsymbol{R}_{0,B})\mathrm{OTF}_a(\boldsymbol{R}) * G_B(\boldsymbol{R}) \\ &+ 2\mu(0)\mathrm{OTF}_a(\boldsymbol{R}) * \delta(\boldsymbol{R}) + \mu(-\boldsymbol{R}_{0,B})\mathrm{OTF}_a(\boldsymbol{R}) * G_B(-\boldsymbol{R})\Big), \end{aligned} \tag{3.71}$$

with $\mu(-\boldsymbol{R}_{0,B}) = \mu^*(\boldsymbol{R}_{0,B})$.

Under these conditions, the quasi-monochromatic approximation provides an expression for the intensity distribution of the fringe pattern, as

$$\begin{aligned} I_{\mathrm{qm}}(\boldsymbol{\alpha}) &= 2I_0\,\mathrm{PSF}_a(\boldsymbol{\alpha}) \\ &\times \Big(1 + g_B(\boldsymbol{\alpha})|\mu(\boldsymbol{R}_{0,B})| \cos\big(\phi(\boldsymbol{R}_{0,B}) - 2\pi \boldsymbol{R}_{0,B} \cdot \boldsymbol{\alpha}\big)\Big), \end{aligned} \tag{3.72}$$

with the visibility function, $\mu(\boldsymbol{R}_{0,B})$, given as the Fourier transform of the object brightness distribution, $O_b(\boldsymbol{\alpha})$, as described by the van Cittert-Zernike theorem, (3.24).

The visibility can be measured in the fringe pattern by applying the ABCD method (2.70) or by deriving it from the Fourier spectrum (3.71) in the uv-plane, obtaining an estimator for $\mu(\boldsymbol{R}_{0,B})$ as

$$\frac{\mu(\boldsymbol{R}_{0,B}) \int \mathrm{OTF}_a(\boldsymbol{R}) * G_B(\boldsymbol{R})\, \mathrm{d}\boldsymbol{R}}{2\mu(0) \int \mathrm{OTF}_a(\boldsymbol{R})\, \mathrm{d}\boldsymbol{R}} = \frac{\mu(\boldsymbol{R}_{0,B})}{2}, \tag{3.73}$$

using $\mu(0) = 1$ and $\int G_B(\boldsymbol{R}) \mathrm{d}\boldsymbol{R} = 1$.

Calculating the power spectrum $|\hat{I}(\boldsymbol{R})|^2$ of the fringe pattern, we obtain an estimator for the squared modulus of the visibility function $|\mu(\boldsymbol{R}_{0,B})|^2$ as

$$\frac{|\mu(\boldsymbol{R}_{0,B}) \int \mathrm{OTF}_a(\boldsymbol{R}) * G_B(\boldsymbol{R})\, \mathrm{d}\boldsymbol{R}|^2}{|2\mu(0) \int \mathrm{OTF}_a(\boldsymbol{R})\, \mathrm{d}\boldsymbol{R}|^2} = \frac{|\mu(\boldsymbol{R}_{0,B})|^2}{4}. \tag{3.74}$$

The power spectrum provides an unbiased measure for the visibility function in case of noisy measurements [224], at the price of losing the phase $\phi(\boldsymbol{R}_{0,B})$ of the visibility function.

3.4 The uv-Plane

In the preceding sections, we have investigated spatial frequency spectra and fringe patterns for individual baselines. Here, we will discuss the combination of observations with different baselines and their impact on the imaging process using the LBT and the VLTI as examples. Aspects of the layout of interferometric arrays will be treated in Sect. 5.4.

The quality of the reconstructed image is governed by the number of observations with different baselines filling the uv-plane. The more baselines are used, the less prominent will be the fringe structure in the image so that the reconstructed point-spread function eventually resembles a point again. It is important to keep in mind that the field of view cannot be increased by adding more observations with different baselines.

In the preceding discussion on the imaging process, we have always assumed that the interferometer baseline $\boldsymbol{R}_B = \boldsymbol{B}/\lambda$ is perpendicular to the optical axis and that the object coordinates, the aperture plane and the image coordinates are fixed with respect to each other.

Sky Coordinates

Observing celestial objects with a telescope or an interferometer, the object coordinates, as seen from the ground, are in permanent motion called *diurnal motion* due to the rotation of the Earth. Therefore, a celestial coordinate system is used, normally the equatorial reference frame with the coordinates declination δ for the angle of latitude above the celestial equator, and right ascension RA, for the longitude. Usually, relative coordinates are used, defined with respect to the *line of sight* pointing at the object. The unit vector along the line of sight is called $\boldsymbol{s}$. A difference in position is given by $\Delta\boldsymbol{s}$ that is also denoted by $\boldsymbol{\theta}$. We use $\boldsymbol{\alpha}' = (\Delta RA, \Delta\delta)$ as the coordinate of the sky plane that is tangential to the celestial sphere and orthogonal on $\boldsymbol{s}$, as illustrated in Fig. 3.22.

The visibility $\mu(\boldsymbol{R})$, as the Fourier transform of the object brightness distribution O_b, is calculated with respect to object coordinates $\boldsymbol{\alpha}'$. This means that the uv-plane is defined with respect to the sky coordinates independent of the rotation of the Earth. The uv-plane is no longer parallel to the ground but it is orthogonal to the line of sight $\boldsymbol{s}$.

Consequently, the physical baseline $\boldsymbol{B}$ of an interferometer with independent telescopes on the ground has to be replaced by the *effective baseline* $\boldsymbol{B}_{\text{eff}}$, which is the projection of the baseline vector onto the uv-plane, orthogonal on $\boldsymbol{s}$. It is $|\boldsymbol{B}_{\text{eff}}| = |\boldsymbol{B}|\sin(\gamma)$, with γ the angle between $\boldsymbol{B}$ and $\boldsymbol{s}$. Depending on the object position, the effective baseline and therefore the OTF rotate during the observation. The baseline $\boldsymbol{R_B}$ that is defined in the uv-plane thus equals $\boldsymbol{B}_{\text{eff}}/\lambda$.

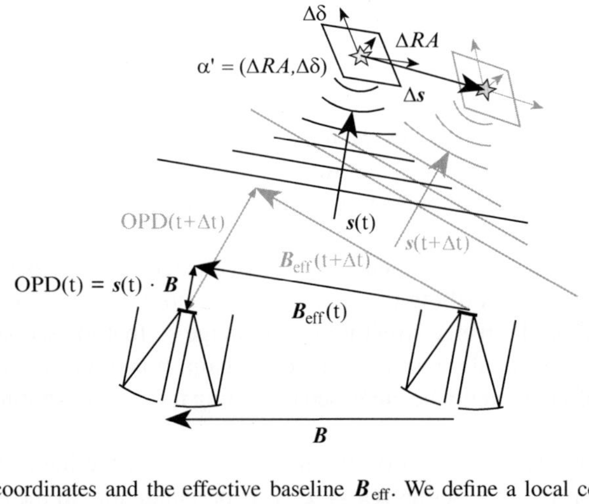

Fig. 3.22 Sky coordinates and the effective baseline $\boldsymbol{B}_{\text{eff}}$. We define a local coordinate system ΔRA, $\Delta\delta$, approximately a plane, that is centred on the object. This plane is orthogonal on the unit vector $\boldsymbol{s}$ pointing at the object. The effective baseline, as seen from the object, is the projection of $\boldsymbol{B}$ onto a plane that is orthogonal on $\boldsymbol{s}$. The situation observing the same object at $t + \Delta t$ is shown in grey. The cover page illustrates this effect. While the sky coordinates are fixed, we see the object at a different position, $\boldsymbol{s} + \Delta\boldsymbol{s}$ and the effective baseline changes in orientation and in length

The OPD is given by $\boldsymbol{s} \cdot \boldsymbol{B}$. The difference of the OPD of an object at $\boldsymbol{s}$ and at $\boldsymbol{s} + \Delta\boldsymbol{s}$ is given by $\Delta\boldsymbol{s} \cdot \boldsymbol{B}$. Note that $\Delta\boldsymbol{s}$ is a vector in three-dimensional space. Projecting $\Delta\boldsymbol{s}$ onto the sky plane, parallel to $\boldsymbol{\alpha}'$, the OPD difference is computed as scalar product with $\boldsymbol{B}_{\text{eff}}$. Staying within thc local coordinate system $\boldsymbol{\alpha}'$ we find that the additional OPD of individual objects at $\boldsymbol{\alpha}'_0$ is given by $\boldsymbol{\alpha}'_0 \cdot \boldsymbol{B}_{\text{eff}}$.

In the following, we will investigate aspects of filling the uv-plane, using two examples of interferometers, the LBT in Fizeau configuration and the VLTI in Michelson configuration, and we will discuss basic principles of aperture synthesis. We shall assume the ideal case that the visibilities are known in modulus and phase.

3.4.1 Large Apertures, Short Baseline: The LBT

We first regard the LBT that, with a baseline of $B = 14.4$ m and a telescope diameter of $D = 8.4$ m, shows only three fringes across the PSF. This means that the three OTF_a – the optical transfer function of the individual sub-apertures – at 0, $\boldsymbol{B}/\lambda$ and $-\boldsymbol{B}/\lambda$, forming the interferometer OTF, slightly overlap since their spacings of B/λ are slightly smaller than their widths of $2D/\lambda$. In Fig. 3.23, interferometer OTF for three different orientations of the effective baseline are displayed for the K-band.

Due to the ratio baseline/telescope diameter, the elongation of the OTF around $\boldsymbol{R}_{0,B} = \boldsymbol{B}/\lambda_0$ is relatively small (see Fig. 3.18). With the K-band spectral bandwidth of $\Delta\lambda = 0.4\,\mu$m, a mean wavelength of $\lambda_0 = 2.2\,\mu$m, and a coherence length of $l_c = \lambda_0^2/\Delta\lambda = 12\,\mu$m we find that the OTF is elongated by a factor of $1 + \frac{B/l_c}{2D/\lambda_0} = 1.16$.

For the LBT, the length of the effective baseline $\boldsymbol{B}_{\text{eff}}$ is the same as that of the physical baseline $\boldsymbol{B}$ since the baseline, defined by the two primary mirrors that are on the same telescope structure, is always perpendicular to the line of sight.

This combination of aperture diameter and baseline permits us to fill the uv-plane reasonably well with three measurements using baselines rotated by 60° as displayed in Fig. 3.23. This can be achieved by observing at three distinct moments in time after rotation of the sky coordinates by 60° with respect to the baseline.

We discuss mainly the properties of the combined OTF and its Fourier transform, the reconstructed PSF OTF – as if observing a point source – keeping in mind that the reconstructed image of an extended object results from combining the measured visibilities of each observation, which is equivalent to multiplying the object visibility with the OTF.

Combining the OTF_i of the individual LBT observations in Fig. 3.23 by simply adding them, we have a situation that is equivalent to adding the intensity distributions PSF_i of the fringe patterns of the three individual observations. However, although there is a central peak, the reconstructed intensity will not display a first minimum of zero due to the different orientations of the added fringe patterns. This means also that the reconstructed PSF is slightly wider than the central fringe of the

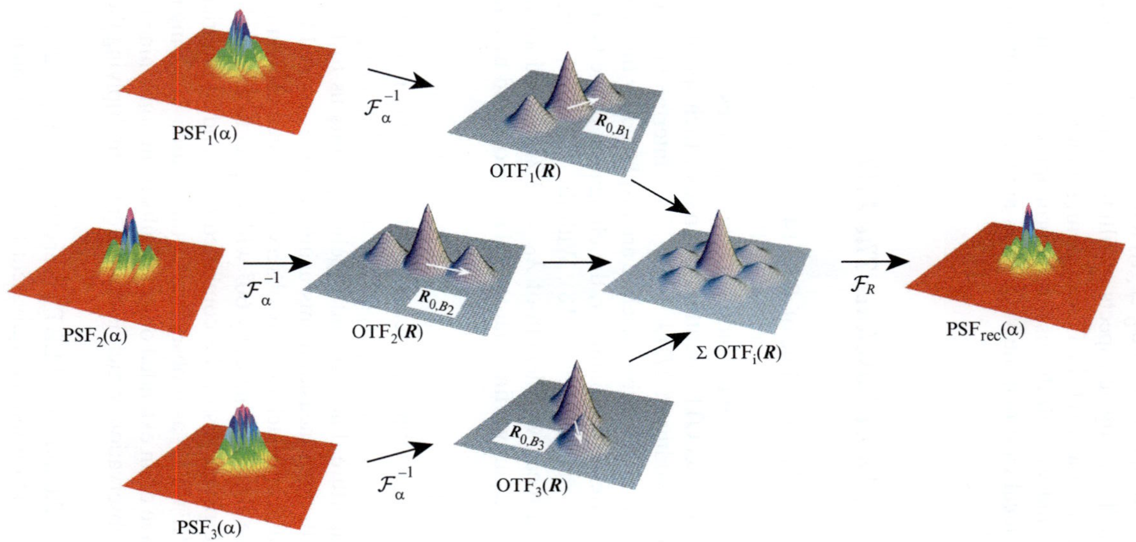

Fig. 3.23 Scenario with three LBT setups. Observing an unresolved star in the K-band, the PSF are determined for three effective baselines $\boldsymbol{R}_{0,B_i}$ that are rotated by 60° with respect to each other. The Fourier transform of PSF$_i\,(\boldsymbol{\alpha})$ is the optical transfer function OTF$_i\,(\boldsymbol{R})$, showing the baseline orientation accordingly. The properly weighted sum of the three OTF provides the spatial frequency spectrum of the reconstructed point-spread function PSF$_{\text{rec}}(\boldsymbol{\alpha})$. Due to the form of the individual OTF the combination of the three OTF fills the uv-plane rather well, resulting in much reduced interference effects: the fringes of the individual point-spread functions are replaced by six small peaks in the reconstructed PSF. This combined OTF is identical to that of a three-aperture interferometer – the three apertures being placed on an equilateral triangle with a side length of $B = 14.4\,\text{m}$ – when the OTF is the autocorrelation function of these three apertures

individual fringe patterns, reducing the image quality and, for instance, the ability to detect very faint objects in the immediate vicinity of the source.

In the uv-plane, this can be attributed to the central OTF_a, at $R = 0$, to which each fringe pattern contributes individually, so that in sum its share is overestimated, and too much weight is given to the contribution of PSF_a in the reconstructed image and too little to the high-frequency content at $\boldsymbol{R}_{B_i}$. We can improve the situation by adjusting the weight of $\mathrm{OTF}_a(0)$.

We introduce the calibration parameter x_{cal} and write the combined interferometer OTF as

$$\mathrm{OTF}_{pc}(\boldsymbol{R}) = \sum_i^3 \mathrm{OTF}_a(\boldsymbol{R}) * \left(G_{B_i}(\boldsymbol{R}) + \frac{2\delta(\boldsymbol{R})}{x_{\mathrm{cal}}} + G_{B_i}(-\boldsymbol{R}) \right). \quad (3.85)$$

Without any weighting, i.e. $x_{\mathrm{cal}} = 1$, we simply add the OTF and the PSF of the individual observations as discussed above. This can be demonstrated by a simple experiment as described in [133].

Reducing the central OTF_a too much means to reduce the contribution from PSF_a, so that the first minimum has negative values that cannot be interpreted as a reconstructed intensity.

In general, it is reasonable to put a positivity constraint on the reconstructed PSF and, thus, on the reconstructed image intensity since they are linked by $I_{\mathrm{rec}}(\boldsymbol{\alpha}) = O_b(\boldsymbol{\alpha}) * \mathrm{PSF}_{\mathrm{rec}}(\boldsymbol{\alpha})$.

One finds that in the case of the LBT a calibration parameter of $x_{\mathrm{cal}} = 2$ corresponds to a reconstructed PSF with a first minimum of zero between the central peak and the six side-peaks – like the first minimum in the individual fringe patterns – as displayed in Fig. 3.23. This reconstructed PSF is real and positive representing the intensity distribution in the image.

Combining more apertures, the emphasis moves from the weighting of the central OTF_a to processing the complete distribution of baselines, involving more complex considerations as will be discussed in Sect. 3.4.3.

It is interesting to look at the positivity constraints from another point of view, picking up the discussion on the autocorrelation theorem in Sect. 3.1.1 (illustrated in Fig. 3.3). There, we argued that a real and positive function, like the reconstructed PSF, has a Fourier transform, like the combined OTF, that is itself the autocorrelation of another function [17, 102, 119].

While this is straightforward in an optical system with a well defined aperture, the autocorrelation of which forms the OTF, what does this mean for our combined interferometer OTF? The basic mathematical principle remains the same but lacking a physical aperture that can be used to calculate the combined OTF, we define a virtual aperture that we call the *meta-aperture*. Then, the reconstructed PSF would have the same form as the PSF of an interferometer with multiple apertures distributed like the meta-aperture, with simultaneous combination of all beams.

Thus, by definition there always exists a meta-aperture as long as the reconstructed PSF is positive.

In our scenario of three distinct visibility measurements with two telescopes, with baselines rotated by 60°, displayed in Fig. 3.23, the combined OTF with a calibration parameter of $x = 2$, and its Fourier transform, the reconstructed PSF, corresponds to the PSF of a three-aperture interferometer – the three apertures being placed on an equilateral triangle with a side length of $B = 14.4$ m forming the meta-aperture – with simultaneous combination of all three apertures.

This meta-aperture is particularly simple. Combining observations with many baselines, it is not straightforward to find a suitable meta-aperture. We will come back to this point in Sect. 3.4.3.

Combining more than three LBT observations, with smaller angles between the baseline vectors the distinct side peaks that show up in the reconstructed PSF in Fig. 3.23 would add up, forming an annulus. The reconstructed PSF would then become similar to the PSF of an annular aperture, see Fig. 2.8.

Additionally taking advantage of the particular OTF shape of the two LBT apertures without zero points between the individual OTF_a, one could amplify the areas of the OTF with low values – and, thus, the measured visibility function – so that the combined OTF resembles that of a filled aperture with a diameter of 22.8 m, and the reconstructed PSF is very similar to the corresponding Airy disk.

The optical design of the LBT permits the operation as a Fizeau interferometer with a large field of view. This requires, as discussed in Sect. 3.3.4, to determine many complex visibility values across each interferometer OTF. Combining many observations with different baseline orientations, the uv-plane can be filled with visibility values while the field of view α_0 is determined by the spacing $1/\alpha_0$ of the visibility values in the uv-plane. The individual values of the spatial frequency spectra then fill a virtual aperture with a diameter of 22.8 m. While the image reconstruction from this combined visibility function is straightforward in principle, we will see later that aspects like atmospheric turbulence and measurement noise require some sophisticated data processing methods [160, 169] to obtain visibility measurements with a good signal-to-noise ratio.

3.4.2 Large Apertures, Long Baselines: The VLTI

The LBT is a special case since it combines large apertures on the same telescope structure with a rather short baseline. Most interferometers combine individual telescopes on the ground with aperture diameters much smaller than their baselines. Then, filling the uv-plane requires many observations. This will be investigated in the remainder of this section.

If the field of view is limited to about an Airy disk – which is still the most common observing situation – the visibility function is invariant over OTF_a, and there is only one single complex visibility point per baseline to be accounted for. This situation will be discussed in the following using illustrations by F. Millour [159].

We regard four scenarios of binary star observations displayed in Fig. 3.24, evolving from single-baseline monochromatic (a), to multi-baseline polychromatic (d) observations. The separation of the binary is 3.65 mas – with each of its components assumed to be a point source – and the separation vector $\boldsymbol{\alpha}_0$ is slightly tilted with respect to the sky coordinate axes as is the visibility function, $\mu(\boldsymbol{R}) = \cos(2\pi \boldsymbol{R} \cdot \boldsymbol{\alpha}_0/2)$, with a period length of $2/\alpha_0 = 548\,\text{arcsec}^{-1}$, the modulus of which is displayed in Fig. 3.24. Keep in mind that these stripes are not the interferometric fringes but the visibility function of the binary!

The interferometric OTF for a single baseline is shown in Fig. 3.24a. The cone-like shape of the OTF of an individual aperture, OTF_a, is displayed by a colour gradient in the circles for the central wavelength $\lambda_0 = 1.1\,\mu\text{m}$ of the J-band ($1.1 \pm 0.15\,\mu\text{m}$). The radius of OTF_a for an 8-m telescope is then $D/\lambda_0 = 35\,\text{arcsec}^{-1}$ and the corresponding FWHM of the Airy disk is 28 mas. The separation of the binary is about eight times smaller than the diameter of the Airy disk so that we can use a spatial filter with the size of an Airy disk to observe the binary. The visibility function $\mu(\boldsymbol{R})$ is approximately invariant over the diameter of OTF_a – as can be seen in Fig. 3.24a – and the visibility can be represented by a single value.

Observing at zenith, the effective baseline length $|\boldsymbol{B}_{\text{eff}}|$ is identical to the physical baseline $|\boldsymbol{B}|$, and its orientation in the uv-plane is determined by the layout of the interferometer on the ground with respect to the sky coordinates.

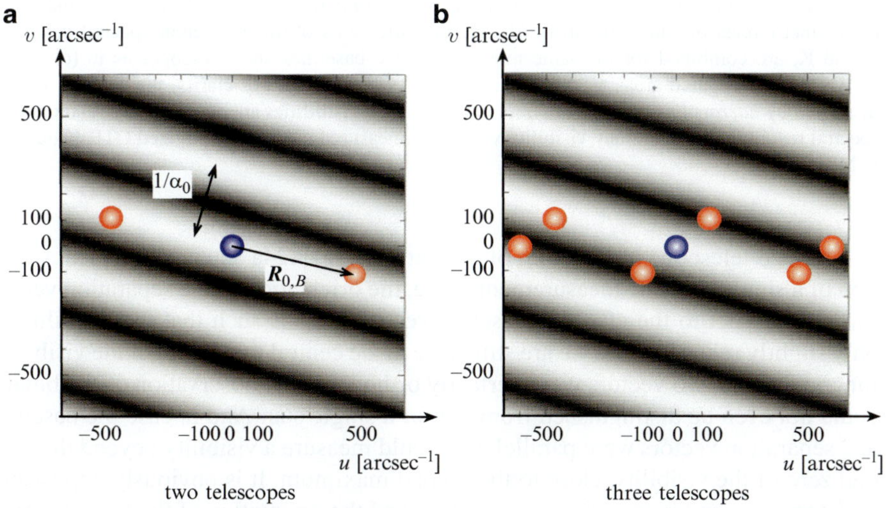

Fig. 3.24 Modelled uv-plane coverage for four different interferometer setups observing a binary star at zenith with a separation of 3.65 mas. The modulus of the visibility function, $\cos(2\pi \boldsymbol{R} \cdot \boldsymbol{\alpha}_0/2)$, is shown by slightly tilted black and white stripes. The separation of its maxima is $1/\alpha_0 = 274\,\text{arcsec}^{-1}$ as displayed. The *coloured circles* indicate the interferometer OTF, like in Fig. 3.16. Up to four 8-m telescopes are used. The telescopes are arranged like at the VLTI. In (**a**), the monochromatic OTF of two 8-m telescopes with a separation of $B = 102$ m and a wavelength of $\lambda_0 = 1.1\,\mu\text{m}$ is displayed, resulting in a uv-plane baseline of $R_{0,B} = B/\lambda_0 = 450\,\text{arcsec}^{-1}$ and a diameter of OTF_a of $2D/\lambda_0 = 70\,\text{arcsec}^{-1}$. In (**b**), the same is shown for three telescopes separated by 102 m, 130 m and 62 m. Parts (**c**) and (**d**) of this figure are displayed on the following page. (Courtesy F. Millour [159])

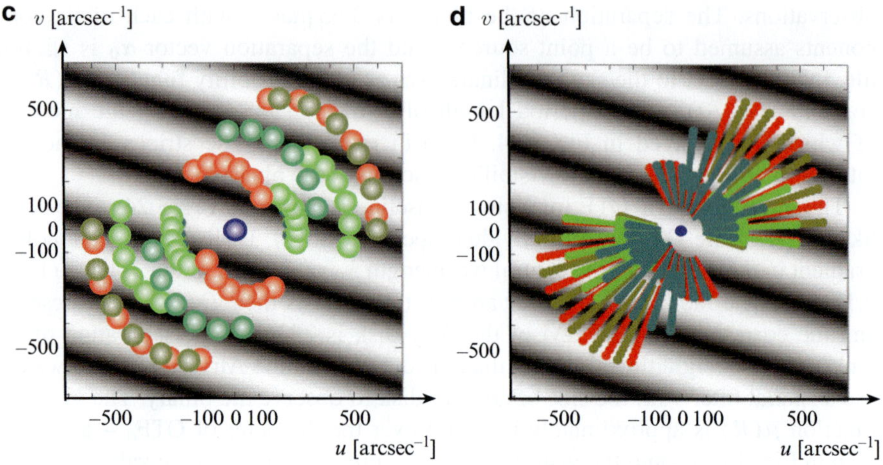

four telescopes, four nights

four telescopes, four nights, three spectral bands (J, H, K)

Fig. 3.24 Observations over four nights, each combining three out of four 8-m telescopes, in (**c**) with the same parameters as in Fig. 3.24a and b. Due to the rotation of the Earth the physical baseline rotates with respect to the uv-plane providing different effective baselines in the uv-plane. Measuring one visibility point per hour, a 6-h observation results in six distinct OTF indicated by distinct circles of equal colour. In (**d**), the measurements of three different spectral bands, J, H and K, are combined for the same mix of effective baselines and telescopes as in (**c**). Now, the OTF has elongated areas as displayed in Figs. 3.18 and 3.19. For clarity, the OTF in (**d**) are displayed by *narrow lines* instead of *circles* as in (**c**). The visibilities are determined with moderate spectral resolution so that many visibility values are measured along the elongated OTF. (Courtesy F. Millour [159])

The two-telescope observation of a binary in Fig. 3.24a shows the typical situation of a single baseline measurement. Here, the orientation of the separation vector $\boldsymbol{\alpha}_0$ with respect to the effective baseline vector $\boldsymbol{R}_{0,B}$ is such that the visibility is only slightly reduced since we are still close to the central maximum of the visibility function. If the two vectors were perfectly orthogonal the observation of the binary could not even be distinguished from that of a single star. And in case the baseline and separation vectors were parallel, we would measure a visibility beyond the second zero of the visibility, close to the second maximum. It is obviously impossible to determine unambiguously the separation and the orientation of the binary with a single measurement.

Combining three telescopes for a single observation as in Fig. 3.24b provides three baselines so that the two parameters of the binary can be determined. However, due to the periodicity of the visibility function one could still encounter situations when the unambiguous measurement of the binary parameters is not possible.

The uv-plane coverage in Fig. 3.24b represents the interferometer OTF, which is the autocorrelation function of the three apertures. Note that the central OTF_a (the blue dot) has a value that is three times larger than that of the six OTF_a (the red dots) of the three baselines, which all have the same value. A similar case, for three pinholes instead of three telescopes, was discussed in Sect. 3.3.1. Figure 3.13 shows the interferometer OTF and the subsequent PSF that is positive since the OTF is the autocorrelation of the three pinholes. Computing the PSF of the interferometer OTF in Fig. 3.24b also yields a positive function but still with too many sidelobes to reconstruct a good quality image of the binary and to determine its parameters. This is still better being done by fitting the parameters of the binary directly to the visibilities.

In Fig. 3.24c we regard the situation when observations last several hours and the coordinate plane in the sky rotates with respect to the baselines on the ground. Four 8-m telescopes are combined in different combinations of three with each triplet observing over 6 h. It is assumed that every hour one set of three visibility points can be determined. Observations over four nights are combined. The colour indices give an idea, which area in the uv-plane is covered during the observation. Here, the binary parameters can be determined unambiguously.

In Sect. 3.3.5, discussing the interferometric field of view and the content of information, we stated that it is required to have visibility measurements spaced by less than $1/\alpha_0$ to obtain a field of view larger than α_0 so that the binary separation of α_0 can be determined unambiguously. We stated also that it is less important that the shortest baseline vector is smaller than $1/\alpha_0$.

Figure 3.24c illustrates that the shortest baselines, creating a gap around $R = 0$, are not very important for determining the binary parameters since the uv-plane is densely filled at larger $\boldsymbol{R}$. However, the situation would be different if we were not observing a binary composed of point sources but of extended sources. Then the visibility function would fall off at larger $\boldsymbol{R}$ and the visibility measurements at shorter spatial frequencies become more important.

The uv-plane coverage in Fig. 3.24c represents the combined OTF that can no longer be expressed as the autocorrelation of another function, a meta-aperture (see Sect. 3.4.1). Therefore, the Fourier transform of the combined OTF, the PSF, is not necessarily positive. However, the visibility measurements filling the uv-plane rather well, the PSF is now reasonably point-like. We will come back to this point in Sect. 3.4.3.

The first three scenarios that are shown in Fig. 3.24a-c assumed narrow band observations in the J-band. Combining observations with three spectral bands, J, H[5] and K, covering a wavelength range between 0.95 and 2.4 μm, the elongation of each polychromatic OTF, (3.64), is enormous, displayed as long narrow lines in Fig. 3.24d. Comparing the length of these lines to the cosine of the visibility function $\mu(\boldsymbol{R})$, it is obvious that the cosine varies over the area covered by OTF_{pc}. Therefore, one cannot use a single visibility value for this polychromatic observation.

Usually, the light is spectrally dispersed before the beam combination, producing multiple narrow-band fringe patterns. The corresponding multiple narrow-band OTF, forming subdivisions of the elongated OTF_{pc}, cover narrow areas of the

[5] The H-band is the atmospheric window between 1.475 and 1.825 μm, see Appendix A.2.

visibility function each with approximately constant values. Thus, instead of a single wide-band visibility value there are many narrow-band fringe pattern and visibility values.

While the uv-plane coverage is defined by the shape of the combination of elongated OTF_{pc}, defining the quality of the reconstructed image, the subdivision in several narrow-band OTF increases the field of view from the width of the white-light fringe to the diameter of the Airy disk of an individual telescope as discussed in Sect. 3.3.6.

Discussing polychromatic observations we have assumed that the visibility function $\mu(\boldsymbol{R})$ is independent of wavelength, which in this case of a binary means that the separation is wavelength independent and that both stars are of the same spectral type so that the intensity ratio is also wavelength independent. For a more complex object, e.g. the surface of a star, this might not be true any more. We would then have to treat every spectral channel individually – the width of which is defined by the actual variation of object shape with wavelength – finally combining the reconstructed object features in a polychromatic image. Each spectral channel would have a narrow-band uv-plane coverage similar to Fig. 3.24c.

Thus, we determine a distribution of visibility points each centred at the corresponding coordinate $\boldsymbol{R}_{0,B_i}$ that can be represented by $\mu(\boldsymbol{R}_{0,B_i})\delta(\boldsymbol{R} - \boldsymbol{R}_{0,B_i})$. Before processing this distribution of visibility peaks to reconstruct the image, we can convolve each δ-peak with the optical transfer function of an individual aperture OTF_a, processing $\mu(\boldsymbol{R}_{0,B_i})OTF_a(\boldsymbol{R} - \boldsymbol{R}_{0,B_i})$. This is equivalent to multiplying the reconstructed image by the point-spread function PSF_a of an individual telescope.

This was done in the PSF reconstruction in Fig. 3.25. These observations were performed as a demonstration for the achievable interferometric point-spread function for a given uv-plane coverage. One can see that, while modelling the real

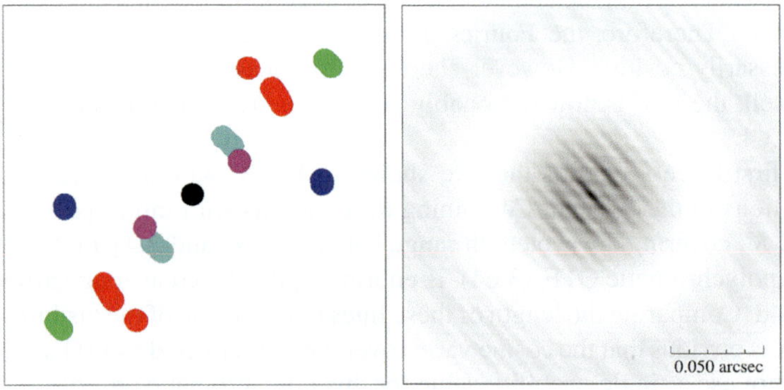

Fig. 3.25 Observing Achernar in the K-band with the VLTI, on September 15/16 and 16/17, 2002, pairwise combining four 8-m telescopes. The uv-plane coverage is shown on the left, and the reconstructed point spread function (PSF) on the right. The width of the central fringe, i.e. the angular resolution limit, is about 3 by 15 mas due to the distribution of baselines used during these observations. On the largest scale, the image is enveloped by the Airy disk of a single 8-m telescope, called the primary beam in radio interferometry. Its first minimum with a radius of 57 mas can be clearly seen

imaging process, the convolution with OTF_a is not strictly required if the observed object is well within the central maximum of the Airy disk of the 8-m aperture, i.e. smaller than about 20 mas in this case. Then it is sufficient to deal with the distribution of δ-peaks representing the visibility measurements.

3.4.3 Image Reconstruction: General Principles

Assuming that we have measured the complex visibilities for a number of baselines, for instance like in Figs. 3.24 or 3.25, how do we reconstruct an image?

The problem of *synthesis imaging*, or *aperture synthesis*, has been treated in radio interferometry during the last decades, and there is a huge body of literature (see e.g. [228, 232]). The straightforward process of Fourier transforming the measured visibility distribution, $\mu(\boldsymbol{R}_{0,B_i})\delta(\boldsymbol{R} - \boldsymbol{R}_{0,B_i})$, has its limitations in the achievable image quality. We have to keep in mind, that the distribution of baselines after several hours of observations (Fig. 3.24c) is distinctly different from the instantaneous distribution (Fig. 3.24b) for a snapshot observation. Filling the uv-plane by combining observations of several hours is called *Earth rotation synthesis*.

Depending on the number of baselines, their distribution and the weighting of the visibility at $|\boldsymbol{R}| = 0$, the reconstructed image is full of sidelobes and possibly negative values. In radio interferometry, this image is called the *dirty map*, and the reconstructed PSF, the Fourier transform of the interferometer OTF given by the distribution of $\delta(\boldsymbol{R} - \boldsymbol{R}_{0,B_i})$, is called the *dirty beam* [22].

Using the OTF of the individual apertures by processing the distribution $\mu(\boldsymbol{R}_{0,B_i}) \times \mathrm{OTF}_a(\boldsymbol{R} - \boldsymbol{R}_{0,B_i})$ instead of the δ-peaks does not affect the image quality as long as the object is much smaller than the PSF of an individual aperture – the primary beam – since the PSF only acts as an envelope for the reconstructed image.

Both the dirty map and the dirty beam are real functions since we set $\mu(-\boldsymbol{R}_{0,B_i}) = \mu^*(\boldsymbol{R}_{0,B_i})$. If the chosen distribution of baselines were such that the combined OTF could also be expressed by an autocorrelation of a meta-aperture (see Sect. 3.4.1), then the dirty map would be positive by definition [102]. Unfortunately, the combination of measured baselines does not usually correspond to the autocorrelation of a multi-aperture interferometer, so that positivity can only be ensured by the image reconstruction process.

Figure 3.26 shows an example of a dirty beam and a dirty map combining measurements of 90 visibility values. It is interesting to see that even this apparently large number of visibilities does not permit us to identify the rather simple object consisting of seven point sources.

The appearance of the sidelobes and negative values can be affected by applying **weighting functions**, for instance the computed OTF of a filled aperture with the diameter of the longest baseline $B_{\max}$. This OTF has a cone-like shape dropping to zero at $B_{\max}/\lambda_0$. In image space, this would be equivalent to convolving the dirty beam with the corresponding PSF, an Airy disk with a FWHM of $\lambda_0/B_{\max}$. Then, any information in the fringe pattern of the longest baseline – corresponding to the

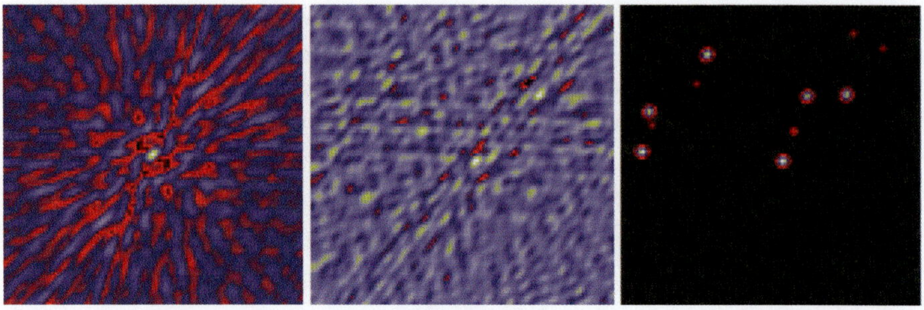

Fig. 3.26 Simulation of an image reconstruction with CLEAN, modelling observations with four 8-m telescopes measuring 18 baselines in the K-band. The K-band is subdivided into 5 spectral channels so that a total of 90 visibility values are processed. The dirty beam on the left and the dirty map in the middle still show a large number of sidelobes, and the reconstructed image on the right displays the enormous improvement. (Courtesy T. Paumard [179])

highest spatial frequency – would be eliminated by putting the visibility to zero. For instance a faint companion of a bright star at a distance of $\lambda/2B_{\max}$ that could be detected in the fringe pattern of the longest baseline would simply be obscured by the convolution with this PSF.

The shape of this OTF is determined by the imaging process with a filled aperture. Here, low frequencies, corresponding to short baselines, are over-emphasised since they appear many times in the telescope aperture, while the longest baselines measuring the highest spatial frequency and, thus, the finest detail, do not contribute much to the imaging process.

By contrast, interferometric observations with individual apertures, measuring the visibility with each baseline only once, are non-redundant (unless there are redundant baselines) so that all visibility measurements should have the same weight. Forcing down the visibilities at long baselines does not make proper use of these measurements.

On the other hand, applying a much wider weighting function does not do much to improve the apparent image quality of the dirty map. The choice of weighting function depends entirely on the imaging task, and more sophisticated functions are sometimes useful [22], applying for instance a weighting function consisting of the reciprocal of the variance of each measured visibility.

However, none of these linear methods of applying weighting functions to the measured visibilities can fundamentally improve the dirty map since the sparse distribution of the visibilities in the uv-plane is not affected. Other, **non-linear methods** are required, extrapolating the measured visibilities in the unmeasured regions of the uv-plane reducing sidelobes and artefacts [169]. If the resulting distribution of visibilities then corresponds to the autocorrelation of a meta-aperture, the reconstructed image intensity would be positive by definition. Thus, the reduction of sidelobes and artefacts, improving the image quality, is linked to forcing positivity without, however, reaching either goal perfectly.

A novel approach was presented in 1974 by J. A. Högbom [102]. He introduced the image reconstruction method CLEAN that has been enormously successful [42]. CLEAN basically performs a deconvolution in image space extracting the dirty beam from the dirty map.

Another method, reconstructing the image by fitting its visibility function to the measured visibilities in the uv-plane is the *maximum entropy method*, MEM, described by J.G. Ables [1] in 1974, based on mathematical research by E.T. Jaynes [108].

Both methods will be discussed in the following. A more detailed discussion can be found in the proceedings of the NRAO summer school [228] and, there, in particular in the contribution by T. Cornwell [42].

CLEAN

CLEAN is an iterative process, starting with the dirty map. Convolving the dirty map with an appropriately chosen PSF – for instance an Airy disk or a Gaussian and the corresponding OTF as weighting function as discussed above – can be advantageous to smooth the noise in the intensity distribution. The dirty beam is then convolved with the same PSF. Due to their appearance we still call the convolution products dirty map and dirty beam.

Based on the assumption that the brightest intensity value in the dirty map is due to an object feature, the first step consists in subtracting the dirty beam from the dirty map, centring the dirty beam at the position of the brightest intensity. The value and the coordinates of this brightest point are stored.

The second step repeats the first step but now subtracting the dirty beam from the maximum of the dirty map remaining after step one. One will find another maximum and subtract again the properly positioned dirty beam from the dirty map, store its value and coordinates, and so on. The iteration ends when the remaining dirty map seems to contain only noise and no real object features.

The image is then reconstructed by adding all previously determined PSF – with proper values and coordinates – to the dirty map remaining after the last iteration, presumably only containing noise. This process ensures that the reconstructed image is positive, apart from remaining negative bits in the dirty map indicating the noise level of the reconstruction. Recipes for the numerical implementation of CLEAN can be found in [31] and [43].

The reconstructed image in Fig. 3.26 displays an impressive quality, free of side-lobes compared to the dirty map. This corresponds to a much denser uv-plane coverage that one obtains by Fourier transforming the reconstructed image. Thus, applying CLEAN in the image plane implies that visibility values are extrapolated in unmeasured parts of the uv-plane. This is the desired result but without being able to establish criteria on how to extrapolate the data in the uv-plane and, in particular, on how to consider measurement noise CLEAN does not make the best use of the data.

Without a clear rule when to stop the iteration in the presence of noise there is an element of arbitrariness in it since the treatment of noise depends on the experience of the user, on the complexity of the object and on the weighting function smoothing the intensity distribution. Continuing the iteration for too long, i.e. picking maxima in the dirty map that are effectively noise, eventually produces fake object detail [42, 102]. The inherent problem is that by working with the dirty beam in the image plane the signal quality of the individual visibility measurements in the uv-plane cannot be taken into account.

Maximum Entropy Method, MEM

In contrast to the procedure described by CLEAN, without a connection between the visibility data and the reconstruction process, MEM directly fits the data providing a link between the data quality and the reconstructed image [42, 169].

MEM is based on the mathematical principle that amongst the infinite number of possible images all agreeing with the visibility measurements, one should choose that image that contains the least amount of information (hence, the maximum entropy), and therefore also the least amount of false information. Here, *information* is defined in analogy to entropy in thermodynamics.

While this seems to be a fundamental physical principle it has created a lot of confusion on how exactly to define entropy in this context and there are many suggestions [169]. The most commonly used definition for this information entropy is given [42] as

$$H = -\sum_k I_k \ln(I_k/O_k), \tag{3.86}$$

with I_k the reconstructed image intensity at pixel k and O_k the image prior, for instance a low resolution image providing a starting point for the reconstruction. This is very similar to the quantity to be maximised using Bayesian statistics [169].

The maximization of the entropy is constrained by the measured visibilities. In the case of noisy measurements this constraint can be expressed by attempting to set the χ^2 sum equal to the number of degrees of freedom Ω, which is the number of independent data [1, 169]. It is

$$\chi^2 = \sum_i \frac{|\mu_{\text{rec}}(\boldsymbol{R}_{0,B_i}) - \mu(\boldsymbol{R}_{0,B_i})|^2}{\sigma_\mu^2(\boldsymbol{R}_{0,B_i})} \approx \Omega\,, \tag{3.87}$$

with μ the measured visibilities, μ_{rec} the reconstructed visibilities, and σ_μ^2 the variance of the visibility measurement at baseline $\boldsymbol{R}_{0,B_i}$.

We then have to maximize a linear combination of the entropy and of the constraints,

$$H' = H - \Lambda(\chi^2 - \Omega)\,, \tag{3.88}$$

with Λ the Lagrange multiplier that, in practice, has to be adjusted so that the residuals have the expected level [169].

Setting the reconstructed visibilities exactly to the measured values is almost always incompatible with the positivity constraint since the reconstructed image would then be identical to the dirty map. Enforcing positivity by maximizing H' can provide artefacts in the image that are related to the fact that the entropy is insensitive to spatial information [42] so that the extrapolation of visibility points in the uv-plane is somewhat random. One should note that the extrapolation of visibility points is not restricted to the region within the longest baseline but it can also go beyond, i.e. beyond the diffraction limited. This super-resolution is not undesirable but it has to be treated carefully.

There have been numerous attempts to improve this method using the entropy [45, 216] or Bayes Theorem [185].

In the optical domain, problems of image reconstruction i.e. blind deconvolution were first treated in speckle interferometry (see Sect. 4.4) with single telescopes [50, 126]. A generalized approach, applicable to interferometry, was presented by E. Thiébaut [230, 231]. He regards the entropy H only as one of the quantities to be optimized, adding other positivity and field size constraints that are linearly combined as in (3.88).

Positivity could also be enforced by searching for a meta-aperture so that the image is positive by definition. However, first attempts treating one-dimensional problems [119] could not be extended to two-dimensional images [169].

Starting in 2004, the optical interferometric community arranged *Imaging Beauty Contests* at the biannual SPIE conferences [46, 131, 132]. The contestants used the methods mentioned above with the maximum entropy method BSMEM [25] and with the MIRA method [230] displaying the best performance over the years. However, since this field is still developing [136] this only presents a snapshot of the current state of the art.

Last but not least, the simple method of fitting model parameters – for instance the orientation and relative brightness of a binary star – to the measured visibilities has been the bread-and-butter of optical interferometrists for many years. Regarding this as an image reconstruction with extremely limiting constraints, it provides the desired parameters with a much higher accuracy than a less constrained general image reconstruction can do. Running against the trend to produce pretty pictures, *model fitting* provides valuable scientific insight that should not be dismissed.

The uv-Plane: Summary

Discussing astronomical observations, the uv-plane is defined with respect to the celestial coordinate system. Using the equatorial reference frame with the declination δ for the angle of latitude above the celestial equator, and the right ascension RA, for the longitude, we define $\boldsymbol{\alpha}' = (\delta, RA)$ as the coordinate vector of the sky plane that is tangential to the celestial sphere and perpendicular to the *line of sight* pointing at the object. Processing interferometric data, relative coordinates can be used, with the line of sight defining the zero position.

The uv-plane is defined with respect to these sky coordinates and is, thus, parallel to the sky plane, independent of the rotation of the Earth. We have to replace the physical baseline $\boldsymbol{B}$ on the ground by the *effective baseline* $\boldsymbol{B}_{\mathrm{eff}}$, which is the projection of the baseline vector onto the uv-plane, perpendicular to the line of sight. It is $|\boldsymbol{B}_{\mathrm{eff}}| = |\boldsymbol{B}|\cos(\gamma)$, with γ the angle between the baseline vector and the coordinate vector $\boldsymbol{\alpha}'$. Depending on the object position, the effective baseline and therefore the OTF rotate during the observation.

The quality of the reconstructed image depends on the density of the data points, the measured visibilities, in the uv-plane. Depending on the number of telescopes, their apertures and their baselines, different strategies can be used to fill the uv-plane.

A rather short baseline combined with large apertures – like on the LBT – permits us to fill the uv-plane smoothly with as little as three observations (see Fig. 3.23). The weighting of the individual measurements, reducing the overemphasis of the central OTF, determines the appearance of the reconstructed image. The weighting can be used as a simple measure to control the positivity in the reconstructed image.

The case of an interferometer composed of individual telescopes on the ground is the most common situation in interferometry. Here, the effective baselines change in length and orientation as a function of the position of the celestial object and of the duration of the observation. Combining several nights of observations with a small number of telescopes – with baselines much longer than the telescope diameter – can provide a reasonable uv-plane coverage (see Fig. 3.24).

Synthesizing images of these measurements largely relies on methods that have been developed in radio interferometry (see e.g. [228, 232]). Simply Fourier transforming the measured distribution of visibilities in the uv-plane usually provides an image full of sidelobes and negative values, called the dirty map in radio interferometry [22].

Trying to improve the dirty map by applying weighting functions in the uv-plane is only of limited effect since the fundamental problem of a patchy distribution of visibilities in the uv-plane is not addressed [22].

Two other, non-linear, methods permit us to extrapolate the visibilities into regions of the uv-plane without measurements, even beyond the longest baselines providing superresolution beyond the diffraction limit [169]. The first approach is called CLEAN performing basically a deconvolution of the dirty map and the related PSF in the image plane [102]. This method has been enormously successful but has its pitfalls when treating noise [42].

The second approach is called the maximum entropy method, MEM, using the fundamental physical principle that amongst the infinite possible reconstructions that with the least amount of information is the most probable one

[1]. Fitting the reconstructed visibilities to the measured visibilities weighted by the measurement variance, the entropy can be maximized providing a positivity constraint through its definition [169].

Here the problems arise from the lack of spatial sensitivity of the image entropy so that artefacts can spoil the reconstructed image [42]. Adding more constraints to this process, eventually can make the entropy obsolete [230]. Combining MEM and image reconstruction methods that were developed for speckle interferometry seems to be a promising path for aperture synthesis in optical interferometry [231].

Having only few baselines it is often impossible to reconstruct an image of useful quality. One can then resort to fitting model parameters to the measured visibilities. Stellar diameters, when the star is modelled as a circular disk and the diameter is the only unknown parameter, or binary stars, with parameters relative brightness and separation vector, are the simplest examples for model fitting. Although not as spectacular as real images, model fitting can provide valuable scientific insight at an angular resolution that cannot be reached by single telescopes.

Chapter 4
Atmospheric Turbulence

The imaging process that we have described in Chap. 3 suffers from atmospheric turbulence perturbing the incoming wave front by random index of refraction fluctuations. This effect is caused by turbulent mixing of air with different temperature so that the image quality in large telescopes is far from the diffraction limit but corresponds to telescopes with a few 10 cm diameter. Operating interferometers on the ground, the fringe pattern is wiped out unless the integration times are limited to a about 100 ms.

In this chapter, we will discuss the statistics of atmospheric turbulence as derived by Kolmogorov [117, 118]. Following from the index of refraction fluctuations, we will investigate the statistical properties of the electromagnetic wave, assuming thin horizontal turbulent layers. Basic temporal and spatial parameters, like the Fried parameter, are given in the second section.

Observing through turbulence – with a single telescope or with an interferometer – the imaging process is determined by the statistics of the turbulence inside the apertures. The random process is affected by the averaging over the apertures when proceeding into the image plane, and we will derive the statistical quantities as a function of turbulence and instrumental parameters in the third section.

We will provide the statistics for OPD fluctuations and for differential OPD fluctuations, affecting differential fringe motion when observing two separate object, for the wave front tilt, and for the perturbed wave front, using Zernike polynomials for the description of Kolmogorov turbulence. In Chap. 6, the derived quantities will be used to investigate the limitations of observing through turbulence.

Finally, the twinkling of the stars, the scintillation will be investigated with the same methods, and speckle interferometry, as one of the early attempts to beat atmospheric turbulence, will be discussed in the last section.

A. Glindemann, *Principles of Stellar Interferometry*, Astronomy and Astrophysics Library, DOI 10.1007/978-3-642-15028-9_4,

4.1 Kolmogorov Turbulence

4.1.1 First Principles

The statistics of the spatial and temporal structure of atmospheric turbulence determines the propagation of light through the atmosphere. Following from the theory of fluid motion, the flow of air becomes turbulent, i.e. unstable and random, if the *Reynolds number* $Re = L_0 v_0 / k_v$ exceeds a critical value. L_0 is the characteristic size of the flow, v_0 is the characteristic velocity and k_v is the kinematic viscosity. With typical values for these parameters, $L_0 \approx 22\,\mathrm{m}$, $v_0 \approx 10\,\mathrm{m\,s^{-1}}$ and $k_v = 15 \times 10^{-6}\,\mathrm{m^2\,s^{-1}}$, it is $Re \approx 10^7$ which corresponds to fully developed turbulence.

A.N. Kolmogorov assumed that the kinetic energy in the largest structures of the turbulence is transferred successively to smaller and smaller structures [117, 118, 225]. He also presumed that the motion of the turbulent structure is both statistically stationary and isotropic, implying that the second and higher order statistical moments of the turbulence depend only on the distance between any two points in the structure.

If the product of the characteristic size L_0 of the small structure and its velocity v_0 is too small to keep the Reynolds number in the turbulent regime, the break up process stops and the kinetic energy is dissipated as heat by viscous friction.

In a stationary state, the energy flow from larger structures L to smaller structures l must be constant, i.e. the amount of energy that is being injected into the largest structure must be equal to the energy that is dissipated as heat. It is $E(l)\mathrm{d}l$ the kinetic energy in a volume element with size $\mathrm{d}l$ between l and $l + \mathrm{d}l$. If the typical transfer time of $E(l)\mathrm{d}l$ through a structure of size l is given by l/v, then the energy flow rate, ϵ_0, can be written as

$$\epsilon_0 = \frac{E(L)\mathrm{d}L}{t(L)} = \frac{E(l)\mathrm{d}l}{t(l)} = \frac{\frac{1}{2}\rho v^2}{l/v} = \text{const}\,, \tag{4.1}$$

and it is

$$v \propto \epsilon_0^{1/3} l^{1/3}\,. \tag{4.2}$$

The kinetic energy $\hat{E}(k)\mathrm{d}k$ in the interval k and $k + \mathrm{d}k$ of spatial wave numbers is proportional to v^2. With the spatial wave number $k \propto 2\pi/l$ one obtains

$$\hat{E}(k)\mathrm{d}k \propto k^{-2/3} \text{ or } \hat{E}(k) \propto k^{-5/3}\,. \tag{4.3}$$

For isotropic turbulence, the kinetic energy spectrum $\hat{E}_p(\boldsymbol{k})$ can be computed by integrating over the unit sphere:

$$\hat{E}(k) = 4\pi k^2 \hat{E}_p(\boldsymbol{k}) \Rightarrow \hat{E}_p(\boldsymbol{k}) \propto k^{-11/3}, \tag{4.4}$$

with $\boldsymbol{k}$ the three-dimensional spatial wave vector, $k = |\boldsymbol{k}|$ and $k \propto 2\pi/l$.

This relationship expresses the *Kolmogorov spectrum.* It holds in the *inertial range* of turbulence for $l_0 \ll 2\pi/k \ll L_0$ where L_0 is the outer scale of turbulence, generally the size of the largest structure that moves with homogeneous speed, and l_0 is the inner scale at which the viscous dissipation starts. The outer scale of turbulence varies between a few metres close to the ground where the largest structure is determined by the height over the ground, and a few hundred metres in the free atmosphere, which is the thickness of the turbulent layer [33, 98, 226]. The inner scale of turbulence is in the range of a few millimetres near the ground to about 1 cm near the tropopause [194].

4.1.2 Index of Refraction Fluctuations

Light travelling through the atmosphere is affected by fluctuations of the index of refraction (or refractive index). The physical source of these fluctuations are temperature inhomogeneities produced by turbulent mixing of air. The refractive index of dry air as a function of wavelength, pressure and temperature is in good approximation [18]

$$n(\boldsymbol{r},\lambda) = 1 + \left(77.49 + \frac{0.435}{(\lambda/\mu\text{m})^2}\right)\frac{P/\text{mbar}}{T/\text{K}}10^{-6}, \tag{4.5}$$

with λ in μm, P the pressure in millibar and T the temperature in Kelvin.

For the refractive index of moist air, the following corrective term has to be applied:

$$n_f(\boldsymbol{r},\lambda) - n(\boldsymbol{r},\lambda) = -\frac{f}{\text{mbar}}\left(3.73 - \frac{0.040}{(\lambda/\mu\text{m})^2}\right)10^{-8}, \tag{4.6}$$

with f the partial pressure of water vapor in mbar.

These empirical formulae are based on the *Edlén formula* from 1966 [66]. Over the years, they have been improved by Owens [175], by Birch and Downs [18] and more recently by Ciddor [29, 30]. The modifications improved the accuracy of the formulae with respect to measurements to below 10^{-7}. However, these authors mostly concentrated on the visible and the near-infrared.

In the mid-infrared, there are only few experimental results and the extrapolation of these expressions beyond 2 μm is accordingly uncertain. An analytical approach computing the effect of absorption bands by accumulating oscillator strengths was presented by Mathar [152]. His results fit the available experimental data up to 20 μm and confirm that in the mid-infrared refractive index fluctuations depend heavily on variations of water vapor.

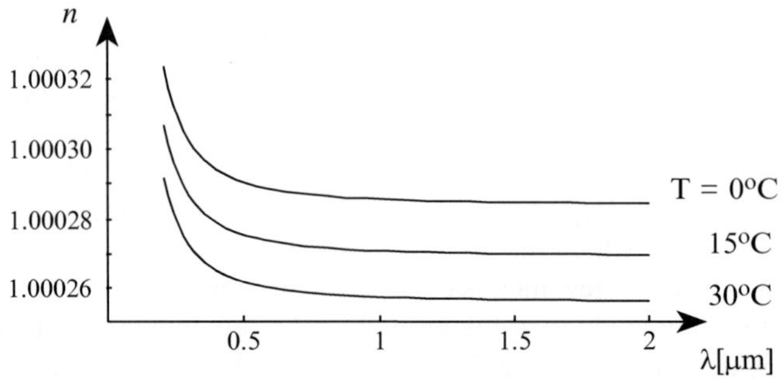

Fig. 4.1 The refractive index of dry air at 0, 15 and 30°C and 1,000 mbar as given by the Edlén formula (4.5). The dependence on temperature is modelled by approximating the refraction as a sum of temperature and wavelength dependent terms. This figure illustrates that temperature variations affect all wavelengths in the same way

The refractive index at different temperatures is displayed in Fig. 4.1. $n(\boldsymbol{r}, \lambda)$ is a function of the three-dimensional position vector $\boldsymbol{r}$ because of the temperature and pressure inhomogeneities in the atmosphere. The variation of the refractive index with wavelength, $n(\lambda)$, is called *dispersion*. It is mainly determined by the second term, $0.435/\lambda^2$, in (4.5). The dispersion is much smaller than the variation of the refractive index due to pressure and temperature changes under typical conditions in the atmosphere that is mainly governed by the first term, 77.49, in (4.5). Thus, the three curves in Fig. 4.1 have approximately identical shape but different absolute values.

This approximation has the consequence that the shape of the wave front in a turbulent layer depends mainly on the local temperature and pressure, and not on the wavelength. Effects that show a wavelength dependence in the image plane of a telescope, like the different number of speckles in the optical and in the infrared, are caused by the different relative impact of the same wave front distortion at different wavelengths.

However, the dispersion, the wavelength dependence of the index of refraction, does have an influence on the imaging process. The apparent angular position of a star depends on the refractive index due to the refraction at the Earth's atmosphere that to a first approximation can be regarded as a thick coplanar plate. Thus, the image of a star is slightly elongated, coloured like a rainbow, unless the star is at zenith. This effect is called *transversal dispersion*.

In interferometers, the position of the white-light fringe that is related to an optical path difference (OPD) of zero varies with wavelength because the optical paths through the two telescopes do not have equal shares of travel in vacuum and in the atmosphere. Unless the star is at zenith, the optical path difference eventually is a difference between an optical path in vacuum and an optical path in the atmosphere. Due to the dispersion, the OPD then varies with wavelength. This *longitudinal dispersion* will be discussed in Sect. 6.2.1. The size of these effects does not depend on the actual turbulence but only on the invariable dispersion.

Kolmogorov and von Kármán Spectrum

It can be shown that the refractive index as a passive, conservative additive, i.e. a quantity that does not affect atmospheric turbulence and that is not affected by the motion of the air, also follows Kolmogorov statistics [172]. Then, the power spectrum $\Phi_n(\boldsymbol{k})$ of $n(\boldsymbol{r})$ that is called the Kolmogorov spectrum has the same spatial frequency dependency as the kinetic energy (4.4) and can be expressed as

$$\Phi_n(\boldsymbol{k}) = 0.033\, C_n^2\, k^{-11/3} \;\; [1/\mathrm{m}^{-3}], \tag{4.7}$$

with $\boldsymbol{k}$ the three-dimensional spatial wave vector, $k = |\boldsymbol{k}|$ and $k \propto 2\pi/l$. The dimension of $\Phi_n(\boldsymbol{k})$, the power spectrum or, more precisely, the power spectral density, is power per volume element $\mathrm{d}\boldsymbol{k}$, i.e. per m^{-3}. Following common practice, we use the term power spectrum keeping in mind that it is always the power per spatial (or temporal) frequency that we refer to.

The quantity C_n^2 is called the *structure constant* of the refractive index fluctuations and has units of $\mathrm{m}^{-2/3}$ so that $\Phi_n(\boldsymbol{k})$ has units of $1/\mathrm{m}^{-3}$. C_n^2 characterises the strength of the fluctuations of n. Measurements of C_n^2 have shown a good agreement with the Kolmogorov theory within the inertial range [32, 106], and with the von Kármán spectrum [235].

Like the statistical distribution of velocity discussed in Sect. 4.1.1, the refractive index distribution is isotropic and homogeneous as long as the spatial frequencies involved are in the inertial range, with $l_0 \ll 2\pi/k \ll L_0$. The Kolmogorov theory predicts a mathematical form for $\Phi_n(\boldsymbol{k})$ only inside the inertial range.

An extension beyond this regime is given by the *von Kármán spectrum* [107], reading as

$$\Phi_{n,vK}(\boldsymbol{k}) = 0.033\, C_n^2 \left(k_{L_0}^2 + k^2\right)^{-11/6} \mathrm{e}^{-k^2/k_{l_0}^2} \;\; [1/\mathrm{m}^{-3}], \tag{4.8}$$

with $k_{l_0} = 5.92/l_0$ and $k_{L_0} = 2\pi/L_0$. Compared to the Kolmogorov spectrum, the power is reduced outside the inertial range (see Fig. 4.2), which is a reasonable assumption. For practical purposes the outer scale is of larger interest than the inner scale since it reduces the pole at $k = 0$ to finite values.

So far, only the power spectrum $\Phi_n(\boldsymbol{k})$ of the refractive index fluctuations has been discussed. The power spectrum of $n(\boldsymbol{r})$ is related to its (dimensionless) autocorrelation $\Gamma_n(\boldsymbol{r}) =< n(\boldsymbol{r}')n(\boldsymbol{r}' + \boldsymbol{r}) >$ by the Wiener–Khinchine theorem (see also Sect. 3.1.1):

$$\Gamma_n(\boldsymbol{r}) = \int \Phi_n(\boldsymbol{k})\mathrm{e}^{-\mathrm{i}\boldsymbol{k}\cdot\boldsymbol{r}}\mathrm{d}\boldsymbol{k}\,. \tag{4.9}$$

The consequence of the form of the power spectrum with its pole at $k = 0$ is that the variance $\Gamma_n(0) =< n^2(\boldsymbol{r}) >$ of the refractive index fluctuations is infinite. However, if the von Kármán spectrum is used there is a large but finite value for the variance of $n(\boldsymbol{r})$.

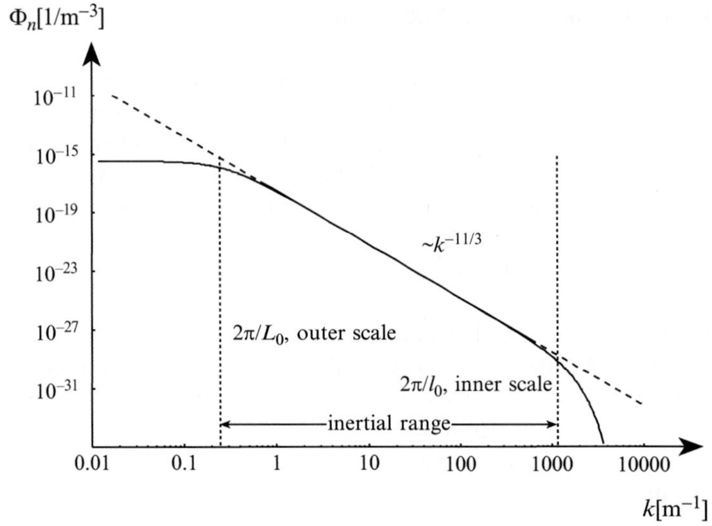

Fig. 4.2 The Kolmogorov spectrum (*dashed line*), valid inside the inertial range of turbulence, and the von Kármán spectrum (*solid line*) for an outer scale of turbulence of $L_0 = 22\,\text{m}$ and an inner scale of $l_0 = 5\,\text{mm}$, and for $C_n^2 = 10^{-16}\,\text{m}^{-2/3}$, a typical value for atmospheric turbulence

Structure Function

Introducing the structure function of the refractive index, which is the mean-square difference of the refractive index at two points separated by $\boldsymbol{r}$ and, thus, dimensionless, the discussion of the inertial range can be avoided, yielding

$$\begin{aligned} D_n(\boldsymbol{r}) &= < |n(\boldsymbol{r}') - n(\boldsymbol{r}' - \boldsymbol{r})|^2 > \\ &= 2(< n(\boldsymbol{r}')^2 > - < n(\boldsymbol{r}')n(\boldsymbol{r}' - \boldsymbol{r}) >) \\ &= 2(\Gamma_n(0) - \Gamma_n(\boldsymbol{r}))\,. \end{aligned} \tag{4.10}$$

While this looks like yet another statistical function we will see in the next section how the structure function helps computing the correlation function of a light wave travelling through atmospheric turbulence.

The structure function for Kolmogorov turbulence was derived by Obukhov [172]:

$$D_n(\boldsymbol{r}) = C_n^2 r^{2/3}\,. \tag{4.11}$$

Like the power spectrum (4.7), the structure function only depends on $r = |\boldsymbol{r}|$ because the turbulence is isotropic.

This form of the structure function of the refractive index is known as *Obukhov's law*. Together with the Kolmogorov spectrum (4.7) it completely characterizes atmospheric turbulence.

Kolmogorov Turbulence: Summary

Refractive index fluctuations in the atmosphere that are caused by natural temperature and pressure fluctuations can be described in good approximation by Kolmogorov statistics. The power spectrum – more precisely the power spectral density – $\Phi_n(\boldsymbol{k})$ of the fluctuations of the refractive index $n(\boldsymbol{r})$ is called the Kolmogorov spectrum [172], given by

$$\Phi_n(\boldsymbol{k}) = 0.033\, C_n^2\, k^{-11/3} \ [1/\mathrm{m}^{-3}], \tag{4.7}$$

with $\boldsymbol{k}$ the three-dimensional spatial wave vector, $k = |\boldsymbol{k}|$ and $k \propto 2\pi/l$. The dimension of $\Phi_n(\boldsymbol{k})$ is power per volume element $\mathrm{d}\boldsymbol{k}$, i.e. per m^{-3}.

The quantity C_n^2 is called the *structure constant* of the refractive index fluctuations and has units of $\mathrm{m}^{-2/3}$. It characterises the strength of the fluctuations of n.

The Kolmogorov spectrum only holds within the inertial range of turbulence with $2\pi/k$, i.e. the distances in the turbulence, smaller than the outer scale of turbulence L_0 and larger than the inner scale of turbulence l_0.

An extension beyond this regime is given by the *von Kármán spectrum* [107] reading as

$$\Phi_{n,vK}(\boldsymbol{k}) = 0.033\, C_n^2 \left(k_{L_0}^2 + k^2\right)^{-11/6} \mathrm{e}^{-k^2/k_{l_0}^2} \ [1/\mathrm{m}^{-3}], \tag{4.8}$$

with $k_{l_0} = 5.92/l_0$ and $k_{L_0} = 2\pi/L_0$. The outer scale of turbulence is of larger practical interest than the inner scale since it reduces the pole of the power spectrum at $k = 0$ to finite values.

The structure function $D_n(\boldsymbol{r})$ of the refractive index variations is the second important function characterizing Kolmogorov turbulence. It is defined by

$$\begin{aligned} D_n(\boldsymbol{r}) &= < |n(\boldsymbol{r}') - n(\boldsymbol{r}' - \boldsymbol{r})|^2 > \\ &= 2(< n(\boldsymbol{r}')^2 > - < n(\boldsymbol{r}')n(\boldsymbol{r}' - \boldsymbol{r}) >)\,. \end{aligned} \tag{4.10}$$

For Kolmogorov turbulence its functional form was derived by Obukhov [172], yielding

$$D_n(\boldsymbol{r}) = C_n^2 r^{2/3}, \tag{4.11}$$

with $r = |\boldsymbol{r}|$. The Kolmogorov spectrum and its structure function form the basis for the description of wave propagation through turbulence.

4.2 Statistical Properties of the Perturbed Complex Wave

For the sake of simplicity, only horizontal monochromatic plane waves are considered, propagating downwards through atmospheric turbulence. Light from an unresolved, i.e. point-like star at zenith approximately fulfils this assumption. The fluctuations of the amplitude are calculated assuming that the turbulence follows Kolmogorov statistics.

4.2.1 Thin Layer Turbulence Model

Using the thin screen approximation [194], the layer thickness is assumed to be large compared to the correlation scale of the fluctuations but small enough to neglect diffraction effects within the layer. Also, the layer is non-absorbing and its statistical properties depend only on its altitude h, i.e. the structure constant C_n^2 does not vary in the horizontal direction.

After propagation through a thin turbulent layer at altitude h, the phase is related to the distribution of the refractive index through

$$\varphi_h(\boldsymbol{\xi}) = \frac{2\pi}{\lambda} \int_h^{h+\delta h} n(\boldsymbol{\xi}, z)\mathrm{d}z\,, \tag{4.12}$$

where δh is the thickness of the layer and $\boldsymbol{\xi} = (\xi, \zeta)$ denotes the horizontal coordinate vector. This definition of the phase ensures that it has the same statistical properties as the refractive index n.

The amplitude of a plane wave immediately after propagation through a layer at altitude h can be written as

$$V_{\varphi,h}(\boldsymbol{\xi}) = V(\boldsymbol{\xi})\mathrm{e}^{\mathrm{i}\varphi_h(\boldsymbol{\xi})}\,, \tag{4.13}$$

with $V(\boldsymbol{\xi}) = V_0 \exp(\mathrm{i}kz)$ the amplitude of the incoming plane wave from a point-like star at zenith. Discussing horizontal layers, we set $V(\boldsymbol{\xi}) = V_0 = 1$ in the following. $V_{\varphi,h}(\boldsymbol{\xi})$ depends on ν through $\varphi_h(\boldsymbol{\xi})$, which is a function of wavelength, (4.12). To describe the statistical properties of the complex wave we need the correlation function of the amplitude $V_{\varphi,h}(\boldsymbol{\xi})$.

In Sect. 2.3, we calculated the mutual coherence function (MCF) as the time average of optical disturbances. There, the optical disturbances were random variables due to the random emission process of thermal light sources. Here, propagating through atmospheric turbulence the amplitude has a random phase $\varphi_h(\boldsymbol{\xi})$ varying with time.

We assume that averages over one turbulent layer (as one possible realization of the random process) are the same as temporal averages at a given point waiting a "long" time. Thus, the process is regarded as ergodic, like the emission process of thermal light in Sect. 2.3. The definition of a "long" period is related to the typical time span of fluctuations of the turbulence, which is of the order of a few milliseconds. This means that – unlike for the fluctuations of the light emitted by an incoherent source – detectors are fast enough to take snapshots of atmospheric turbulence. We will come back to this in Sect. 4.4.

Thus, we calculate the correlation function $\Gamma_{\varphi,h}(\boldsymbol{\xi})$ for atmospheric turbulence, equivalent to the MCF, as time average:

$$\begin{aligned}\Gamma_{\varphi,h}(\boldsymbol{\xi}) &= < V_{\varphi,h}(\boldsymbol{\xi}')V^*_{\varphi,h}(\boldsymbol{\xi}'-\boldsymbol{\xi}) > \\ &= \left\langle \mathrm{e}^{\mathrm{i}[\varphi_h(\boldsymbol{\xi}')-\varphi_h(\boldsymbol{\xi}'-\boldsymbol{\xi})]} \right\rangle . \end{aligned} \tag{4.14}$$

In principle, the correlation of the amplitudes has to be computed not only at two different points but also at two different moments in time, like the MCF. However, for all practical purposes in the context of imaging through turbulence, the time difference is much smaller than the typical time span of the fluctuations. We assume the two amplitudes to be taken at the same moment, with zero time difference,[1] so that $\Gamma_{\varphi,h}(\boldsymbol{\xi})$ is only a function of coordinate difference $\boldsymbol{\xi}$, like the visibility function, which is the MCF for time difference $\tau = 0$, see (2.51).

Discussing coherence functions in Sect. 2.3, the optical disturbance was the random variable with a Gaussian random process. Here, in the context of atmospheric turbulence, the refractive index and, subsequently, the phase $\varphi_h(\boldsymbol{\xi})$ are the random variables. Following the central-limit theorem [176] it is reasonable to assume Gaussian statistics with zero mean for the phase. The average of the amplitude $\exp(\mathrm{i}\varphi_h(\boldsymbol{\xi}))$ can then be calculated using the *characteristic function* that – for an arbitrary random variable g – reads as

$$< e^{\mathrm{i}g} > = \int \mathrm{e}^{\mathrm{i}g}\, p(g) \mathrm{d}g = \mathrm{e}^{-\frac{1}{2}<g^2>} , \tag{4.15}$$

where $p(g)$ denotes the Gaussian probability density function of the random variable g. Setting $g = \varphi_h(\boldsymbol{\xi})$ we find that the average amplitude is $< \exp(\mathrm{i}\varphi_h(\boldsymbol{\xi})) >= \exp(-\frac{1}{2} < \varphi_h^2(\boldsymbol{\xi}) >)$. The statistical properties of the phase being the same as those of the refractive index (4.12), the variance of the phase $< \varphi_h^2(\boldsymbol{\xi}) >$ is also infinite, and the averaged amplitude is zero.

Regarding a turbulent layer with a finite outer scale of turbulence L_0 and, thus, with finite variance, the averaged amplitude has values slightly larger than zero. This is in contrast to the case of the random amplitude emitted by a thermal source when the amplitude itself follows a Gaussian distribution and its average is always zero.

Returning to Kolmogorov turbulence, the difference $\varphi_h(\boldsymbol{\xi}') - \varphi_h(\boldsymbol{\xi}' - \boldsymbol{\xi})$ also follows a Gaussian distribution with zero mean and we can calculate the correlation function in (4.14) using the characteristic function (4.15) with $g = \varphi_h(\boldsymbol{\xi}') - \varphi_h(\boldsymbol{\xi}' - \boldsymbol{\xi})$. $\Gamma_{\varphi,h}(\boldsymbol{\xi}, \nu)$ can now be written as

$$\Gamma_{\varphi,h}(\boldsymbol{\xi}) = \left\langle \mathrm{e}^{\mathrm{i}[\varphi_h(\boldsymbol{\xi}')-\varphi_h(\boldsymbol{\xi}'-\boldsymbol{\xi})]} \right\rangle = \mathrm{e}^{-\frac{1}{2}<[\varphi_h(\boldsymbol{\xi}')-\varphi_h(\boldsymbol{\xi}'-\boldsymbol{\xi})]^2>} . \tag{4.16}$$

[1] Note that, despite the time difference between the amplitudes being zero, the temporal average is taken over a long time.

In consequence of the Gaussian distribution of the phase difference having zero mean, $\Gamma_{\varphi,h}(\boldsymbol{\xi})$ is a real function.

Phase Structure Function

With the phase structure function $D_{\varphi,h}(\boldsymbol{\xi}) =< [\varphi_h(\boldsymbol{\xi}') - \varphi_h(\boldsymbol{\xi}' - \boldsymbol{\xi})]^2 >$, we obtain

$$\Gamma_{\varphi,h}(\boldsymbol{\xi}) = \mathrm{e}^{-\frac{1}{2}D_{\varphi,h}(\boldsymbol{\xi})} \,. \tag{4.17}$$

Thus, the problem of determining the correlation function $\Gamma_{\varphi,h}(\boldsymbol{\xi})$ of the amplitude is shifted to calculating the phase structure function $D_{\varphi,h}(\boldsymbol{\xi})$.

The relation between the three-dimensional distribution of the refractive index $n(\boldsymbol{\xi}, z)$ and the two-dimensional distribution of the phase is given by (4.12). A similar reasoning leads from the structure function $D_n(\boldsymbol{r})$ of the refractive index (4.11) to that of the phase $D_{\varphi,h}(\boldsymbol{\xi})$, which is a function of the two-dimensional vector $\boldsymbol{\xi}$ since we use the approximation of a thin layer. Assuming also that δh is much larger than the correlation scale of the fluctuations, one can show that for a horizontal wave front entering the layer i at altitude h_i, the phase structure function at the exit of the layer is [72]

$$D_{\varphi,h_i}(\boldsymbol{\xi}) = 2.91 \left(\frac{2\pi}{\lambda}\right)^2 \delta h_i C_{n_i}^2 \xi^{5/3} \ \ [\mathrm{rad}^2], \tag{4.18}$$

with $\xi = |\boldsymbol{\xi}|$.

Inserting this phase structure function into (4.17) we find the correlation function of the monochromatic plane wave after propagating through a turbulent layer. The finite width of $\Gamma_{\varphi,h}(\boldsymbol{\xi})$, as a consequence of the shape of $D_{\varphi,h}(\boldsymbol{\xi})$, means that the correlation width of the amplitude is now limited while the unperturbed plane wave before entering the turbulent layer was perfectly correlated. A suitable correlation length that is determined by the parameters of the turbulence will be defined in the next section.

D_{φ,h_i} is the structure function of the phase in rad^2. If the phase is given in the dimension of metre it describes the physical shape of the turbulent wave front. It is $D^m_{\varphi,h_i} = D_{\varphi,h_i} \times (\frac{\lambda}{2\pi})^2 = 2.91 \delta h_i C_{n_i}^2 \xi^{5/3}$. Thus, the structure function of the phase in m^2, i.e. the shape of the phase, is approximately independent of wavelength. Therefore a wave front sensor can be operated in the visible determining the shape of the wave front and steering the deformable mirror for observations in the infrared.

4.2.2 Multiple Layers, the Fried Parameter

The real atmosphere can be regarded as a composition of many turbulent layers, each of them fulfilling the thin screen approximation. The distance between the layers, the typical size of the turbulence cell and the strength of the turbulence are such that the propagation of the amplitude can be described by Fresnel diffraction [79, 194].

A simple model for weak turbulence gives an impression of the effects [194]. We regard the turbulent layer i at altitude h_i that gives rise to a Fresnel diffraction pattern on layer $i+1$ at altitude h_{i+1}, with $h_{i+1} < h_i$. If the phase is regarded as sufficiently weak, $\varphi_{h_i}(\boldsymbol{\xi}) \ll 1$, we use an approximation for the amplitude, $V_{h_i}(\boldsymbol{\xi}) = \exp(\mathrm{i}\varphi_{h_i}(\boldsymbol{\xi})) \approx 1 + \mathrm{i}\varphi_{h_i}(\boldsymbol{\xi})$, and the Fresnel diffraction between layer i and layer $i+1$ is approximately given by [155]:

$$\begin{aligned} V^{-}_{\varphi,h_{i+1}}(\boldsymbol{\xi}) &\approx V_{\varphi,h_i}(\boldsymbol{\xi}) - \mathrm{i}\frac{\Delta h_i}{2k}\frac{\partial^2 V_{\varphi,h_i}(\boldsymbol{\xi})}{\partial \boldsymbol{\xi}^2} \\ &\approx 1 + \mathrm{i}\varphi_{h_i}(\boldsymbol{\xi}) + \frac{\Delta h_i}{2k}\frac{\partial^2 \varphi_{h_i}(\boldsymbol{\xi})}{\partial \boldsymbol{\xi}^2}, \end{aligned} \tag{4.19}$$

with $V^{-}_{\varphi,h_{i+1}}(\boldsymbol{\xi})$ the amplitude of the Fresnel diffraction pattern entering layer $i+1$ and $\Delta h_i = h_i - h_{i+1}$.

The imaginary part of the amplitude, $\mathrm{i}\varphi_{h_i}(\boldsymbol{\xi})$, is unaffected by Fresnel diffraction, and the real part has an additional term proportional to the curvature of the phase $\partial^2\varphi_{h_i}(\boldsymbol{\xi})/\partial\boldsymbol{\xi}^2$ and to the distance Δh_i between the layers. This additional term causes intensity fluctuations called the *scintillation*. We will come back to this in Sect. 4.3.4.

The curvature of the phase affecting the real part also affects the resulting phase of the amplitude $V^{-}_{\varphi,h_{i+1}}(\boldsymbol{\xi})$. Thus, the phase of the wave entering layer $i+1$ is sightly different from the perturbation $\varphi_{h_i}(\boldsymbol{\xi})$ in layer i, modifying also its statistical properties like the phase structure function and the power spectrum [194].

Usually the curvature terms are neglected in the **near-field approximation** assuming that the distance Δh_i between the turbulent layers is very small. Then, the phase of the originally plane wave, being perturbed by the first turbulent layer in its propagation path, remains unaffected until it hits the next turbulent layer. After passage through multiple layers the phase can be approximated by the sum of the individual perturbations. For very weak turbulence this is intuitively understandable.

While this seems like a rather crude approximation – given that scintillation is an observable and measurable effect – the derived quantities, in particular the correlations functions, provide a very good description of the real process.

The amplitude $V_{\varphi,h_{i+1}}(\boldsymbol{\xi})$ at the exit of a layer at altitude h_{i+1} is related to that at the entrance of that layer, $V^{-}_{\varphi,h_{i+1}}(\boldsymbol{\xi})$, by multiplication with the phase disturbance of layer $i+1$,

$$V_{\varphi,h_{i+1}}(\boldsymbol{\xi}) = V^{-}_{\varphi,h_{i+1}}(\boldsymbol{\xi})\mathrm{e}^{\mathrm{i}\varphi_{h_{i+1}}(\boldsymbol{\xi})}, \tag{4.20}$$

and, the correlation function can be determined by

$$\begin{aligned}< V_{\varphi,h_{i+1}}(\boldsymbol{\xi}')V^*_{\varphi,h_{i+1}}(\boldsymbol{\xi}'-\boldsymbol{\xi}) > &= < V^-_{\varphi,h_{i+1}}(\boldsymbol{\xi}')V^{*-}_{\varphi,h_{i+1}}(\boldsymbol{\xi}'-\boldsymbol{\xi}) > \\ &\times < \mathrm{e}^{\mathrm{i}[\varphi_{h_{i+1}}(\xi')-\varphi_{h_{i+1}}(\xi'-\xi)]} > \\ &= < V^-_{\varphi,h_{i+1}}(\boldsymbol{\xi}')V^{*-}_{\varphi,h_{i+1}}(\boldsymbol{\xi}'+\boldsymbol{\xi}) > \mathrm{e}^{-\frac{1}{2}D_{\varphi,h_{i+1}}(\xi)} .\end{aligned} \tag{4.21}$$

In near-field approximation, the correlation function of $V^-_{\varphi,h_{i+1}}(\boldsymbol{\xi}')$ is the same as that of $V_{\varphi,h_i}(\boldsymbol{\xi}')$, unaffected by Fresnel diffraction.

Thus, the propagation of the correlation function through the atmosphere is reduced to a simple product of the correlation functions of the single layers, since they are statistically independent [194].

Calculating the correlation function iteratively for all layers, one obtains the correlation function on the ground after propagation through N turbulent layers as

$$\begin{aligned}< V_\varphi(\boldsymbol{\xi}')V^*_\varphi(\boldsymbol{\xi}'-\boldsymbol{\xi}) > &= \mathrm{e}^{-\frac{1}{2}D_\varphi(\xi)}, \text{ with} \\ D_\varphi(\boldsymbol{\xi}) &= 2.91\left(\frac{2\pi}{\lambda}\right)^2 \sum_{i=1}^{N} \delta h_i C^2_{n_i}\, \xi^{5/3} .\end{aligned} \tag{4.22}$$

In the more realistic case of a continuous distribution of turbulence and of a source at a distance $\zeta = |\boldsymbol{\zeta}|$ from zenith one obtains

$$D_\varphi(\boldsymbol{\xi}) = 2.91\left(\frac{2\pi}{\lambda}\right)^2 (\cos\zeta)^{-1}\int C^2_n(h)\mathrm{d}h\, \xi^{5/3} \ [\mathrm{rad}^2]. \tag{4.23}$$

Fried simplified the expression by introducing the correlation length r_0, called the *Fried parameter* [71], which is defined by

$$r_0 := \left(0.423\left(\frac{2\pi}{\lambda}\right)^2 (\cos\zeta)^{-1}\int C^2_n(h)\mathrm{d}h\right)^{-3/5} . \tag{4.24}$$

The numerical parameter, 0.423, defines r_0 such that the variance of the phase over a circle with a diameter of r_0 is about 1 rad^2. In this sense, the Fried parameter defines the size of a turbulence cell. The wavelength dependence of r_0 is given by $r_0 \propto \lambda^{6/5}$ and the dependence on zenith angle is $r_0 \propto (\cos\zeta)^{3/5}$. Typical values of r_0 are 0.6 m in the K-band and, correspondingly, 0.11 m in the visible.

The phase structure function, which is the mean-square difference of the phase at two points on the wave front separated by $\boldsymbol{\xi}$, can now be expressed by

$$D_\varphi(\boldsymbol{\xi}) = 6.88\left(\frac{\xi}{r_0}\right)^{5/3} \quad [\mathrm{rad}^2], \tag{4.25}$$

with $\xi = |\boldsymbol{\xi}|$, and the correlation function of the amplitudes is

$$\Gamma_\varphi(\boldsymbol{\xi}) =< V_\varphi(\boldsymbol{\xi}')V_\varphi^*(\boldsymbol{\xi}' - \boldsymbol{\xi}) >= \mathrm{e}^{-3.44\left(\frac{\xi}{r_0}\right)^{5/3}}. \tag{4.26}$$

Note that $\Gamma_\varphi(\boldsymbol{\xi})$ is a real function since we made the assumption that the phase difference follows a Gaussian distribution with zero mean.

Very often the power spectrum of the phase fluctuations is needed for analysis. Similar to the calculation that relates the structure function of the refractive index (4.11) to the Kolmogorov spectrum of refractive index fluctuations (4.7) for three-dimensional coordinates, the phase structure function (4.25) as a function of the two-dimensional coordinate $\boldsymbol{\xi}$ can be used to compute the Kolmogorov spectrum of the phase fluctuations [40, 107, 194], yielding

$$\begin{aligned}\Phi(\kappa) &= 0.033\,(2\pi)^{-2/3}\left(\frac{2\pi}{\lambda}\right)^2 (\cos\zeta)^{-1}\int C_n^2(h)\mathrm{d}h\,\kappa^{-11/3}\\ &= 0.0229\,r_0^{-5/3}\,\kappa^{-11/3}\;\;[\mathrm{rad}^2/\mathrm{m}^{-2}],\end{aligned} \tag{4.27}$$

with $\kappa = |\boldsymbol{\kappa}|$ and $\boldsymbol{\kappa}$ the two-dimensional spatial frequency vector. Therefore, the dimension of $\Phi(\boldsymbol{\kappa})$, as power per surface element $\mathrm{d}\boldsymbol{\kappa}$, is rad^2/m^{-2}.

Outer Scale of Turbulence

Note that the Kolmogorov spectrum of the phase fluctuations $\Phi_n(\boldsymbol{k})$, (4.7), was given as a function of the spatial wave vector $\boldsymbol{k} \propto 2\pi/\boldsymbol{r}$, while, here, for the power spectrum of the phase fluctuations we use the spatial frequency vector $\boldsymbol{\kappa} \propto 1/\boldsymbol{\xi}$.

The integral over the Kolmogorov spectrum (the power spectrum) provides the variance of the phase. As noted above, the integral over $\Phi(\kappa) \propto \kappa^{-11/3}$ is infinite. This means, that the variance of the turbulent phase is infinite, which is a well known property of Kolmogorov turbulence [225]. However, this is only true in the theoretical limit of an infinite outer scale of turbulence. For a finite outer scale, L_0, the von Kármán spectrum of the phase fluctuations can be derived from (4.8) (see also Fig. 4.2),

$$\Phi(\kappa)_{vK} = 0.0229\,r_0^{-5/3}\,(L_0^{-2} + \kappa^2)^{-11/6}\;\;[\mathrm{rad}^2/\mathrm{m}^{-2}]. \tag{4.28}$$

The inner scale of turbulence (see Fig. 4.2) does not need to be considered for our purposes [145]. Integrating over $\Phi(\kappa)_{vK}$ yields a finite value for the phase variance.

According to (4.26), the correlation between two amplitudes V_φ goes to zero if their distance ξ is much larger than the Fried parameter r_0. This could be regarded as a complete loss of coherence of the originally plane wave by the process of propagation through turbulence.

However, the terms coherence and coherence function, defined in Sect. 2.3, should be reserved for statistical effects related to the nature of the light source.

The correlation function $\Gamma_\varphi(\boldsymbol{\xi})$, which is defined as a time average (4.14) over time scales that are much longer than the typical time scale of atmospheric turbulence, is independent of the nature of the illuminating light.

To illustrate the consequence we regard again **Young's Experiment**. Light that has passed through atmospheric turbulence is used to illuminate the two pinholes that are separated by a distance larger than r_0. An exposure time of several seconds would show a homogeneous intensity distribution without any interference pattern. If the exposure time were shorter than a few milliseconds the turbulence would be quasi frozen creating a random but fixed phase difference between the pinholes, and we would find fringes.

The presence of fringes with a non-zero visibility indicates that the light has a non-zero degree of coherence. A sequence of snapshots would show fringes at varying positions depending on the random phase difference. Thus, after passing through turbulence the measured fringe visibility depends on the exposure time and not only on the coherence of the light. This deterioration of the measured visibility due to atmospheric turbulence will be discussed in great detail in Chap. 6.

Turbulence Profile

The atmospheric turbulence profile is the distribution of turbulence with altitude, given by $C_n^2(h)$. Based on measurements, Hufnagel together with Valley [106, 247] suggested a model for the turbulence profile called the *Hufnagel–Valley model* (HV-model). This was later modified [178] yielding

$$C_n^2(h) = 1.9 \times 10^{-15} \mathrm{e}^{-h/0.1} + 3.02 \times 10^{-17} \mathrm{e}^{-h/1.5} \\ + 8.16 \times 10^{-24} h^{10} \mathrm{e}^{-h}, \tag{4.29}$$

with h in km. This profile corresponds to an r_0 of 0.18 m at $\lambda = 0.5\,\mu$m. It is displayed in Fig. 4.3.

The three terms describe the strength of the turbulence in three altitude regimes. The first exponential term affects the turbulence in the *ground layer* a few hundred metres above the ground. In the HV-model, the structure constant $C_n^2(h)$ has values about 100 times larger than at higher altitudes. The second exponential term dominates the turbulence in the first few kilometres, and the third term describes a general increase of the turbulence at about 9–12 km altitude due to the jet stream. Investigating this in more detail, one finds distinct layers of turbulence between 6 and 18 km that are about 1 km thick [250].

Recent measurements using a variety of methods [236] show a good agreement at altitudes below 1,000 m, confirming the exponential law as in the first and second term in (4.29). The numerical constants are different depending not only on r_0 but also on random fluctuations of the turbulence profile.

There is no detailed model for the variation of the outer scale of turbulence L_0 with altitude. While the integrated value is about 20–50 m [260], L_0 varies between a few metres in the ground layer and a few hundred metres above [33, 98, 226].

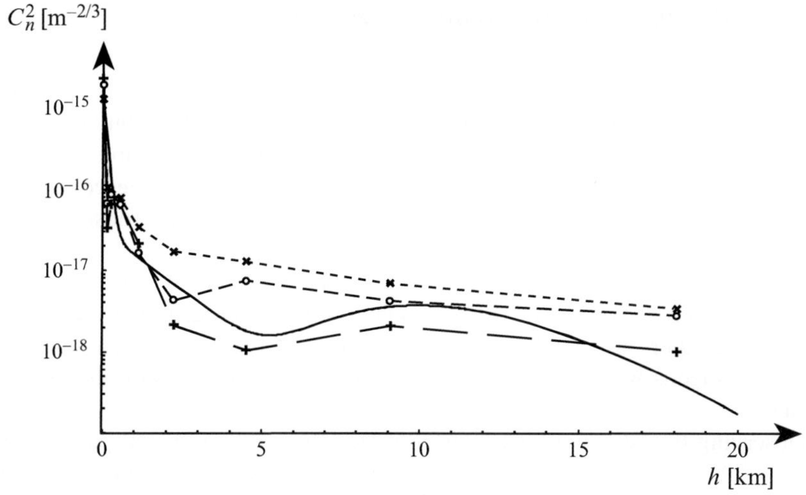

Fig. 4.3 Turbulence profiles as a function of altitude in km corresponding to an integrated r_0 of 0.18 m at $\lambda = 0.5\,\mu$m. The *solid line* displays the modified Hufnagel–Valley model (4.29). The individual points (marked by *times*, *plus* and *circle*) show median values and the lowest and highest quartiles of C_n^2 measurements at each altitude. The measured profiles represented by the three *dashed curves*, all result in an overall r_0 of 0.18 m. This large variance of 50–100% of the C_n^2 values has to be kept in mind when using the turbulence profiles for numerical simulations

Figure 4.3 displays the modified HV-model together with measurements of C_n^2 at individual altitudes. These measurements were collected at ESO's VLT Observatory on Cerro Paranal, Chile, in 58 nights between 2005 and 2007, combining a total of 1851 measurements of 10-min averages [141]. Here, only profiles with an overall r_0 of 0.18 m at $\lambda = 0.5\,\mu$m are accounted for.

The altitudes of the individual C_n^2 values do not correspond to any particular turbulent layer but they are determined by instrumental constraints. For numerical simulations of propagation through turbulence and of the performance of adaptive optics systems, the continuous turbulence profile is represented by these individual C_n^2 values at discrete altitudes, weighted by the layer thickness that they represent.

The spread of C_n^2 profiles in Fig. 4.3 represents profiles for the low and high quartiles of the measured C_n^2 values in the ground layer below 500 m (Kolb, 2009, Private communication). Thus, if due to random fluctuations the C_n^2 values in the ground layer are smaller than the median value, the corresponding values at higher altitudes are larger since, for this comparison, we only regard profiles with the same integrated r_0.

This means, that there is no typical profile that is simply scaled according to the value of the overall r_0 but there are fluctuations of 50–100% of the C_n^2 values at individual altitudes. The turbulence profiles vary from moment to moment and from site to site, and any model can only give a rough idea about the profile structure to expect for a given observation.

4.2.3 Anisoplanatic and Temporal Effects

The discussions in the last two sections were about turbulent layers of unlimited extent, describing general statistical properties of atmospheric turbulence. Observing with a telescope through atmospheric turbulence, only a subsample of each layer contributes to the image. The shape of the wave front in the telescope aperture, observing a star at position $\boldsymbol{s}$, is determined by the sum of the subsamples of all turbulent layers in a cylindrical volume in this direction.

The light of a star at a slightly different angular position $\boldsymbol{s} + \Delta\boldsymbol{s}$ travels through a slightly different volume of the atmosphere – the more different the higher the contributing layers are – and displays a wave front of slightly different shape. Thus, the instantaneous wave front in the telescope aperture depends on the position of the star.

Observing two stars with an interferometer, the random optical path difference (OPD) of the light from the star at position $\boldsymbol{s}$ will be slightly different from the OPD of the star at $\boldsymbol{s} + \Delta\boldsymbol{s}$ for the same reason.

Long time averages of the random phase distribution, like the phase structure function, are the same for both star positions $\boldsymbol{s}$ and $\boldsymbol{s} + \Delta\boldsymbol{s}$ if statistical parameters do not change. This means that long exposure images look the same.

However, we are now interested in the resemblance of the instantaneous wave fronts coming from the two stars. This is important for adaptive optics and fringe tracking systems that rely on the measurement of a reference star to correct for the aberrations and the fringe motion of the scientific object. These techniques will be discussed in detail in Chap. 6.

The acceptable angular distance between the scientific object and the reference star – still ensuring a good resemblance between the aberrations – is called the *isoplanatic angle*. Another important quantity, which describes the temporal evolution of the atmosphere and, thus, the permitted time difference between measuring the wave front and applying this information, is given by the typical time scale of the decorrelation, the *atmospheric coherence time*. These two quantities are the most important limiting factors for the performance of adaptive optics and fringe tracking systems.

Isoplanatic Angle

Simplifying atmospheric turbulence by replacing the turbulence profile by a single layer permits a simple computation for the isoplanatic angle. We write $\varphi_s(\boldsymbol{\xi})$ for the phase in the observing direction $\boldsymbol{s}$, and $\varphi_{s+\Delta s}(\boldsymbol{\xi})$ in direction $\boldsymbol{s} + \Delta\boldsymbol{s}$. The difference angle $\Delta\boldsymbol{s}$ can be replaced by the lateral displacement $\Delta\boldsymbol{s}\, h$ of the single layer at altitude h [194], writing

$$\varphi_s(\boldsymbol{\xi} - \Delta\boldsymbol{s}\, h) = \varphi_{s+\Delta s}(\boldsymbol{\xi})\,. \tag{4.30}$$

The average resemblance of the phase values can be quantified by rewriting the phase structure function for the phase distribution in the directions $\boldsymbol{s}$ and $\boldsymbol{s} + \Delta\boldsymbol{s}$, which is a function of coordinate difference $\Delta\boldsymbol{s}$ only:

$$D_\varphi(\Delta \boldsymbol{s} h) = <[\varphi_s(\boldsymbol{\xi}) - \varphi_s(\boldsymbol{\xi} - \Delta \boldsymbol{s} h)]^2> = 6.88 \left(\frac{\Delta s\, h}{r_0}\right)^{5/3} . \quad (4.31)$$

Note that, as customary, we will denote the difference vector $\Delta\boldsymbol{s}$ by $\boldsymbol{\theta}$ when discussing variations of the apparent star position due to turbulence.

If the isoplanatic angle is defined such that the rms phase difference between two points at $\boldsymbol{\xi}$ and at $\boldsymbol{\xi} - \boldsymbol{\theta} h$ is 1 rad, we write the isoplanatic angle as $\theta_0 = 0.314\, r_0/h$ so that we obtain

$$D_\varphi(\boldsymbol{\theta} h) = \left(\frac{\theta}{\theta_0}\right)^{5/3} . \quad (4.32)$$

The influence of different layers or of a continuous turbulence distribution can be accounted for by replacing r_0 in $\theta_0 = 0.314\, r_0/h$ by its definition (4.24) and then by weighting C_n^2 by $h^{5/3}$ (compare to 4.23). We introduce the effective altitude, $\bar{h}$, as [74]

$$\bar{h} = \left(\frac{\int C_n^2(h) h^{5/3} \mathrm{d}h}{\int C_n^2(h) \mathrm{d}h}\right)^{3/5} , \quad (4.33)$$

ensuring that turbulence at high altitudes turbulence has more weight than at low altitudes, and we write the isoplanatic angle as [73]

$$\theta_0 = 0.314 \left(0.423 \left(\frac{2\pi}{\lambda}\right)^2 \int C_n^2(h) h^{5/3} \mathrm{d}h\right)^{-3/5} = 0.314 \frac{r_0}{\bar{h}} . \quad (4.34)$$

If the effective altitude is $\bar{h} = 2.5$ km and $r_0 = 0.6$ m – a typical value in the K-band – it is $\theta_0 = 16$ arcsec. Since the isoplanatic angle scales with r_0 it also depends on wavelength $\propto \lambda^{6/5}$.

This definition of the isoplanatic angle is based on the comparison of individual points on the wave front. In practical cases, the value is often larger due to averaging effects over the telescope aperture. For adaptive optics systems, the position of the deformable mirror, the correction degree and the acceptable error affect the value [28,234], and for interferometers, correcting the OPD fluctuations due to turbulence, the required performance determines the value of the isoplanatic angle.

However, the definition in (4.34) is still useful as a lower value of the isoplanatic angle independent of telescope and instrumental parameters.

Atmospheric Coherence Time

Using the *Taylor hypothesis of frozen turbulence* the temporal evolution can be estimated in perfect analogy [194]. The assumption is that a static piece of turbulence moves with constant speed $|\boldsymbol{v}|$ in front of the telescope aperture, which is a good approximation as long as the observed time differences are small. Then the phase at point $\boldsymbol{\xi}$ at time $t+\tau$ can be written as

$$\varphi_{t+\tau}(\boldsymbol{\xi}) = \varphi_t(\boldsymbol{\xi} - \boldsymbol{v}\tau)\,, \tag{4.35}$$

and the temporal phase structure function is

$$D_\varphi(\boldsymbol{v}\tau) = < [\varphi_t(\boldsymbol{\xi}) - \varphi_t(\boldsymbol{\xi} - \boldsymbol{v}\tau)]^2 >= 6.88 \left(\frac{v\tau}{r_0}\right)^{5/3}. \tag{4.36}$$

The temporal difference is thus transformed to a difference in spatial coordinates with the difference being $v\tau$. A simple estimate of the correlation time similar to the isoplanatic angle above yields the coherence time $\tau_0 = 0.314\, r_0/v$ [24] and the temporal phase structure function

$$D_\varphi(\boldsymbol{v}\tau) = \left(\frac{\tau}{\tau_0}\right)^{5/3}. \tag{4.37}$$

Multiple layers with different wind speeds are considered equivalently to the case of anisoplanacy by applying individual wind speeds $v^{5/3}$ to individual layers with structure constants C_n^2, defining the effective wind speed, $\bar{v}$, as

$$\bar{v} = \left(\frac{\int C_n^2(h) v^{5/3} \mathrm{d}h}{\int C_n^2(h) \mathrm{d}h}\right)^{3/5}, \tag{4.38}$$

so that the atmospheric coherence time is written as

$$\tau_0 = 0.314 \left(0.423 \left(\frac{2\pi}{\lambda}\right)^2 \int C_n^2(h) v^{5/3} \mathrm{d}h\right)^{-3/5} = 0.314\,\frac{r_0}{\bar{v}}\,. \tag{4.39}$$

We then obtain a coherence time of $\tau_0 = 19\,\mathrm{ms}$ for an effective wind speed of $\bar{v} = 10\,\mathrm{m\,s^{-1}}$ and a Fried parameter of $r_0 = 0.6\,\mathrm{m}$. Often, the coherence time is given for the visible when, with $r_0 \propto \lambda^{6/5}$, it is $r_0 = 0.11\,\mathrm{m}$, obtaining $\tau_0 = 3.8\,\mathrm{ms}$.

As for the isoplanatic angle, this definition of the coherence time is based on the bare atmosphere, comparing the phase at a single point on the wave front at two moments in time with a fixed time difference, permitting a root-mean-square difference of these two values of 1 rad.

In interferometry, a suitable measure for the coherence time is the variance of the fringe motion. Regarding the two sub-apertures of an interferometer, one has to compute the mean-square phase difference of two points separated by the baseline and not of a single point as in (4.36), taking not only the temporal but also the spatial correlation into account. Simply assuming that the two points are separated far enough so that they are spatially uncorrelated, the variances add and the value of the phase structure function doubles so that the constant has to be set at 2×6.88. Then, equivalent to (4.37), the **interferometric coherence time** can be defined as [38]

$$\tau_{0,2} = 0.207 \, \frac{r_0}{\bar{v}} \, . \tag{4.40}$$

This definition of the coherence time is related to the phase variance for a specified time difference τ, (4.37). The situation is different if a temporal integration is performed, for instance during a limited exposure time, when the phase variance during the exposure is of interest. One might be interested in the average smearing of the fringes during the exposure time – resulting in a loss of fringe contrast – or in the residual fringe motion when stabilising the fringe position on a close by reference star. And one might use real apertures instead of pinholes performing an averaging process over the apertures. For this type of analysis, the power spectrum of the phase fluctuations is a more suitable tool than the phase structure function.

Temporal Power Spectrum

The temporal power spectrum of the phase fluctuations can be calculated from the spatial power spectrum $\Phi(\boldsymbol{\kappa})$ (4.27). With $\boldsymbol{v}$ being e.g. parallel to the ξ axis, one sets $\kappa_\xi = f/v$ and performs an integration over κ_ζ to obtain the temporal power spectrum $\Phi_t(f)$ [24],

$$\Phi_t(f) = \frac{1}{\bar{v}} \int \Phi\left(\frac{f}{\bar{v}}, \kappa_\zeta\right) \mathrm{d}\kappa_\zeta = 0.077 \left(\frac{\bar{v}}{r_0}\right)^{5/3} f^{-8/3} \;\; [\mathrm{rad}^2/\mathrm{Hz}] \, . \tag{4.41}$$

It is important to note that, unlike the spatial power spectra, we define temporal power spectra from 0 to $+\infty$. If for some reason it is required to define them from $-\infty$ to $+\infty$, the constant 0.077 has to be replaced by half its value.

The variance of the temporal phase fluctuations is the integral over $\Phi_t(f)$. As there is a pole at $f = 0$ this integral is infinite as in the case of spatial phase fluctuations (4.27). Using the von Kármán spectrum with finite outer scale L_0 (4.8) the integration provides a finite value for the variance.

In this section, we had an observing situation in mind, discussing the resemblance of wave fronts in different directions, but we regarded the bare wave front without taking telescope apertures or averaging processes over apertures and baselines into account. This will be done in the following section computing power spectra of quantities that are derived from the phase, like for instance the fringe motion or the image motion. We shall see that the variances of these quantities – the integrals over their power spectra – are finite, and that atmospheric coherence times can be defined according to individual performance criteria.

Statistical Properties of the Perturbed Complex Wave: Summary

The Fried parameter r_0 [71] provides the diameter of a circular patch of the wave front in atmospheric turbulence that has a phase variance of about 1 rad^2. It is defined as

$$r_0 = \left(0.423\left(\frac{2\pi}{\lambda}\right)^2 (\cos\zeta)^{-1}\int C_n^2(h)\mathrm{d}h\right)^{-3/5}. \qquad (4.24)$$

r_0 is proportional to $\lambda^{6/5}$ and to $(\cos\zeta)^{3/5}$, with ζ the zenith angle.

The phase structure function $D_\varphi(\boldsymbol{\xi}) =< [\varphi(\boldsymbol{\xi}') - \varphi(\boldsymbol{\xi}' + \boldsymbol{\xi})]^2 >$, which is the mean-square difference of the phase at two points on the wave front separated by $\boldsymbol{\xi}$, reads as

$$D_\varphi(\boldsymbol{\xi}) = 6.88\left(\frac{\xi}{r_0}\right)^{5/3} \quad [\mathrm{rad}^2], \qquad (4.25)$$

with $\xi = |\boldsymbol{\xi}|$, and $\boldsymbol{\xi}$ the two-dimensional coordinate vector.

Assuming a plane wave – originating from a point-like star at a very large distance – enters the turbulent atmosphere, the correlation function of its amplitude $V_\varphi(\boldsymbol{\xi}) = \exp(\mathrm{i}\varphi_h(\boldsymbol{\xi}))$ on the ground is given by

$$\Gamma_\varphi(\boldsymbol{\xi}) =< V_\varphi(\boldsymbol{\xi}')V_\varphi^*(\boldsymbol{\xi}' - \boldsymbol{\xi}) >= \mathrm{e}^{-\frac{1}{2}D_\varphi(\boldsymbol{\xi})} = \mathrm{e}^{-3.44\left(\frac{\xi}{r_0}\right)^{5/3}}, \qquad (4.26)$$

assuming the near-field approximation, and neglecting scintillation. $\Gamma_{\varphi,h}(\boldsymbol{\xi})$ is a real function since we made the assumption that the phase difference follows a Gaussian distribution with zero mean.

Thus, the correlation goes to zero if the positions of two amplitudes are separated by more than r_0. This could be regarded as a complete loss of coherence of the originally coherent plane wave.

However, with the typical time scale of atmospheric turbulence of a few milliseconds, snapshot images of the fringe pattern in an interferometer would still provide very high visibility values, preserving the coherence of the light, and, thus, the opportunity to measure visibilities and derive high-angular resolution images.

Very often the power spectrum of the phase fluctuations is needed for analysis. Using the phase structure function (4.25), the Kolmogorov spectrum of the phase fluctuations can be computed [40, 171], reading

$$\Phi(\boldsymbol{\kappa}) = 0.0229\, r_0^{-5/3}\kappa^{-11/3} \quad [\mathrm{rad}^2/\mathrm{m}^{-2}], \qquad (4.27)$$

with $\boldsymbol{\kappa}$ the two-dimensional spatial frequency vector, $\kappa = |\boldsymbol{\kappa}|$ and, hence, $\Phi(\boldsymbol{\kappa})$ in units of power per surface element $\mathrm{d}\boldsymbol{\kappa}$, i.e. rad^2/m^{-2}.

For a finite outer scale L_0, the von Kármán spectrum of the phase fluctuations can be derived from (4.8), yielding

$$\Phi(\kappa)_{vK} = 0.0229\, r_0^{-5/3}\, (L_0^{-2} + \kappa^2)^{-11/6}\ \ [\mathrm{rad}^2/\mathrm{m}^{-2}]. \tag{4.28}$$

The inner scale of turbulence (see Fig. 4.2) does not need to be considered for our purposes [145].

The isoplanatic angle indicates the angular distance of two sources in the sky for which the rms phase fluctuations on the ground are about 1 rad, and the atmospheric coherence time is the time difference when the rms phase fluctuations at a single point have a value of 1 rad. The two quantities are derived in perfect analogy [194] using the phase structure function (4.25).

We define an isoplanatic angle θ_0 [73] such that it is

$$D_\varphi(\theta \bar{h}) = \left(\frac{\theta}{\theta_0}\right)^{5/3}, \tag{4.32}$$

with

$$\theta_0 = 0.314\, \frac{r_0}{\bar{h}} = \left(2.91 \left(\frac{2\pi}{\lambda}\right)^2 \int C_n^2(h) h^{5/3} \mathrm{d}h\right)^{-3/5}. \tag{4.34}$$

and $\bar{h} = \left(\int C_n^2(h) h^{5/3} \mathrm{d}h / \int C_n^2(h) \mathrm{d}h\right)^{3/5}$ the effective altitude defined in (4.33) [74].

In analogy, we define the atmospheric coherence time τ_0 [24] so that

$$D_\varphi(v\tau) = \left(\frac{\tau}{\tau_0}\right)^{5/3}, \tag{4.37}$$

with

$$\tau_0 = 0.314\, \frac{r_0}{\bar{v}} = \left(2.91 \left(\frac{2\pi}{\lambda}\right)^2 \int C_n^2(h) v^{5/3} \mathrm{d}h\right)^{-3/5}, \tag{4.39}$$

and $\bar{v} = \left(\int C_n^2(h) v^{5/3} \mathrm{d}h / \int C_n^2(h) \mathrm{d}h\right)^{3/5}$ the effective wind speed defined in (4.38).

The atmospheric coherence time for interferometry is based on the mean-square phase difference of two points taking not only the temporal but also the spatial correlation into account. Equivalent to (4.39) we define [38]

$$\tau_{0,2} = 0.207\, \frac{r_0}{\bar{v}}. \tag{4.40}$$

The temporal power spectrum of the phase fluctuations can be computed from the spatial power spectrum $\Phi(\kappa)$ (4.27). With $\boldsymbol{v}$ being e.g. parallel to the ξ axis, one sets $\kappa_\xi = f/v$ and performs an integration over κ_ζ to obtain the temporal power spectrum $\Phi_t(f)$ [24],

$$\Phi_t(f) = 0.077 \left(\frac{\bar{v}}{r_0}\right)^{5/3} f^{-8/3} \; [\mathrm{rad}^2/\mathrm{Hz}]. \tag{4.41}$$

It is important to note that, unlike the spatial power spectra, we define temporal power spectra from 0 to $+\infty$.

4.3 Propagation Through Optical Systems

In the previous section, the statistical properties of the propagating turbulent wave front were described. When it comes to analysing the impact of turbulence on the imaging process in telescopes and interferometers, some assumptions have to be made about the phase distribution in the telescope aperture. We assume that the turbulent atmosphere can be represented by a single thin layer in the telescope aperture neglecting Fresnel diffraction, as discussed in Sect. 4.2. This is called the near-field approximation [194, 255]. Only when discussing scintillation in Sect. 4.3.4, the Fresnel diffraction will be considered.

We will start discussing fringe motion, differential fringe motion and image motion employing the Taylor hypothesis of frozen turbulence as in Sect. 4.2.3. Anisoplanatic effects will also be discussed under this hypothesis. For the sake of simplicity, we will always use the effective altitude, $\bar{h}$, of turbulence (4.33) and the effective wind speed, $\bar{v}$, (4.38) to avoid bulky integrals in the formulae. For accurate results, however, it is required to compute the integrals using measured turbulence profiles and wind speeds (and directions). Since these are often not available or, as discussed following Fig. 4.3, highly variable so that a conclusion is difficult to reach, the simplifications permit to understand the principle interrelation of the parameters and to obtain useful numerical estimates.

For the interpretation of the results one always has to be aware not only of its approximative character but also of the natural variability of atmospheric turbulence that evades accurate predictions.

4.3.1 Fringe Motion

The fringe motion in an interferometer is caused by random OPD variations between the apertures. These fluctuations reduce the measured visibility if the exposure time is longer than the atmospheric coherence time. Freezing the fringe motion with a *fringe tracker* using a bright reference star permits us to integrate a long time and, thus, to observe fainter sources in the vicinity of the reference star.

In order to specify the performance parameters of a fringe tracker, the power spectrum of the fringe motion needs to be known. In addition, we have to discuss the differential fringe motion between the reference star and the faint object so that we can quantify the isoplanatic angle.

The first quantity to be investigated is the variance of the phase difference $\Delta\varphi$ between two points – for instance the pinholes in Young's experiment – separated by the baseline $\boldsymbol{B}$, $< |\varphi(\boldsymbol{\xi}) - \varphi(\boldsymbol{\xi} - \boldsymbol{B})|^2 >$, which is the phase structure function $D_\varphi(\boldsymbol{B})$, (4.25).

In order to arrive at the power spectrum of the phase difference, we write this difference as a convolution of the phase $\varphi(\boldsymbol{\xi})$ with the two δ-functions representing the pinholes,

$$\Delta\varphi(\boldsymbol{\xi}) = \varphi(\boldsymbol{\xi}) * (\delta(\boldsymbol{\xi} - \boldsymbol{B}/2) - \delta(\boldsymbol{\xi} + \boldsymbol{B}/2)), \tag{4.42}$$

with $\boldsymbol{B}$, the baseline of the interferometer.

In Fourier space, the convolution on the right hand side transforms into the product of the individual Fourier transforms, yielding

$$\mathcal{F}_\xi(\Delta\varphi(\boldsymbol{\xi})) = \mathcal{F}_\xi(\varphi(\boldsymbol{\xi})) \times \mathcal{F}_\xi\,(\delta(\boldsymbol{\xi} - \boldsymbol{B}/2) - \delta(\boldsymbol{\xi} + \boldsymbol{B}/2)). \tag{4.43}$$

The power spectrum of the phase difference fluctuations is the square of $\mathcal{F}_\xi(\Delta\varphi(\boldsymbol{\xi}))$. Computing the square of the right hand side of the equation, we have the power spectrum $\mathcal{F}_\xi^2(\varphi(\boldsymbol{\xi})) = |\hat{\varphi}(\boldsymbol{\kappa})|^2 = \Phi(\boldsymbol{\kappa})$ of the phase fluctuations, (4.27), and $4\sin^2(\pi\boldsymbol{B}\cdot\boldsymbol{\kappa})$, the square of the Fourier transform of $\delta(\boldsymbol{\xi} - \boldsymbol{B}/2) - \delta(\boldsymbol{\xi} + \boldsymbol{B}/2)$.

Since we are interested in the OPD in m, we use $\Delta\varphi(\boldsymbol{\xi})\lambda/(2\pi)$, forming the power spectrum $\Phi_{\mathrm{OPD}}(\boldsymbol{\kappa})$ of the OPD fluctuations as[2]

$$\Phi_{\mathrm{OPD}}(\boldsymbol{\kappa}) = \frac{\lambda^2}{4\pi^2}\,0.0229\,r_0^{-5/3}\,\kappa^{-11/3} 4\sin^2(\pi\boldsymbol{B}\cdot\boldsymbol{\kappa}) \;\; [\mathrm{m}^2/\mathrm{m}^{-2}], \tag{4.44}$$

with $\kappa = |\boldsymbol{\kappa}|$. The power spectrum is no longer radial-symmetric but it depends on the orientation of the baseline vector $\boldsymbol{B}$. At low spatial frequencies, in the direction parallel to $\boldsymbol{B}$, the square of the sine-function can be approximated by $\pi^2 B^2 \kappa^2$ so that $\Phi_{\mathrm{OPD}}(\boldsymbol{\kappa})$ is proportional to $\kappa^{-5/3}$. At high frequencies – the equivalent of very small values of $\boldsymbol{\xi}$ – $\sin^2(\pi\boldsymbol{B}\cdot\boldsymbol{\kappa})$ can be replaced by 0.5, its average value, so that the proportionality to $\kappa^{-11/3}$ remains as in the power spectrum of the phase $\Phi(\boldsymbol{\kappa})$. In the orthogonal direction, $\Phi_{\mathrm{OPD}}(\boldsymbol{\kappa})$ is always proportional to $\kappa^{-11/3}$.

The spatial power spectrum $\Phi_{\mathrm{OPD}}(\boldsymbol{\kappa})$ will be used in the following to derive the temporal power spectrum. We can also use it to compute the variance σ^2_{OPD} of the

---

[2] The notation of the dimension as $\mathrm{m}^2/\mathrm{m}^{-2}$ is chosen since we describe the power in m^2 per surface element in frequency space, $d\boldsymbol{\kappa}$, in m^{-2}.

OPD fluctuations by integrating $\Phi_{\text{OPD}}(\kappa)$ from $-\infty$ to $+\infty$ although it is easier to determine its value directly through the phase structure function $D_\varphi(\boldsymbol{B})$ that, by definition, gives the variance of the phase difference. Then, the **OPD variance** for a baseline length of B is given by [194]

$$\sigma^2_{\text{OPD}} = \frac{\lambda^2}{(2\pi)^2} D_\varphi(\boldsymbol{B}) = 0.17\,\lambda^2 \left(\frac{B}{r_0}\right)^{5/3} \quad [\text{m}^2], \tag{4.45}$$

displayed in Fig. 4.4. Due to the definition of the Fried parameter (4.24), r_0 scales with wavelength $\propto \lambda^{6/5}$ and it is $\lambda^2\, r_0^{-5/3} = 16.7 \int C_n^2(h)\mathrm{d}h$ so that the optical path length fluctuations are independent of wavelength and scale linear with the integral over C_n^2.

However, the impact of path length fluctuations on observations is determined by the phase fluctuations, i.e. by $\sigma_{\text{OPD}} \frac{2\pi}{\lambda}$, so that, at short wavelengths the fringe visibility is more heavily affected than at long wavelengths.

Outer Scale of Turbulence

With baselines from several 10 up to a few hundred metres, the phase structure function has to be corrected for the outer scale of turbulence, L_0, that is of the order of several 10 m [26,260]. Useful estimates for the phase structure function, $D_{\varphi,vK}(\boldsymbol{B})$, based on the von Kármán spectrum (4.28), can be found in [41,246]. The OPD variance with finite outer scale can be computed as $\sigma^2_{\text{OPD},vK} = \lambda^2/(2\pi)^2\, D_{\varphi,vK}(\boldsymbol{B})$

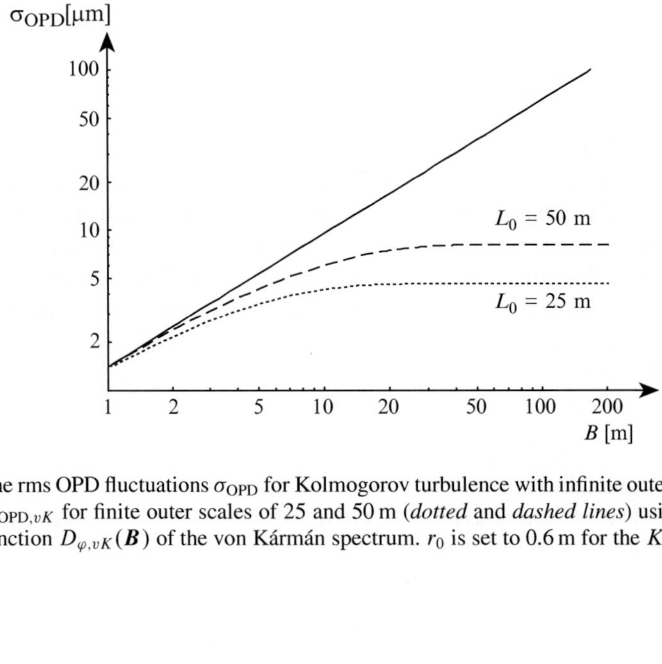

Fig. 4.4 The rms OPD fluctuations σ_{OPD} for Kolmogorov turbulence with infinite outer scale (*solid line*), and $\sigma_{\text{OPD},vK}$ for finite outer scales of 25 and 50 m (*dotted* and *dashed lines*) using the phase structure function $D_{\varphi,vK}(\boldsymbol{B})$ of the von Kármán spectrum. r_0 is set to 0.6 m for the K-band

and we obtain

$$\sigma^2_{\mathrm{OPD},vK} = 0.00876\,\lambda^2 \left(\frac{L_0}{r_0}\right)^{5/3} \left(1 - \left(\frac{2\pi B}{L_0}\right)^{5/6} K_{5/6}\left(\frac{2\pi B}{L_0}\right)\right), \quad (4.46)$$

with $K_{5/6}(.)$ the Macdonald function. This formula provides a useful approximation of the real situation although it is based on the assumption that the outer scale is constant at all altitudes, which is not confirmed by measurement [33, 98, 226].

The important conclusion is that the outer scale has a very strong influence on the OPD fluctuations, as displayed in Fig. 4.4. If the baseline is comparable in length to the outer scale of turbulence, the fluctuations are reduced by a factor of about five (!). For a 50-m baseline and infinite outer scale for instance, the rms OPD fluctuations σ_{OPD} are about 37 μm, according to (4.45) for an r_0 of 0.6 m in the K-band. If the outer scale is $L_0 = 50$ m, the rms OPD fluctuations are reduced to about 8 μm, using (4.46). The order of magnitude of this effect is confirmed by measurements [52, 57].

Differential Fringe Motion

The differential fringe motion can be derived in the same way as in (4.42). We write the difference in fringe positions of an object at $\boldsymbol{s}$ and an object at position $\boldsymbol{s} + \Delta\boldsymbol{s}$ as $\big(\varphi(\boldsymbol{\xi}) - \varphi(\boldsymbol{\xi} - \boldsymbol{B})\big) - \big(\varphi(\boldsymbol{\xi} - \Delta\boldsymbol{s}\,\bar{h}) - \varphi(\boldsymbol{\xi} - \Delta\boldsymbol{s}\,\bar{h} - \boldsymbol{B})\big)$ when the difference in angular position is represented by a linear shift of the turbulent layer at the effective altitude $\bar{h}$ as in (4.30).

Then, the convolution in (4.42) reads as

$$\Delta\varphi_{\Delta s}(\boldsymbol{\xi}) = \varphi(\boldsymbol{\xi}) * \left(\delta\left(\boldsymbol{\xi} - \frac{\Delta\boldsymbol{s}\,\bar{h}}{2} - \frac{\boldsymbol{B}}{2}\right) - \delta\left(\boldsymbol{\xi} - \frac{\Delta\boldsymbol{s}\,\bar{h}}{2} + \frac{\boldsymbol{B}}{2}\right)\right.$$
$$\left. - \delta\left(\boldsymbol{\xi} + \frac{\Delta\boldsymbol{s}\,\bar{h}}{2} - \frac{\boldsymbol{B}}{2}\right) + \delta\left(\boldsymbol{\xi} + \frac{\Delta\boldsymbol{s}\,\bar{h}}{2} + \frac{\boldsymbol{B}}{2}\right)\right). \quad (4.47)$$

As in (4.44) we move to the power spectrum $\Phi_{\mathrm{dOPD}}(\boldsymbol{\kappa})$ of the differential OPD fluctuations, dOPD, transforming the convolution of the phase with the δ-functions into the product of the power spectrum $\Phi(\boldsymbol{\kappa})$ of the phase, (4.27), with their Fourier transforms – in this case $4\sin^2(\pi\boldsymbol{B}\cdot\boldsymbol{\kappa})4\sin^2(\pi\bar{h}\Delta\boldsymbol{s}\cdot\boldsymbol{\kappa})$ – yielding

$$\Phi_{\mathrm{dOPD}}(\boldsymbol{\kappa}) = \frac{\lambda^2}{4\pi^2}\,0.0229\,r_0^{-5/3}\,\kappa^{-11/3}$$
$$\times 16\sin^2(\pi\boldsymbol{B}\cdot\boldsymbol{\kappa})\sin^2(\pi\bar{h}\Delta\boldsymbol{s}\cdot\boldsymbol{\kappa})\ \ [\mathrm{m}^2/\mathrm{m}^{-2}]. \quad (4.48)$$

Like Φ_{OPD} in (4.44), this power spectrum is not radial-symmetric, but it depends on the orientation of the baseline vector $\boldsymbol{B}$ and on the separation $\Delta\boldsymbol{s}$. There are now three regimes to be distinguished: At low frequencies, in the direction parallel to $\boldsymbol{B}$ and to Δs, the sine-functions can be approximated by $\pi^2 B^2 \kappa^2$ and by $\pi^2 \Delta s^2 \bar{h}^2 \kappa^2$ so that $\Phi_{\text{dOPD}}(\boldsymbol{\kappa})$ is proportional to $\kappa^{1/3}$. At medium frequencies, making the assumption that $\Delta s\,\bar{h}$ is much smaller than the baseline B, the sine-function of $\boldsymbol{B}$ can be regarded as constant while $\pi^2 \Delta s^2 \bar{h}^2 \kappa^2$ remains so that $\Phi_{\text{dOPD}}(\boldsymbol{\kappa}) \propto \kappa^{-5/3}$, and at high frequencies, both sine functions are effectively constant and the dependency on $\kappa^{-11/3}$ remains. In the orthogonal direction, $\Phi_{\text{dOPD}}(\boldsymbol{\kappa})$ is proportional to $\kappa^{-11/3}$ as $\Phi(\boldsymbol{\kappa})$. We will come back to these properties when discussing the temporal evolution.

Finite Apertures

Moving on to real apertures with diameter d, we express the interferometer aperture as in Sect. 3.3.2, by a convolution of the individual telescope aperture $a(\boldsymbol{\xi})$ with two δ-functions, putting the centres of the apertures at $-\boldsymbol{B}/2$ and $\boldsymbol{B}/2$.

We investigate the difference of the phases that are averaged over each individual sub-aperture. First, we look at the average phase in a single aperture that can be written as a convolution of $a(\boldsymbol{\xi})$ with the phase,

$$\varphi^d(\boldsymbol{\xi}) = \frac{1}{a_0} \int \varphi(\boldsymbol{\xi}') a(\boldsymbol{\xi} - \boldsymbol{\xi}') \mathrm{d}\boldsymbol{\xi}' , \qquad (4.49)$$

with a_0 the area of the sub-aperture and the superscript d indicating the averaged quantity.

The power spectrum of the averaged phase, $\mathcal{F}^2_{\xi}(\varphi^d(\boldsymbol{\xi}))$, can now be computed as the product of $\mathcal{F}^2_{\xi}(\varphi(\boldsymbol{\xi})) = \Phi(\boldsymbol{\kappa})$ and of $\mathcal{F}^2_{\xi}(a(\boldsymbol{\xi})) = \text{Besinc}^2(\pi d \kappa)$, yielding

$$\Phi^d_{\Delta\varphi}(\boldsymbol{\kappa}) = 0.0229\, r_0^{-5/3}\, \kappa^{-11/3} \text{Besinc}^2(\pi d \kappa) \;\; [\text{rad}^2/\text{m}^{-2}]. \qquad (4.50)$$

The squared Besinc function, that – as squared Fourier transform of the telescope aperture – has the form of an Airy disc, acts as a low pass filter on the power spectrum $\Phi(\boldsymbol{\kappa})$ of the phase fluctuations. The contributions at high frequencies corresponding to small distances in the turbulent wave front are reduced, as the averaging process smooths the gradients. At low frequencies, i.e. for distances larger than the aperture diameter the effect of the pupil averaging is much reduced and the power spectrum is approximately unaffected.

Combining the averaging of the phase over a single sub-aperture and the phase difference between two sub-apertures, we write the spatial power spectrum of OPD fluctuations between two apertures of diameter d and separated by the baseline $\boldsymbol{B}$ as

$$\Phi^d_{\text{OPD}}(\boldsymbol{\kappa}) = \frac{\lambda^2}{4\pi^2} \Phi^d_{\Delta\varphi}(\boldsymbol{\kappa}) 4 \sin^2(\pi \boldsymbol{B} \cdot \boldsymbol{\kappa}) \;\; [\text{m}^2/\text{m}^{-2}], \qquad (4.51)$$

and finally, by forming the product with $4\sin^2(\pi\bar{h}\Delta\boldsymbol{s}\cdot\boldsymbol{\kappa})$, the power spectrum of the differential OPD fluctuations can be written as

$$\begin{aligned}\Phi^d_{\mathrm{dOPD}}(\boldsymbol{\kappa}) = &\frac{\lambda^2}{4\pi^2}\Phi^d_{\Delta\varphi}(\boldsymbol{\kappa})\\ &\times 16\sin^2(\pi\boldsymbol{B}\cdot\boldsymbol{\kappa})\sin^2(\pi\bar{h}\Delta\boldsymbol{s}\cdot\boldsymbol{\kappa}) \quad [\mathrm{m}^2/\mathrm{m}^{-2}]. \end{aligned} \tag{4.52}$$

Computing the OPD variance of the phase fluctuations from (4.51) yields a slightly smaller value than in the case of point-like sub-apertures given in (4.45) since $\Phi^d_{\mathrm{OPD}}(\boldsymbol{\kappa})$ is filtered by $\mathrm{Besinc}^2(\pi d\kappa)$. This reflects the averaging effect over the individual sub-apertures. However, the effect is rather small since most of the power is at low frequencies due to the proportionality to $k^{-11/3}$, (4.50), so that the OPD variance in (4.45) can be used as a good, conservative approximation.

Temporal Evolution of Fringe Motion

The Taylor hypothesis of frozen turbulence is used again to estimate the effect of moving turbulence. The temporal power spectrum of the OPD fluctuations can be calculated similar to the one of the phase (4.41) by integrating over the direction perpendicular to the wind speed.

First, we discuss the case of two pinhole sized apertures with the power spectrum $\Phi_{\mathrm{OPD}}(\kappa_\xi,\kappa_\zeta)$, (4.44). Assuming that the effective wind direction is parallel to the ξ-axis, we set $\kappa_\xi = f/\bar{v}$, obtaining

$$\Phi_{\mathrm{OPD},t}(f) = \frac{1}{\bar{v}}\int \Phi_{\mathrm{OPD}}(f/\bar{v},\kappa_\zeta)\mathrm{d}\kappa_\zeta \quad [\mathrm{m}^2/\mathrm{Hz}]. \tag{4.53}$$

This integral cannot be solved in closed form but one can use approximations for low and high frequencies, averaged for all orientations $\bar{\boldsymbol{v}}$ vs. $\boldsymbol{B}$ [37,224] as

$$\begin{aligned}\Phi_{\mathrm{OPD},t}(f) &= 0.096\,\lambda^2\left(\frac{\bar{v}}{r_0}\right)^{5/3}\left(\frac{B}{\bar{v}}\right)^2 f^{-2/3} & f \ll 0.2\,\bar{v}/B\\ \Phi_{\mathrm{OPD},t}(f) &= 0.00392\,\lambda^2\left(\frac{\bar{v}}{r_0}\right)^{5/3} f^{-8/3} & f \gg 0.2\,\bar{v}/B\,, \end{aligned} \tag{4.54}$$

when the numerical values are averaged with respect to wind direction [139]. Note that temporal power spectra are defined for positive frequencies.

Compared to the temporal power spectrum $\Phi_t(f)$ of the phase, (4.41), we see the same $f^{-8/3}$ proportionality of the power spectrum at high frequencies but a rather flat spectrum $\propto f^{-2/3}$ at low frequencies. Converting the high frequency approximation of $\Phi_{\mathrm{OPD},t}$ to the power spectrum of the phase difference by multiplying it by $4\pi^2/\lambda^2$, we obtain $0.154(\bar{v}/r_0)^{5/3}f^{-8/3}$, which is twice the temporal power spectrum, $2\Phi_t(f)$, of the phase (4.41).

This can be explained by assuming a very long baseline with very small transient frequency $0.2\,\bar{v}/B$ so that the power spectrum is almost completely described by the high frequency approximation. Computing the variance of the phase difference from this power spectrum, we find a value that is twice as large as the variance of the phase since the phases that are separated by B are no longer correlated and, in order to compute the variance of the phase difference, twice its individual variance has to be taken. A similar argument was used when deriving the atmospheric coherence time $\tau_{0,2}$ in Sect. 4.2.3.

For shorter baselines the variance of the phase difference is smaller than twice that of the phase alone since the phases are more and more correlated. Formally this is expressed by the small frequency approximation taking more and more effect.

Using the power spectrum, we can calculate the **OPD variance** that we measure during a time period T. This describes the fringe jitter – and the related loss in visibility – during an observation.

The effect of the rectangular time window of width T is considered in frequency space by its Fourier transform, $\mathrm{sinc}(\pi T f)$. Since we are interested in the variance $<(\Delta\varphi - \overline{\Delta\varphi})^2>$, with $\overline{\Delta\varphi}$ the average OPD during time T, we have to compute the integral of $\Phi_{\mathrm{OPD},t}(f)$ filtered by $(1-\mathrm{sinc}^2(\pi T f))$, to obtain the OPD variance [224]. For very long integration times, the OPD averages to zero as discussed in Sect. 4.2. Sometimes, this computation is simplified by integrating the power spectrum from $f = 1/T$ to ∞, ignoring the contribution of frequencies smaller than $1/T$. However, due to the exponentially increasing power, the contribution of the small frequencies – the slow fringe motion – is substantial and cannot be neglected, as illustrated by Fig. 4.5.

Applying the filter function to the power spectrum in (4.54), we obtain an approximation for the rms OPD variations for integration times shorter than a few 100 ms as [224]

$$\sigma_{\mathrm{OPD},T} = 0.19\,\lambda \left(\frac{\bar{v}}{r_0}\right)^{5/6} T^{5/6}\ [\mathrm{m}]. \tag{4.55}$$

Writing $\sigma_{\mathrm{OPD},T}$ as a function of atmospheric coherence time, we replace $\bar{v}/r_0$ by $0.314/\tau_0$ (4.39) or, if the interferometric coherence time $\tau_{0,2}$ (4.40) is preferred, by $0.207/\tau_{0,2}$.

Under typical conditions, for instance an r_0 of 0.6 m, at a wavelength of 2.2 μm and an effective wind speed of $\bar{v} = 10\,\mathrm{m\,s^{-1}}$, we find that the rms OPD variations over 100 ms are about $\sigma_{\mathrm{OPD},T} = 600$ nm.

Note that due to the scaling of r_0 with $\lambda^{6/5}$, OPD variations are independent of wavelength and the product $\lambda r_0^{-5/6}$ could be replaced, see (4.24), by $4.09(\int C_n^2(h)\mathrm{d}h)^{1/2}$ if the turbulence profile is available.

It should be noted that this approach, regarding the OPD variations during a time period T, is different from the one used to define the atmospheric coherence time $\tau_{0,2}$ in (4.40). There, we investigated the mean-square phase difference of two points in time, at t and at $t+\tau$, and in space, separated by the baseline, and we found that an rms phase difference of 1 rad, corresponding to an OPD of $\lambda/(2\pi)$, was obtained

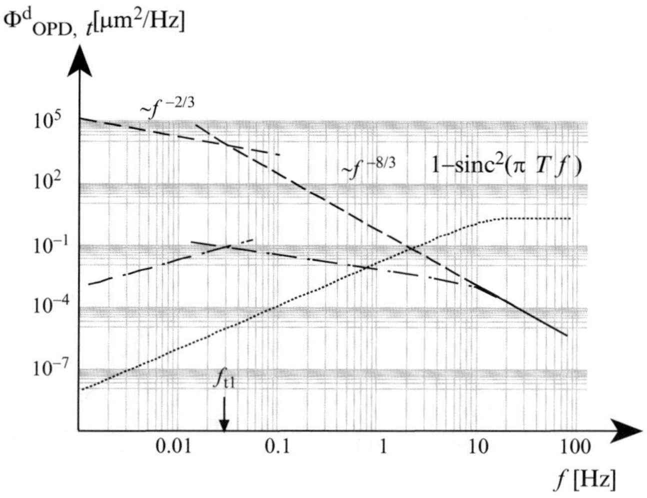

Fig. 4.5 Temporal power spectrum of the OPD fluctuations for an interferometer with baseline $B = 100\,\mathrm{m}$, pinhole sized apertures and an effective wind speed $\bar{v} = 10\,\mathrm{m\,s^{-1}}$, using the approximation given in (4.54). The *dashed lines* represent the two asymptotes $\propto f^{-2/3}$ and $\propto f^{-8/3}$ with transient frequency $f_{t1} = 0.2\,\bar{v}/B$, and the *dotted line* shows the filter function $(1 - \mathrm{sinc}^2(\pi T f))$ for an integration time of $T = 50\,\mathrm{ms}$. The filtered power spectrum is displayed by the *dashed-dotted lines* in the respective frequency regimes. The OPD variance over time T is computed by integrating the filtered power spectrum. The substantial contribution of the frequencies below $f = 1/T = 20\,\mathrm{Hz}$ is readily apparent

for a time difference of $\tau_{0,2}$. Here, we integrate over a time period T and compute the accumulated OPD variations so that T can be much longer than $\tau_{0,2}$ before the rms OPD variations are $\lambda/(2\pi)$. Using (4.55), we find that after an integration time of $T_0 = 0.81\,r_0/\bar{v}$ we have OPD variations of $\lambda/(2\pi)$, thus about four times longer than $\tau_{0,2}$ [224].

Finite Apertures

Expanding the discussion to sub-apertures with diameter d, we use the spatial power spectrum of OPD fluctuations, Φ^d_{OPD}, in (4.51) that is reduced at high spatial frequency due to the averaging over the sub-aperture. We apply the same integration perpendicular to the wind speed as for the power spectrum of pinhole sized apertures in (4.54), obtaining a third asymptote for frequencies larger than $0.3\,\bar{v}/d$ as

$$\Phi^d_{\mathrm{OPD},t}(f) = \quad 0.000105\lambda^2 \left(\frac{\bar{v}}{r_0}\right)^{5/3} \left(\frac{d}{\bar{v}}\right)^{-3} f^{-17/3} \qquad f \gg 0.3\,\bar{v}/d\,. \qquad (4.56)$$

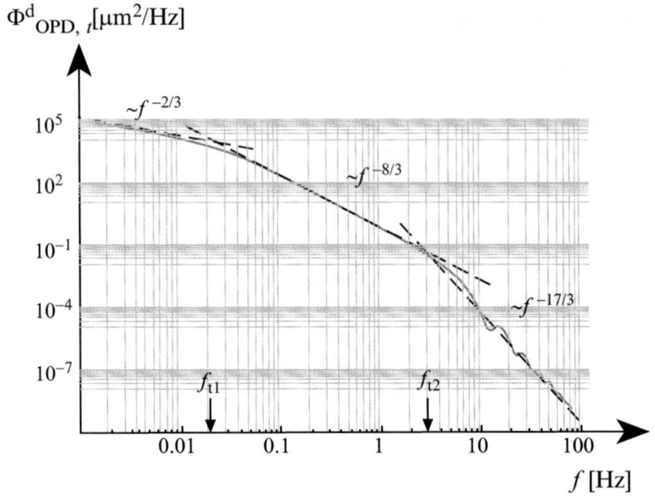

Fig. 4.6 Temporal power spectrum of the OPD fluctuations for an interferometer with baseline $B = 100\,\mathrm{m}$, sub-aperture diameter $d = 1\,\mathrm{m}$, effective wind speed $\bar{v} = 10\,\mathrm{m\,s^{-1}}$, and an r_0 of 0.6 m in the K-band. One can distinguish three regimes indicated by *dashed lines*: $\propto f^{-2/3}$ for $f \ll f_{t1} = 0.2\,\bar{v}/B$, $\propto f^{-8/3}$ for $f_{t1} \ll f \ll f_{t2} = 0.3\,\bar{v}/d$, and $\propto f^{-17/3}$ for $f \gg f_{t2}$. Approximative formulae for the three asymptotes are given in (4.54) and in (4.56)

The effect of the finite aperture diameter is, thus, that beyond $0.3\,\bar{v}/d$ the power spectrum has reduced values, dropping with the power of $-17/3$ instead of $-8/3$. However, the OPD variance is still dominated by the asymptote $\propto f^{-8/3}$, due to the filtering with $(1 - \mathrm{sinc}^2(\pi T f))$ as illustrated in Fig. 4.5 so that $\sigma_{\mathrm{OPD},T}$ in (4.55) provides a good, conservative estimate for the OPD variance.

The temporal power spectrum $\Phi^d_{\mathrm{OPD},t}$ with its three regimes is displayed in Fig. 4.6.

Measurements of OPD power spectra show in general the same behaviour as the theoretical models. The power spectrum can be subdivided in regimes with different asymptotes, and the proportionality to $f^{-8/3}$ has been found, with the tendency to have a smaller exponent, between $-7.5/3$ [26, 52] and $-6.5/3$ (Gitton, 2009, Private communication). The sharp decay $\propto f^{-17/3}$, however, has not been confirmed by measurements yet. Fitting measured spectra by asymptotes, one finds that especially the areas around the transient frequencies display deviations as shown in Fig. 4.7. Thus, the theoretical models and the derived approximations provide a general understanding of the effects, when it comes to details however, measurements have to be used.

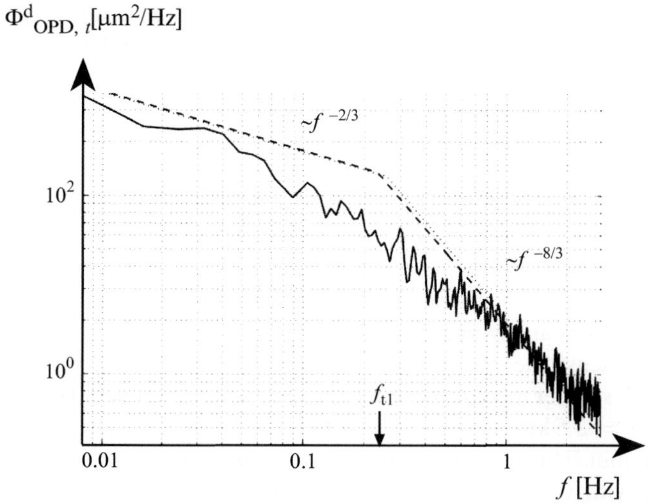

Fig. 4.7 Example of a measured power spectrum of OPD fluctuations, with interferometer baseline $B = 16\,\mathrm{m}$, sub-aperture diameter $d = 0.4\,\mathrm{m}$ and effective wind speed $\bar{v} = 20\,\mathrm{m\,s^{-1}}$. Two asymptotes, $\propto f^{-2/3}$ and $\propto f^{-8/3}$, with transient frequency $f_{t1} = 0.25\,\mathrm{Hz}$, are displayed by *dashed lines*. While at low and at high frequencies the asymptotes fit the curve rather well, there are substantial deviations around f_{t1}. Courtesy E. di Folco [57]

Outer Scale of Turbulence

For a finite outer scale of turbulence L_0 and for $\bar{\boldsymbol{v}} \parallel \boldsymbol{B}$, the following approximation [37] can be used for $f \ll 0.2\,\bar{v}/B$

$$\Phi_{\mathrm{OPD},t}^{L_0}(f) = 0.8\,\Phi_{\mathrm{OPD},t}(f)\left(1 + \left(\frac{\bar{v}}{f\,L_0}\right)^2\right)^{-4/3} \quad [\mathrm{m^2/Hz}]. \qquad (4.57)$$

This power spectrum is proportional to f^2, thus, dropping drastically towards lower frequencies. However, for typical wind speeds of $10\,\mathrm{m\,s^{-1}}$ and baselines of 100 m, this approximation is valid at frequencies below $f = 0.02\,\mathrm{Hz}$, i.e. for time differences longer than 50 s. It is questionable if the turbulence remains frozen over this period of time [52, 139], so that in practice, the turbulence decorrelates faster and the power spectrum is not reduced that much.

For the case of a wind direction perpendicular to the baseline, $\bar{\boldsymbol{v}} \perp \boldsymbol{B}$, one obtains a flat spectrum, that can be used to define the spectrum averaged for wind directions as $\propto f^0$ and $\propto L_0^{2/3}$ [37]. The flattening at very low frequencies has been confirmed by measurements, indicating at outer scales of turbulence of the order of 50 m [26, 52, 57].

One should note, that while the outer scale of turbulence has a large effect on the long term OPD variance (see Fig. 4.4), it has no effect on the OPD variance over a few seconds. This is not only intuitively understandable but also expressed by the correction to the power spectrum, (4.57), that only applies to low frequencies.

Temporal Evolution of Differential Fringe Motion

We use again the integral of the spatial power spectrum over the direction perpendicular to the wind speed as in (4.53) to compute the temporal power spectrum. A detailed analysis for all possible combinations of baseline $\boldsymbol{B}$, wind speed $\bar{v}$ and separation vectors $\Delta\boldsymbol{s}$ can be found in [51]. In the following, we will present the results when all directions are averaged [139].

We will discuss the case when the separation of beams $\Delta s\,\bar{h}$ in the turbulent layer is much smaller than the baseline. For a baseline of $B = 50$ m and an effective altitude of $\bar{h} = 2.5$ km, this means that the separation has to be much smaller than about 50 arcmin. This regime is referred to as *very narrow angle regime* [212].

Starting with very small apertures, the spatial power spectrum of differential OPD fluctuations, $\Phi_{\mathrm{dOPD}}(\boldsymbol{\kappa})$ is given in (4.48) and, integrating over κ_ζ as in (4.53), one obtains three asymptotes for different frequency regimes of the temporal power spectrum of differential fringe motion as

$$\begin{aligned}
\Phi_{\mathrm{dOPD},t}(f) &= 0.628\lambda^2\left(\frac{\bar{v}}{r_0}\right)^{5/3}\left(\frac{\Delta s\,\bar{h}}{\bar{v}}\right)^2\left(\frac{B}{\bar{v}}\right)^{2/3} \\
&\quad\times\left(1+4.84\,f^{4/3}\left(\frac{B}{\bar{v}}\right)^{4/3}\right) && f \ll 0.12\frac{\bar{v}}{B} \\
\Phi_{\mathrm{dOPD},t}(f) &= 0.193\lambda^2\left(\frac{\bar{v}}{r_0}\right)^{5/3}\left(\frac{\Delta s\,\bar{h}}{\bar{v}}\right)^2 f^{-2/3} && 0.12\frac{\bar{v}}{B} \ll f \ll 0.2\frac{\bar{v}}{\Delta s\,\bar{h}} \\
\Phi_{\mathrm{dOPD},t}(f) &= 0.0078\lambda^2\left(\frac{\bar{v}}{r_0}\right)^{5/3} f^{-8/3} && 0.2\frac{\bar{v}}{\Delta s\,\bar{h}} \ll f\,.
\end{aligned} \tag{4.58}$$

At very low frequencies the power spectrum is basically constant, slightly increasing with $f^{4/3}$ towards the transient frequency $0.12\,\bar{v}/B$. At medium and high frequencies, there is the same dependency $\propto f^{-2/3}$ resp. $f^{-8/3}$ as for fringe motion (4.54), but now the transient frequency between the two regimes is $0.2\,\bar{v}/(\Delta s\,\bar{h})$ instead of $0.2\,\bar{v}/B$.

Using the spatial power spectrum of differential OPD fluctuations (4.48) with the two sine functions of $\boldsymbol{B}\cdot\boldsymbol{\kappa}$ and of $\bar{h}\Delta\boldsymbol{s}\cdot\boldsymbol{\kappa}$, the function of $\boldsymbol{B}$ can be regarded as constant for $\kappa_\xi = f/\bar{v} \gg 0.12/B$, and only $\sin^2(\pi\bar{h}\Delta\boldsymbol{s}\cdot\boldsymbol{\kappa})$ takes effect. Therefore, in this regime, the effect of the separation vector for differential fringe motion is the same as that of the baseline for fringe motion, which is the consequence of computing the difference of phase differences.

Finite Apertures

Considering the effect of finite sub-apertures, we use the spatial power spectrum Φ^d_{dOPD} in (4.52) to compute the temporal power spectrum. Similar to the power spectrum for fringe motion, we can keep the low frequency approximations and replace the asymptote for high frequencies that is $\propto f^{-8/3}$ in (4.58) by two asymptotes, assuming that $\Delta s\,\bar{h}$ is much smaller than the sub-aperture diameter d, i.e. Δs is of the order of some 10 arcsec. Then, the two high frequency asymptotes of the temporal power spectrum read as

$$\begin{aligned}\Phi^d_{\mathrm{dOPD},t}(f) &= 0.00263\lambda^2\left(\frac{\bar{v}}{r_0}\right)^{5/3}\left(\frac{d}{\bar{v}}\right)^{-3} \\ &\quad\times\left(\frac{\Delta s\,\bar{h}}{\bar{v}}\right)^2 f^{-11/3} \qquad 0.24\frac{\bar{v}}{d} \ll f \ll 0.28\frac{\bar{v}}{\Delta s\,\bar{h}} \\ \Phi^d_{\mathrm{dOPD},t}(f) &= 0.00021\lambda^2\left(\frac{\bar{v}}{r_0}\right)^{5/3}\left(\frac{d}{\bar{v}}\right)^{-3} f^{-17/3} \qquad f \gg 0.28\frac{\bar{v}}{\Delta s\,\bar{h}}\,.\end{aligned} \tag{4.59}$$

The high frequency approximation $\propto f^{-17/3}$ is the same as that for fringe motion (4.56). However, the values are twice as large since we compute the differential fringe motion, so that, with no correlation between the individual phase values contributing to the differences at high frequencies, we have to add up the phase variances. We found the same behaviour for the high frequency approximation of fringe motion compared to phase fluctuations as discussed following (4.54).

The **variance of the differential fringe motion** is very important when discussing fringe tracking (see Sect. 6.2.1). Assuming that we keep the fringe position of one object stable, the fringe motion of the second object at distance Δs is given by the variance of the differential fringe motion. Thus, this permits to quantify the isoplanatic angle and to identify temporal effects of the turbulence on the visibility measurement.

Using again the filter function $(1 - \mathrm{sinc}^2(\pi T f))$ to consider the limited integration time T – as for $\sigma_{\mathrm{OPD},T}$ in (4.55) – we integrate the filtered temporal power spectrum $\Phi^d_{\mathrm{dOPD},t}$ to obtain the variance over time T. The two asymptotes in (4.59) provide approximations for $1/T \gg 0.24\bar{v}/d$, which means that T should be shorter than 1 s. Since power spectra $\propto f^{-17/3}$ have not been confirmed by measurements yet , we use only the asymptote $\propto f^{-11/3}$, obtaining a conservative estimate for the variance.

Then, the approximation for rms fluctuations of the differential fringe motion reads as

$$\sigma^d_{\mathrm{dOPD},T} = 0.1\,\lambda r_0^{-5/6}\,\bar{v}^{8/6}\bar{h}\,d^{-3/2}\,\Delta s\,T^{4/3}\ [\mathrm{m}]. \tag{4.60}$$

Apart from the atmospheric parameters r_0, $\bar{v}$ and $\bar{h}$, the differential fringe motion decreases with increasing telescope diameter, $d^{-3/2}$, it is linear with separation of the objects Δs and it is almost linear with exposure time, $T^{4/3}$, within the range of validity of this approximation, which is $\Delta s \ll d/\bar{h}$ and $T \ll 4.17\, d/\bar{v}$. If we go to longer integration times, the asymptote $\propto f^{-2/3}$ in (4.58) has to be used additionally so that the further increase of the variance with exposure time is slower but the linearity with Δs remains.

For atmospheric parameters of $r_0 = 0.6\,\mathrm{m}$ at $\lambda = 2.2\,\mu\mathrm{m}$, (remembering that the product $\lambda r_0^{-5/6}$ is independent of wavelength), $\bar{v} = 10\,\mathrm{m\,s^{-1}}$, $\bar{h} = 2.5\,\mathrm{km}$, and assuming that the telescope aperture has a diameter of $d = 2\,\mathrm{m}$, that the objects are separated by $\Delta s = 10\,\mathrm{arcsec}$ and that the exposure time is $T = 1\,\mathrm{s}$, we obtain a value of $\sigma^d_{\mathrm{dOPD},T} = 310\,\mathrm{nm}$ for the differential fringe motion.

The differential OPD variance for infinite exposure time has to be computed with the complete power spectrum adding the contributions of all asymptotes. The result shows that the fluctuations saturate at about 10 s integration time reaching about four times the value for a 1 s integration [51]. For the atmospheric parameters that we used above, the rms fluctuations are of the order of $1.2\,\mu\mathrm{m}$.

4.3.2 Image Motion

The average gradient of the phase distribution in the telescope aperture determines the position of the image in the telescope focus. The fluctuations of the image position are also referred to as *angle of arrival fluctuations*. The image motion determines the requirements for wave front sensors like the Shack-Hartmann sensor, see Sect. 6.3.1 that rely on reconstructing the wave front from gradient measurements in the sub-apertures.

First we discuss the statistical properties of the gradient $\nabla\varphi$ of the wave front without averaging over the telescope aperture. The two components $\nabla\varphi_\xi$ and $\nabla\varphi_\zeta$ as a function of the horizontal coordinate $\boldsymbol{\xi} = (\xi, \zeta)$ are [194, 224]

$$\nabla\varphi_\xi(\xi,\zeta) = -\frac{\lambda}{2\pi}\frac{\partial}{\partial\xi}\varphi(\xi,\zeta) \text{ and } \nabla\varphi_\zeta(\xi,\zeta) = -\frac{\lambda}{2\pi}\frac{\partial}{\partial\zeta}\varphi(\xi,\zeta)\,. \quad (4.61)$$

Instead of computing the first derivative, we use its Fourier transform, $-\mathrm{i}2\pi\kappa_{\xi,\zeta}\hat{\varphi}(\boldsymbol{\kappa})$, with $\hat{\varphi}(\boldsymbol{\kappa})$ the Fourier transform of $\varphi(\boldsymbol{\xi})$. We denote the Fourier transforms of the gradients by $\widehat{\nabla\varphi_\xi}(\boldsymbol{\kappa})$ and $\widehat{\nabla\varphi_\zeta}(\boldsymbol{\kappa})$.

Then, the power spectra of the vector components of the two gradients, $\Phi_{\nabla,\xi}(\boldsymbol{\kappa}) = |\widehat{\nabla\varphi_\xi}(\boldsymbol{\kappa})|^2$ and $\Phi_{\nabla,\zeta}(\boldsymbol{\kappa}) = |\widehat{\nabla\varphi_\zeta}(\boldsymbol{\kappa})|^2$, are related to the power spectrum of the phase $\Phi(\boldsymbol{\kappa}) = |\hat{\varphi}(\boldsymbol{\kappa})|^2$, (4.27), by

$$\Phi_{\nabla,\xi}(\boldsymbol{\kappa}) = \lambda^2\kappa_\xi^2\Phi(\boldsymbol{\kappa}) \text{ and } \Phi_{\nabla,\zeta}(\boldsymbol{\kappa}) = \lambda^2\kappa_\zeta^2\Phi(\boldsymbol{\kappa})\,, \quad (4.62)$$

and the power spectrum of the phase gradient $\Phi_\nabla(\boldsymbol{\kappa})$ is obtained by adding $\Phi_{\nabla,\xi}$ and $\Phi_{\nabla,\zeta}$, yielding

$$\Phi_\nabla(\boldsymbol{\kappa}) = 0.0229\,\lambda^2\left(\kappa_\xi^2 + \kappa_\zeta^2\right) r_0^{-5/3}\kappa^{-11/3} = 0.0229\,\lambda^2 r_0^{-5/3}\kappa^{-5/3}, \quad (4.63)$$

with $\kappa = |\boldsymbol{\kappa}|$.

Averaging the gradient over the **telescope aperture** provides the angular coordinate $\boldsymbol{\theta}$ of the image centroid. The averaging process is accounted for by convolving the gradient in (4.61) with the aperture function $A(\boldsymbol{\xi})$, writing

$$\boldsymbol{\theta}(\boldsymbol{\xi}) = \frac{1}{A_0}\int \nabla\varphi(\boldsymbol{\xi}')A(\boldsymbol{\xi} - \boldsymbol{\xi}')\mathrm{d}\boldsymbol{\xi}', \quad (4.64)$$

with A_0 the surface area of the aperture. $\boldsymbol{\theta}(\boldsymbol{\xi})$ is also referred to as *tip-tilt of the wave front*. For a point like aperture the averaging process collapses yielding $\boldsymbol{\theta} = \nabla\varphi(\boldsymbol{\xi})$.

As for (4.50) when we computed the averaged phase, we replace the convolution by a multiplication of the individual Fourier transforms in Fourier space, using $\mathcal{F}_\xi(\nabla\varphi(\boldsymbol{\xi})) = \widehat{\boldsymbol{\nabla}\boldsymbol{\varphi}}(\boldsymbol{\kappa})$ and $\mathcal{F}_\xi(A(\boldsymbol{\xi})) = \mathrm{Besinc}(\pi D\kappa)$.

The power spectrum of the phase gradient after averaging with the telescope aperture can then be written as the product of $\Phi_\nabla(\boldsymbol{\kappa}) = |\widehat{\boldsymbol{\nabla}\boldsymbol{\varphi}}(\boldsymbol{\kappa})|^2$, (4.63), and $\mathrm{Besinc}^2(\pi D\kappa)$ [40, 151], yielding

$$\Phi_\theta(\boldsymbol{\kappa}) = 0.0229\,\lambda^2 r_0^{-5/3}\kappa^{-5/3}\mathrm{Besinc}^2(\pi D\kappa)\ [\mathrm{rad}^2/\mathrm{m}^{-2}]. \quad (4.65)$$

This is also referred to as the power spectrum of the image motion or of the tip-tilt.

As in (4.50), the Besinc function acts as a low pass filter on the power spectrum. Since we are now regarding the phase gradient, the power spectrum decreases with $\kappa^{-5/3}$, slower than the power spectrum of the phase that is $\propto \kappa^{-11/3}$. But as for the power spectrum of the averaged phase, the averaging process only affects the small distances – the high spatial frequencies κ – and not the long distances.

The **variance $\boldsymbol{\sigma_\theta^2}$ of the image motion**, i.e. of the angle of arrival, can be computed as the integral over the power spectrum $\Phi_\theta(\boldsymbol{\kappa})$. Half of this value yields the one-axis variance of the position $\boldsymbol{\theta}$ of the image centroid as [243]

$$\sigma_\theta^2 = 0.17\left(\frac{\lambda}{D}\right)^2\left(\frac{D}{r_0}\right)^{5/3}\ [\mathrm{rad}^2]. \quad (4.66)$$

If λ/D, the FWHM of the Airy disk, is given in arcsec, one obtains the variance in arcsec^2. The two-axis variance is $2\sigma_\theta^2$.

NB 8. *The relationship between the variance of the phase, σ_{tt}^2, over the telescope aperture and the variance of the tip-tilt, σ_θ^2, can be computed fairly simple, yielding*

$$\sigma_{\mathrm{tt}}^2 = \frac{\pi^2}{4}\frac{\sigma_\theta^2}{(\lambda/D)^2}\ [\mathrm{rad}^2]. \quad (4.67)$$

Using the value for the variance of the image motion in (4.66), we obtain the variance of the aberrations in the telescope aperture due to one-axis image motion as

$$\sigma_{\mathrm{tt}}^2 = 0.419 \left(\frac{D}{r_0}\right)^{5/3} \ [\mathrm{rad}^2]. \tag{4.68}$$

We will come back to this point in the Sect. 4.3.3 discussing Zernike polynomials.

For a numerical example, we use again an 8-m telescope and the K-band ($\lambda = 2.2\,\mu\mathrm{m}$). With a typical value of $r_0 = 0.6\,\mathrm{m}$ at $\lambda = 2.2\,\mu\mathrm{m}$, we find the one-axis rms image motion as 0.2 arcsec, or 3.6 times the Airy disk diameter.

The dependence of σ_θ^2 on $D^{-1/3}$ means that the image motion in terms of arcsec decreases with increasing telescope diameter. However, the variance of the relative image motion – relative to the size of the diffraction limited PSF – increases with $D^{5/3}$.

It is important to note that, due to $r_0 \propto \lambda^{5/6}$, the product $\lambda^2\, r_0^{-5/3}$ in (4.66) and, thus, σ_θ^2 is independent of wavelength; the image motion in arcsec – like the OPD fluctuations in m – is the same at all wavelengths. Thus, wave front sensors like the Shack-Hartmann sensor measuring the wave front gradient can be operated in the visible for corrections at all wavelengths.

In (4.45), the variance of the OPD between two apertures was given. Converting the OPD of two points separated by $B = D$ into a tilt angle of the wave front, $\theta = \mathrm{OPD}/B$, we see that it is exactly $\sigma_\theta^2 = \sigma_{\mathrm{OPD}}^2/B^2$. Thus, the one-axis variance of the image motion in a telescope with diameter D corresponds to the variance of the OPD fluctuations of an interferometer with baseline $B = D$.

The influence of the **outer scale of turbulence L_0** on image motion can be estimated by computing the power spectrum of the phase gradient (4.65) by using the von Kármán spectrum of the phase fluctuations (4.28). As for fringe motion, the impact of L_0 on image motion can be substantial if the telescope diameter approaches the outer scale of turbulence. If the two quantities are about the same, the image motion is reduced by more than 80%, and even if the telescope is 100 times smaller than L_0 the reduction on the rms image motion is still about 15% [207]. These results are comparable to those for fringe motion using (4.46).

For our numerical example, this means that on an 8-m telescope with an outer scale of approximately 25–50 m, [260], the image motion is reduced by about 50%, while for larger telescopes of e.g 42 m diameter, the reduction of image motion can be as much as a factor of 8–10. Then the image motion can be as small as the diameter of the diffraction limited Airy disk in the K-band, which is 11 mas on a 42-m telescope.

Temporal Evolution of Image Motion

Using the Taylor hypothesis of frozen turbulence as for the previous computations of temporal power spectra, we determine the temporal power spectrum of the averaged phase gradient as

$$\Phi_{\theta,t}(f) = \frac{1}{\bar{v}} \int \Phi_\theta(f/\bar{v}, \kappa_\zeta) \mathrm{d}\kappa_\zeta \ [\mathrm{rad}^2/\mathrm{Hz}]. \tag{4.69}$$

Tyler [243] gave low and high frequency approximations for this integral that can be simplified by using the effective wind speed $\hat{v}$ [81]. Then, the one-axis power spectrum of the image motion in the two regimes is

$$\Phi_{\theta,t}(f) = 0.096\,\lambda^2\,\bar{v}^{-2}\left(\frac{\bar{v}}{r_0}\right)^{5/3} f^{-2/3} \ [\mathrm{rad}^2/\mathrm{Hz}] \quad f \ll 0.24\,\bar{v}/D \tag{4.70}$$

$$\Phi_{\theta,t}(f) = 0.0013\,\lambda^2\,\bar{v}^{-2}\left(\frac{D}{\hat{v}}\right)^{-3}\left(\frac{\bar{v}}{r_0}\right)^{5/3} f^{-11/3} \ [\mathrm{rad}^2/\mathrm{Hz}] \quad f \gg 0.24\,\bar{v}/D\,.$$

The value of the transient frequency, $f_t = 0.24\,\bar{v}/D$, agrees well with the value given by Conan [40]. $\Phi_{\theta,t}(f)$ is displayed in Fig. 4.8. Note that the temporal power spectrum is defined from 0 to $+\infty$.

In the low frequency domain the power spectrum decreases with $f^{-2/3}$, and it is independent of the size of the aperture D. Compared to the low frequency approximation for OPD fluctuations (4.54), we have exactly the same formula considering that the tilt angle $\boldsymbol{\theta}$ is converted into an OPD by OPD $= \boldsymbol{\theta} \cdot \boldsymbol{B}$, so that it is $\Phi_{\mathrm{OPD},t}(f) = B^2 \Phi_{\theta,t}(f)$.

In the high frequency domain the spectrum is proportional to $f^{-11/3}$ and to D^{-3}, decreasing for increasing telescope diameter. This illustrates the influence of the Bessel function as a low pass filter that leaves the low frequency domain unaffected

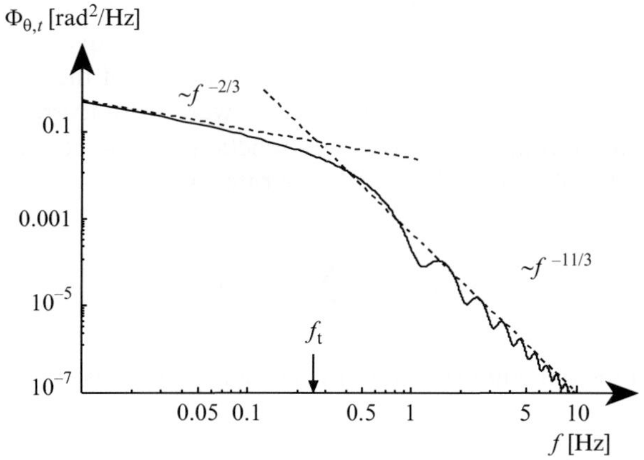

Fig. 4.8 One-axis power spectrum of the image motion defined for positive frequencies. The dashed lines show the approximations for the low frequency domain ($\propto f^{-2/3}$) and the high frequency domain ($\propto f^{-11/3}$). The transient frequency is approximately $f_t = 0.24\,\bar{v}/D$. The solid curve shows the exact calculation. The parameters are $D = 8$ m and $\hat{v} = 8\,\mathrm{m\,s^{-1}}$, so that $f_t = 0.24$ Hz

by the aperture and that takes effect as soon as the frequency is beyond the transient frequency f_t.

Using the von Kármán spectrum to account for a finite outer scale of turbulence L_0, the power spectrum (4.70) is reduced at low frequencies so that the variance of the image motion is reduced, too. The consequence is that the seeing disk has a smaller FWHM, as discussed in Sect. 4.3.5 (4.88).

However, one has to be careful when applying the Taylor hypothesis at very low frequencies, corresponding to very long times [139]. The turbulence might not be frozen any more. For many applications it is more important to properly model the high frequency behaviour since these frequencies remain after correction of the low frequencies e.g. with adaptive optics.

Because of the steep slope ($\propto f^{-11/3}$) of the power spectrum at frequencies beyond f_t the contributions to the image motion are small, only 10% of the power is at frequencies larger than f_t. However, adaptive optics systems have to be operated at substantially larger bandwidth if a good correction is to be achieved. One should note, too, that the outer scale of turbulence, affecting the low frequencies only, has no effect on the performance of adaptive optics systems working with a bandwidth larger than f_t. In Chap. 6, the bandwidth requirements will be discussed in more detail.

4.3.3 Zernike Representation of Atmospheric Turbulence

We introduced Zernike polynomials in Sect. 3.1.2 discussing optical aberrations. Their convenience lies in the property that, following from the Kolmogorov statistics, one can determine individually the power in every single mode like tip-tilt, astigmatism or coma. One can then immediately calculate the residual aberration after correcting a specified number of modes with an adaptive optics system. This computation was done by R.J. Noll [171]. The variance of the residual aberration is expressed as the variance of the difference between the uncorrected phase and of the removed modes. If the aberration – as a function of polar coordinates ρ, θ – that is due to the first J Zernike polynomials is written as

$$\varphi_J(\rho, \theta) = \sum_{i=1}^{J} a_i Z_i(\rho, \theta), \tag{4.71}$$

the variance of the remaining aberrations can be expressed as

$$\sigma_J^2 = \iint_{\text{aperture}} < [\varphi(\rho, \theta) - \varphi_J(\rho, \theta)]^2 > \rho \mathrm{d}\rho \mathrm{d}\theta. \tag{4.72}$$

As already noted, the variance of the phase fluctuations $< \varphi^2(\rho, \theta) >$ is infinite. The analysis in terms of Zernike polynomials shows that the infinity lies in the piston

Table 4.1 The residual variances Δ_j of Kolmogorov turbulence after the first j Zernike modes are removed. The difference in the right column illustrates the strength of the single modes demonstrating that modes of equal radial order contribute the same amount to the variance

Δ_1	$= 1.030(D/r_0)^{5/3}$		
Δ_2	$= 0.582(D/r_0)^{5/3}$	$\Delta_2 - \Delta_1$	$= 0.448(D/r_0)^{5/3}$
Δ_3	$= 0.134(D/r_0)^{5/3}$	$\Delta_3 - \Delta_2$	$= 0.448(D/r_0)^{5/3}$
Δ_4	$= 0.111(D/r_0)^{5/3}$	$\Delta_4 - \Delta_3$	$= 0.023(D/r_0)^{5/3}$
Δ_5	$= 0.0880(D/r_0)^{5/3}$	$\Delta_5 - \Delta_4$	$= 0.023(D/r_0)^{5/3}$
Δ_6	$= 0.0648(D/r_0)^{5/3}$	$\Delta_6 - \Delta_5$	$= 0.023(D/r_0)^{5/3}$
Δ_7	$= 0.0587(D/r_0)^{5/3}$	$\Delta_7 - \Delta_6$	$= 0.0062(D/r_0)^{5/3}$
Δ_8	$= 0.0525(D/r_0)^{5/3}$	$\Delta_8 - \Delta_7$	$= 0.0062(D/r_0)^{5/3}$
Δ_9	$= 0.0463(D/r_0)^{5/3}$	$\Delta_9 - \Delta_8$	$= 0.0062(D/r_0)^{5/3}$
Δ_{10}	$= 0.0401(D/r_0)^{5/3}$	$\Delta_{10} - \Delta_9$	$= 0.0062(D/r_0)^{5/3}$
Δ_{11}	$= 0.0377(D/r_0)^{5/3}$	$\Delta_{11} - \Delta_{10}$	$= 0.0024(D/r_0)^{5/3}$
Δ_{12}	$= 0.0352(D/r_0)^{5/3}$	$\Delta_{12} - \Delta_{11}$	$= 0.0024(D/r_0)^{5/3}$

Table 4.2 Covariance matrix between the first 11 Zernike modes as given by N. Roddier [199]. For an optimal reconstruction of the turbulent wave front the covariance matrix should be diagonal

$$\begin{pmatrix}
0.448 & 0 & 0 & 0 & 0 & 0 & -0.0141 & 0 & 0 & 0 & 0 \\
0 & 0.448 & 0 & 0 & 0 & -0.0141 & 0 & 0 & 0 & 0 & 0 \\
0 & 0 & 0.0232 & 0 & 0 & 0 & 0 & 0 & 0 & -0.0039 & 0 \\
0 & 0 & 0 & 0.0232 & 0 & 0 & 0 & 0 & 0 & 0 & 0 \\
0 & 0 & 0 & 0 & 0.0232 & 0 & 0 & 0 & 0 & 0 & -0.0039 \\
0 & -0.0141 & 0 & 0 & 0 & 0.0062 & 0 & 0 & 0 & 0 & 0 \\
-0.0141 & 0 & 0 & 0 & 0 & 0 & 0.0062 & 0 & 0 & 0 & 0 \\
0 & 0 & 0 & 0 & 0 & 0 & 0 & 0.0062 & 0 & 0 & 0 \\
0 & 0 & 0 & 0 & 0 & 0 & 0 & 0 & 0.0062 & 0 & 0 \\
0 & 0 & -0.0039 & 0 & 0 & 0 & 0 & 0 & 0 & 0.0024 & 0 \\
0 & 0 & 0 & 0 & -0.0039 & 0 & 0 & 0 & 0 & 0 & 0.0024
\end{pmatrix}$$

term. Removing the piston term gives a finite value for the variance of the residual aberration. The residual variances in Table 4.1 are given in terms of $(D/r_0)^{5/3}$ as the Zernike polynomials are defined in the telescope aperture D. The right column of the table shows the differential improvement. It shows that the differences are constant for the same radial degree n.

In Sect. 4.3.2, discussing image motion, we found the variance of the phase in the telescope aperture due to one-axis image motion as $0.419\,(D/r_0)^{5/3}$, (4.68). This value, based on the gradient of the wave front, is slightly smaller than that of the Zernike tip-tilt, $\Delta_2 - \Delta_1$, of $0.448(D/r_0)^{5/3}$ in Table 4.1. This is due to the fact that higher order Zernike modes like coma, $Z_{7,8}$ are balanced by a linear term (see Sect. 3.1.2) so that the contribution of this aberration to the overall aberration is minimized. The consequence is that the phase gradient is not only determined by the linear tip-tilt mode but it is also sensitive to coma. In Table 4.2, one finds that the covariance between tip-tilt and coma has a negative value, so that the variance of the phase gradient is smaller than the variance of the pure tip-tilt [187, 243].

For the removal of higher orders, Noll [171] gave the approximation for the phase variance as

$$\Delta_J \approx 0.2944 J^{-\sqrt{3}/2} (D/r_0)^{5/3} \, [\mathrm{rad}^2] \,. \tag{4.73}$$

Expressing the wave front as a Zernike polynomial the covariance matrix of the expansion coefficients $< a_i a_{i'} >$ plays an important role. This matrix can be calculated using the power spectrum of the phase fluctuations [171, 199]. It turns out that the covariance matrix is not perfectly diagonal. This means that when describing Kolmogorov turbulence with Zernike polynomials the Zernike modes are not statistically independent with the consequence that the wave front reconstruction from the wave front sensor data is sub-optimal. Karhunen-Loève functions are more appropriate as they have a diagonal covariance matrix. They cannot be obtained in closed form, but there is an approximation for Zernike polynomials [199].

Temporal Evolution of Zernike Modes

Using the same formalism as for the analysis of the temporal characteristics of the image motion (Sect. 4.3.2) the temporal evolution of Zernike modes can be calculated [40, 171, 198]. The results are important for the specification of the bandwidth requirements of adaptive optics systems.

In the last section, the covariance matrix of the Zernike coefficients $< a_i a_{i'} >$ was calculated. Now we are interested in the temporal correlation of single Zernike coefficients $< a_i(t') a_i(t'+t) >$. Thus, equivalent to the calculation of the power spectrum of the image motion (Sect. 4.3.2) we determine a Zernike coefficient as the convolution

$$a_i(\rho, \theta) = \int_{\mathrm{aperture}} \varphi(\rho', \theta') Z_i(\rho - \rho', \theta - \theta') \rho \mathrm{d}\rho' \mathrm{d}\theta' \,. \tag{4.74}$$

At $\rho = 0$ and $\theta = 0$ this equation is identical to the calculation of the Zernike coefficient a_i (3.19) that can be used, [171], to calculate the variance of the Zernike modes (see Table. 4.1). The temporal covariance follows from the spatial covariance $< a_i(\rho', \theta') \, a_i(\rho' + \rho, \theta' + \theta) >$ by using the frozen turbulence hypothesis similar to the calculation of the image motion (4.69).

The resulting power spectra cannot be given in closed form but numerical results were discussed in [40, 198]. The spectra show a dependence on the radial degree of the Zernike polynomial at low frequencies and a high frequency behaviour proportional to $f^{-17/3}$ that is independent of the Zernike mode. In the low frequency domain, polynomials with a radial degree of $n = 1$, Zernike tip and tilt, decrease with $f^{-2/3}$. Higher order polynomials have a slightly different characteristic depending on their azimuthal dependence; all radially symmetric polynomials go with f^0, all others with f^0, with $f^{4/3}$ or with f^2 depending on the wind direction.

The transient frequency between the high and the low frequency domains can be approximated by

$$f_t^n \approx 0.3(n+1)\hat{v}/D, \tag{4.75}$$

where n is the radial degree of the Zernike polynomials. The transient frequency is approximately equal to the bandwidth required to correct for the Zernike mode in an adaptive optics system.

Averaging the Zernike spectra for a given radial degree shows the mean behaviour for this degree. This behaviour can also be modelled when using a multi layer model with different wind directions that is more realistic than the single layer approach [40]. The curves are displayed in Fig 4.9. It shows that for $n > 1$ the power spectra are all proportional to f^0 at low frequencies. The curves are scaled in order to give the proper variance of the single modes (see Table 4.1). It is interesting to note that at high frequencies the power spectra increase only very slowly with n. The increase in transient frequency f_t, and thus in bandwidth is partially compensated by the decrease in variance at higher radial degrees. For a given degree of correction all modes have to be corrected with approximately the same bandwidth [40].

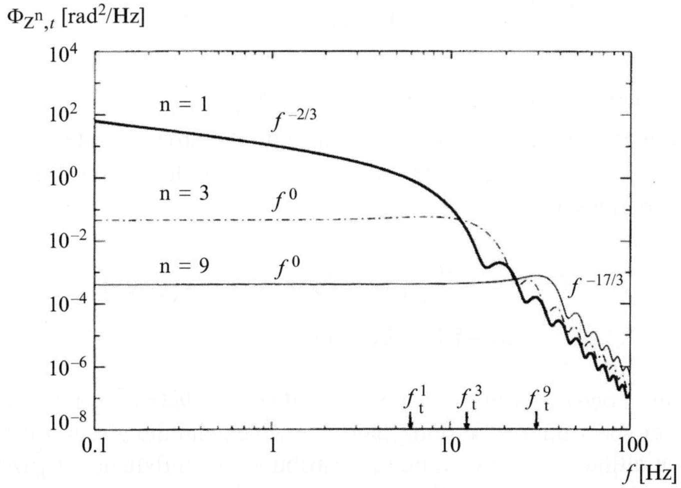

Fig. 4.9 Zernike polynomial mean temporal power spectrum in a given radial degree n for $n = 1$, 3, 9. The spectra are normalised to the turbulence variance of one polynomial of the considered radial degree: $\hat{v}/D = 10$ Hz. The asymptotic power laws and the cutoff frequencies are indicated. (Courtesy J.-M. Conan [40])

4.3.4 Scintillation

The scintillation, the twinkling of the stars, is easily observable with the naked eye. These intensity variations are due to random focussing effects of atmospheric turbulence that depend on the local curvature $\frac{\partial^2 \varphi(\xi)}{\partial \xi^2}$ of the phase. We will see that the scintillation is strong when observing with the naked eye, averaging the intensity over the small aperture of the eye's pupil, and it is very weak when using large telescopes. An exhaustive summary of scintillation effects and measurements was given by Dravins et al. [61–63].

The intensity variations are caused by Fresnel diffraction at the turbulent layer at some altitude above the telescope. Thus, we cannot apply the near-field approximation assuming that the Fresnel diffraction is negligible and that the atmospheric turbulence is placed in the telescope aperture, as we did in the previous sections, but we have to investigate the Fresnel diffraction between the layer and the ground.

In Sect. 4.2.2, we discussed the propagation of the wave front through individual turbulent layers, using the Fresnel approximation. We modify (4.19) to obtain the amplitude $V_\varphi(\boldsymbol{\xi})$ on the ground for a layer at altitude h as approximately $1 + \mathrm{i}\varphi(\boldsymbol{\xi}) + h\lambda/(4\pi)\frac{\partial^2 \varphi(\xi)}{\partial \xi^2}$. The intensity can then be written as

$$I(\boldsymbol{\xi}) \approx 1 + \frac{h\lambda}{2\pi}\left(\frac{\partial^2 \varphi(\xi,\zeta)}{\partial \xi^2} + \frac{\partial^2 \varphi(\xi,\zeta)}{\partial \zeta^2}\right), \tag{4.76}$$

neglecting quadratic terms like $\varphi^2(\boldsymbol{\xi})$ and $\left(\frac{\partial^2 \varphi(\xi)}{\partial \xi^2}\right)^2$.

The second derivative of $\varphi(\boldsymbol{\xi})$ transforms into the product $-\mathrm{i}4\pi^2\kappa_{\xi,\zeta}^2\hat{\varphi}(\boldsymbol{\kappa})$, with $\hat{\varphi}(\boldsymbol{\kappa})$ the Fourier transform of $\varphi(\boldsymbol{\xi})$, and we can write the Fourier transform of the intensity variations as

$$\begin{aligned}\hat{I}(\boldsymbol{\kappa}) &= -\mathrm{i}\frac{h\lambda}{2\pi}\left(4\pi^2\kappa_\xi^2\hat{\varphi}(\boldsymbol{\kappa}) + 4\pi^2\kappa_\zeta^2\hat{\varphi}(\boldsymbol{\kappa})\right) \\ &= -\mathrm{i}\,2\pi\, h\lambda\, \kappa^2\hat{\varphi}(\boldsymbol{\kappa}).\end{aligned} \tag{4.77}$$

Using the power spectrum of the phase $\Phi(\boldsymbol{\kappa}) = |\hat{\varphi}(\boldsymbol{\kappa})|^2$, (4.27), we proceed to the power spectrum of the scintillation. If we extend the discussion from a single layer at altitude h to a continuous distribution of turbulence, $C_n^2(h)$, we have to compute the integral over C_n^2 weighted by h^2, obtaining the power spectrum $\Phi_I(\boldsymbol{\kappa}) = |\hat{I}(\boldsymbol{\kappa})|^2$ of the scintillation as

$$\Phi_I(\boldsymbol{\kappa}) = 4\pi^2\,\lambda^2\bar{h}_2^2\kappa^4\,0.0229\,r_0^{-5/3}\kappa^{-11/3} = 0.904\,\lambda^2\bar{h}_2^2\,r_0^{-5/3}\kappa^{1/3}, \tag{4.78}$$

with the effective altitude $\bar{h}_2$ defined by $\bar{h}_2 = \left(\int C_n^2(h)h^2\mathrm{d}h / \int C_n^2(h)\mathrm{d}h\right)^{1/2}$.

Unlike the power spectra of fringe motion, Φ_{OPD} (4.44), and of image motion, Φ_∇ (4.63), that decrease with κ, Φ_I is proportional to $\kappa^{1/3}$, slightly increasing with spatial frequency. This means that there is more power, i.e. larger intensity fluctuations, on small scales than on large scales, whereas the fringe and image motion have smaller values on small scales.

Computing the variance of the scintillation requires to integrate over the power spectrum. The Fresnel approximation for small phase variations that we have used to derive Φ_I is not sufficiently accurate for large spatial frequencies. Roddier [194] presents an approximation that is valid over the complete frequency space, yielding the intensity variance as

$$\begin{aligned}\sigma_I^2 &= 19.12\lambda^{-7/6} \int C_n^2(h) h^{5/6} \mathrm{d}h \\ &= 1.145\, r_0^{-5/3} (\lambda \bar{h}_{5/6})^{5/6}\,. \end{aligned} \qquad (4.79)$$

Here, the effective altitude is defined by $\bar{h}_{5/6} = \left(\int C_n^2(h) h^{5/6} \mathrm{d}h / \int C_n^2(h) \mathrm{d}h\right)^{6/5}$.

σ_I^2 is the variance of the scintillation on the ground without any optical systems. Observing through small apertures this approximation can be used as long as the apertures are smaller than $(\lambda \bar{h}_{5/6})^{1/2}$. With $\bar{h}_{5/6}$ typically several kilometres, (4.79) is valid for apertures of some cm.

For **larger apertures** with diameter D, the power spectrum Φ_I in (4.78) has to be multiplied by $\mathrm{Besinc}^2(\pi D \kappa)$ – the equivalent of the convolution in the aperture plane – as we did computing the power spectrum of image motion, Φ_∇^D (4.65). If D is much larger than $(\lambda \bar{h}_{5/6})^{1/2}$, the filtering effect of the aperture is sufficiently strong so that the approximation leading to the power spectrum in (4.78) is appropriate [194], and we obtain the intensity power spectrum in the focal plane of a telescope as

$$\Phi_I^D(\kappa) = 0.904\, \lambda^2 \bar{h}_2^2\, r_0^{-5/3} \kappa^{1/3}\, \mathrm{Besinc}^2(\pi D \kappa). \qquad (4.80)$$

The slightly increasing power $\propto \kappa^{1/3}$ in $\Phi_I(\kappa)$ is turned into a decreasing function $\Phi_I^D(\kappa)$ for larger spatial frequencies due to the averaging effect over spatial frequencies larger than $1/D$.

The intensity variance in the focal plane of telescopes with diameter D is then given by

$$\begin{aligned}\sigma_{I,D}^2 &\propto D^{-7/3} \int C_n^2(h) h^2 \mathrm{d}h \\ &\propto D^{-7/3}\, r_0^{-5/3} (\lambda \bar{h}_2)^2\,. \end{aligned} \qquad (4.81)$$

The expression for $\sigma_{I,D}^2$ can also be derived from geometrical optics [191]. It is valid for telescope apertures larger than $(\pi \lambda \bar{h}_2)^{1/2}$. For telescope apertures of several metres the scintillation is extremely small since $\sigma_{I,D}^2$ decreases strongly with increasing D.

Like the variances of image motion and fringe motion, the variance of the scintillation $\sigma_{I,D}^2$ is independent of wavelength, whereas the expression σ_I^2 for very small apertures, (4.79), decreases with longer wavelengths.

It is interesting to note that the effective altitude $\bar{h}_2$, which affects the intensity variance (4.81) when observing through a telescope, is more sensitive to high altitude turbulence than $\bar{h}_{5/6}$ that determines the intensity variance (4.79) in very small apertures. This can be explained by considering that small scale intensity variations of low altitude turbulence are reduced due to the averaging over a large aperture so that high altitude turbulence is relatively more important.

Temporal Evolution of Scintillation

The Taylor hypothesis of frozen turbulence is used again to determine the temporal power spectrum of scintillation that we obtain by integrating over the component of the spatial frequency vector $\boldsymbol{\kappa}$ that is perpendicular to the wind direction, yielding

$$\Phi_{I,t}(f) = \frac{1}{\bar{v}} \int \Phi_I^D(f/\bar{v}, \kappa_\zeta) \mathrm{d}\kappa_\zeta \, , \tag{4.82}$$

obtaining a temporal power spectrum that is constant, $\propto f^0$, for frequencies up to about $0.4\bar{v}/D$ and decreasing $\propto f^{-5/3}$ for larger frequencies [40].

4.3.5 Speckle Pattern and Seeing Disk

Assuming a point source at infinity, a plane wave, $V(\boldsymbol{\xi}) = V_0 \exp(\mathrm{i}kz)$ from a point-like star at zenith enters the atmosphere. After passage through atmospheric turbulence we find the amplitude in the telescope aperture as $V_\varphi(\boldsymbol{\xi}) = V(\boldsymbol{\xi}) \exp(\mathrm{i}\varphi(\boldsymbol{\xi}))$, with random phase $\varphi(\boldsymbol{\xi})$. Usually, we set $V(\boldsymbol{\xi}) = V_0 = 1$, and we write the instantaneous amplitude $V(\boldsymbol{\alpha})$ in the image plane of a telescope, as in (2.22), as the Fourier transform of $V_\varphi(\boldsymbol{\xi})$ in the telescope aperture,

$$V(\boldsymbol{\alpha}) = \frac{1}{\mathrm{i}\lambda F} \int A(\boldsymbol{\xi}) V_\varphi(\boldsymbol{\xi}) \mathrm{e}^{-\mathrm{i}\frac{2\pi}{\lambda}\boldsymbol{\xi}\cdot\boldsymbol{\alpha}} \mathrm{d}\boldsymbol{\xi} \, , \tag{4.83}$$

with $A(\boldsymbol{\xi})$, the telescope aperture.

The image intensity distribution in the image plane is given by $I(\boldsymbol{\alpha}) = |V(\boldsymbol{\alpha})|^2$ and it shows the characteristic *speckle pattern* that is displayed in Fig. 4.10. There are approximately as many speckles in the image as there are turbulence cells of diameter r_0 in the telescope aperture.

Thus, after passage through atmospheric turbulence, the instantaneous image of a point-like star is scattered into a speckle cloud when each speckle has the size of an Airy disk.

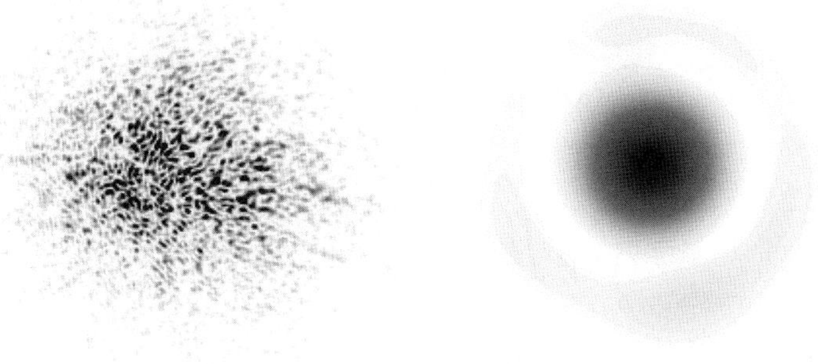

Fig. 4.10 Speckle images of a single point-like star in the visible (*on the left*) and in the mid infrared at 10 μm (*on the right*) under identical atmospheric conditions on an 8-m telescope. In the 10 μm image, parts of the first diffraction ring can be seen. The envelope of the speckle image has about the size of the seeing disk. A simulation for atmospheric turbulence was used to produce the images

Fig. 4.11 Speckle pattern in the image plane of an interferometer when observing a point-like star through atmospheric turbulence. The interference fringes are now distributed irregularly due to the random phase distribution in the interferometer aperture

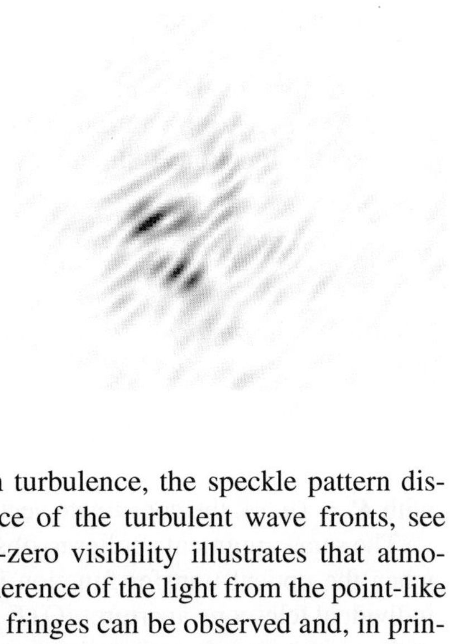

Observing with an interferometer through turbulence, the speckle pattern displays irregular fringes due to the interference of the turbulent wave fronts, see Fig. 4.11. The presence of fringes with non-zero visibility illustrates that atmospheric turbulence does not extinguish the coherence of the light from the point-like star. As long as the exposure time is short the fringes can be observed and, in principle, the visibility can be determined by processing the Fourier transform of the speckle pattern as discussed in Sect. 2.4.3.

Seeing Disk

If the exposure time of the speckle image is longer than the time scale of the evolution of the speckles, i.e. longer than the coherence time of atmospheric turbulence, the speckle start to smear until they merge into the so-called *seeing disk* for an exposure time of some seconds.

We write the time averaged image intensity as

$$
\begin{aligned}
< I(\boldsymbol{\alpha}) > &= < \frac{1}{(\lambda F)^2} \int A(\boldsymbol{\xi}_1) V_\varphi(\boldsymbol{\xi}_1) \mathrm{e}^{-\mathrm{i}\frac{2\pi}{\lambda}\boldsymbol{\xi}_1 \cdot \boldsymbol{\alpha}} \mathrm{d}\boldsymbol{\xi}_1 \\
&\quad \times \int A^*(\boldsymbol{\xi}_2) V_\varphi^*(\boldsymbol{\xi}_2) \mathrm{e}^{\mathrm{i}\frac{2\pi}{\lambda}\boldsymbol{\xi}_2 \cdot \boldsymbol{\alpha}} \mathrm{d}\boldsymbol{\xi}_2 > \\
&= \frac{1}{(\lambda F)^2} \iint A(\boldsymbol{\xi}_1) A^*(\boldsymbol{\xi}_1 - \boldsymbol{\xi}) \\
&\quad \times < V_\varphi(\boldsymbol{\xi}_1) V_\varphi^*(\boldsymbol{\xi}_1 - \boldsymbol{\xi}) > \mathrm{d}\boldsymbol{\xi}_1 \, \mathrm{e}^{-\mathrm{i}\frac{2\pi}{\lambda}\boldsymbol{\xi} \cdot \boldsymbol{\alpha}} \mathrm{d}\boldsymbol{\xi} \, . \\
&= \frac{1}{(\lambda F)^2} \iint A(\boldsymbol{\xi}_1) A^*(\boldsymbol{\xi}_1 - \boldsymbol{\xi}) \mathrm{d}\boldsymbol{\xi}_1 \, \Gamma_\varphi(\boldsymbol{\xi}) \, \mathrm{e}^{-\mathrm{i}\frac{2\pi}{\lambda}\boldsymbol{\xi} \cdot \boldsymbol{\alpha}} \mathrm{d}\boldsymbol{\xi} \, , \qquad (4.84)
\end{aligned}
$$

with $\boldsymbol{\xi} = \boldsymbol{\xi}_1 - \boldsymbol{\xi}_2$.

The time averaging process is applied to the amplitude V_φ in the turbulent layer only since the aperture function A is time-invariant. Using (4.26), we write the correlation of the amplitude as $\Gamma_\varphi(\boldsymbol{\xi}) = \exp(-\frac{1}{2} D_\varphi(\boldsymbol{\xi})) = \exp(-3.44(\xi/r_0)^{5/3})$, with $D_\varphi(\boldsymbol{\xi})$ the phase structure function (4.25).

In Sect. 3.1.1, discussing Fourier Optics, we defined the telescope OTF as the autocorrelation of the aperture function $A(\boldsymbol{\xi})$ in the notation with spatial frequencies $\boldsymbol{R} = \boldsymbol{\xi}/\lambda$, (3.6). Introducing spatial frequencies into (4.84), we obtain the point-spread function[3] PSF_φ after passage through turbulence as

$$
\mathrm{PSF}_\varphi(\boldsymbol{\alpha}) =< I(\boldsymbol{\alpha}) > = \int \mathrm{OTF}(\boldsymbol{R}) \Gamma_\varphi(\boldsymbol{R}\lambda) \mathrm{e}^{-\mathrm{i}2\pi \boldsymbol{R} \cdot \boldsymbol{\alpha}} \mathrm{d}\boldsymbol{R} \, , \qquad (4.85)
$$

introducing the *atmospheric transfer function* OTF_φ for imaging through turbulence as

$$
\mathrm{OTF}_\varphi(\boldsymbol{R}) = \mathrm{OTF}(\boldsymbol{R}) \Gamma_\varphi(\boldsymbol{R}\lambda) \, , \qquad (4.86)
$$

with $\boldsymbol{R} = (u, v)$, the coordinate vector in the uv-plane (see Sect. 3.1.1).

The same argument applies to observations with an interferometer (see Sect. 6.1) when the optical transfer function is composed of three transfer functions of the individual telescope apertures, OTF, centred at $|\boldsymbol{R}| = 0$ and at $\boldsymbol{R} = \pm \boldsymbol{R}_B$ (3.52).

If r_0 is larger than the telescope aperture D but smaller than the baseline B, then $\Gamma_\varphi(\boldsymbol{R}\lambda)$ is much wider than the central $\mathrm{OTF}(\boldsymbol{R})$, but zero at $\boldsymbol{R}_B\lambda$, so that OTF_φ is dominated by the central telescope OTF, and the resulting image is very similar to the Airy disk of an individual telescope, without fringes. This is typical in the mid-infrared, displayed on the right hand side of Fig. 4.10, when r_0 is about

[3] The PSF as dimensionless quantity requires to divide the intensity by V_0 (3.3). Since we set $V_0 = 1$ this will not be noted explicitly.

10 m – larger than most telescopes – so that diffraction rings can be seen already in short time exposures.

In the near-infrared, r_0 is typically < 1 m, much smaller than the telescope diameter and than the interferometer baseline. Then, $\Gamma_\varphi(\boldsymbol{R}\lambda)$ is much narrower than the OTF of an individual telescope so that the influence of the OTF on OTF_φ can be neglected, and the size of the seeing disk is determined by the Fourier transform of $\Gamma_\varphi(\boldsymbol{R}\lambda)$ alone.

Using (4.26) and $\boldsymbol{\xi} = \boldsymbol{R}\lambda$, we obtain the seeing disk, i.e. the point-spread function PSF_φ, for $r_0 < D$ as

$$\begin{aligned}\mathrm{PSF}_\varphi(\boldsymbol{\alpha}) &= \int \Gamma_\varphi(\boldsymbol{\xi}) \mathrm{e}^{-\mathrm{i}\frac{2\pi}{\lambda}\boldsymbol{\xi}\cdot\boldsymbol{\alpha}} \mathrm{d}\boldsymbol{\xi} \\ &= \int \mathrm{e}^{-3.44\left(\frac{\xi}{r_0}\right)^{5/3}} \mathrm{e}^{-\mathrm{i}\frac{2\pi}{\lambda}\boldsymbol{\xi}\cdot\boldsymbol{\alpha}} \mathrm{d}\boldsymbol{\xi} \end{aligned} \tag{4.87}$$

The seeing disk is displayed in Fig. 4.12. The seeing disk is usually many times larger than the Airy disk of an individual telescope and than the fringe spacing of an interferometer, so that the angular resolution of long exposure images is limited to the diameter of the seeing disk.

A Gaussian function models the seeing disk reasonably well, using the approximation $(\xi/r_0)^{5/3} \approx (\xi/r_0)^2$ in (4.87). However, with the Gaussian approximation the seeing disk converges to zero faster than measured seeing profiles that are better described by the Kolmogorov model. The full width at half maximum (FWHM) of the seeing disk is $0.98\lambda/r_0$ corresponding in good approximation to a telescope with diameter r_0.[4] With $r_0 \propto \lambda^{6/5}$ the seeing is $\lambda/r_0 \propto \lambda^{-1/5}$, i.e. it is decreasing slowly with increasing wavelength.

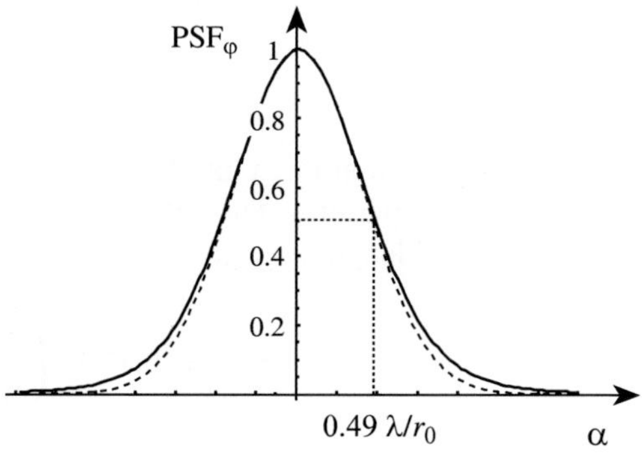

Fig. 4.12 The intensity distribution in a seeing disk computed from the phase structure function $D_\varphi(\xi) = 6.88\left(\frac{\xi}{r_0}\right)^{5/3}$. The full width at half maximum (FWHM) is $0.98\lambda/r_0$. The *dashed line* shows a Gaussian function for comparison

[4] One should note that it was Fried's original intention to define r_0 as the equivalent telescope diameter, [71].

Outer Scale of Turbulence

The above computation of the seeing disk relied on an infinite outer scale L_0. For finite L_0 values, an empirical formula, approximating the FWHM of the seeing disk in a large telescope for the regime $L_0/r_0 > 20$, is given by [233]

$$\mathrm{FWHM}_{vK} \approx \sqrt{1 - 2.183\left(\frac{r_0}{L_0}\right)^{0.356}}\; 0.98\frac{\lambda}{r_0}\,, \tag{4.88}$$

with the subscript vK indicating that the von Kármán spectrum with outer scale L_0 was used.

Assuming an outer scale of $L_0 = 22\,\mathrm{m}$ and an r_0 of 0.6 m, corresponding to 0.76 arcsec seeing at 2.2 μm, the FWHM of the seeing disk is 38% smaller than the Kolmogorov seeing disk, i.e. 0.48 arcsec instead of 0.76 arcsec. In the visible, with $r_0 = 0.11$ m under the same seeing conditions, the FWHM is reduced by 19%. Thus, deriving r_0 values from measurements of the FWHM of the seeing disks on large telescopes leads to substantially wrong results if the outer scale of turbulence is not considered or cannot be estimated reliably. Therefore, dedicated *seeing monitors* with smaller apertures that are less prone to the outer scale are used to measure r_0 [206].

Propagation Through Optical Systems: Summary

The impact of atmospheric turbulence on the imaging process in interferometers and telescopes can best be described by the spatial and temporal power spectra of the disturbance, e.g. fringe or image motion, since they are related to their variances. Their mean values approach a constant value after a few seconds and do not affect the imaging process.

Fringe Motion

The fringe motion in an interferometer is caused by random OPD variations between the apertures. The OPD variance for two point-like sub-apertures and a baseline of length B is given by

$$\sigma^2_{\mathrm{OPD}} = 0.17\,\lambda^2\left(\frac{B}{r_0}\right)^{5/3} \quad [\mathrm{m}^2]. \tag{4.45}$$

For a finite outer scale of turbulence, L_0, the fringe motion is reduced substantially if the baseline is comparable or larger than L_0. A useful estimate for the OPD variance is based on the von Kármán spectrum (4.28), [41,246], and reads as

$$\sigma^2_{\mathrm{OPD},vK} = 0.00876\,\lambda^2 \left(\frac{L_0}{r_0}\right)^{5/3} \left(1 - \left(\frac{2\pi B}{L_0}\right)^{5/6} K_{5/6}\left(\frac{2\pi B}{L_0}\right)\right), \quad (4.46)$$

with $K_{5/6}(x)$ the Macdonald function.

For a 50-m baseline and infinite outer scale for instance, the rms OPD fluctuations σ_{OPD} are about 37 μm, according to (4.45) for an r_0 of 0.6 m in the K-band. If the outer scale is $L_0 = 50$ m, the rms OPD fluctuations are reduced to about 8 μm, using (4.46).

Using temporal power spectra we can compute the OPD variance of fringe motion for a given exposure time and estimate the requirements for fringe trackers.

The spectra can be approximated by asymptotes for different frequency domains. The temporal power spectrum of fringe motion for pinhole sized apertures is

$$\Phi_{\mathrm{OPD},t}(f) = 0.096\,\lambda^2 \left(\frac{\bar{v}}{r_0}\right)^{5/3} \left(\frac{B}{\bar{v}}\right)^2 f^{-2/3}\,[\mathrm{m}^2/\mathrm{Hz}] \quad f \ll 0.2\,\bar{v}/B$$

$$\Phi_{\mathrm{OPD},t}(f) = 0.00392\,\lambda^2 \left(\frac{\bar{v}}{r_0}\right)^{5/3} f^{-8/3}[\mathrm{m}^2/\mathrm{Hz}] \quad f \gg 0.2\,\bar{v}/B. \quad (4.54)$$

Moving to sub-apertures of diameter d, we have to add a third asymptote $\propto f^{-17/3}$ for frequencies larger than $0.3\,\bar{v}/d$ (4.56).

The variance of the OPD variations for short exposure times can be computed using (4.54), yielding an approximation for the rms fluctuations during time period T of

$$\sigma_{\mathrm{OPD},T} = 0.19\,\lambda \left(\frac{\bar{v}}{r_0}\right)^{5/6} T^{5/6}\,[\mathrm{m}]. \quad (4.55)$$

The differential fringe motion, dOPD, describes the difference in fringe positions of two objects separated by Δs. This quantity is essential to estimate anisoplanatic and temporal effects of turbulence in a fringe tracker. If the fringes of one object are kept stable the second object displays a fringe motion that is described by the differential fringe motion.

The temporal power spectrum of differential OPD fluctuations with pinhole sized apertures reads as

$$\Phi_{\mathrm{dOPD},t}(f) = 0.628\lambda^2 \left(\frac{\bar{v}}{r_0}\right)^{5/3} \left(\frac{\Delta s\,\bar{h}}{\bar{v}}\right)^2 \left(\frac{B}{\bar{v}}\right)^{2/3} \times \left(1 + 4.84\,f^{4/3} \left(\frac{B}{\bar{v}}\right)^{4/3}\right) \quad f \ll 0.12\frac{\bar{v}}{B}$$

$$\Phi_{\text{dOPD},t}(f) = 0.193\lambda^2 \left(\frac{\bar{v}}{r_0}\right)^{5/3} \left(\frac{\Delta s\, \bar{h}}{\bar{v}}\right)^2 f^{-2/3} \qquad 0.12\frac{\bar{v}}{B} \ll f \ll 0.2\frac{\bar{v}}{\Delta s\, \bar{h}}$$

$$\Phi_{\text{dOPD},t}(f) = 0.0078\lambda^2 \left(\frac{\bar{v}}{r_0}\right)^{5/3} f^{-8/3} \qquad 0.2\frac{\bar{v}}{\Delta s\, \bar{h}} \ll f, \tag{4.58}$$

when the high frequency asymptote $\propto f^{-8/3}$ has to be replaced in case of finite apertures with diameter d by two additional asymptotes $\propto f^{-11/3}$ and $\propto f^{-17/3}$ (4.59) for frequencies larger than $0.3\,\bar{v}/d$.

For short exposure times $T \ll 4.17\,d/\bar{v}$ and small separations $\Delta s \ll d/\bar{h}$, the approximation for rms variations of the differential fringe motion is expressed by

$$\sigma^d_{\text{dOPD},T} = 0.1\,\lambda r_0^{-5/6}\,\bar{v}^{8/6}\bar{h}\,d^{-3/2}\,\Delta s\,T^{4/3}\ [\text{m}]. \tag{4.60}$$

Image Motion

The average gradient of the phase distribution in the telescope aperture determines the position of the image in the telescope focus, also referred to as *angle of arrival*.

The one-axis variance of the image motion, i.e. of the position $\boldsymbol{\theta}$ of the image centroid is given by [243]

$$\sigma_\theta^2 = 0.17 \left(\frac{\lambda}{D}\right)^2 \left(\frac{D}{r_0}\right)^{5/3} [\text{rad}^2]. \tag{4.66}$$

If λ/D, the FWHM of the Airy disk, is given in arcsec, one obtains the variance in arcsec^2. The two-axis variance is $2\sigma_\theta$.

For a numerical example, we use again an 8-m telescope and the K-band ($\lambda = 2.2\,\mu$m). For a typical value of $r_0 = 0.6$ m at $\lambda = 2.2\,\mu$m we find the one-axis rms image motion as 0.2 arcsec, or 3.6 times the Airy disk diameter.

Converting the tilt angle of the wave front into the OPD of two points separated by $B = D$, using $\theta = \text{OPD}/B$, we see that it is exactly $\sigma_\theta^2 = \sigma_{\text{OPD}}^2/B^2$. Thus, the one-axis variance of the image motion (4.66) in a telescope with diameter D corresponds to the variance of the OPD fluctuations (4.45) of an interferometer with baseline $B = D$.

In analogy to the reduction of OPD variance if the length of the baseline approaches the outer scale of turbulence, L_0, the image motion is reduced by more than 80% if the telescope diameter is comparable to the outer scale of turbulence. However, this reduction is due exclusively to the lower contribution at low frequencies.

For the one-axis power spectrum of the image motion we obtain

$$\Phi_{\theta,t}(f) = 0.096\,\lambda^2\,\bar{v}^{-2}\left(\frac{\bar{v}}{r_0}\right)^{5/3} f^{-2/3}[\mathrm{rad}^2/\mathrm{Hz}] \quad f \ll 0.24\,\bar{v}/D \quad (4.70)$$

$$\Phi_{\theta,t}(f) = 0.0013\,\lambda^2\,\bar{v}^{-2}\left(\frac{D}{\hat{v}}\right)^{-3}\left(\frac{\bar{v}}{r_0}\right)^{5/3} f^{-11/3}\,[\mathrm{rad}^2/\mathrm{Hz}] \quad f \gg 0.24\,\bar{v}/D$$

Note that the temporal power spectrum is defined from 0 to $+\infty$.

Scintillation

The twinkling of the stars, the scintillation, is rather strong when observing with small apertures like the naked eye, and very weak for telescopes larger than 1 m in diameter. It is related to the curvature of the wave front $\frac{\partial^2\varphi(\xi)}{\partial\xi^2}$.

The variance of the scintillation in the image plane of telescopes smaller than $(\lambda\bar{h}_{I_{5/6}})^{1/2}$, typically a few cm, is given by [194]:

$$\sigma_I^2 = 1.145\,r_0^{-5/3}(\lambda\bar{h}_{I_{5/6}})^{5/6}. \quad (4.79)$$

The effective altitude $\bar{h}_{I_{5/6}}$ is defined by $\bar{h}_{I_{5/6}}^{5/6} = \int C_n^2(h)h^{5/6}\mathrm{d}h / \int C_n^2(h)\mathrm{d}h$.

For apertures, D of several 10 cm and larger, the intensity variance is given by [191, 194]

$$\sigma_{I,D}^2 \propto D^{-7/3} r_0^{-5/3}(\lambda\bar{h}_{I_2})^2, \quad (4.81)$$

displaying a steep decrease with increasing telescope diameter. The effective altitude $\bar{h}_{I_2}$ is defined by $\bar{h}_{I_2}^2 = \int C_n^2(h)h^2\mathrm{d}h / \int C_n^2(h)\mathrm{d}h$. Both $\bar{h}_{I_2}$ and $\bar{h}_{I_{5/6}}$ have values of typically several kilometres.

Like the variances of fringe motion, σ_{OPD}^2, and image motion, σ_α^2, the variance of the scintillation $\sigma_{I,D}^2$ is independent of wavelength, whereas the expression σ_I^2 for very small apertures, (4.79), decreases with longer wavelengths.

The temporal power spectrum $\Phi_{I,t}(f)$ is proportional to f^0 for frequency smaller than $0.4\bar{v}/D$ and $\propto f^{-5/3}$ for larger frequencies [40].

Speckle Pattern and Seeing Disk

Observing through atmospheric turbulence, the image intensity distribution of a point-like star at zenith shows the characteristic speckle pattern that is displayed in Fig. 4.10. Increasing the exposure time, the speckles smear, merging eventually into the *seeing disk*.

The time averaged intensity of the seeing disk, PSF_φ is written as

$$\mathrm{PSF}_\varphi(\boldsymbol{\alpha}) = \int \mathrm{OTF}(\boldsymbol{R})\Gamma_\varphi(\boldsymbol{R}\lambda)\mathrm{e}^{-\mathrm{i}2\pi\boldsymbol{R}\cdot\boldsymbol{\alpha}}\mathrm{d}\boldsymbol{R}, \tag{4.85}$$

with $\mathrm{OTF}(\boldsymbol{R})$ the optical transfer function either of an individual telescope or of an interferometer, and $\Gamma_\varphi(\boldsymbol{R}\lambda)$ is the correlation of the amplitude, with $\Gamma_\varphi(\boldsymbol{\xi}) = \exp(-\frac{1}{2}D_\varphi(\boldsymbol{R}\lambda))$, and $D_\varphi(\boldsymbol{R}\lambda)$ the phase structure function (4.25).

Then, the atmospheric transfer function OTF_φ for imaging through turbulence is given by

$$\mathrm{OTF}_\varphi(\boldsymbol{R}) = \mathrm{OTF}(\boldsymbol{R})\Gamma_\varphi(\boldsymbol{R}\lambda). \tag{4.86}$$

$\boldsymbol{R} = (u, v)$ is the coordinate vector in the uv-plane (see Sect. 3.1.1).

In the near-infrared, r_0 is typically < 1 m, much smaller than the telescope diameter. Then, $\Gamma_\varphi(\boldsymbol{R}\lambda)$ is much narrower than the OTF so that the influence of telescope and interferometer OTF on OTF_φ can be neglected, and the size of the seeing disk is determined by the Fourier transform of $\Gamma_\varphi(\boldsymbol{R}\lambda)$ alone, yielding

$$\mathrm{PSF}_\varphi(\boldsymbol{\alpha}) = \mathcal{F}_\xi\left(\mathrm{e}^{-\frac{1}{2}D_\varphi(\boldsymbol{\xi})}\right) = \int \mathrm{e}^{-3.44\left(\frac{\xi}{r_0}\right)^{5/3}}\mathrm{e}^{-\mathrm{i}\frac{2\pi}{\lambda}\boldsymbol{\xi}\cdot\boldsymbol{\alpha}}\mathrm{d}\boldsymbol{\xi}, \tag{4.87}$$

The seeing disk is displayed in Fig. 4.12, reasonably well modelled by a Gaussian function.

The full width half maximum (FWHM) of the seeing disk is $0.98\lambda/r_0$ corresponding in good approximation to a telescope with diameter r_0. For a finite outer scale L_0, an empirical formula , approximating the FWHM of the seeing disk in a large telescope for the regime $L_0/r_0 > 20$, is given by [233]

$$\mathrm{FWHM}_{vK} \approx \sqrt{1 - 2.183\left(\frac{r_0}{L_0}\right)^{0.356}}\, 0.98\frac{\lambda}{r_0}. \tag{4.88}$$

4.4 Speckle Interferometry

Speckle interferometry was originally proposed by A. Labeyrie in 1970 [120] for single telescopes. Although formally computing the correlation of intensities like in the intensity interferometer (see Sect. 2.5) the motivation and the formalism are very different. The motivation was driven by the desire to overcome the degradation of the angular resolution by atmospheric turbulence and to reach the diffraction limit of the telescope.

After passage through atmospheric turbulence the image of a point-like star is scattered into a speckle cloud when each speckle has the size of an Airy disk as displayed in Fig. 4.10. The exposure time for a speckle image has to be shorter than the time scale of the evolution of the speckles, i.e. shorter than the atmospheric coherence time (4.39), that is in the millisecond range. Exposure times of a few seconds merge the individual speckle images into the seeing disk with a diameter of approximately the envelope of the speckle image in Fig. 4.10. The seeing disk is usually many times larger than the Airy disk. The angular resolution is then limited to the diameter of the seeing-disk.

The principle of speckle interferometry is based on the fact that two close-by objects, e.g. a binary star, form two identical speckle images, slightly shifted with respect to each other, if the distance between the two objects is much smaller than the size of a speckle image. The intensities add up, displaying a new speckle image that, to the naked eye, is indistinguishable from the speckle image of a single point source. Thus, one cannot decide if a single star or a binary are observed.

However, calculating the autocorrelation function of the intensity pattern one finds a peak, i.e. a strong correlation, symmetrically at plus and minus the difference coordinate corresponding to distance and orientation of the objects. Due to the symmetry of the autocorrelation function one cannot determine, which one of the two stars is the brighter one. Unfortunately, there can be many more peaks due to random correlations of the speckle pattern. Their random positions average out when calculating the mean autocorrelation function of many instantaneous speckle images. The advantage of using the autocorrelation function is illustrated in Fig. 4.13 by displaying the Fourier transforms of the speckle images.

Computing the autocorrelation of intensities in the image plane after imaging through turbulence requires to deal with two different correlation widths and times since we have two random processes. The first process is the emission process of a thermal light source introduced in Sect. 2.3 that is characterised by the coherence function as a function of the form of the light source and its spectrum. The second one is determined by the statistics of atmospheric turbulence. The most important difference between the two processes is the temporal scale at which changes take place. While the coherence time as the reciprocal of the spectral bandwidth is of the order of $10^{-12} - 10^{-14}$ sec, the correlation time of atmospheric turbulence is in the millisecond range.

The measured intensity is the result of an averaging process over a time span that is limited by the temporal resolution of the detector, which is of the order of 10^{-9} s and, thus, considerably longer than the coherence time. If we choose the averaging time to be shorter than the atmospheric coherence time we take snapshots of speckle images. The image intensity $I(\boldsymbol{\alpha}, t)$ is then a function of time taking a different speckle image at every exposure.

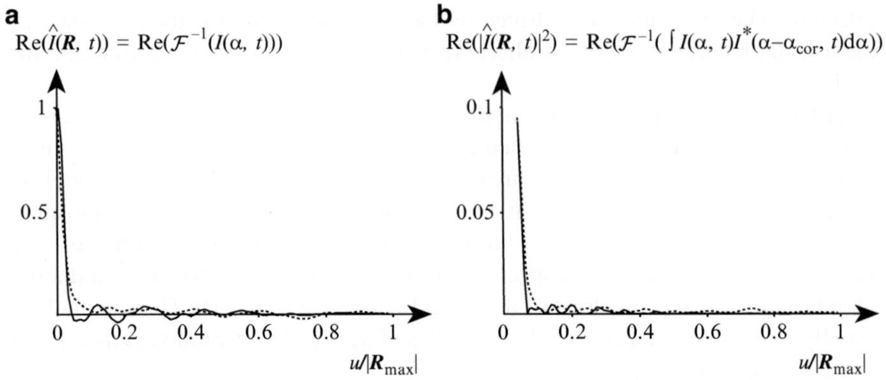

Fig. 4.13 Illustration of the averaging process of individual speckle images and of their autocorrelations. The effect of the averaging process is discussed by regarding the Fourier transform of the intensities. In (**a**), one-dimensional slices through the real parts of image intensity spectra – the Fourier transforms – of two speckle images are displayed (solid line and dashed line), and in (**b**), their power spectra – the Fourier transforms of the autocorrelation functions. The scale of the rather small values in (**b**) is enhanced by a factor of 10. It is $\boldsymbol{R} = (u, v)$ and $|\boldsymbol{R}_{\max}| = D/\lambda$ with D the aperture diameter. With random positive and negative values the average of many intensity spectra like in (**a**) averages to zero beyond the central peak. The remaining central peak is the Fourier transform of the seeing disk. The power spectra (**b**) having only positive values by definition add up constructively. The theoretical limit for this averaging process is shown in Fig. 4.14

In analogy to (3.4) in Sect. 3.1.1, we write down the monochromatic intensity distribution in the image plane of a telescope as

$$I(\boldsymbol{\alpha}, t) = O(\boldsymbol{\alpha}) * \mathrm{PSF}_{\varphi}(\boldsymbol{\alpha}, t)\,, \tag{4.89}$$

with $O(\boldsymbol{\alpha})$ the object intensity distribution that is convolved with the time dependent, instantaneous speckle pattern $\mathrm{PSF}_{\varphi}(\boldsymbol{\alpha}, t)$ displayed in Fig. 4.10.

The instantaneous image intensity spectrum $\hat{I}(\boldsymbol{R}, t)$, which is the Fourier transform of $I(\boldsymbol{\alpha}, t)$ (3.10), can be expressed (3.11) as

$$\hat{I}(\boldsymbol{R}, t) = \hat{O}(\boldsymbol{R})\,\mathrm{OTF}_{\varphi}(\boldsymbol{R}, t) = I_0\,\mu(\boldsymbol{R})\,\mathrm{OTF}_{\varphi}(\boldsymbol{R}, t), \tag{4.90}$$

with $\mathrm{OTF}_{\varphi}(\boldsymbol{R}, t)$, the instantaneous atmospheric transfer function and $\hat{O}(\boldsymbol{R})$ the spatial frequency spectrum of the object intensity. This is the common notation in the literature on speckle interferometry. In the monochromatic case, we can replace $\hat{O}(\boldsymbol{R})$ by the product of average intensity I_0 and visibility function $\mu(\boldsymbol{R})$ (3.28). In the following, we will only use the visibility function $\mu(\boldsymbol{R})$, to emphasise the common formal basis of the imaging process in speckle and amplitude interferometry, using $\hat{O}(\boldsymbol{R}) = I_0\,\mu(\boldsymbol{R})$ as given by (3.28).

First, we look at the time average of the speckle image using the formalism above. We stated that the long time average of the speckle pattern produces the **seeing disk** $\mathrm{PSF}_{\varphi}(\boldsymbol{\alpha})$ (4.85) that can be many times wider than the diffraction limited Airy disk.

Accordingly, transforming this into Fourier space, we find that the time averaged atmospheric transfer function, $\mathrm{OTF}_\varphi(\boldsymbol{R})$, is narrower than the diffraction limited OTF.

Denoting the time average by < . >, we obtain

$$< I(\boldsymbol{\alpha}) > = O(\boldsymbol{\alpha}) * \mathrm{PSF}_\varphi(\boldsymbol{\alpha})$$
$$< \hat{I}(\boldsymbol{R},t) > = I_0\,\mu(\boldsymbol{R})\,\mathrm{OTF}_\varphi(\boldsymbol{R})\,, \tag{4.91}$$

with $\mathrm{OTF}_\varphi(\boldsymbol{R}) = \mathrm{OTF}(\boldsymbol{R})\Gamma_\varphi(\boldsymbol{R}\lambda)$ (4.86), and Γ_φ the correlation function of the amplitude due to turbulence (4.26).

Depending on the atmospheric parameters and on the properties of the observed object, the autocorrelation function of atmospheric turbulence can go to zero much faster, i.e. the time averaged $\mathrm{OTF}_\varphi(\boldsymbol{R})$ can be narrower, than the diffraction limited OTF, so that $\mathrm{OTF}_\varphi(\boldsymbol{R}) \approx \Gamma_\varphi(\boldsymbol{R}\lambda)$. Then the seeing disk is wider than the diffraction limited Airy disk. This occurs frequently even at good observing sites. Figure 4.14a displays examples for time averaged OTFs.

Thus, the result of (4.91) is the time averaged image intensity spectrum as the product of the visibility function, $\mu(\boldsymbol{R})$, which is the coherence function of the incoming light, with $\mathrm{OTF}_\varphi(\boldsymbol{R}) \approx \Gamma_\varphi(\boldsymbol{R}\lambda)$, which is the correlation function of

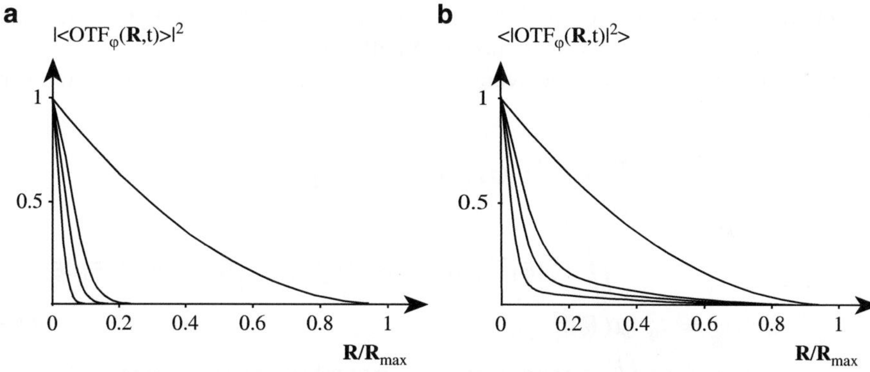

Square of the time averaged OTF_φ — Time average of the squared OTF_φ

Fig. 4.14 The squared modulus of the time averaged atmospheric transfer function, OTF_φ, (**a**) and the time average of the squared OTF_φ (**b**) for a variety of seeing conditions. The transfer functions in (**a**) correspond to the long time average of the image intensity i.e. the seeing disk that is many times wider than the diffraction limited Airy disk. The size of the seeing disk increases with the strength of the turbulence and the corresponding transfer function becomes narrower. Curves for three different seeing conditions are given. For comparison the diffraction limited OTF is also shown (*top curve*). In (**b**) speckle transfer functions are displayed when the Fourier transform of the image correlation, the power spectrum, is averaged. Here, atmospheric turbulence also causes a rapid drop of the transfer function but there is slowly decreasing part almost up to the diffraction limit at $\boldsymbol{R}/\boldsymbol{R}_{\max} = 1$ that allows to determine the square of the visibility function $|\mu(\boldsymbol{R})|^2$, i.e. the square of the spatial frequency spectrum $|\hat{O}_b(\boldsymbol{R})|^2$, of the object at large spatial frequencies recovering the information on small detail in the object

atmospheric turbulence in the telescope aperture. The two random processes – due to coherence and due to atmospheric turbulence – are cleanly separated as a product of their respective correlation functions.

Power Spectrum

Discussing now the spatial autocorrelation of speckle images we transform the calculation again into Fourier space, replacing the autocorrelation of $I(\boldsymbol{\alpha}, t)$ by the square of the individual Fourier transforms $\mu(\boldsymbol{R})\mathrm{OTF}_\varphi(\boldsymbol{R}, t)$ yielding the *power spectrum* of the individual speckle image,

$$\begin{aligned} \mathcal{F}_\alpha^{-1}\left(\int I(\boldsymbol{\alpha}', t) I^*(\boldsymbol{\alpha}' - \boldsymbol{\alpha}, t)\mathrm{d}\boldsymbol{\alpha}'\right) &= |\hat{I}(\boldsymbol{R}, t)|^2 \\ &= I_0^2\, |\mu(\boldsymbol{R})|^2\, |\mathrm{OTF}_\varphi(\boldsymbol{R}, t)|^2 \,. \end{aligned} \quad (4.92)$$

The autocorrelation of an individual speckle image does not reveal too much about the observed object since there are many random correlations in the image. The time average of the autocorrelation or of the power spectrum reduces the random contributions replacing them by their mean values.

We write the amplitude in the aperture as $V_\varphi(\xi) = \exp(\mathrm{i}\varphi(\xi))$. For the sake of simplicity, we omit the aperture function $A(\xi)$, and we keep in mind that the phase $\varphi(\xi)$ is a time varying function without explicitly mentioning it.

Then, in analogy to the time averaged intensity (4.84), the time averaged power spectrum is written as

$$\begin{aligned} < |\hat{I}(\boldsymbol{R}, t)|^2 > &= I_0^2\, |\mu(\boldsymbol{R})|^2 < |\mathrm{OTF}_\varphi(\boldsymbol{R}, t)|^2 > \\ &= I_0^2\, |\mu(\boldsymbol{R})|^2 \\ &\quad \times < \iint V_\varphi(\boldsymbol{\xi}_1) V_\varphi^*(\boldsymbol{\xi}_1 - \boldsymbol{R}\lambda) V_\varphi^*(\boldsymbol{\xi}_2) V_\varphi(\boldsymbol{\xi}_2 - \boldsymbol{R}\lambda)\mathrm{d}\boldsymbol{\xi}_1 \mathrm{d}\boldsymbol{\xi}_2 > \\ &= I_0^2\, |\mu(\boldsymbol{R})|^2 \qquad (4.93) \\ &\quad \times \iint < V_\varphi(\boldsymbol{\xi}_1) V_\varphi^*(\boldsymbol{\xi}_1 - \boldsymbol{R}\lambda) V_\varphi^*(\boldsymbol{\xi}_2) V_\varphi(\boldsymbol{\xi}_2 - \boldsymbol{R}\lambda) > \mathrm{d}\boldsymbol{\xi}_1 \mathrm{d}\boldsymbol{\xi}_2 \,. \end{aligned}$$

The fourth order moment of the amplitude $V_\varphi(\boldsymbol{\xi})$ can be reduced to groups of second order moments like for the optical disturbances in intensity interferometry. Similar to the result of intensity correlations in (2.84) we obtain two main groups of functions. The first group contains mainly correlation functions like in (4.91) that are narrower than the diffraction limited OTF. The second group consists of integrals over correlation functions $< V_\varphi(\boldsymbol{\xi}_1) V_\varphi^*(\boldsymbol{\xi}_2) >, < V_\varphi^*(\boldsymbol{\xi}_1 - \boldsymbol{R}\lambda) V_\varphi(\boldsymbol{\xi}_2 - \boldsymbol{R}\lambda) >$ that are independent of $\boldsymbol{R}$ and provide a slowly decreasing value up to the diffraction limit.

While the first group represents the seeing limited transfer function, the second group provides the required non-zero values of the transfer function, increasing the angular resolution beyond the seeing limit up to the diffraction limit. Examples of time averaged power spectra, also called the *speckle transfer function*, under different seeing conditions are given in Fig. 4.14b.

Thus, averaging the power spectra of a large number of speckle images, the square of the visibility function, $|\mu(\boldsymbol{R})|^2$, can be recovered up to the diffraction limit. However, the exact form of the visibility function can only be determined if the speckle transfer function is known. The speckle transfer function can be measured by observing a source, the reference source, that is smaller than the diffraction limit of the telescope so that the spatial frequency spectrum is approximately constant up to diffraction limit $\boldsymbol{R}_{\max} = D/\lambda$. Computing its time averaged power spectrum $< |\hat{I}_{\text{ref}}(\boldsymbol{R},t)|^2 >$ yields a good estimator for the speckle transfer function $< |\text{OTF}_\varphi(\boldsymbol{R},t)|^2 >$, assuming that atmospheric turbulence is a stationary random process. The power spectrum of the object of interest can now be determined in detail by dividing its averaged power spectrum by the averaged power spectrum of the reference source, obtaining the squared visibility function as

$$|\mu(\boldsymbol{R})|^2 = \frac{< |\hat{I}(\boldsymbol{R},t)|^2 >}{< |\hat{I}_{\text{ref}}(\boldsymbol{R},t)|^2 >} = \frac{< |\hat{I}(\boldsymbol{R},t)|^2 >}{< |\text{OTF}_\varphi(\boldsymbol{R},t)|^2 >}\,. \tag{4.94}$$

Thus, the power spectrum of the speckle pattern – just as the power spectrum of the fringe pattern in Sect. 2.4.3 – yields $|\mu(\boldsymbol{R}))|^2$, loosing the phase $\phi(\boldsymbol{R})$ of the visibility function and with it all the information on the asymmetric content of the intensity distribution. Binary stars appear as triple stars with a bright central "star" properly reflecting the correlation of a binary but not very helpful when determining which star might be the brightest of the two.

In the following, we will present a technique – triple correlation, or bispectrum – that permits to reconstruct the phase of the visibility function eliminating the effects of atmospheric turbulence through the usage of the closure phase. In Sect. 6.2.3, we will discuss the equivalent formalism for stellar interferometry.

Bispectrum

In analogy to the autocorrelation of the speckle pattern (4.92), G. Weigelt suggested to compute the time average of the *triple correlation* of the intensity distribution [140,252]. We regard its four-dimensional Fourier transform, called the *bispectrum*, reading

$$\begin{aligned}
&< \mathcal{F}^{-1}_{\alpha_1,\alpha_2}\left(\int I(\boldsymbol{\alpha},t)I(\boldsymbol{\alpha}-\boldsymbol{\alpha}_1,t)I(\boldsymbol{\alpha}-\boldsymbol{\alpha}_2,t)\mathrm{d}\boldsymbol{\alpha}\right) > \\
&=< \hat{I}(\boldsymbol{R}_1)\hat{I}(\boldsymbol{R}_2)\hat{I}(-\boldsymbol{R}_1-\boldsymbol{R}_2) > \\
&= I_0^3\mu(\boldsymbol{R}_1)\mu(\boldsymbol{R}_2)\mu^*(\boldsymbol{R}_1+\boldsymbol{R}_2) \\
&\quad \times < \text{OTF}_\varphi(\boldsymbol{R}_1,t)\text{OTF}_\varphi(\boldsymbol{R}_2,t)\text{OTF}^*_\varphi(\boldsymbol{R}_1+\boldsymbol{R}_2,t) >\,.
\end{aligned} \tag{4.95}$$

As a triple product of complex functions, the bispectrum is a complex function, too. Its phase, $\psi(\boldsymbol{R}_1, \boldsymbol{R}_2)$, is determined by the sum of phases of the visibility function,

$$\psi(\boldsymbol{R}_1, \boldsymbol{R}_2) = \phi(\boldsymbol{R}_1) + \phi(\boldsymbol{R}_2) - \phi(\boldsymbol{R}_1 + \boldsymbol{R}_2), \quad (4.96)$$

which is called the *closure phase* since it is the sum of phases around a closed loop of uv-plane vectors $\boldsymbol{R}_i$. There is a large variety of methods to reconstruct the phase distribution $\phi(\boldsymbol{R})$ from the bispectrum closure phase [7, 9, 84, 115].

The phase of the triple product of transfer functions $\mathrm{OTF}_\varphi(\boldsymbol{R}_i)$ is in good approximation zero [140]. This can be understood when writing each $\boldsymbol{R}_i$ as a difference of spatial frequencies $\boldsymbol{R}_i = \boldsymbol{\xi}_i/\lambda - \boldsymbol{\xi}_j/\lambda$ and expressing the phase of the OTF by the difference $\varphi(\boldsymbol{\xi}_i) - \varphi(\boldsymbol{\xi}_j)$ of phase values in the aperture plane (see Sect. 4.3.5) so that the sum of the phases at $\boldsymbol{R}_1$ and $\boldsymbol{R}_2$ is identical to the phase at $\boldsymbol{R}_1 + \boldsymbol{R}_2$ and the total sum cancels.

In amplitude interferometry, this technique is also used albeit for individual combinations of three telescopes so that the three baselines $\boldsymbol{R}_i$ form a closed loop and a single closure phase can be determined (see Sect. 6.2.3). Formally, these methods are identical.

Speckle vs. Intensity Interferometry

At first glance, both methods seem to be very similar, since both use intensity correlations to provide the square of the visibility function. However, despite this formal similarity the methods have nothing in common.

In intensity interferometry – as discussed in Sect. 2.5 – the correlation of two instantaneous intensities $i(\boldsymbol{\xi})$ at two different points in the aperture plane is computed. For practical reasons, these intensities are measured in the foci of individual telescopes that are being used as light buckets, averaging over the telescope apertures, to collect more photons. The distance vector between the two aperture centres is the baseline vector $\boldsymbol{B}$, and the correlation function permits to determine the square of the visibility function, $|\mu(\boldsymbol{R_B})|^2$, for that particular baseline vector. The instantaneous intensities need to be measured with extreme temporal resolution (10^{-9} s) using a very narrow spectral bandwidth in order to be sensitive to the coherence function of the amplitudes. Atmospheric turbulence has a negligible effect on this procedure. The triple correlation of intensities is a real function containing the real part of the product of visibility functions when the closure phase appears in the cosine term (2.87).

In speckle interferometry, we compute the correlation function of the image intensity distribution $I(\boldsymbol{\alpha}, t)$, which is a function of t due to atmospheric turbulence. Thus, firstly, we do this computation in the image plane instead of the aperture plane, and secondly, we investigate the intensity and not the instantaneous intensity. Therefore, we do not determine the visibility function, i.e. the coherence function of the fast varying amplitudes, in the image plane. However, the relatively slowly fluctuating effects of atmospheric turbulence on the amplitude are temporally resolved and affect the correlation function.

Instead of investigating the correlation function in the image plane, we move into the aperture plane by Fourier transforming the intensity correlation, yielding the square of the image intensity spectrum, the power spectrum $|\hat{I}(\boldsymbol{R},t)|^2$. Due to the imaging process, the visibility function in the aperture plane is directly available, since the Fourier transform, $\hat{I}(\boldsymbol{R},t)$, of the image intensity is the product of the visibility function, $\mu(\boldsymbol{R})$, in the aperture plane multiplied by the optical transfer function, $\mathrm{OTF}_\varphi(\boldsymbol{R},t)$. The time averaged power spectrum is then the product of $|\mu(\boldsymbol{R})|^2$ with the speckle transfer function $< |\mathrm{OTF}_\varphi(\boldsymbol{R},t)|^2 >$. The latter has non zero values up to the diffraction limit permitting to recover the visibility function up to the diffraction limit. The Fourier transform of the triple correlation of intensities, the bispectrum, is a complex function when the phase is given by the closure phase (4.96).

Thus, in intensity interferometry, the visibility function is determined by the correlation of the intensities, whereas in speckle interferometry, we determine the speckle transfer function by processing the image intensity correlation, more precisely by computing the time average of the power spectrum. Here, the visibility function is always available as the Fourier transform of the image intensity, independent of the temporal averaging process. The effect of speckle interferometry is that, by applying the speckle transfer function, the visibility function can be determined up to the diffraction limit, improving the angular resolution accordingly.

The formal similarity between the two methods is limited to the fact that both compute the second order moment of intensities and, thus, the fourth order moments of amplitudes. In intensity interferometry, determining the instantaneous intensities in the aperture plane with maximum temporal resolution and a very narrow spectral bandwidth, this is the fourth order moments of the fast varying amplitudes providing the visibility function. In speckle interferometry, measuring the image intensities with sufficient temporal resolution to freeze atmospheric turbulence, we determine the fourth order moments of the slowly varying amplitudes providing the speckle transfer function.

Returning to amplitude interferometry, we can combine two telescopes separated by the baseline vector $\boldsymbol{R_B}$, and we will find a speckle pattern with irregular fringes (see Fig. 4.11). This speckle pattern can be treated like that of individual telescopes, providing the visibility function $\mu(\boldsymbol{R_B})$ eventually. We will come back to this in Sect. 6.1.

Chapter 5
Instrumental Techniques

The beam combination scheme and the layout of the interferometer array determine the characteristics of the interferometer. If the reimaged telescope apertures in the beam combining instrument are a downscaled replica of the interferometer array, i.e. if they are mapped homothetically, the interferometer can be treated like a single telescope with a masked aperture, as discussed in Sect. 3.3.2, and the interferometric field of view is as large as that of the individual telescopes, limited by the optical design only. This is called the *Fizeau configuration*, after Hippolyte Fizeau who was one of the first to publish the idea of stellar interferometry in 1868 [69].

While the Fizeau configuration permits to discuss the imaging process in close analogy to single aperture systems, there is another configuration playing a major role in interferometry, which is called *Michelson configuration* after Albert A. Michelson who together with Francis G. Pease was the first to measure stellar diameters with an interferometer on Mount Wilson, California, in 1920 [158]. In contrast to the Fizeau configuration, the telescope apertures are no longer mapped homothetically but their distances in the beam combining instrument are chosen independently of the interferometer baseline. The consequence is that the interferometric field of view in Michelson configuration is limited in size to less than the Airy disk of the individual apertures.

If the telescope apertures are projected even parallel to and on top of each other with a beam splitter – the so-called *co-axial combination* – then, unlike all configurations with separate exit pupils, Airy disks without fringes form in the image plane. By temporally modulating the optical path length in one of the two beams, constructive and destructive interference can then be produced, "switching" the Airy disk on and off. Thus, instead of a fringe pattern as a function of diffraction angle α there is a fringe pattern as a function of OPD or of time difference τ.

The term Michelson configuration is often restricted to co-axial combination (see Fig. 5.3), since the optical layout is similar to the classical Michelson interferometer [157]. However, this restriction leaves out those configurations with separate telescope apertures in the exit pupil that are still not a Fizeau configuration. Given that Michelson in his experiment on Mount Wilson did not use co-axial combination but had the apertures projected side by side in the beam combining instrument, we apply the term Michelson configuration to *all* designs without homothetic mapping of the telescope apertures.

A. Glindemann, *Principles of Stellar Interferometry*, Astronomy and Astrophysics Library, DOI 10.1007/978-3-642-15028-9_5,

So far, we have centred the discussion on fringe patterns, their visibilities and on forming images through aperture synthesis, having a rather modest number of telescopes – typically fewer than 10 – in mind. However, combining a very large number of telescopes simultaneously and homothetically, we could produce a PSF with a narrow central core and few bumps and sidelobes, similar to an Airy disk, so that we instantaneously have a usable image. This is a straightforward process in Fizeau configuration, forming an image with a field of view that is limited by the optical design of the telescopes only.

Combining many telescopes in Michelson configuration by maximising the re-imaged telescope apertures in the exit pupil so that the gaps are minimised, we form a *densified pupil* providing a PSF with a bright central core. This concept was proposed by A. Labeyrie in 1996 [122] who called it *Hypertelescope*. While it has advantages over the Fizeau configuration, its field of view is severely limited, deteriorating very quickly when going off-axis.

Rearranging the telescope apertures similar to the Hypertelescope but forming an image by projecting the apertures (!) on top of each other was suggested by F. Vakili in 2004 [245]. The acronym for this configuration is IRAN, for *Interferometric Remapped Array Nulling*. The IRAN concept is intriguing since the image is formed in the pupil plane, apparently contradicting all principles of classical image formation. The field of view is similar to that of the Hypertelescope, but its PSF is of constant quality over the field of view.

Last but not least, we will discuss the concept of the *Nulling interferometer* [20] that was developed for a very specific scientific programme, that of searching for extra-solar planets. Here, the number of telescopes is modest but the very faint companion of the star can be detected by extinguishing, *nulling*, the light of an on-axis point source.

In Sect. 3.4 we investigated the impact of given interferometric arrays on the imaging process. Here, we will discuss criteria for their design based on the studies that have been performed for radio interferometry in the last decades [104]. While the underlying theory is the same as in the optical some practical issues drive us to different conclusions when it comes to designing an array.

5.1 Combination of Two Telescopes

In this section, we discuss basic aspects of telescope combination taking into account the reimaging of the telescopes apertures, the entrance pupil, into the exit pupil inside the beam combining instrument.

5.1.1 Fizeau Configuration

In Fizeau configuration, the exit pupil, i.e. the image of the telescope apertures inside the beam combining instrument, is a downscaled replica of the interferometer array so that the telescope apertures and their distances are reduced by the same factor,

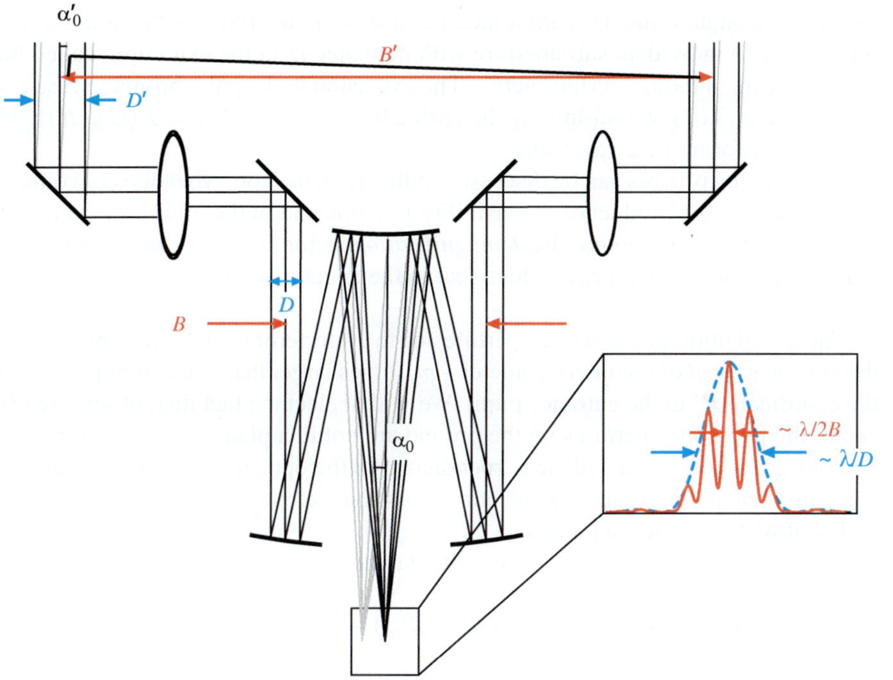

Fig. 5.1 Principle of the Fizeau configuration. The two lenses represent the optical system, reimaging the telescope apertures from the entrance pupil into the exit pupil. In Fizeau configuration it is $B/D = B'/D'$ so that the interferometer behaves like a masked aperture and fringes are found everywhere in the image plane. This figure illustrates also that downscaling the entrance pupil corresponds to an angular magnification $\alpha_0/\alpha_0' > 1$

as sketched in Fig. 5.1.[1] Therefore, the imaging process can be computed like that of a large telescope, with an aperture mask containing sub-apertures according to size and position of the individual telescopes. Like in a single telescope, the object-image relationship is given by a convolution between the object intensity and a PSF that now consists of the Airy disk of an individual sub-aperture with a narrow fringe pattern. This was discussed in great detail in Sect. 3.3 for the case that both entrance and exit pupil are a 1:1 image of the telescope apertures. Here, we will concentrate on the impact of downscaling the telescope apertures in the exit pupil.

We write the convolution product in the image plane as in (3.54), relating the object brightness distribution $O_b(\boldsymbol{\alpha}')$ to the image intensity by

$$\begin{aligned} I(\boldsymbol{\alpha}) &= O_b(\boldsymbol{\alpha}/m_p) * \mathrm{PSF}(\boldsymbol{\alpha}) \\ &= 2\, O_b(\boldsymbol{\alpha}/m_p) * (\mathrm{PSF}_a(\boldsymbol{\alpha})\, (1 + \cos(2\pi \boldsymbol{R}_B \cdot \boldsymbol{\alpha}))) \qquad (5.1) \\ &= 2 \int O_b(\boldsymbol{\alpha}')\mathrm{PSF}_a(\boldsymbol{\alpha} - m_p\boldsymbol{\alpha}')\, \big(1 + \cos(2\pi \boldsymbol{R}_B \cdot (\boldsymbol{\alpha} - m_p\boldsymbol{\alpha}'))\big)\, \mathrm{d}\boldsymbol{\alpha}', \end{aligned}$$

[1] This figure, as well as Figs. 5.2, 5.3 and 5.7 were inspired by illustrations in the PhD Thesis of R. Wilhelm [253].

with m_p the angular magnification factor and $\boldsymbol{\alpha} = m_p\boldsymbol{\alpha}'$. PSF_a is the point-spread function of an individual sub-aperture with diameter D in the exit pupil, called the primary beam in radio interferometry . The extension to the polychromatic case can be done as in (3.67) by multiplying the cosine function by $g_B(\boldsymbol{\alpha}) = \mathcal{F}(G_B(\boldsymbol{R}))$, the Fourier transform of the spectrum.

The imaging process can be discussed in the uv-plane, too, when the convolution transforms into the product of the visibility function and of the interferometer OTF. Following elementary optics, the *demagnification* factor D'/D, by which we downscale the interferometer array, is identical to the angular *magnification* factor m_p in the image.

The distribution of telescope apertures of the interferometer array is now called the *entrance pupil* of the interferometer – m_p times larger than the exit pupil – using the coordinate $\boldsymbol{R}'$ in the entrance pupil. We will neglect the fact that, observing off-zenith the telescope apertures on the ground are not in a plane perpendicular to the line of sight – and not in a plane perpendicular to the optical axis in the exit pupil – since this has no consequence for our discussion except for an additional optical path that will be treated separately.

Starting in the entrance pupil, we write the interferometer OTF (3.52) as

$$\mathrm{OTF}(\boldsymbol{R}') = \mathrm{OTF}_{a'}(\boldsymbol{R}') * \left(\delta(\boldsymbol{R}' - \boldsymbol{R}'_{B'}) + 2\delta(\boldsymbol{R}') + \delta(\boldsymbol{R}' + \boldsymbol{R}'_{B'})\right), \quad (5.2)$$

when $\mathrm{OTF}_{a'}$ is the monochromatic optical transfer function of a single telescope with aperture D'.

The OTF is multiplied by the visibility function $\mu(\boldsymbol{R}')$, the Fourier transform of the object brightness distribution $O_b(\boldsymbol{\alpha}')$, and by I_0, the integral over the object brightness, obtaining the image intensity spectrum $\hat{I}(\boldsymbol{R}')$, (3.55), as

$$\begin{aligned}\hat{I}(\boldsymbol{R}') &= I_0\mu(\boldsymbol{R}') \\ &\quad \times \left(\mathrm{OTF}_{a'}(\boldsymbol{R}') * \left(\delta(\boldsymbol{R}' - \boldsymbol{R}'_{B'}) + 2\delta(\boldsymbol{R}') + \delta(\boldsymbol{R}' + \boldsymbol{R}'_{B'})\right)\right). \quad (5.3)\end{aligned}$$

Replacing $\boldsymbol{R}'$ by $m_p\boldsymbol{R}$ yields the downscaled image intensity spectrum in the exit pupil,

$$\begin{aligned}\hat{I}(\boldsymbol{R}) &= I_0\mu(m_p\boldsymbol{R}) \\ &\quad \times \left(\mathrm{OTF}_{a'}(m_p\boldsymbol{R}) * \left(\delta(m_p\boldsymbol{R} - \boldsymbol{R}'_{B'}) + 2\delta(m_p\boldsymbol{R}) + \delta(m_p\boldsymbol{R} + \boldsymbol{R}'_{B'})\right)\right) \\ &= I_0\mu(m_p\boldsymbol{R}) \quad (5.4) \\ &\quad \times (\mathrm{OTF}_a(\boldsymbol{R}) * (\delta(\boldsymbol{R} - \boldsymbol{R}_B) + 2\delta(\boldsymbol{R}) + \delta(\boldsymbol{R} + \boldsymbol{R}_B))),\end{aligned}$$

with OTF_a the optical transfer function of a reimaged telescope aperture with diameter $D = D'/m_p$, and the baseline in the exit pupil of $\boldsymbol{R}_B = \boldsymbol{R}'_{B'}/m_p$. The visibility function in the exit pupil is also downscaled by m_p, so that, at $\boldsymbol{R} = \boldsymbol{R}_B$, we have the visibility $\mu(m_p\boldsymbol{R}_B) = \mu(\boldsymbol{R}'_{B'})$, i.e. the visibility determined by the baseline in the entrance pupil.

In Sect. 3.3.4, the polychromatic case was discussed providing the polychromatic OTF in (3.64) when the individual δ-functions at $\pm \boldsymbol{R}_B$ are replaced by the spectra $G_B(\boldsymbol{R})$. All discussions and conclusions on the imaging process in Sect. 3.3.4 apply here.

The Fourier transform of the image intensity spectrum in the exit pupil, $\hat{I}(\boldsymbol{R})$, provides the image intensity distribution, $I(\boldsymbol{\alpha})$, which is the convolution product of the individual Fourier transforms $O_b(\boldsymbol{\alpha}/m_p)$ and the interferometer PSF, as given in (5.1). There is no fundamental limit for the field of view and the convolution holds for any object size.

A true Fizeau configuration is difficult to build since, observing a celestial source with telescopes on the ground, the effective baselines – the projections of the baseline vector onto the sky plane perpendicular to the line of sight – are moving constantly due to the rotation of the Earth so that the projected telescope apertures in the entrance pupil are in permanent motion, both laterally and longitudinally (see Sect. 3.4). Providing homothetic mapping of the entrance pupil, the exit pupil also undergoes permanent baseline changes so that the fringe pattern has variable fringe spacing.

This poses considerable practical problems [11]. First, one has to ensure that the optical systems reimaging the telescope apertures in the exit pupil place them dynamically in exactly the correct lateral and longitudinal position to properly define the baseline as $\boldsymbol{R}_B = \boldsymbol{R}'_{B'}/m_p$. Second, scanning the fringe pattern in order to measure the visibility requires a pixel scale that is suitable for all baselines during the observation. This means that the smallest pixel scale, suitable for the narrowest fringe pattern of the longest baseline, is not optimal for shorter baselines since measuring the visibility, the wider fringes of the shorter baselines are scanned with more pixels than necessary so that the sensitivity is reduced. Therefore, to date there has been no serious effort to build a Fizeau interferometer with separate telescopes.

Both problems are alleviated using aperture masks on single telescopes when the optical layout ensures the proper reimaging of the entrance pupil, and when the baselines are fixed. This technique is also known as ***sparse aperture masking***.

The advantage of sparse aperture masking over observations with the full aperture lies in the fact that the individual spatial frequencies defined by the baselines are measured with much higher accuracy than when they are part of the spatial frequency spectrum of the Airy disk of the full aperture [99, 242]. The advantages of observations with sparse aperture masking were demonstrated very impressively by P. Tuthill [241].

Almost like an aperture mask, albeit with only two apertures, the Large Binocular Telescope (LBT), [101], with its two 8.4-m primary mirrors mounted on the same telescope structure, comes very close to a Fizeau configuration because the baseline is by design perpendicular to the line of sight so that only static aperture remapping is required.

5.1.2 Michelson Configuration

The unrestrained arrangement of the telescope apertures in the exit pupil characterizes the Michelson configuration, named after Michelson's original experiment [158]. Then, the telescope apertures are not mapped homothetically, and, in general, the object-image relationship can no longer be described by a convolution since the positions of the Airy disks in the field no longer coincide with equal optical path lengths in the interferometer (see Fig. 5.2). On axis, the PSF is an Airy disk with fringes while elsewhere in the field of view, the PSF does not show fringes. This means that the PSF is no longer shift-invariant. This argument applies to all realistic cases when the spectral bandwidth is finite. However, in strictly monochromatic

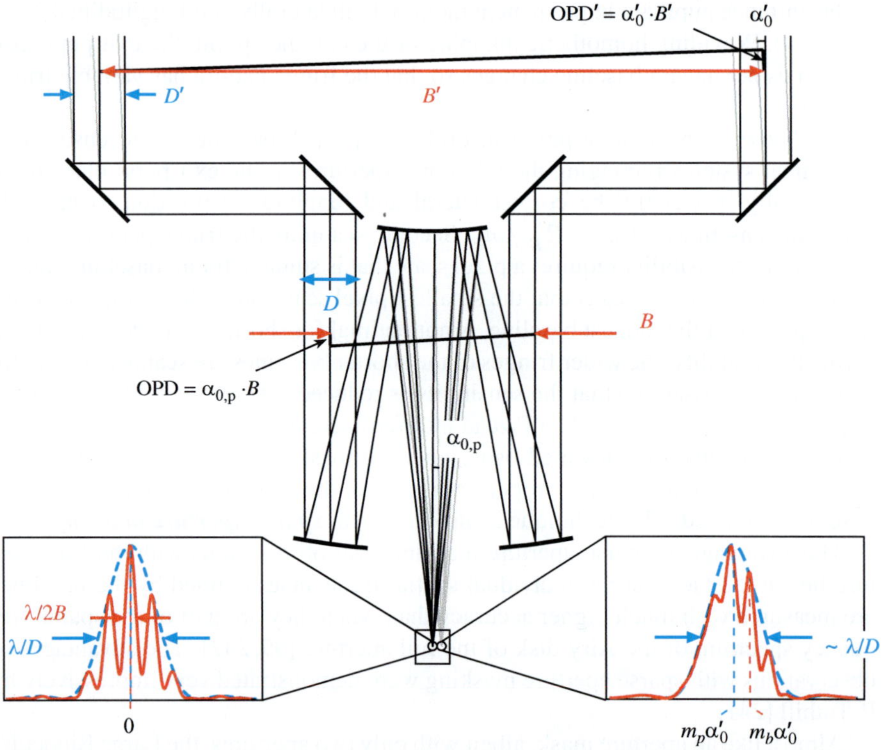

Fig. 5.2 Principle of the Michelson configuration in multi-axial combination. For simplicity, the telescope apertures of diameter D' are reimaged by a periscope so that $D = D'$ and $m_p = 1$, as in the original Michelson experiment [158]. This figure illustrates that, in the entrance pupil, the OPD for an object at position $\boldsymbol{\alpha}'_0$, OPD $= \boldsymbol{\alpha}'_0 \cdot \boldsymbol{B}'$, is larger than the OPD corresponding to the center of the Airy disc, $\boldsymbol{\alpha}_{0,p} \cdot \boldsymbol{B}$, due to the downscaled baseline $\boldsymbol{B}$. While the image is at position $\boldsymbol{\alpha}_{0,p} = m_p\boldsymbol{\alpha}'_0$, the white-light fringe is at position $\boldsymbol{\alpha}_{0,b} = m_b\boldsymbol{\alpha}'_0$, for $\boldsymbol{\alpha}'_0 || \boldsymbol{B}$, when the OPD in the entrance pupil and that in the exit pupil cancel. Then, the fringe package is no longer at the center of the Airy disk, $m_p\boldsymbol{\alpha}'_0$, unless $m_b = m_p$ as in Fizeau configuration

illumination with extremely long coherence length, fringes can be found everywhere in the field of view.

We base our definition of the Michelson configuration on the imaging process. If the PSF is shift-invariant, a convolution describes the imaging process, requiring homothetic mapping as in Fizeau configuration. In Michelson configuration, without homothetic mapping, the PSF is not shift-invariant and, without further assumptions, the imaging process cannot be described by a convolution.

As in Fizeau configuration, we call m_p the demagnification factor of the telescope apertures, and we introduce the demagnification factor $m_b = \boldsymbol{R}'_{B'}/\boldsymbol{R}_B$ for the baselines.

Therefore, we have to write the interferometer PSF as a function not only of the image coordinate $\boldsymbol{\alpha}$ but also of the object coordinate $\boldsymbol{\alpha}'$ since it varies with the object position. For an object at position $\boldsymbol{\alpha}'_0$, we have the point-spread function of an individual aperture, PSF_a, centred at position $\boldsymbol{\alpha}_{0,p} = m_p\boldsymbol{\alpha}'_0$ in the image plane, as in (5.1).

With $m_p \neq m_b$, the white-light fringe, i.e. the position of zero OPD, is at position $\boldsymbol{\alpha}_{0,b} = m_b\boldsymbol{\alpha}'_0 \neq \boldsymbol{\alpha}_{0,p}$. Then, the OPD in the entrance pupil, $\boldsymbol{\alpha}'_0 \cdot \boldsymbol{B}'$, is identical to the OPD in the exit pupil, $\boldsymbol{\alpha}_{0,b} \cdot \boldsymbol{B} = m_b\boldsymbol{\alpha}'_0 \cdot \boldsymbol{B} = \boldsymbol{\alpha}'_0 \cdot \boldsymbol{B}'$, as displayed in Fig. 5.2. This case of separate apertures in the exit pupil is also called *multi-axial combination.*

Starting with the monochromatic case and using (3.53), the resulting interferometer PSF in Michelson configuration reads as

$$\mathrm{PSF}_M(\boldsymbol{\alpha},\boldsymbol{\alpha}') = 2\mathrm{PSF}_a(\boldsymbol{\alpha} - m_p\boldsymbol{\alpha}')\left(1 + \cos(2\pi \boldsymbol{R}_B \cdot (\boldsymbol{\alpha} - m_b\boldsymbol{\alpha}'))\right), \quad (5.5)$$

and the image intensity distribution is

$$\begin{aligned} I(\boldsymbol{\alpha}) &= \int O_b\left(\boldsymbol{\alpha}'\right) \mathrm{PSF}_M(\boldsymbol{\alpha},\boldsymbol{\alpha}')\mathrm{d}\boldsymbol{\alpha}' \quad (5.6)\\ &= 2\int O_b\left(\boldsymbol{\alpha}'\right) \mathrm{PSF}_a(\boldsymbol{\alpha} - m_p\boldsymbol{\alpha}')\left(1 + \cos(2\pi \boldsymbol{R}_B \cdot (\boldsymbol{\alpha} - m_b\boldsymbol{\alpha}'))\right)\mathrm{d}\boldsymbol{\alpha}'\,. \end{aligned}$$

Due to different demagnification factors m_p for the telescope aperture and m_b for the baseline PSF_M is not shift-invariant and the integral cannot be written as a convolution. However, in the monochromatic case this does not restrict the imaging capabilities even for large objects since fringes are found for any coordinate $\boldsymbol{\alpha}$ in the image plane, and the object brightness distribution can be recovered from the visibility in the image intensity spectrum [221].

Small Objects

Preparing the case of polychromatic objects, we restrict the object size to the diameter of the Airy disk of an individual telescope, PSF_a, as discussed in Sect. 3.3.2, so that we can extract PSF_a from the integral and the integral describes a convolution again:

$$\begin{aligned} I(\boldsymbol{\alpha}) &= 2\,\mathrm{PSF}_a(\boldsymbol{\alpha}) \int O_b(\boldsymbol{\alpha}') \left(1 + \cos(2\pi \boldsymbol{R}_B \cdot (\boldsymbol{\alpha} - m_b \boldsymbol{\alpha}'))\right) \mathrm{d}\boldsymbol{\alpha}' \\ &= 2\,\mathrm{PSF}_a(\boldsymbol{\alpha}) \big(O_b(\boldsymbol{\alpha}/m_b) * (1 + \cos(2\pi \boldsymbol{R}_B \cdot \boldsymbol{\alpha}))\big). \end{aligned} \tag{5.7}$$

Thus, the image intensity distribution of small objects is described by the convolution of the object brightness distribution with the fringe pattern, similar to the Fizeau configuration, (5.1).

Note that the scale factor of the object brightness distribution is given by m_b since the object position is defined by the position of the white-light fringe at zero OPD so that the object size for the application of a convolution (5.7) is limited to

$$\begin{aligned} m_b \alpha'_{\max} - m_p \alpha'_{\max} &\ll \frac{\lambda}{D} \\ \alpha'_{\max} &\ll \frac{\lambda}{m_b D - m_p D}\,. \end{aligned} \tag{5.8}$$

Usually, combining apertures smaller than 10 m in diameter that are much smaller than their baselines of several 10 m, the demagnification factor m_b of the baselines is much stronger than that of the apertures, m_p, in order to have a compact exit pupil. In this approximation, we can set $m_b - m_p \approx m_b$, so that the object size is limited to $\alpha'_{\max} \ll \lambda/(m_b D'/m_p)$, which is the FWHM of the Airy disk, demagnified by m_b/m_p.

In the uv-plane, we take the subsequent assumption that the visibility function $\mu(\boldsymbol{R}')$ is invariant over the width of the OTF of an individual aperture since $O_b(\boldsymbol{\alpha}')$ is much smaller than PSF_a as discussed in Sect. 3.3.2.

The image intensity spectrum in the entrance pupil is given by the product of the visibility function with the interferometer OTF, as in Fizeau configuration, (5.3). However, with the visibility function invariant over $\mathrm{OTF}_{a'}$, we can replace $\mu(\boldsymbol{R}')$ by its values at $\boldsymbol{R}' \pm \boldsymbol{R}'_{B'}$ as in (3.57), yielding

$$\begin{aligned} \hat{I}(\boldsymbol{R}') &= I_0 \mathrm{OTF}_{a'}(\boldsymbol{R}') * \big(\mu(\boldsymbol{R}')(\delta(\boldsymbol{R}' - \boldsymbol{R}'_{B'}) + 2\delta(\boldsymbol{R}') + \delta(\boldsymbol{R}' + \boldsymbol{R}'_{B'}))\big) \\ &= I_0 \mathrm{OTF}_{a'}(\boldsymbol{R}') \\ &\quad * \big(\mu(\boldsymbol{R}'_{B'})\delta(\boldsymbol{R}' - \boldsymbol{R}'_{B'}) + 2\mu(0)\delta(\boldsymbol{R}') + \mu(-\boldsymbol{R}'_{B'})\delta(\boldsymbol{R}' + \boldsymbol{R}'_{B'})\big), \end{aligned} \tag{5.9}$$

with $\mu(-\boldsymbol{R}'_{B'}) = \mu^*(\boldsymbol{R}'_{B'})$.

The image intensity spectrum in the exit pupil is obtained by rescaling the coordinates, using m_p for the OTF of the individual apertures and m_b for the baseline $\boldsymbol{R}'_{B'}$, as

$$\begin{aligned}\hat{I}(\boldsymbol{R}) &= I_0 \mathrm{OTF}_{a'}(m_p \boldsymbol{R}) \\ &\quad * \big(\mu(\boldsymbol{R}'_{B'})\delta(m_b \boldsymbol{R} - \boldsymbol{R}'_{B'}) + 2\mu(0)\delta(m_b \boldsymbol{R}) + \mu^*(\boldsymbol{R}'_{B'})\delta(m_b \boldsymbol{R} + \boldsymbol{R}'_{B'})\big) \\ &= I_0 \mathrm{OTF}_a(\boldsymbol{R}) \\ &\quad * \big(\mu(m_b \boldsymbol{R}_B)\delta(\boldsymbol{R} - \boldsymbol{R}_B) + 2\mu(0)\delta(\boldsymbol{R}) + \mu^*(m_b \boldsymbol{R}_B)\delta(\boldsymbol{R} + \boldsymbol{R}_B)\big),\end{aligned} \tag{5.10}$$

OTF_a is the OTF of aperture D'/m_p, and the baseline in the exit pupil is $\boldsymbol{R}_B = \boldsymbol{R}'_{B'}/m_b$.

Moving into the image plane by Fourier transforming $\hat{I}(\boldsymbol{R})$, we simplify the convolution in the image intensity distribution in (5.7) by using the visibility $\mu(m_b \boldsymbol{R}_B)$ to obtain

$$I(\boldsymbol{\alpha}) = 2\mathrm{PSF}_a(\boldsymbol{\alpha}) I_0 \big(1 + |\mu(m_b \boldsymbol{R}_B)| \cos(\phi(m_b \boldsymbol{R}_B) - 2\pi \boldsymbol{R}_B \cdot \boldsymbol{\alpha})\big), \tag{5.11}$$

with $\phi(m_b \boldsymbol{R}_B)$ the phase of the visibility function. The contrast of the fringe pattern with fringe spacing $1/\boldsymbol{R}_B$ is given by the modulus of the visibility function μ for the baseline $m_b \boldsymbol{R}_B = \boldsymbol{R}'_{B'}$ in the entrance pupil.

Thus, whatever the demagnification of the baseline, the visibility is always determined by the baseline in the telescope array and not by the baseline in the exit pupil.

Observing in Polychromatic Light

Observing with finite spectral bandwidth, we found that the contrast of the fringes decreases with increasing OPD. This is accounted for by multiplying the cosine function in (5.7) by $g_B(\boldsymbol{\alpha})$, the Fourier transform of the spectrum, $G_B(\boldsymbol{R})$, yielding the polychromatic PSF as in (3.61), as

$$\mathrm{PSF}_{pc}(\boldsymbol{\alpha}) = 2\mathrm{PSF}_a(\boldsymbol{\alpha})\big(1 + g_B(\boldsymbol{\alpha}) \cos(2\pi \boldsymbol{R}_{0,B} \cdot \boldsymbol{\alpha})\big).$$

The half-width of the fringe package in the image plane is $(\lambda_0/\Delta\lambda)\lambda_0/B = l_c/B$, with l_c the coherence length, as displayed in Fig. 3.17. We assume in the following, that the fringe package is narrower than the Airy disk, λ/D. Thus, if the off-axis distance $\alpha'_0 = |\boldsymbol{\alpha}'_0|$ of the object, with $\boldsymbol{\alpha}_0 || \boldsymbol{B}$, is so large that the coordinate difference between the centre of the Airy disk and the white-light fringe at $m_b\alpha'_0 - m_p\alpha'_0$ is larger than l_c/B (see Fig. 5.2), the fringe patterns of an on-axis object and of an object at α'_0 no longer overlap as discussed in Sect. 3.3.3.

Therefore, we write the **field of view** in Michelson configuration as [222]

$$\begin{aligned} m_b\alpha'_{\max} - m_p\alpha'_{\max} &\ll \frac{l_c}{B} \\ \alpha'_{\max} &\ll \frac{l_c}{m_b B - m_p B} \approx \frac{l_c}{B'}, \end{aligned} \tag{5.12}$$

using the approximation $m_b - m_p \approx m_b$ (see 5.8) and $m_b B = B'$, the baseline in the entrance pupil, we find that the object size is limited to the half-width of the fringe package, l_c/B', which we assumed to be smaller than the Airy disk.

Quasi-Monochromatic Approximation

The field of view in (5.12) is exactly the same as in the quasi-monochromatic approximation (Sect. 3.3.4) when the limitation was motivated by representing the fringe package by a single visibility value determined by the object shape through the van Cittert–Zernike theorem (2.50).

With an object size, smaller than the half-width of the fringe package – and smaller than the Airy disk – the visibility function is invariable not only over OTF_a but also over the width of the polychromatic optical transfer function that is elongated due to the convolution of OTF_a with the spectrum $G_B(\boldsymbol{R})$. This was investigated in detail in Sect. 3.3.4 when discussing small objects and the quasi-monochromatic approximation.

Similar to (3.71), the polychromatic image intensity spectrum is now written as

$$\hat{I}(\boldsymbol{R}) = I_0 \mathrm{OTF}_a(\boldsymbol{R}) * \big(\mu(m_b \boldsymbol{R}_{0,B}) G_B(\boldsymbol{R}) + 2\mu(0)\delta(\boldsymbol{R}) + \mu^*(m_b \boldsymbol{R}_{0,B}) G_B(-\boldsymbol{R})\big), \tag{5.13}$$

with $\mu(m_b \boldsymbol{R}_{0,B})$ the visibility function at the mean baseline $m_b \boldsymbol{R}_{0,B} = \boldsymbol{R}'_{0,B'} = \boldsymbol{B}'/\lambda_0$, and $G_B(\boldsymbol{R})$ the spectrum as a function of spatial frequency, centred at $\boldsymbol{B}/\lambda_0$, (3.64).

Proceeding into the image plane, we obtain the image intensity distribution (see 3.72), as

$$I_{\mathrm{qm}}(\boldsymbol{\alpha}) = 2 I_0 \, \mathrm{PSF}_a(\boldsymbol{\alpha}) \times \left(1 + g_B(\boldsymbol{\alpha}) |\mu(\boldsymbol{R}'_{0,B'})| \cos(\phi(\boldsymbol{R}'_{0,B'}) - 2\pi \boldsymbol{R}_{0,B} \cdot \boldsymbol{\alpha})\right), \tag{5.14}$$

with $\boldsymbol{R}'_{0,B'} = m_b \boldsymbol{R}_{0,B}$.

The contrast of the fringe pattern around the white-light fringe is determined by the modulus of the visibility at the entrance pupil baseline $\boldsymbol{R}'_{0,B'}$, and the position of the fringe pattern with respect to the position of zero OPD, $|\boldsymbol{\alpha}| = 0$, provides its phase $\phi(\boldsymbol{R}'_{0,B'})$.

5.1.3 Co-Axial Combination

In the limiting case of a Michelson configuration when the individual telescope apertures are imaged on top of each other, it is $|\boldsymbol{R}_B| = 0$ and $m_b \to \infty$. The exit pupil appears as a single aperture, and the PSF is an Airy disk without fringes.

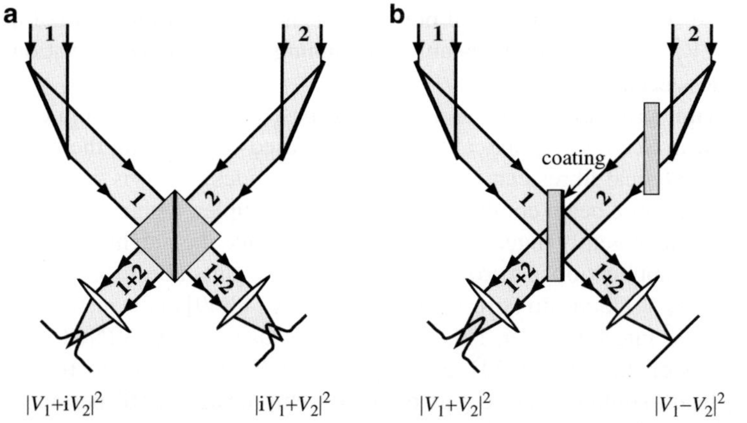

Fig. 5.3 Principle of coaxial beam combination with a beam splitter cube (**a**) and a plane-parallel plate with semi-reflective coating (**b**), assuming a 50:50 split of the incoming intensities. We denote the incoming intensities by $2V_1^2$ and $2V_2^2$, and assume $2V_1^2 = 2V_2^2$. The exit pupils are represented by single lenses in each output arm of the beam splitter. The phase shift of $\pi/2$ of the transmitted beams in the beam splitter cube (**a**) is indicated by i in the sum of the amplitudes. For OPD $= 0$ between the input beams, the total intensity in each output arm is equal to the intensity in each input. For the plane-parallel plate (**b**), the reflection of beam 2 at the optically denser medium causes a phase shift of π, so that the intensity in the right output arm of (**b**) is zero. In the left output arm, the intensity is maximum, equal to the sum of the intensities in beam 1 and 2. The additional plane-parallel plate in incoming beam 2 is required to compensate for the optical path of beam 1 in the beam splitter

This is called the *co-axial combination*. The Airy disk is either bright if the phase difference between the beams from the two telescopes is zero, or it is dark if the phase difference is π. Thus, there is no spatial fringe pattern in the image plane but an Airy disk with an intensity that varies with the OPD.

Figure 5.3 displays the principle of co-axial combination, when the single lens in each output of the beam splitter represents the exit pupil where the superimposed images of the telescope apertures form. The intensities of the Airy disks, drawn in the focal planes of the lenses, depend on the phase differences between the beams that are determined by the OPD in the input beams and by beam splitting schemes.

In the case of a beam splitter cube (Fig. 5.3a), the intensity split of the incoming intensities is based on an effect called *frustrated total internal reflection*, when the gap between two prisms forming the cube is so narrow that the total reflection at the optically thinner medium filling the gap is disturbed and a fraction of the light, as a function of the thickness of the gap, passes through.

Then, the phase of each transmitted beam is shifted by $\pi/2$ so that, for an optical path difference of zero between the incoming beams, the phase difference between the amplitudes V_1 and V_2 of the exiting beams is $\pi/2$ in one output arm of the beam splitter and $-\pi/2$ in the other arm. It can be shown that in symmetric beam splitters, the symmetry of phase shifts occurs for fundamental reasons [143, 257], independent of the material of the beam splitter. The intensity in each output arm is

$I = V_1^2 + V_2^2$ so that the sum of both outputs is the same as the sum of intensities, $2V_1^2 + 2V_2^2$, of the incoming beams maintaining the energy conservation principle for lossless beam splitters.

The same principle applies to a technique based on monomode optical fibres. As discussed in Sect. 3.3.6, optical fibres can be used for masking the image in each telescope. Optical fibres can also serve as waveguides, and it is possible to design a symmetric fibre beam combiner, a so-called X-coupler, with monomode fibres by squeezing them gently together so that light leaks into the neighbour fibre.

The first application in astronomy, using fluoride glass fibres for the K-band, was presented by V. Coudé du Foresto and S. Ridgway [48,49] in the instrument FLUOR at the IOTA[2] interferometer. This technique was also used for VINCI, the first-light instrument of the VLTI [111]. Moving from individual fibres to fully integrated optics, beam splitters can be part of the on-chip design [110] permitting a large flexibility for the number of input and output beams [12].

The case of a plane-parallel plate is shown in Fig. 5.3b when one surface has a semi-reflective coating to provide the split of the intensities. The reflection of beam 2 at the optically denser medium causes a phase shift of π while the transmitted beam 1 undergoes no phase shift, so that for zero OPD in the incoming beams – and assuming $V_1 = V_2$ – there is zero intensity in the right output of the beam splitter. In the left output, both the reflection of beam 1 and the transmission of beam 2 do not suffer from a phase shift so that we have constructive interference, and the intensity is $I = (V_1 + V_2)(V_1 + V_2)^* = V_1^2 + 2V_1V_2 + V_2^2 = 4V^2$, using $V = V_1 = V_2$, which is the same as the sum of the incoming intensities so that the energy is conserved.

We can account for the phase shifts in the individual output arms of the beam splitters by adding the appropriate phase φ_{bs} in the interferometer OTF, yielding the image intensity spectrum in the exit pupil as in (5.10), for objects substantially smaller than PSF_a, now for $|\boldsymbol{R}_B| = 0$,

$$\begin{aligned}\hat{I}(\boldsymbol{R}) = {}& I_0 \mathrm{OTF}_a(\boldsymbol{R}) \\ & * \left(\mu(\boldsymbol{R}'_{B'})\delta(\boldsymbol{R} - 0)\mathrm{e}^{-\mathrm{i}\varphi_{\mathrm{bs}}} + 2\mu(0)\delta(\boldsymbol{R}) + \mu^*(\boldsymbol{R}'_{B'})\delta(\boldsymbol{R} + 0)\mathrm{e}^{\mathrm{i}\varphi_{\mathrm{bs}}}\right),\end{aligned} \tag{5.15}$$

with $\boldsymbol{R}'_{B'}$ the baseline in the entrance pupil.

Assuming that the OPD between the incoming beams is zero, we have $\varphi_{\mathrm{bs}} = \pm\pi/2$ in the outputs of the beam-splitter cube, and $\varphi_{\mathrm{bs}} = 0$ and π in the outputs of the plane-parallel plate [209].

Temporal Modulation

Modulating the optical path length in one arm of the interferometer is equivalent to introducing a time varying optical path difference $\mathrm{OPD}(t)$ between the incoming

[2] IOTA is the Infrared-Optical Telescope Array on Mt. Hopkins in Arizona, USA.

beams so that the phase in the interferometer OTF is $\psi(t) = \frac{2\pi}{\lambda}\text{OPD}(t) + \varphi_{\text{bs}}$. This provides the time dependent image intensity spectrum as

$$\hat{I}(\boldsymbol{R},t) = I_0\text{OTF}_a(\boldsymbol{R}) * \Big(\mu(\boldsymbol{R}'_{B'})\delta(\boldsymbol{R}-0)\text{e}^{-\text{i}\psi(t)} + 2\mu(0)\delta(\boldsymbol{R}) + \mu^*(\boldsymbol{R}'_{B'})\delta(\boldsymbol{R}+0)\text{e}^{\text{i}\psi(t)}\Big). \tag{5.16}$$

The image intensity distribution, $I(\boldsymbol{\alpha},t) = \mathcal{F}_R(\hat{I}(\boldsymbol{R},t))$ is now also a function of time t. The quantity that is usually discussed is the integral of the image intensity $I_{\text{mod}}(t) = \int I(\boldsymbol{\alpha},t)\text{d}\boldsymbol{\alpha}$, obtaining

$$\begin{aligned} I_{\text{mod}}(t) &= \int 2\text{PSF}_a(\boldsymbol{\alpha})I_0\big(1+|\mu(\boldsymbol{R}'_{B'})|\cos(\phi(\boldsymbol{R}'_{B'})-\psi(t))\big)\text{d}\boldsymbol{\alpha} \\ &= 2I'_0\big(1+|\mu(\boldsymbol{R}'_{B'})|\cos(\phi(\boldsymbol{R}'_{B'})-\psi(t))\big), \end{aligned} \tag{5.17}$$

with $I'_0 = I_0 \int \text{PSF}_a(\boldsymbol{\alpha})\text{d}\boldsymbol{\alpha}$.

We have, thus, a signal that oscillates between its maximum value $1+|\mu(\boldsymbol{R}'_{B'})|$ and its minimum $1-|\mu(\boldsymbol{R}'_{B'})|$, displaying a complete fringe for a modulation of $\text{OPD} = \lambda$, i.e. of $\psi = 2\pi$. Restricting the modulation to λ usually is sufficient if only the fringe position is of interest, for instance for a fringe tracker. If the amplitude of the modulation is larger than the coherence length l_c, we scan the complete fringe package, as displayed in Fig. 5.5, and the visibility is derived from the Fourier spectrum of $I_{\text{mod}}(t)$ as discussed in Sect. 2.4.3.

With $\varphi_{\text{bs}} = +\pi/2$ and $-\pi/2$ in the two output arms of the beam splitter cube, as well as with $\varphi_{\text{bs}} = 0$ and π in the outputs of the plane-parallel plate, the phases of the fringe patterns in the two outputs differ by π in both cases, sometimes referred to as fringes in anti-phase, so that if one arm displays maximum intensity the other arm has zero intensity.

Spatial Modulation

Instead of a temporal modulation of the OPD to cover a phase shift range of 2π, it is possible to provide this range instantaneously by spatial phase modulation. Introducing in one of the incoming beams of the beam splitter cube an achromatic phase shift of $\pi/2$ between the orthogonal p and s polarization, the phase shift in the output beams of the beam splitter is now polarization dependent. For one of the polarization components we still have phases $\varphi_{\text{bs}} = \pm\pi/2$ in the two output arms, but for the second component we find $\varphi_{\text{bs}} = \pi/2 + \pi/2 = \pi$ and $\varphi_{\text{bs}} = -\pi/2 + \pi/2 = 0$. Disentangling the s and p polarization in the exiting beams by polarizing beam splitters provides four signals with $0, \pi/2, \pi$ and $3\pi/2$ phase difference between the interfering beams as displayed in Fig. 5.4.

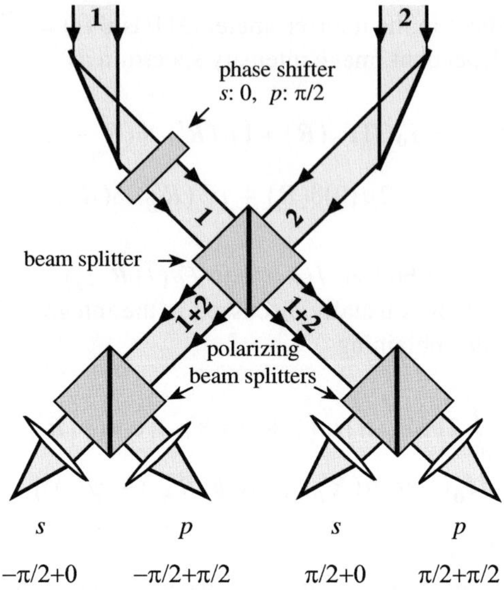

Fig. 5.4 Principle of spatial phase modulation. The achromatic phase shifter in one of the input arms provides a phase shift of $\pi/2$ for the p polarization with respect to the s polarization. Then, the phase shifts between beams 1 and 2 in the output arms are different for s and p. The polarizing beam splitters separate the s and p polarization and we find the indicated phase shifts between the beams in the four output arms. This scheme instantaneously provides four phase shifts, $-\pi/2, 0, \pi/2$ and π so that the ABCD method can be applied to measure the visibility

In astronomy, spatial phase modulation was implemented for the first time for the fringe tracker of the PRIMA[3] facility of the VLTI [204].

Integrating over the Airy disk eliminates all information on the spatial distribution of the image intensity and yields the same value as the integral of the intensity in the exit pupil. Therefore, co-axial combination is sometimes also labelled *pupil plane combination*. However, one has to be aware that the fringe pattern does not form in the pupil plane – although this could be done by slightly tilting one of the reimaged telescope apertures and observe the fringes in the exit pupil as will be discussed in Sect. 5.3.3 – but that a single value of the intensity is used, and fringes form exclusively through OPD modulation, modulating the PSF intensity.

Quasi-Monochromatic Approximation

In co-axial combination, we write the modulated intensity (5.17) in quasi-monochromatic approximation (compare to 5.14) as

$$I_{\mathrm{mod,qm}}(t) = 2I_0'\big(1 + g_p(\psi(t))|\mu(\boldsymbol{R}'_{0,B'})|\cos(\phi(\boldsymbol{R}'_{0,B'}) - \psi(t))\big), \quad (5.18)$$

---

[3] PRIMA (Phase Referenced Imaging and Micro-arcsecond Astrometry) is the dual-feed facility of the VLTI.

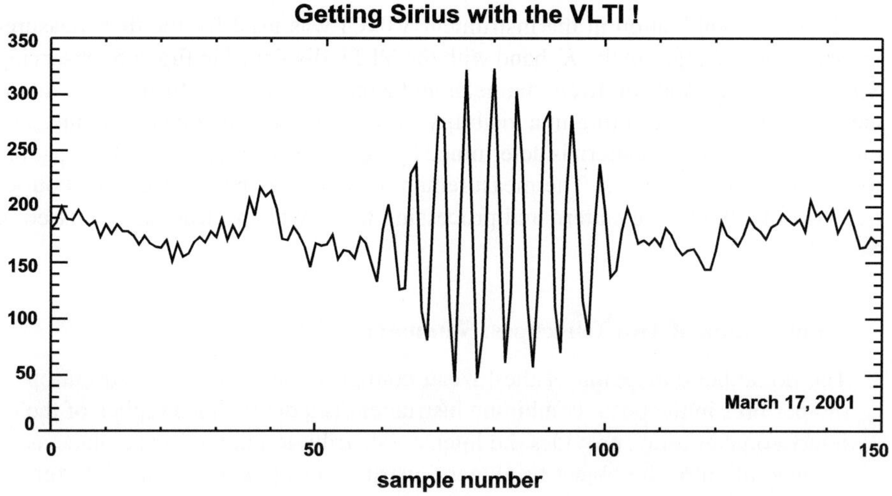

Fig. 5.5 The first fringe pattern of the VLTI observing Sirius on March 17, 2001. The spectral band is the K-band (2.2±0.2 μm), and $\lambda/\Delta\lambda = 5.5$. The finite spectral bandwidth allows for about 11 fringes similar to Fig. 2.16

with $I_0' = I_0 \int \mathrm{PSF}_a(\boldsymbol{\alpha})\mathrm{d}\boldsymbol{\alpha}$ and $g_p(\psi(t)) = g(\psi(t)/(2\pi\nu_0)) = g(\tau)$ so that the time delay τ between the interfering beams is replaced by the dependence on the modulated phase, using $\tau = \mathrm{OPD}/c = \psi(t)/(2\pi\nu_0)$.

For the case of the K-band, with $\lambda/\Delta\lambda = 5.5$ we find that $g(\tau)$ drops to zero if the OPD is larger than the coherence length of 5.5λ, i.e. after 5.5 fringes. If the modulation $\psi(t)$ of the fringe pattern is larger than $\pm 5.5 \times 2\pi$ we will see the complete fringe package, as displayed in Fig. 5.5. This relates the properties of $g(\tau)$ to those of the phase modulation. Using the notation $g_p(\psi(t))$ we can write the intensity distribution as function of time for a given modulation $\psi(t)$.

The contrast of the white-light fringe is determined by $|\mu(\boldsymbol{R}'_{0,B'})|$ when $g_p(\psi(t))$ is basically constant and the ABCD method can be used to determine both modulus and phase of the visibility.

Regarding the Fourier spectrum of the modulated intensity distribution, we should note that $I_{\mathrm{mod,\,qm}}(t)$ is formally identical to the fringe pattern in Young's experiment (2.63). Discussing its Fourier transform in Sect. 2.4.3, we can apply the result to obtain the image intensity spectrum as

$$\hat{I}_{\mathrm{mod,\,qm}}(\nu) = I_0'\left(2\delta(\nu) + G(\nu)\mu\left(\frac{\boldsymbol{B}'}{c}\nu_0\right) + G(-\nu)\mu^*\left(-\frac{\boldsymbol{B}'}{c}\nu_0\right)\right), \quad (5.19)$$

when the spatial frequency $\boldsymbol{R}'_{0,B'}$ is written as $\frac{\boldsymbol{B}'}{c}\nu_0$.

Due to the integration over the PSF, there is no longer a convolution with OTF_a as in (5.16), and we obtain the Fourier spectrum as a function frequency ν due to $I_{\mathrm{mod,qm}}$ being a function of time.

Co-axial combination in the instrument VINCI was used for the first measurement of stellar fringes in the K-band with the VLTI, displayed in Fig. 5.5, observing Sirius with siderostats of 40 cm diameter and a baseline of 16 m. Sirius has a diameter of about 6 mas resulting in a visibility of about 0.9 for the given baseline. The envelope of the fringe pattern is determined by the spectrum, here the K-band, going to zero beyond 11 fringes. The measurement of $|\mu|^2$ was performed after Fourier transforming the fringe pattern and processing the power spectrum as discussed in Sect. 2.4.3 [111].

Combination of Two Telescopes: Summary

The homothetic mapping of the Fizeau configuration, when the distribution of apertures in the beam combining instrument is a downscaled replica of the interferometer array, provides the intensity distribution in the image plane as a convolution of the object brightness distribution $O_b(\boldsymbol{\alpha}')$ with the interferometer point-spread function, which is the Airy disk with fringes. The image coordinate $\boldsymbol{\alpha}$ is related to the object coordinate by $\boldsymbol{\alpha}' = \boldsymbol{\alpha}/m_p$, yielding the image intensity distribution as

$$I(\boldsymbol{\alpha}) = 2\, O_b(\boldsymbol{\alpha}/m_p) * (\mathrm{PSF}_a(\boldsymbol{\alpha})\,(1 + \cos(2\pi \boldsymbol{R}_B \cdot \boldsymbol{\alpha})))\,. \tag{5.1}$$

The image intensity spectrum is formed by the Fourier transform of the image intensity distribution, which is the product of the visibility function $\mu(m_p \boldsymbol{R})$ with the interferometer OTF. The spatial frequency vectors in the entrance and exit pupil are related by $\boldsymbol{R}' = m_p \boldsymbol{R}$, so that the baseline in the interferometer array $\boldsymbol{R}'_{B'}$ is downscaled by m_p to $\boldsymbol{R}_B$ in the exit pupil, and we find the image intensity spectrum in the exit pupil as

$$\hat{I}(\boldsymbol{R}) = I_0 \mu(m_p \boldsymbol{R}) \\ \times (\mathrm{OTF}_a(\boldsymbol{R}) * (\delta(\boldsymbol{R} - \boldsymbol{R}_B) + 2\delta(\boldsymbol{R}) + \delta(\boldsymbol{R} + \boldsymbol{R}_B)))\,, \tag{5.4}$$

with OTF_a the optical transfer function of a downscaled telescope aperture with diameter $D = D'/m_p$.

The visibility function in the exit pupil is also demagnified by m_p, so that, at $\boldsymbol{R} = \boldsymbol{R}_B$, we have the visibility $\mu(m_p \boldsymbol{R}_B) = \mu(\boldsymbol{R}'_{B'})$, i.e. the visibility is determined by the baseline in the entrance pupil.

Note that the effective baselines – the projections of the baseline vectors onto the sky plane perpendicular to the line of sight – are moving constantly due to the rotation of the Earth so that the projected telescope apertures in the entrance pupil are in permanent motion, both laterally and longitudinally. Thus, in Fizeau configuration, one has to ensure that the reimaged telescope apertures in the exit pupil are placed dynamically in exactly the correct lateral and longitudinal position to properly define the baseline as $\boldsymbol{R}_B = \boldsymbol{R}'_{B'}/m_p$.

In Michelson configuration, there is no homothetic mapping of the interferometer array, i.e. the distance and size of the apertures in the exit pupil do not form a downscaled replica of the entrance pupil. The demagnification factor m_p is now used for the demagnification of the apertures only, and we introduce m_b for the demagnification of the baseline. This implies that the position of zero OPD no longer coincides with the centre of the Airy disk, PSF_a of an individual aperture (see Fig. 5.2), and the field of view is restricted. PSF_a is also called the primary beam. Therefore the convolution only holds for objects smaller than the central core of the Airy disk, writing

$$I(\boldsymbol{\alpha}) = 2\,\mathrm{PSF}_a(\boldsymbol{\alpha})\big(O_b(\boldsymbol{\alpha}/m_b) * (1 + \cos(2\pi \boldsymbol{R}_B \cdot \boldsymbol{\alpha}))\big). \tag{5.7}$$

Note that the demagnification of the object is now determined by m_b.

Consequently, the image intensity spectrum consists of the optical transfer functions of the individual apertures, OTF_a, convolved with the product of the visibility functions $\mu(m_b \boldsymbol{R}_B)$ with the δ-peaks at $\pm\boldsymbol{R}_B$ and 0, yielding

$$\begin{aligned}\hat{I}(\boldsymbol{R}) = {} & I_0 \mathrm{OTF}_a(\boldsymbol{R}) \\ & * \big(\mu(m_b \boldsymbol{R}_B)\delta(\boldsymbol{R} - \boldsymbol{R}_B) + 2\mu(0)\delta(\boldsymbol{R}) + \mu^*(m_b \boldsymbol{R}_B)\delta(\boldsymbol{R} + \boldsymbol{R}_B)\big).\end{aligned} \tag{5.10}$$

In polychromatic light, the field of view is restricted [222] such that fringe package remains within PSF_a, to

$$\alpha'_{\max} \ll \frac{l_c}{m_b B - m_p B} \approx \frac{l_c}{B'}\,, \tag{5.12}$$

with $m_b B = B'$, the baseline in the entrance pupil.

Restricting the object size to $\alpha'_{\max}$ we can apply the quasi-monochromatic approximation, writing the image intensity distribution as

$$\begin{aligned}I_{\mathrm{qm}}(\boldsymbol{\alpha}) = {} & 2 I_0\,\mathrm{PSF}_a(\boldsymbol{\alpha}) \\ & \times \big(1 + g_B(\boldsymbol{\alpha})|\mu(\boldsymbol{R}'_{0,B'})| \cos(\phi(\boldsymbol{R}'_{0,B'}) - 2\pi \boldsymbol{R}_{0,B} \cdot \boldsymbol{\alpha})\big)\,,\end{aligned} \tag{5.14}$$

with $\boldsymbol{R}'_{0,B'} = m_b \boldsymbol{R}_{0,B}$, and $g_B(\boldsymbol{\alpha})$ – the Fourier transform of the spectrum, $G_B(\boldsymbol{R})$ – is the envelope function of the fringe package. $g_B(\boldsymbol{\alpha})$ is a function of delay τ between the beams, using $g_B(\boldsymbol{\alpha}) = g(\boldsymbol{\alpha} \cdot \boldsymbol{B}/c) = g(\tau)$.

In co-axial combination, we write the modulated intensity as

$$I_{\mathrm{mod,qm}}(t) = 2 I'_0\big(1 + g_p(\psi(t))|\mu(\boldsymbol{R}'_{0,B'})| \cos(\phi(\boldsymbol{R}'_{0,B'}) - \psi(t))\big), \tag{5.18}$$

with $I'_0 = I_0 \int \mathrm{PSF}_a(\boldsymbol{\alpha})\mathrm{d}\boldsymbol{\alpha}$, $g_p(\psi(t)) = g(\frac{\psi(t)}{2\pi\nu_0}) = g(\tau)$, and $\psi(t)$ the temporal phase modulation, related to the OPD modulation through $\frac{\lambda_0}{2\pi}\psi(t)$.

The image intensity spectrum is

$$\hat{I}_{\text{mod,qm}}(\nu) = I'_0 \left(2\delta(\nu) + G(\nu)\mu\left(\frac{\boldsymbol{B}'}{c}\nu_0\right) + G(-\nu)\mu^*\left(-\frac{\boldsymbol{B}'}{c}\nu_0\right)\right), \tag{5.19}$$

when the spatial frequency $\boldsymbol{R}'_{0,\boldsymbol{B}'}$ is written as $\frac{\boldsymbol{B}'}{c}\nu_0$.

5.2 Multi-Aperture Combination: Michelson Configuration

Observing with more than two apertures, one can follow two strategies. Either the light from all homothetically mapped apertures is combined in a single beam combiner, directly forming an image – hence called direct imaging – or the beams are combined such that the visibilities of individual baselines can be determined individually. The latter will be discussed in this section while direct imaging will be treated in Sect. 5.3.

The determination of visibilities of individual baselines is not limited to the pairwise combination of apertures but can also be done *all-in-one*, simultaneously combining all apertures. For this all-in-one combination it is not necessarily required that the apertures in the exit pupil are mapped homothetically if a restricted field of view is acceptable. While the interference fringe of many apertures is an increasingly indistinguishable mix of fringe spacings and orientations (see for instance Fig. 5.12), we can discern the individual visibilities in the image intensity spectrum in the uv-plane.

5.2.1 *Multi-Axial and Co-Axial Combination*

Combining several telescopes, it is advantageous to ensure a non-redundant baseline distribution in the telescope array since – still discussing a moderate number of apertures – we want to maximize the number of baselines for a given number of telescopes. This means that there is only one telescope pair per baseline vector.

In a redundant distribution, different telescope pairs with the same distance and orientation would contribute to the visibility of the same baseline vector, i.e. to the same fringe, so that the independent visibility measurement of each telescope pair is replaced by the averaged visibility. OPD inaccuracies due to turbulence or to instrumental effects could result in destructive interference between telescope pairs and subsequently in errors in the measured visibility.

Taking this argument further to the highly redundant combination of baselines in individual apertures of single telescopes – when "baseline" refers to all possible difference vectors between points in the aperture – there is a substantial influence of the phase errors between redundant baselines, which are determined by the aberrations of the imaging process. The result of these aberrations is a PSF that is irregularly shaped so that the features of the PSF cannot be distinguished from the structure of extended objects.

However, having the freedom in Michelson configuration to arrange the baselines in the exit pupil as suitable for the measurement process, one could provide a non-redundant baseline distribution in the exit pupil even if the telescope array has redundant baselines. The advantage of the independent measurement of visibilities of redundant baselines lies in the individual averaging and calibration of visibility modulus and phase that can be done in post-processing, not suffering from possibly destructive interference when averaging the complex amplitudes from identical baseline vectors in the fringe pattern [166, 182].

The interferometer OTF is the autocorrelation of the distribution of all apertures in the array (3.6), as in (5.2) for two-telescope combination. For a non-redundant distribution of N_T telescopes, there are $N_B = N_T(N_T - 1)/2$ baselines, and we have replica of the optical transfer functions, $\mathrm{OTF}_{a'}$, of the individual apertures at positions $\pm \boldsymbol{R}'_{B'j}$ as displayed in Fig. 5.6. The central $\mathrm{OTF}_{a'}$ at $|\boldsymbol{R}'| = 0$ has a value that is N_T times larger than the value at the positions $\boldsymbol{R}'_{B'j}$ since, due to the autocorrelation process, all telescopes contribute to the value at the origin.

Reimaging the telescope apertures in the exit pupil, the closest packing of the apertures is given by twice the aperture diameter D, i.e. the width of OTF_a as displayed in Fig. 5.7. Arranging the reimaged apertures linearly in a row, we write the

Fig. 5.6 Multi-aperture OTF in the entrance pupil for six randomly distributed telescopes resulting in 15 baselines. The cones represent the optical transfer function $\mathrm{OTF}_{a'}$ of the individual apertures, with the cone at $|\boldsymbol{R}'| = 0$ six times larger than the values at $\boldsymbol{R}'_{B'_j}$

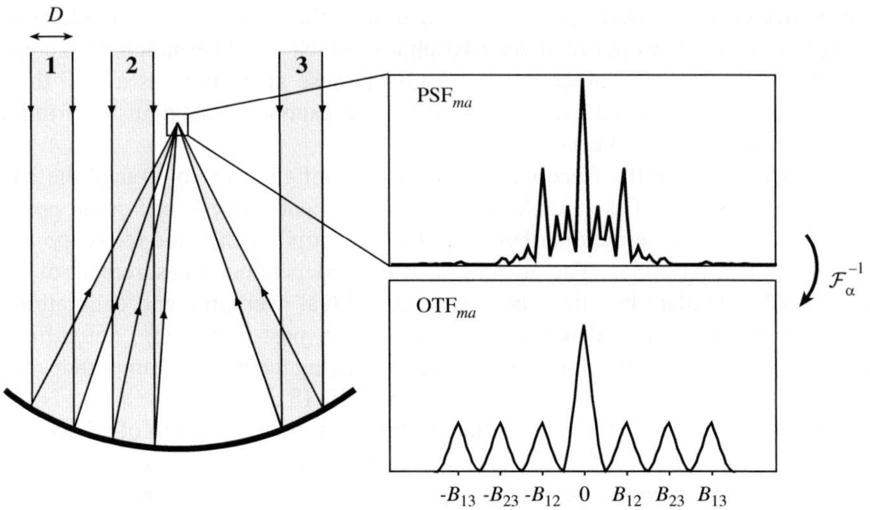

Fig. 5.7 Concept of a Michelson configuration with three beams and all-in-one combination. The three baselines B_{12}, B_{23} and B_{13} in the exit pupil are fixed and lined-up in a row, independent of the real baselines in the entrance pupil. The shortest baseline has a length of $2D$, and their lengths are related like 1:2:3. *On the right*, the fringe pattern PSF_{ma}, (5.21), and OTF_{ma}, (5.20), are displayed containing the three spatial frequencies B_{12}/λ, B_{23}/λ and B_{13}/λ. Barely recognizable, the envelope of the fringe pattern is the Airy disk of an individual aperture given by the diameter D of the individual beams

multi-aperture interferometer OTF in the exit pupil as

$$\begin{aligned}\text{OTF}_{ma}(\boldsymbol{R}) &= \text{OTF}_a(\boldsymbol{R}) \\ &\quad * \Big(N_T\,\delta(\boldsymbol{R}) + \sum_{j=1}^{N_B}\big(\delta(\boldsymbol{R} - j\,\boldsymbol{R}_{B_0}) + \delta(\boldsymbol{R} + j\,\boldsymbol{R}_{B_0})\big)\Big), \qquad (5.20)\end{aligned}$$

with $\boldsymbol{R}_{B_0} = (2D, 0)$ the shortest baseline with a length corresponding to the width of OTF_a. Note that the baseline demagnification factor m_b is now different for each baseline.

The point-spread function of the multi-aperture interferometer is given by the Fourier transform of OTF_{ma}, yielding

$$\begin{aligned}\text{PSF}_{ma}(\boldsymbol{\alpha})|_{\alpha'=0} &= \text{PSF}_a(\boldsymbol{\alpha})\Big(N_T + \sum_{j=1}^{N_B}\big(\mathrm{e}^{\mathrm{i}2\pi j\,\boldsymbol{R}_{B_0}\cdot\boldsymbol{\alpha}} + \mathrm{e}^{-\mathrm{i}2\pi j\,\boldsymbol{R}_{B_0}\cdot\boldsymbol{\alpha}}\big)\Big) \\ &= \text{PSF}_a(\boldsymbol{\alpha})\Big(N_T + \sum_{j=1}^{N_B} 2\cos(2\pi j\,\boldsymbol{R}_{B_0}\cdot\boldsymbol{\alpha})\Big), \qquad (5.21)\end{aligned}$$

describing the interferometer PSF as the product of the Airy disk of an individual aperture in the exit pupil, PSF_a, with the fringe pattern. The validity of (5.21) is limited explicitly to on-axis objects at $\alpha' = 0$, reminding us that due to different demagnification factors m_p for the aperture and m_b for the baselines, the PSF is not shift-invariant, as discussed in Sect. 5.1.2.

Therefore, we have to reduce the field of view to the half-width of the fringe package (5.12), much smaller than PSF_a so that we can write the image intensity distribution as a convolution of the object brightness distribution $O_b(\boldsymbol{\alpha}')$ with the fringe pattern, as in (5.7), yielding

$$\begin{aligned} I(\boldsymbol{\alpha}) &= \mathrm{PSF}_a(\boldsymbol{\alpha})\Big(O_b(\boldsymbol{\alpha}/m_p) * \big(N_T + \sum_{j=1}^{N_B} 2\cos(2\pi j\,\boldsymbol{R}_{B_0}\cdot\boldsymbol{\alpha})\big)\Big) \\ &= \mathrm{PSF}_a(\boldsymbol{\alpha}) I_0\Big(N_T + \sum_{j=1}^{N_B} 2|\mu(\boldsymbol{R}'_{B'_j})|(\cos(\phi(\boldsymbol{R}'_{B'_j}) - 2\pi j\,\boldsymbol{R}_{B_0}\cdot\boldsymbol{\alpha}))\Big), \end{aligned} \tag{5.22}$$

in the first line in the notation as convolution, and in the second line as the sum of products of the visibility function with the cosine functions at individual spatial frequencies $\boldsymbol{R}'_{B'j}$, as in (5.11). $\phi(\boldsymbol{R}'_{B'j})$ is the phase of the visibility function.

The image intensity spectrum in the exit pupil is the Fourier transform of the image intensity distribution as in (5.10), reading

$$\begin{aligned} \hat{I}(\boldsymbol{R}) &= I_0 \mathrm{OTF}_a(\boldsymbol{R}) \\ &* \Big(N_T\mu(0)\delta(\boldsymbol{R}) + \sum_{j=1}^{N_B}\big(\mu(\boldsymbol{R}'_{B'_j})\delta(\boldsymbol{R}-j\,\boldsymbol{R}_{B_0}) + \mu^*(\boldsymbol{R}'_{B'_j})\delta(\boldsymbol{R}+j\,\boldsymbol{R}_{B_0})\big)\Big). \end{aligned} \tag{5.23}$$

Unlike in Fizeau configuration, there is no fixed relationship between the baselines in the exit pupil $j\,\boldsymbol{R}_{B_0}$ and those in the entrance pupil $\boldsymbol{R}'_{B'j}$ for which we determine the visibility. Formally, we could also project baselines that are identical in the entrance pupil but that are realised by separate telescope pairs with the same separation and orientation, on different baselines in the exit pupil so that the visibilities can be determined independently.

Figure 5.7 shows an example of a multi-axial three-beam combination. The fringe pattern shows a mix of three different spatial frequencies – according to the three baselines – in the image. In the image intensity spectrum, the Fourier spectrum of the fringe pattern with the contributions of the three baselines can be disentangled, providing the visibilities of the entrance pupil baselines (5.23).

The linear arrangement of the reimaged apertures has the advantage that the signals can be dispersed spectroscopically in the direction orthogonal to the line of apertures as displayed in Fig. 5.8. This has the additional advantage that *differential phase*, i.e. the phase differences between the spectral channels can be measured, providing valuable information for the image reconstruction [184].

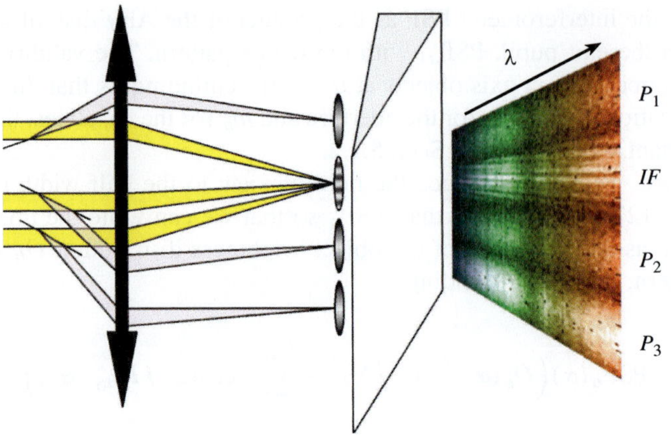

Fig. 5.8 Principle of spectral dispersion in a three-beam Michelson configuration as implemented in the instrument AMBER at the VLTI [184]. A cylindrical lens in the exit pupil forms the interference pattern that is then dispersed by a spectrograph providing narrow band fringe patterns (*IF*) as a function of λ on the detector. Fractions of the light from each beam are sent into the spectrally resolved photometric channels P_i that are required to calibrate the raw visibility (Courtesy F. Millour)

We will see in Sect. 6.2.3 that the combination of three apertures is interesting in the case of atmospheric turbulence. Usually, the fringe position is not stable but subject to random motion due to random phase variations caused by turbulence. Thus, the phase of the visibility function cannot be determined in the fringe pattern without, for instance, measuring the phase of a reference star with known phase simultaneously (see Sect. 6.2.1). However, the sum of the phases of three baselines, when the three baseline vectors in the entrance pupil form a closed triangle, is independent of phase disturbances at each aperture. This technique is called *closure phase* and will be discussed in Sect. 6.2.3. It is also applied in intensity interferometry, see Sect. 2.5, and in speckle interferometry, see Sect. 4.4.

Combining several telescopes co-axially can be done by choosing different temporal modulation frequencies f_j to disentangle the contributions of the different baselines in the temporal Fourier spectrum of the modulated signal. Similar to (5.17) we obtain the multi-aperture intensity distribution as

$$I_{\mathrm{mod}}(t) = \int 2\mathrm{PSF}_a(\boldsymbol{\alpha}) I_0 \Big(N_T + \sum_{j=1}^{N_B} 2|\mu(\boldsymbol{R}'_{B'})| \cos(\phi(\boldsymbol{R}'_{B'}) - \varphi_{\mathrm{mod}_{f_j}}(t)) \Big) \mathrm{d}\boldsymbol{\alpha}$$

$$= 2I'_0 \Big(N_T + \sum_{j=1}^{N_B} 2|\mu(\boldsymbol{R}'_{B'})| \cos(\phi(\boldsymbol{R}'_{B'}) - \varphi_{\mathrm{mod}_{f_j}}(t)) \Big), \quad (5.24)$$

with $I'_0 = I_0 \int \mathrm{PSF}_a(\boldsymbol{\alpha}) \mathrm{d}\boldsymbol{\alpha}$.

5.2.2 Aspects of Beam Combination

In practice, the simultaneous, all-in-one combination of a large number of telescopes that we have discussed in this section can pose considerable problems both in multi-axial and in co-axial combination since it becomes increasingly difficult to distinguish the visibilities of the individual baselines. In multi-axial combination the baselines in the exit pupil become very long very quickly if non-redundancy is to be ensured, and in co-axial combination very high frequencies with increasing modulation amplitudes are required to safely distinguish the visibilities.

On the other hand, pairwise combining the telescopes of an interferometer array has other disadvantages since the stable relationship between the fringe positions of different baselines is lost due to the limited optomechanical stability of the beam combiners. Then, techniques like closure phase could not be applied. However, if one is interested exclusively in the relative position of the fringes, to freeze their motion, and if a separate reference star is used to define the zero phase position (see Sect. 6.2.1), the pairwise combination is a viable solution.

A schematic layout for pairwise combination of three beams in a plane is shown in Fig. 5.9. This layout can be expanded for more telescopes but is then restricted to adjacent beam pairs, like AB, BC, CD, DE, etc.

Another scheme by E. Ribak [192], providing pairwise combination of all beams, combines N linearly arranged beams in an $N \times N$ matrix of interferograms between all $N(N-1)/2$ beam pairs. The diagonal elements contain the interference of each beam with itself which is the intensity of each beam that can be used for calibration of the visibility. After a first demonstration in the laboratory this scheme needs still to be tested on the sky.

One should note, that the **signal-to-noise ratio** (SNR) of all-in-one and of pairwise combination of all beams is not very different. Regarding the visibility of a single pair of telescopes, we find that the signal is proportional to the amplitude of the cosine of the fringe pattern, $2\,I_0|\mu|$, with the intensity I_0 giving the number of photons per telescope (see for instance 5.22). The background signal on the detector is given by $2\,I_0$, contributing a photon noise proportional to $\sqrt{2\,I_0}$. We then have an SNR for the measurement of the visibility of $|\mu|\sqrt{2\,I_0}$, or, in the notation of visibility $\mathcal{V} = |\mu|$ and number of photons N_{ph} in the fringe pattern, the SNR of bright objects is $\mathcal{V}\sqrt{N_{ph}}$ [211,224].

Combining N_T telescopes pairwise, the light from each telescope has to be combined with $N_T - 1$ other telescopes reducing the intensity of the fringe pattern to $2\,I_0/(N_T-1)$, and we obtain

$$\mathrm{SNR}_{pw} = |\mu|\sqrt{\frac{2\,I_0}{N_T - 1}}. \tag{5.25}$$

In simultaneous all-in-one combination, the visibility of each individual baseline is determined by an individual pair of telescopes. The signal, in this case the fringe amplitude, is $2\,I_0|\mu|$, but all N_T telescopes contribute to the background signal,

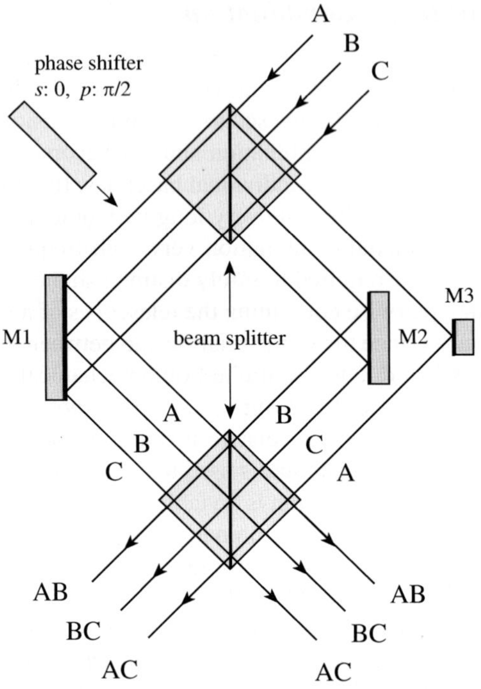

Fig. 5.9 Schematic layout of pairwise beam combination for three telescopes. The top beam splitter distributes the light to a single mirror on the left and to an arrangement of two mirrors on the right, permuting the beams from ABC to BCA, so that the pairwise combination appears in the exits of the bottom beam splitter. The phase difference between the beams in each pair is 0 in the left output and π in the right output. Inserting a phase shifter in the light path, as in Fig. 5.4, the s and p polarization components in the output arms would display phase differences of 0 and $\pi/2$ in the left arm, and of π and $3\pi/2$ in the right arm, providing spatial phase modulation. More telescopes can be accommodated by enlarging M1 and M2, maintaining the permutation of beams by M3 so that pairs of adjacent beams (AB, BC, CD, DE etc.) are formed

$N_T\, I_0$, with noise contribution $\sqrt{N_T\, I_0}$. We then find [24]

$$\mathrm{SNR}_{sim} = 2|\mu|\sqrt{\frac{I_0}{N_T}}. \tag{5.26}$$

Thus, the advantage for all-in-one over pairwise combination is a factor of $\sqrt{2}$ when N_T becomes large. Given that pairwise combination can hardly be done without beamsplitters, creating two sets of outputs as in Fig. 5.9, the light from one telescope is distributed to $2(N_T - 1)$ fringe pattern – thus, to 4 in Fig. 5.9 – and SNR_{pw} is reduced by another factor of $\sqrt{2}$.

However, if pairwise combination is used for fringe tracking, the minimum number of combinations is AB, BC, CD, DE... – since this provides a "chain" between all telescopes – so that the light from each telescope is only distributed to two fringe patterns. Then, the SNR is $|\mu|\sqrt{I_0}$, which is a factor of $\sqrt{N_T}/2$ larger than in all-in-one combination, albeit for the price of having only N_T instead of N_B baselines. It is an open question if N_T baselines are sufficient for fringe tracking since losing the fringes on one baseline interrupts the process. Experience with fringe trackers like CHAMP[4], using a scheme as in Fig. 5.9 expanded to six telescopes and providing six beam pairs [14], will help to determine the best solution. Thus, there are many issues to consider before choosing a beam combination scheme.

To date, in the laboratory a maximum of six beams has been combined all-in-one with the instrument MIRC using optical fibres [163]. MIRC, the Michigan InfraRed Combiner, was designed and built for the CHARA interferometer[5] on Mount Wilson, California. So far, four telescopes of the interferometer have been combined with MIRC, obtaining excellent astronomical results [114] and demonstrating the advantage that even as few as four telescopes with six baselines provide over observations with two telescopes.

Arranging the fibres in MIRC side by side in a silicon V-groove array – as customary for integrated optics applications [15] – a non-redundant baseline distribution can be arranged and, due to the flexibility of the setup, modified if necessary (see Fig. 5.10). Although this technique is applicable to more than six beams, there is a limit, and for a very large number of telescopes the concept of a hybrid beam combiner could be applied, grouping the telescope beams in manageable quantities [165].

The Michelson configuration is by far the most common choice for interferometric instruments. Although the field of view is restricted, depending on the diameter

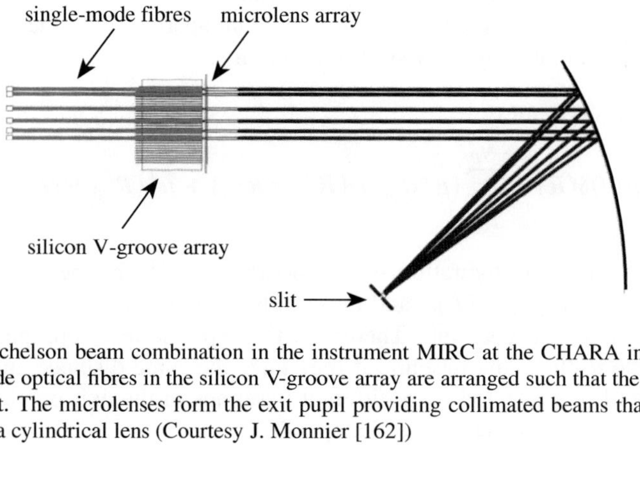

Fig. 5.10 Michelson beam combination in the instrument MIRC at the CHARA interferometer. The monomode optical fibres in the silicon V-groove array are arranged such that the baselines are non-redundant. The microlenses form the exit pupil providing collimated beams that are focused on a slit with a cylindrical lens (Courtesy J. Monnier [162])

[4] CHAMP was built for the CHARA interferometer on Mt. Wilson, California.

[5] CHARA is the Center for High Angular Resolution Astronomy at the Georgia State University, Georgia, USA.

of the individual telescopes, down to a few 10 mas (5.12), this disadvantage is outweighed by far by the stable opto-mechanical configuration in the instrument – no need to permanently adjust baselines as in the Fizeau configuration – and by the possibility to implement spectrography. The direct access to the visibility of each baseline, and the possibility to calibrate the visibilities individually adds a further advantage, arguably outperforming even direct imaging techniques for a large number of apertures if field of view is not an issue [166].

Multi-Aperture Combination: Summary

We call N_T the number of telescopes in the array and N_B the number of baselines with $N_B = N_T(N_T - 1)/2$ if the distribution of telescopes is non-redundant.

In Michelson configuration, we write the image intensity distribution of an object that is smaller than the Airy disk, PSF_a, of an individual aperture as a convolution of the object brightness distribution $O_b(\boldsymbol{\alpha}')$ with the interference pattern:

$$I(\boldsymbol{\alpha}) = \text{PSF}_a(\boldsymbol{\alpha})\Big(O_b(\boldsymbol{\alpha}/m_p) * \big(N_T + \sum_{j=1}^{N_B} 2\cos(2\pi j\,\boldsymbol{R}_{B_0}\cdot\boldsymbol{\alpha})\big)\Big), \quad (5.22)$$

when the exit pupil baselines are multiples of the shortest baseline $\boldsymbol{R}_{B_0}$, lined-up linearly (see Fig. 5.7). The linear arrangement is not required for the Michelson configuration but it has the advantage that spectroscopy can be implemented easily (see Fig. 5.8).

Accordingly, the image intensity spectrum, the Fourier transform of $I(\boldsymbol{\alpha})$, is the convolution of the OTF of an individual aperture with the distribution of δ-peaks weighted by the visibility function μ:

$$\hat{I}(\boldsymbol{R}) = I_0\text{OTF}_a(\boldsymbol{R}) \quad (5.23)$$
$$* \Big(N_T\mu(0)\delta(\boldsymbol{R}) + \sum_{j=1}^{N_B}\big(\mu(\boldsymbol{R}'_{B'_j})\delta(\boldsymbol{R} - j\,\boldsymbol{R}_{B_0}) + \mu^*(\boldsymbol{R}'_{B'_j})\delta(\boldsymbol{R} + j\,\boldsymbol{R}_{B_0})\big)\Big).$$

Unlike in Fizeau configuration, there is no fixed relationship between the baselines in the exit pupil $j\,\boldsymbol{R}_{B_0}$ and those in the entrance pupil $\boldsymbol{R}'_{B'j}$ for which we determine the visibility. Therefore, the separate telescope pairs, which contribute to the same baseline if they have the same distance and orientation, could be projected onto different baselines in the exit pupil, and their visibilities could be determined independently.

In co-axial combination, different temporal frequencies f_j can be used to disentangle the contributions of the different baselines in the temporal Fourier spectrum of the modulated signal, writing

$$I_{\mathrm{mod}}(t) = 2I_0'\Big(N_T + \sum_{j=1}^{N_B} |\mu(\boldsymbol{R}'_{B'})| \cos(\phi(\boldsymbol{R}'_{B'}) - \varphi_{\mathrm{mod}_{f_j}}(t))\Big), \quad (5.24)$$

with $I_0' = I_0 \int \mathrm{PSF}_a(\boldsymbol{\alpha})\mathrm{d}\boldsymbol{\alpha}$.

The actual beam combination can be done all-in-one as in Fig. 5.7, forming a single fringe pattern mixing fringes of different period $1/R_{B_j}$, or in pairwise combination as in Fig. 5.9, forming as many pairs as baselines. The signal-to-noise ratio for the visibility measurement for both options differs only by a factor of $\sqrt{2}$, see (5.25) and (5.26).

In all cases, the visibility is determined for each baseline and an image is formed by processing the visibilities, i.e. by aperture synthesis.

5.3 Multi-Aperture Combination: Direct Imaging

Direct imaging instead of visibility measurement is very appealing but it completely depends on the ability to phase the telescope array, i.e. to ensure that the phase differences between all apertures in the exit pupil are very small, ideally zero. This is the equivalent of a single telescope with very small aberrations and, thus, a diffraction limited image. While this can be done in principle, it adds complexity to the interferometer so that the seeming simplicity of direct imaging is gone.

Direct imaging is also substantially different from aperture synthesis since we rely on the instantaneous interference of the amplitudes in the image plane to provide the image intensity, while in aperture synthesis the image is formed by processing the measured and calibrated visibilities, combining them in the uv-plane and applying image reconstruction methods to obtain the image intensity as discussed in Sect. 3.4.3.

The Fizeau configuration, in direct analogy to a masked telescope, provides a straightforward direct imaging mode with large field of view, with the possibility of processing the image intensity distribution instead of visibilities. Unless the interferometer array is completely and seamlessly filled with apertures, the resulting point-spread function, which is the interference pattern of the ensemble of apertures, consists of fringes and sidelobes with a narrow central core with a diameter $\propto 1/B_{\mathrm{max}}$. This interference pattern is enveloped by the PSF of an individual telescope, with a diameter $\propto 1/D$, as displayed in Fig. 5.11.

For some applications, it is desirable to increase the intensity that goes into the central core by reducing the intensity in the halo of sidelobes accordingly. The Hypertelescope and the IRAN configuration provide methods to improve the peak intensity albeit at the cost of field of view. These will be discussed in Sects. 5.3.2 and 5.3.3.

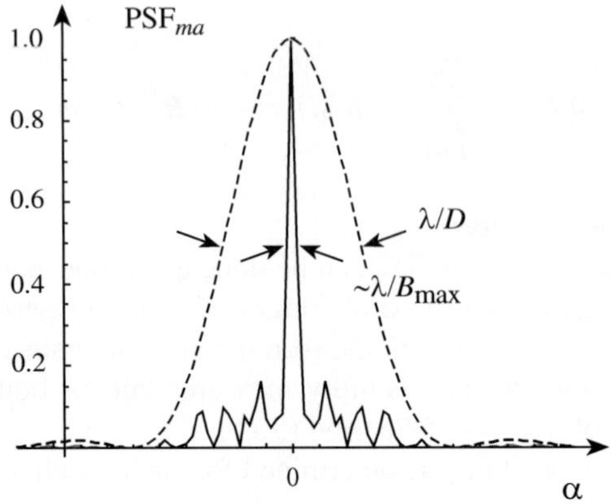

Fig. 5.11 Profile of PSF_{ma} in Fizeau configuration with $N_T = 10$ randomly distributed apertures. *The dashed line* shows the Airy disk, PSF_a, of an individual aperture in the exit pupil. The longest baseline B_{max} is about 10 times longer than the diameter D of an individual aperture so that the core peak is about 10 times narrower than PSF_a. The average intensity in the halo of sidelobes around the core is proportional to N_T while the peak intensity is proportional to N_T^2, in this case 10 times larger

5.3.1 Fizeau Configuration

Combining several telescopes in Fizeau configuration, we develop the formalism by expanding the two-telescope case discussed in Sect. 5.1.1 to many apertures.

Unlike in Michelson configuration, apertures and baselines are demagnified by the same factor m_p in order to obtain a downscaled image of the interferometer array in the exit pupil. This is called homothetic mapping. Then, the multi-aperture interferometer OTF in the exit pupil is the downscaled OTF of the telescopes array displayed in Fig. 5.6, reading

$$\mathrm{OTF}_{ma}(\boldsymbol{R}) = \mathrm{OTF}_a(\boldsymbol{R}) * \left(N_T \delta(\boldsymbol{R}) + \sum_{j=1}^{N_B} \left(\delta(\boldsymbol{R} - \boldsymbol{R}_{B_j}) + \delta(\boldsymbol{R} + \boldsymbol{R}_{B_j}) \right) \right), \tag{5.27}$$

with $\boldsymbol{R}_{B_j} = \boldsymbol{R}'_{B'j}/m_b$ the baseline vectors in the exit pupil. In case of a partially redundant configuration, the number of baselines N_B would be reduced, and the weight at the redundant baselines would be increased accordingly.

Similar to (5.21), we write the multi-aperture PSF in Fizeau configuration as

$$\begin{aligned} \mathrm{PSF}_{ma}(\boldsymbol{\alpha}) &= \mathrm{PSF}_a(\boldsymbol{\alpha}) \Big(N_T + \sum_{j=1}^{N_B} \left(\mathrm{e}^{\mathrm{i}2\pi \boldsymbol{R}_{B_j} \cdot \boldsymbol{\alpha}} + \mathrm{e}^{-\mathrm{i}2\pi \boldsymbol{R}_{B_j} \cdot \boldsymbol{\alpha}} \right) \Big) \\ &= \mathrm{PSF}_a(\boldsymbol{\alpha}) \Big(N_T + \sum_{j=1}^{N_B} 2\cos(2\pi \boldsymbol{R}_{B_j} \cdot \boldsymbol{\alpha}) \Big). \end{aligned} \tag{5.28}$$

Combining many telescopes, the point-spread function displays a narrow core with diameters between $\lambda/(\sqrt{2}B_{\max})$ and $\lambda/B_{\max}$, depending on the distribution of telescopes. If the telescopes are distributed on a circle with diameter $m_p B_{\max}$, the core is $\sqrt{2}$ times narrower than in case of a homogeneously filled circle with diameter $m_p B_{\max}$ (compare to Fig. 2.8). The envelope of the interference pattern is given by PSF_a the Airy disk of an individual aperture in the exit pupil with an FWHM of λ/D, as displayed in Fig. 5.11. If the telescopes are distributed on a regular grid, the point-spread function has a regular distribution of peaks as will be discussed in Sect. 5.4.

One should note that (5.28) describes the multi-aperture PSF for perfectly phased apertures. In addition to requiring dynamic homothetic mapping of all apertures as discussed at the end of Sect. 5.1.1, any kind of influence by atmospheric turbulence or by instrumental effects, perturbing the plane wave, has to be corrected so that the phase differences between the apertures in the exit pupil are minimised, ideally to zero. Adding coherently the contributions from all perfectly phased apertures, the intensity on axis, at $|\boldsymbol{\alpha}| = 0$, is then $\propto N_T^2$. The intensity level in the halo, within PSF_a, is $\propto N_T \times \mathrm{PSF}_a(\boldsymbol{\alpha})$ modulated by the sum of cosine functions.

In case that the apertures are not perfectly phased, we have to add random phase terms to all cosine functions, so that the peak intensity on axis does not add up to N_T^2 but to considerably smaller values, and the central core is possibly broadened. We have the equivalent situation on a single telescope when the aberrations of the wave front reduce the peak intensity of the point-spread function and eventually widen the diffraction limited core.

In Fizeau configuration, we obtain the image intensity distribution as a convolution of the object brightness distribution $O_b(\boldsymbol{\alpha}')$ with the shift-invariant interferometer PSF, writing

$$I(\boldsymbol{\alpha}) = O_b(\boldsymbol{\alpha}/m_p) * \Big(\mathrm{PSF}_a(\boldsymbol{\alpha})\big(N_T + \sum_{j=1}^{N_B} 2\cos(2\pi \boldsymbol{R}_{B_j} \cdot \boldsymbol{\alpha})\big)\Big), \tag{5.29}$$

with $\boldsymbol{\alpha}/m_p = \boldsymbol{\alpha}'$, and, accordingly, the baselines in the exit pupil m_p times smaller than the baselines in the entrance pupil, $\boldsymbol{R}_{B_j} = \boldsymbol{R}'_{B'_j}/m_p$.

The Fourier transform of the image intensity provides the spectrum in the exit pupil, $\hat{I}(\boldsymbol{R})$, as given by (5.4), now with the multi-aperture interferometer OTF (5.27) as

$$\begin{aligned}\hat{I}(\boldsymbol{R}) &= I_0\mu(m_p\boldsymbol{R})\mathrm{OTF}_{ma}(\boldsymbol{R}) \\ &= I_0\mu(m_p\boldsymbol{R}) \\ &\quad\times\Big(\mathrm{OTF}_a(\boldsymbol{R}) * \Big(N_T\delta(\boldsymbol{R}) + \sum_{j=1}^{N_B}\big(\delta(\boldsymbol{R} - \boldsymbol{R}_{B_j}) + \delta(\boldsymbol{R} + \boldsymbol{R}_{B_j})\big)\Big)\Big).\end{aligned} \tag{5.30}$$

If the large field of view is not exploited and the object brightness distribution is smaller than PSF_a, then the visibility function $\mu(m_p \boldsymbol{R})$ is invariant over OTF_a and we can replace the product $\mu(m_p \boldsymbol{R}) \times \mathrm{OTF}_{ma}$ by the products $\mu(m_p \boldsymbol{R}_{B_j}) \times \delta(\boldsymbol{R}_{B_j})$ inside the sum.

Using $\mu(m_p \boldsymbol{R}_{B_j}) = \mu(\boldsymbol{R}'_{B'j})$, the visibility is determined by the baseline vectors $\boldsymbol{R}'_{B'j}$ in the interferometer array, the entrance pupil. Thus, the image intensity spectrum contains the visibilities at the individual baselines in the array.

For large telescopes, and consequently a wide optical transfer function OTF_a, the visibility of an extended object varies over OTF_a as discussed in Sect. 3.3.5. Then we have to consider the shape of the visibility function inside all N_B OTF_a centred at positions $\boldsymbol{R}_{B_j}$ instead of processing a single value for each baseline. For the two apertures of the LBT, this was discussed in detail in Sect. 3.4.1.

5.3.2 *Hypertelescope*

Labeyrie's proposal [122] of a Hypertelescope is based on the idea of a densified pupil, enlarging the individual apertures in the exit pupil of the interferometer so that the enveloping Airy disk, PSF_a, shrinks while the interference pattern, forming the fringes and sidelobes, keeps its shape.

Starting from the Fizeau configuration (5.28), we write the PSF for an on-axis object as

$$\mathrm{PSF}_{ma}(\boldsymbol{\alpha})|_{\alpha'=0} = \mathrm{PSF}_{a'}(\boldsymbol{\alpha}/m_p)\Big(N_T + \sum_{j=1}^{N_B} 2\cos(2\pi \boldsymbol{R}_{B_j} \cdot \boldsymbol{\alpha})\Big), \qquad (5.31)$$

with $\mathrm{PSF}_{a'}(\boldsymbol{\alpha}/m_p) = \mathrm{PSF}_a(\boldsymbol{\alpha})$, the point-spread function of the aperture with diameter D in the exit pupil. The demagnification factors m_p for the telescope apertures and m_b for the baselines define the diameter of the enveloping PSF_a and of the distribution of baselines $\boldsymbol{R}_{B_j}$ independently. For telescope diameter D', we have an aperture diameter of $D = D'/m_p$ in the exit pupil, and baselines $\boldsymbol{R}'_{B'j}$ are downscaled in the exit pupil, with $\boldsymbol{R}_{B_j} = \boldsymbol{R}'_{B'j}/m_b$. Figure 5.12 displays the impact of different m_p on the interferometric PSF.

The central core of PSF_a, the Airy disk of the individual aperture in the exit pupil, has the diameter λ/D, so that a smaller demagnification factor m_p – leading to a larger D – provides a narrower Airy disk.

We see from (5.31) that the interference pattern only depends on the distribution of baselines $\boldsymbol{R}_{B_j}$, and that any modification of the pupil demagnification factor, m_p, affects the width of PSF_a and, thus, the width of the envelope only.

As discussed for the Fizeau configuration and following from (5.31), the peak intensity in the central core is $\propto N_T^2$, and the intensity level in the interference pattern surrounding the peak is $\propto N_T$. With a peak diameter of about $\lambda/B_{\max}$, the intensity in the central core is $\propto N_T^2/B_{\max}^2$, and the integrated intensity in the halo is

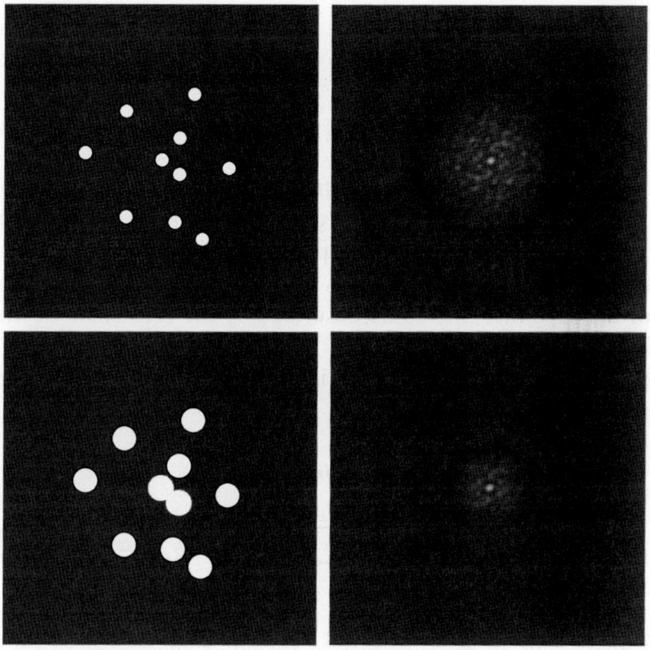

Fig. 5.12 Random distribution of apertures in the exit pupil on the *left*, and the resulting point-spread functions on the *right*. The central core of the PSF is surrounded by a halo with irregular fringes and sidelobes. The envelope of the interference pattern is given by the Airy disk of an individual aperture. Doubling the diameter of the individual apertures decreases the enveloping PSF and with it the diameter of the halo without affecting the distribution of sidelobes inside the halo. Then, the central core obtains a larger share of the total intensity

$\propto N_T/D^2$ for a PSF_a diameter of λ/D. Assuming that the number of telescopes is moderate, e.g. limited to 12 telescopes each 12 m in diameter, and if the maximum baseline is 1 km, we find about 2/1,000 of the total intensity in the core of the PSF.

Enlarging the aperture diameter D in the exit pupil by decreasing m_p, the fraction of light in the central core increases quadratically, by $N_T D^2/B_{\max}^2$. This is illustrated in Fig. 5.12: Doubling the diameter of the individual apertures decreases the diameter of PSF_a by a factor of two so that the area containing the interference pattern is reduced quadratically. Thus, the relative intensity of the central core is increased by a factor of four.

We introduce a densification factor $\gamma_d = m_b/m_p$, with $\gamma_d = 1$ in case of a Fizeau configuration when it is $m_p = m_b$, i.e. without densification of the pupil, and with $\gamma_d > 1$ if the aperture diameter in the exit pupil is increased [122]. For instance, doubling the aperture diameter in the exit pupil with respect to the Fizeau configuration, it is $m_p = m_b/2$ and $\gamma_d = 2$, so that the fraction of light in the core is increased by $\gamma_d^2 = 4$ compared to the Fizeau configuration. We call this increase the gain, g, with

$$g_{\text{hype}} = \gamma_d^2 = \left(\frac{m_b}{m_p}\right)^2. \tag{5.32}$$

As discussed for the Michelson configuration in Sect. 5.1.2, the object-image relationship cannot be described by a convolution if the telescope array is not mapped homothetically into the exit pupil. A densified pupil, when the telescope apertures are downscaled by a factor m_p that is different from the demagnification factor m_b of the baselines, has to be described by the formalism of the Michelson configuration.

We adopt (5.5) for the case of a densified pupil by writing the position dependent point-spread function as

$$\begin{aligned}\text{PSF}_{\text{hype}}(\boldsymbol{\alpha}, \boldsymbol{\alpha}') = {} & \text{PSF}_a\left(\boldsymbol{\alpha} - m_p\boldsymbol{\alpha}'\right) \\ & \times\left(N_T + \sum_{j=1}^{N_B} 2\cos(2\pi \boldsymbol{R}_{B_j} \cdot (\boldsymbol{\alpha} - m_b\boldsymbol{\alpha}'))\right).\end{aligned} \tag{5.33}$$

For an object at $\boldsymbol{\alpha}'_0$, PSF_a is centred at $\boldsymbol{\alpha}_{0,p} = m_p\boldsymbol{\alpha}'_0$ and the sum of cosine functions forming the narrow core of the interference pattern is centred at $\boldsymbol{\alpha}_{0,b} = m_b\boldsymbol{\alpha}'_0$ as discussed for Michelson configuration in (5.12) and as displayed in Fig. 5.13. This means in particular that the peak intensity of the central core is reduced by the

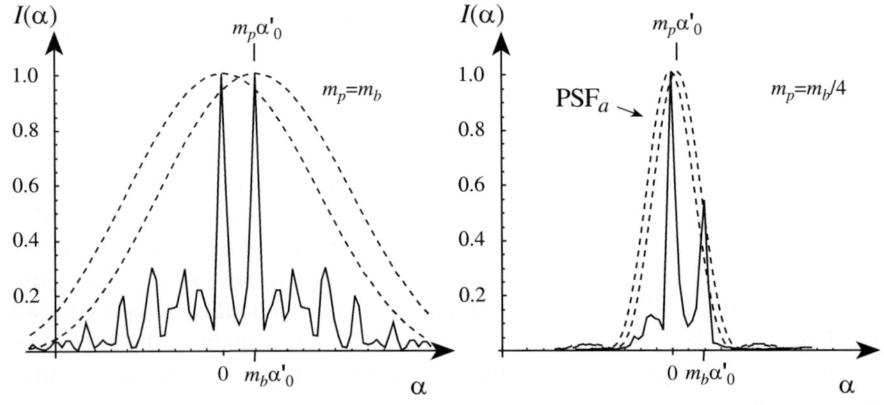

Fig. 5.13 Profile of the image intensity distribution of a binary star with separation $\boldsymbol{\alpha}'_0$ in Fizeau configuration (*left*) and with a four-fold densified pupil (*right*) with $N_T = 10$ randomly distributed apertures using (5.33). *The dashed line* shows the point-spread function, PSF_a, of an individual aperture in the exit pupil. In Fizeau configuration with $m_p = m_b$ (*left*), the PSF is shift-invariant, and the peaks of PSF_a and of the interference pattern are at the same position for each component of the binary. With a pupil densification of $\gamma_d = m_b/m_p > 1$ (*right*), PSF_a is narrower and has its peak at a different position than the interference pattern when moving off-axis (compare to Fig. 5.2). As a consequence, the peak intensity of the interference pattern of the component at $\boldsymbol{\alpha}'_0$ is reduced

multiplication with PSF_a by the value of $\mathrm{PSF}_a(m_b\boldsymbol{\alpha}'_0 - m_p\boldsymbol{\alpha}'_0) = \mathrm{PSF}_{a'}(\gamma_d\boldsymbol{\alpha}'_0 - \boldsymbol{\alpha}'_0) \approx \mathrm{PSF}_{a'}(\gamma_d\boldsymbol{\alpha}'_0)$ for strong densification $\gamma_d \gg 1$.

For circular apertures, $\mathrm{PSF}_{a'}$ is given by a Besinc function with an FWHM of λ/D'. Thus, for $\alpha'_0 = \frac{1}{2}\lambda/(\gamma_d D')$ off-axis, the peak intensity is at half of its on-axis value. Assuming a pupil densification of $\gamma_d = 10$ and 8-m telescopes in the K-band, we find that the peak intensity is reduced to half its value if the object in the sky is 2.8 mas off-axis.

The restriction of the **field of view** is identical to that in Michelson configuration discussed in Sect. 5.1.2, when it was requested (5.8) that the field of view is limited to $\alpha'_{\max} \ll \lambda/(m_b D'/m_p) = \lambda/(\gamma_d D')$.

We write the image intensity distribution for small objects as a convolution of the object brightness distribution $O_b(\boldsymbol{\alpha}')$ with the interference pattern,

$$I(\boldsymbol{\alpha}) = \mathrm{PSF}_a(\boldsymbol{\alpha}) \times O_b(\boldsymbol{\alpha}/m_b) * \Big(N_T + \sum_{j=1}^{N_B} 2\cos(2\pi \boldsymbol{R}_{B_j}\cdot\boldsymbol{\alpha})\Big). \tag{5.34}$$

The image formation can be discussed as well as a function of the object coordinate $\boldsymbol{\alpha}'$ in the sky [4].

Observing with **polychromatic light** in Michelson configuration leads to a further reduction of the field of view, down to the width of the fringe package, l_c/B', (5.12). Determining visibilities, we relied on an overlap of the central part of the fringe package for all objects in the (very small) field of view, so that the visibility of the common fringe pattern was determined by the object shape.

Here, in the case of a Hypertelescope, the imaging process is barely affected since, processing the image intensity distribution, the overlap of the interference patterns is of no interest. Each object point forms an interference pattern with narrow core, the ensemble of which forming the image. The only effect of polychromatic light is its influence on the diameter, $\propto \lambda/B_{\max}$, of the core, and the smearing of the sidelobes in the halo, as given by $\mathrm{PSF}_{\mathrm{hype}}$ in (5.31) when using a finite spectral bandwidth.

The densification of the pupil is only one aspect of the Hypertelescope and of direct imaging in general. The other important aspect is the distribution of the telescopes in the array. Using a regular grid for the telescope positions provides point-spread functions with interesting characteristics as will be discussed in Sect. 5.4.

5.3.3 Interferometric Remapped Array Nulling: IRAN

Starting from masked apertures, the formation of the intensity distribution in the image plane was an intuitively comprehensible process, displaying fringes or, as we saw in the preceding sections, a point-spread function with a narrow central core.

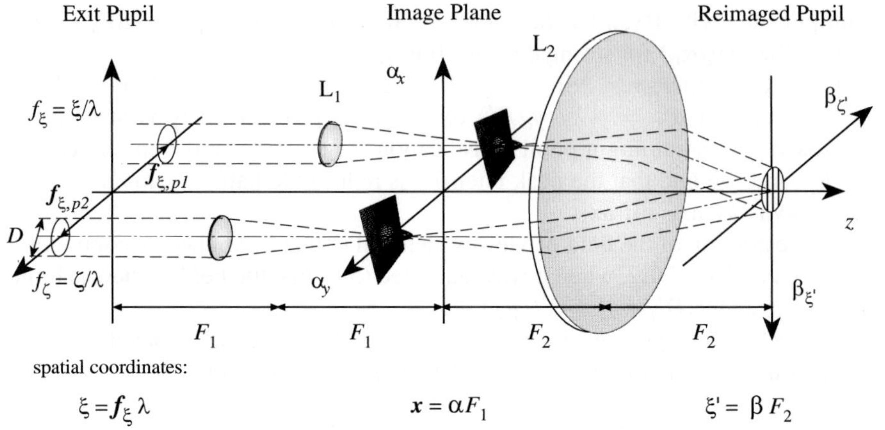

Fig. 5.14 The geometry of the imaging process in IRAN configuration for two telescopes. The exit pupil is in the front focal plane of two lenses L_1 with focal length F_1, centred on each individual aperture with diameter D. They form point-spread functions in the image plane with separation and orientation exactly like the apertures in the exit pupil. The lens L_2 with focal length F_2 is placed such that the image plane is in its front focal plane, and the reimaged pupil plane at distance F_2 behind the image plane so that an image of the exit pupil forms at distance F_2 behind the lens. Due to the process of forming individual PSF in the image plane – instead of a single PSF with fringes – all sub-apertures are superposed in the reimaged pupil plane forming a single image of the aperture with fringes

The fascinating idea of F. Vakili [245] to form an image in the pupil plane of an interferometer takes more imagination to grasp, but we will see that it can be explained in a formally similar way to other direct imaging methods.

The idea is based on forming individual images behind each aperture in the exit pupil, so that the image plane is filled with point-spread functions that show the same distribution as the apertures in the exit pupil (see Fig. 5.14). Adding a large lens behind the image plane, all apertures are reimaged in one common position.

Starting with two apertures and a point source at infinity as in Fig. 5.14, we write the aperture in the exit pupil as the sum of individual aperture functions – circular apertures with diameter D – at positions $\boldsymbol{\xi}_{p1}$ and $\boldsymbol{\xi}_{p1}$. Using spatial frequencies in the pupil plane, with $\boldsymbol{f}_\xi = \boldsymbol{\xi}/\lambda$, we express the sum as a convolution of the individual aperture function $a(\boldsymbol{f}_\xi)$ with δ-functions at spatial frequency positions $\boldsymbol{f}_{\xi,p_j} = \boldsymbol{\xi}_{p_j}/\lambda$,

$$A(\boldsymbol{f}_\xi) = a(\boldsymbol{f}_\xi) * \big(\delta(\boldsymbol{f}_\xi - \boldsymbol{f}_{\xi,p1}) + \delta(\boldsymbol{f}_\xi - \boldsymbol{f}_{\xi,p2})\big). \tag{5.35}$$

Individual images form behind each of the lenses L_1, and we denote the amplitudes of individual point-spread functions on the individual optical axes by $a_{\mathrm{PSF}}(\boldsymbol{\alpha})$. Using (2.23), the aperture in the exit pupil, $a(\boldsymbol{f}_\xi)$, in the front focal plane of the lenses and $a_{\mathrm{PSF}}(\boldsymbol{\alpha})$ in the back focal plane are connected through a Fourier transform,

$$a_{\text{PSF}}(\boldsymbol{\alpha}) = \frac{\lambda}{\text{i}F_1} \int a(\boldsymbol{f}_\xi) \text{e}^{-\text{i}2\pi \boldsymbol{f}_\xi \cdot \boldsymbol{\alpha}} \, \text{d}\boldsymbol{f}_\xi, \tag{5.36}$$

omitting the factor $\lambda/(\text{i}F_1)$ in the following.

The amplitude $V(\boldsymbol{\alpha})$ of the propagating wave in the image plane is now the sum of all $a_{\text{PSF}}(\boldsymbol{\alpha})$, multiplied by the constant amplitude V_0 of the incoming plane wave (see Sect. 3.1), reading

$$V(\boldsymbol{\alpha}) = V_0 \, a_{\text{PSF}}(\boldsymbol{\alpha}) * \big(\delta(\boldsymbol{\alpha} - \boldsymbol{\alpha}_{p1}) + \delta(\boldsymbol{\alpha} - \boldsymbol{\alpha}_{p2})\big). \tag{5.37}$$

It is $\boldsymbol{\alpha}_{pj} = \boldsymbol{f}_{\xi,pj}\lambda/F_1$ the position of the PSF in the image plane so that the spatial coordinates, $\boldsymbol{\alpha}_{pj} F_1$, of the PSF in the image plane and those, $\boldsymbol{f}_{\xi,pj}\lambda$, of the apertures in the exit pupil are identical.

Placing the ensemble of point-spread functions in the front focal plane of a large lens, we find the amplitude in its back focal plane as the Fourier transform of $V(\boldsymbol{\alpha})$. Since, eventually, we want to measure polychromatic intensities in the reimaged pupil plane, the spatial frequency $\boldsymbol{f}'_\xi$ is a bad choice as a coordinate because the shape of the aperture function and of the intensity would then be wavelength dependent as discussed in Sect. 3.1. With the spatial coordinate $\boldsymbol{\xi}' = \boldsymbol{f}'_\xi\lambda$ in the reimaged pupil, we introduce the angle coordinate $\boldsymbol{\beta} = \boldsymbol{\xi}'/F_2$, obtaining

$$V(\boldsymbol{\beta}) = V_0 \, a(\boldsymbol{\beta}) \times \left(\text{e}^{-\text{i}2\pi \boldsymbol{f}_{\xi,p1}\cdot\boldsymbol{\beta}} + \text{e}^{-\text{i}2\pi \boldsymbol{f}_{\xi,p2}\cdot\boldsymbol{\beta}}\right), \tag{5.38}$$

when the convolution of $a_{\text{PSF}}(\boldsymbol{\alpha})$ with the δ-functions in (5.37) becomes a multiplication of the reimaged aperture function $a(\boldsymbol{\beta})$ with exponential functions. The diameter of the reimaged aperture is $D\,F_2/F_1$, which, expressed as angular coordinate, is $\beta_D = D/F_1$. For $F_1 = F_2$, the diameter is identical to the aperture diameter in the exit pupil.

The intensity in the reimaged pupil plane is the modulus squared of the amplitude, yielding

$$\begin{aligned} I(\boldsymbol{\beta}) &= V_0^2 |a(\boldsymbol{\beta})|^2 \left|\text{e}^{-\text{i}2\pi \boldsymbol{f}_{\xi,p1}\cdot\boldsymbol{\beta}} + \text{e}^{-\text{i}2\pi \boldsymbol{f}_{\xi,p2}\cdot\boldsymbol{\beta}}\right|^2 \\ &= V_0^2 |a(\boldsymbol{\beta})|^2 \left(2 + \text{e}^{\text{i}2\pi \boldsymbol{R}_B\cdot\boldsymbol{\beta}} + \text{e}^{-\text{i}2\pi \boldsymbol{R}_B\cdot\boldsymbol{\beta}}\right) \\ &= V_0^2 |a(\boldsymbol{\beta})|^2 \, 2\big(1 + \cos(2\pi \boldsymbol{R}_B \cdot \boldsymbol{\beta})\big), \end{aligned} \tag{5.39}$$

with $\boldsymbol{R}_B = \boldsymbol{f}_{\xi,p1} - \boldsymbol{f}_{\xi,p2}$ the baseline vector.

Thus, reimaging the individual apertures on top of each other, the wave fronts interfere under slightly different angles, $\boldsymbol{f}_{\xi,p1}\lambda/F_2$ and $\boldsymbol{f}_{\xi,p2}\lambda/F_2$, forming the familiar fringe pattern $\propto 1 + \cos(.)$. Here the fringe pattern is confined to $|a(\boldsymbol{\beta})|^2$, which has the same shape as the aperture itself, $|a(\boldsymbol{\beta})|^2 = |a(\boldsymbol{\beta})|$, independent of telescope aberrations.

Comparing the result in (5.39) to that of two apertures forming fringes in the image plane (3.53), we found the same fringe pattern with fringe spacing $1/\boldsymbol{R}_B$, but the envelope was the Airy disk of an individual aperture while it is the aperture function in (5.39). Thus, the Fourier transform of two separate apertures forming fringes in the image plane has been replaced by the Fourier transform of two separate point-spread functions forming fringes in the reimaged pupil plane.

Off-Axis Object

If we regard a point source at angular position $\boldsymbol{\alpha}_0$ instead of on-axis, the ensemble of point-spread functions in the image plane is shifted accordingly, but the baseline and the spacing of the fringe pattern in the reimaged pupil plane do not change.

We have to consider the off-axis object position and the subsequently tilted plane wave by adding a phase term in the exit pupil,

$$A_{\alpha_0}(\boldsymbol{f}_\xi) = A(\boldsymbol{f}_\xi)\mathrm{e}^{\mathrm{i}2\pi\boldsymbol{\alpha}_0\cdot\boldsymbol{f}_\xi}. \tag{5.40}$$

This transforms into a shift of $\boldsymbol{\alpha}_0$ of all point-spread functions in the image plane, and into phase delays $2\pi\boldsymbol{\alpha}_0\cdot\boldsymbol{f}_{\xi,pj}$ at each position $\boldsymbol{\alpha}_{pj} = \boldsymbol{f}_{\xi,pj}\lambda/F_1$, reading

$$\begin{aligned} V_{\alpha_0}(\boldsymbol{\alpha}) = {} & V_0\, a_{\mathrm{PSF}}(\boldsymbol{\alpha}) \\ & * \Big(\delta(\boldsymbol{\alpha}-\boldsymbol{\alpha}_0-\boldsymbol{\alpha}_{p1})\mathrm{e}^{\mathrm{i}2\pi\boldsymbol{\alpha}_0\cdot\boldsymbol{f}_{\xi,p1}} + \delta(\boldsymbol{\alpha}-\boldsymbol{\alpha}_0-\boldsymbol{\alpha}_{p2})\mathrm{e}^{\mathrm{i}2\pi\boldsymbol{\alpha}_0\cdot\boldsymbol{f}_{\xi,p1}}\Big). \end{aligned} \tag{5.41}$$

Now, the Fourier transform into the reimaged pupil plane yields

$$V(\boldsymbol{\beta},\alpha_0) = V_0\, a(\boldsymbol{\beta})\,\mathrm{e}^{\mathrm{i}2\pi\alpha_0\cdot\boldsymbol{\beta}F_1/\lambda}\Big(\mathrm{e}^{\mathrm{i}2\pi f_{\xi,p1}\cdot(\beta-\alpha_0)} + \mathrm{e}^{\mathrm{i}2\pi f_{\xi,p2}\cdot(\beta-\alpha_0)}\Big), \tag{5.42}$$

when the phase delays of the point-spread functions convert into a simple linear shift of the exponential functions by $\boldsymbol{\alpha}_0$. The shift of the ensemble of PSF creates the overall phase delay $2\pi\boldsymbol{\alpha}_0\cdot\boldsymbol{\beta}F_1/\lambda$ in the reimaged pupil plane – describing a tilted plane wave – that is identical to that in the exit pupil if $F_1 = F_2$.

Computing the intensity in the reimaged pupil plane, this overall phase term disappears, obtaining

$$I(\boldsymbol{\beta},\boldsymbol{\alpha}_0) = V_0^2|a(\boldsymbol{\beta})|^2\, 2\big(1+\cos(2\pi\boldsymbol{R}_B\cdot(\boldsymbol{\beta}-\boldsymbol{\alpha}_0))\big). \tag{5.43}$$

Thus, the fringe pattern in the reimaged pupil plane is shifted by $\boldsymbol{\alpha}_0$, exactly like the fringe pattern in an image plane when combining two apertures. If the baselines in the interferometer array are downscaled by a factor m_b, the angular position $\boldsymbol{\alpha}_0'$ in the sky converts into a shift of the fringe pattern by $\boldsymbol{\alpha}_0 = m_b\boldsymbol{\alpha}_0'$.

The shift by $m_b\boldsymbol{\alpha}_0'$ is due to the phase delays, $\exp(\mathrm{i}2\pi\boldsymbol{\alpha}_0\cdot\boldsymbol{f}_{\xi,p1})$, of the point-spread functions in (5.41). The actual phase delay that we have to consider is determined by the product $\boldsymbol{\alpha}_0'\cdot\boldsymbol{f}_{\xi,p1}'$ in the entrance pupil. Using the coordinates in the exit pupil we find $\boldsymbol{\alpha}_0' \cdot \boldsymbol{f}'_{\xi,p1} m_b/m_b = m_b\boldsymbol{\alpha}_0' \cdot \boldsymbol{f}_{\xi,p1}$ so that $\boldsymbol{\alpha}_0$ is replaced by $m_b\boldsymbol{\alpha}_0'$. A similar argument lead to the shift of the fringe pattern in Michelson configuration (5.5).

The envelope of the fringe pattern is given by the individual aperture function $|a(\boldsymbol{\beta})|^2$ with diameter $\beta_D = D/F_1$. If the telescope apertures are downscaled by m_p, the diameter of the apertures in the exit pupil is $D = D'/m_p$, and that of $|a(\boldsymbol{\beta})|^2$ in the reimaged pupil plane is $\beta_D = D'/(m_p F_1)$.

Multi-Aperture Combination

The combination of N_T apertures is now straightforward, describing the exit pupil as the sum of δ-functions

$$A(\boldsymbol{f}_\xi) = a(\boldsymbol{f}_\xi) * \sum_{j=1}^{N_T} \delta(\boldsymbol{f}_\xi - \boldsymbol{f}_{\xi,p_j}),$$

and the distribution of PSF amplitudes in the image plane by

$$V(\boldsymbol{\alpha}) = V_0\, a_{\mathrm{PSF}}(\boldsymbol{\alpha}) * \sum_{j=1}^{N_T} \delta\left(\boldsymbol{\alpha} - \boldsymbol{\alpha}_{p_j}\right).$$

In the reimaged pupil plane we find the amplitude as Fourier transform of $V(\boldsymbol{\alpha})$ as

$$V(\boldsymbol{\beta}) = V_0\, a(\boldsymbol{\beta}) \times \sum_{j=1}^{N_T} \mathrm{e}^{\mathrm{i}2\pi \boldsymbol{f}_{\xi,p_j} \cdot \boldsymbol{\beta}}.$$

The intensity distribution is the modulus squared of $V(\boldsymbol{\beta})$ when the sum over all N_T apertures is replaced by the sum over all N_B baselines as in (5.27), reading

$$I(\boldsymbol{\beta}) = V_0^2 |a(\boldsymbol{\beta})|^2 \Big(N_T + \sum_{j=1}^{N_B} 2\cos(2\pi \boldsymbol{R}_{B_j} \cdot \boldsymbol{\beta})\Big). \tag{5.44}$$

We have the sum of a constant term, N_T, and of the sum of cosine functions exactly like in (5.28) at the beginning of this section when discussing the point-spread function in Fizeau configuration. There, the shape of the PSF of many randomly distributed apertures displayed an irregular distribution of fringes and sidelobes with a narrow core that is created by the sum of cosine functions, and the PSF of an individual aperture as an envelope.

For the IRAN configuration in (5.44), we find precisely the same distribution with a narrow core, now inside an envelope that is given by $|a(\boldsymbol{\beta})|^2$ the reimaged aperture function. This intensity distribution can, thus, be regarded as the point-spread function of the system, since it is the response of the system to a point source. The only difference to the Fizeau configuration is that the IRAN PSF is a function of $\boldsymbol{\beta}$, the angle coordinate in the reimaged pupil plane.

We now use the general object coordinate $\boldsymbol{\alpha}'$ in the sky, and we assume the case that apertures and baselines are downscaled by m_p and m_b, respectively. Using (3.3), the PSF in IRAN configuration is introduced by dividing the intensity distribution by V_0^2, yielding

$$\mathrm{PSF}_{\mathrm{IRAN}}(\boldsymbol{\beta}, \boldsymbol{\alpha}') = |a(\boldsymbol{\beta})|^2 \Big(N_T + \sum_{j=1}^{N_B} 2\cos(2\pi \boldsymbol{R}_{B_j} \cdot (\boldsymbol{\beta} - m_b \boldsymbol{\alpha}')) \Big). \quad (5.45)$$

Therefore, the angle coordinate $\boldsymbol{\beta}$ can be regarded as an image coordinate, mapping each point in the sky onto the position $m_b \boldsymbol{\alpha}'$.

In contrast to the PSF in Fizeau configuration, (5.28), and to the PSF of the Hypertelescope, (5.33), we have an envelope function $|a(\boldsymbol{\beta})|^2$ with a diameter of $\beta_D = D'/(m_p F_1)$ that is independent of wavelength and object position, but that depends on the pupil demagnification factor m_p and on the focal length F_1 of the pupil reimaging system displayed in Fig. 5.14. Thus, the primary beam has the shape of the aperture function. Using an aperture with central obscuration would obviously create zero intensity in the center of the reimaged aperture but, moving the object off-axis, would not change the general principle of the method.

For a numerical example, we assume an observing wavelength of $\lambda = 2.2\,\mu$m, a maximum baseline of $B'_{\max} = 130$ m and an aperture diameter of $D' = 8$ m. Downscaling the array by $m_b = 2{,}000$ and the aperture by $m_p = 8{,}000$, it is $B_{\max} = 65$ mm and $D = 1$ mm in the exit pupil. For the reimaging system we choose $F_1 = 2.4$ m, so that the diameter of the narrow core is $\lambda / B_{\max} = 7$ arcsec and the diameter of the envelope $|a(\boldsymbol{\beta})|^2$ is $D/F_1 = 86$ arcsec, i.e. 12.3 times wider than the core. This situation is displayed in Fig. 5.15.

The imaging process is now described by the convolution of the object brightness distribution $O_b(\boldsymbol{\alpha}')$ with $\mathrm{PSF}_{\mathrm{IRAN}}$ as

$$\begin{aligned} I(\boldsymbol{\beta}) &= O_b(\boldsymbol{\alpha}') * \mathrm{PSF}_{\mathrm{IRAN}}(\boldsymbol{\beta}, \alpha') \\ &= |a(\boldsymbol{\beta})|^2 \times \Big(O_b(\boldsymbol{\beta}/m_b) * \big(N_T + \sum_{j=1}^{N_B} 2\cos(2\pi \boldsymbol{R}_{B_j} \cdot \boldsymbol{\beta})\big)\Big). \quad (5.46) \end{aligned}$$

The convolution holds inside the envelope of diameter D/F_1. Thus, an increase of the aperture diameter in the exit pupil increases the **field of view**, while for the Hypertelescope the enveloping PSF_a, (5.33), shrinks with increasing aperture diameter. However, the idea behind the IRAN concept was not to have a large field of view – this is provided in Fizeau configuration already – but to increase the intensity in the central core as for the Hypertelescope.

We find an increase of the intensity in the central core compared to the Fizeau configuration if the size $D/F_1 = D'/(m_p F_1)$ of the envelope $|a(\boldsymbol{\beta})|^2$ is smaller

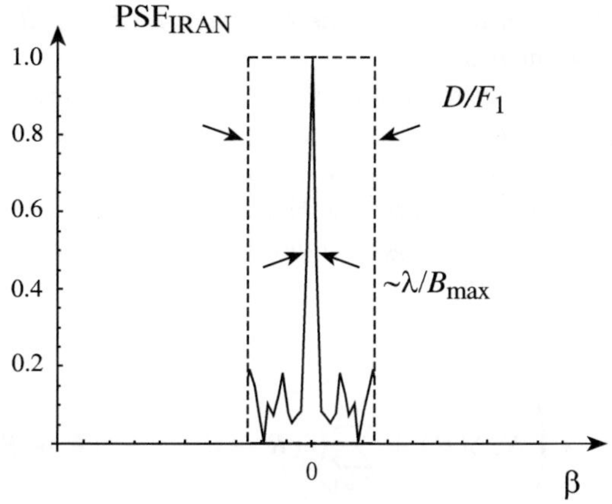

Fig. 5.15 Profile of $\mathrm{PSF_{IRAN}}$, (5.45), with $N_T = 10$ randomly distributed apertures. *The dashed line* shows the envelope, $|a(\boldsymbol{\beta})|^2$, which is the reimaged aperture of width $\beta_D = D/F_1$. The longest baseline $B_{\max}$ determines the width of the narrow central core, which in this case is 12.5 times narrower than the envelope. The average intensity in the halo of sidelobes around the core is proportional to N_T while the peak intensity is proportional to N_T^2, in this case 10 times larger. This figure should be compared to Figs. 5.11 and 5.13 for Fizeau configuration and for the Hypertelescope

than the equivalent width λ/D of the Airy disk which is the envelope in Fizeau configuration [127]. Considering, that in Fizeau configuration, pupil and baselines are downscaled by the same factor $m_p = m_b$, we write the width of the Airy disk as $\lambda m_b/D'$. The gain of the peak intensity of the IRAN over the Fizeau configuration is determined by the area of the envelope functions and we obtain

$$g_{\mathrm{IRAN}} = \left(\frac{\lambda m_b/D'}{D'/(m_p F_1)}\right)^2 = \left(\frac{m_b m_p \lambda F_1}{D'^2}\right)^2 . \tag{5.47}$$

For the numerical example above that is displayed in Fig. 5.15 the gain is 1.75, since for the chosen parameter set, the equivalent width λ/D of the PSF is about 30% wider than the aperture diameter D/F_1. Modifying the parameters in (5.47) one can adjust the instrumental setup for the needs of the particular observing situation.

Before we will investigate the characteristics of the PSF when the telescopes are distributed on a regular grid in Sect. 5.4, we finish this section by discussing the optical transfer function of the IRAN configuration.

NB 9. *Despite the fact that we do not have an image plane in the classical sense of geometrical optics, we can define an optical transfer function as the Fourier transform of the point-spread function* $\mathrm{PSF_{IRAN}}$ *(5.45). The coordinate of this optical transfer function is the spatial frequency* $\boldsymbol{R}$ *since* $\mathrm{OTF_{IRAN}}$ *describes the spatial frequency content of the point-spread function. We formally set* $\boldsymbol{\alpha}' = 0$ *in (5.45)* *since we restrict the field of view to the diameter D/F_1 of the aperture function when the PSF is shift invariant.*

OTF_{IRAN} as Fourier transform of PSF_{IRAN} is then the convolution of the Fourier transform of $|a(\boldsymbol{\beta})|^2$ with the Fourier transform of the sum of cosine functions, which is the sum of δ-peaks. Considering that $|a(\boldsymbol{\beta})|^2 = |a(\boldsymbol{\beta})|$ is the aperture function of an individual telescope, its Fourier transform is the amplitude of the PSF of the individual aperture, a_{PSF}, so that we have replica of a_{PSF} at all $\boldsymbol{R}_{B_j}$, as displayed in Fig. 5.16, writing

$$\text{OTF}_{\text{IRAN}}(\boldsymbol{R}) = a_{\text{PSF}}(\boldsymbol{R}) * \left(N_T \delta(\boldsymbol{R}) + \sum_{j=1}^{N_B} \left(\delta(\boldsymbol{R} - \boldsymbol{R}_{B_j}) + \delta(\boldsymbol{R} + \boldsymbol{R}_{B_j}) \right) \right). \tag{5.48}$$

Restricting the field of view to D/F_1 is equivalent to having a visibility function that is invariant over $a_{\text{PSF}}(\boldsymbol{R})$ so that in the image intensity spectrum the individual δ-peaks in OTF_{IRAN} are multiplied by $\mu(m_b \boldsymbol{R}_{B_j})$ similar to the Michelson configuration (5.23).

However, increasing the field of view by using a very large aperture D, we find that $a_{\text{PSF}}(\boldsymbol{R})$ is a very narrow function so that eventually, the OTF is described in good approximation by the sum of δ-functions. One would obtain exactly the same OTF in a standard interferometric setup if the aperture array consisted of pinholes.

As displayed in Fig. 5.6, the interferometer OTF in Fizeau configuration consists of replica of optical transfer functions, OTF_a, of the individual apertures. The

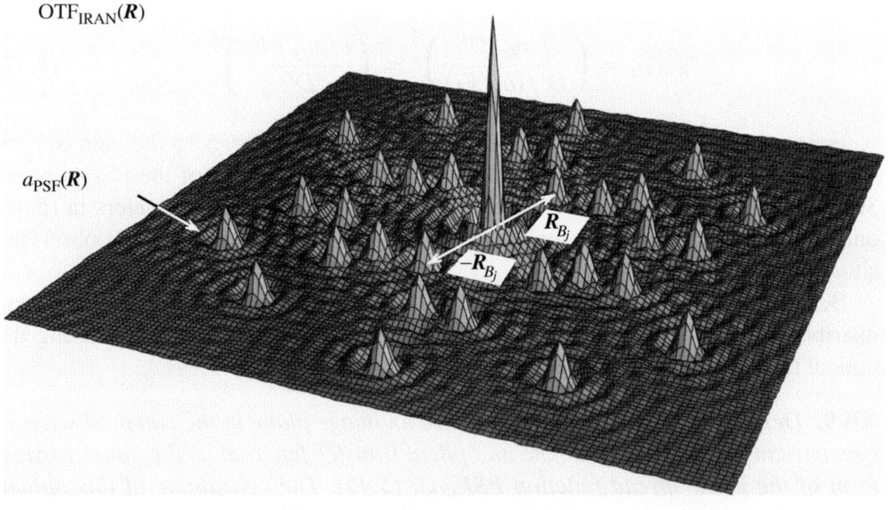

Fig. 5.16 Multi-aperture OTF for six randomly distributed telescopes in IRAN configuration, composed of replica of the PSF amplitude a_{PSF}. In the image intensity spectrum, each PSF amplitude is multiplied by the appropriate visibility function, $\mu(\boldsymbol{R}_{B_j})$

PSF of this array displays an interference pattern with a narrow central core and with a rather narrow envelope, PSF_a, increasing the relative intensity in the central core for larger apertures. A narrow envelope reduces the background in the image and helps to reduce the confusion when imaging complex objects. This can also be interpreted as due to the increasing number of visibility values inside the wider OTF_a for larger apertures.

In the IRAN configuration with large apertures, we have a very wide envelope, but we have the same interference pattern with a narrow central core since the latter is determined by the distribution of the δ-functions only. The relative intensity in the central core is lower than in the Fizeau configuration, and the background level in the image is higher so that the image has a higher degree of confusion and a lower content of information. The latter is related to the form of OTF_{IRAN} permitting only a single visibility values at each frequency $\boldsymbol{R}_{B_j}$ independent of the size of the aperture.

Thus, the peculiar imaging process with reversed roles of image and pupil plane leads to inverse effects when increasing the aperture size and the field of view.

However, both the Hypertelescope and the IRAN configuration were not developed for large fields of view but for rather small fields with well defined properties.

5.3.4 Nulling Interferometer

The basic principle of the Nulling interferometer is very simple, introducing an achromatic phase shift of π in one of the two arms of an interferometer so that the fringe pattern of an on-axis object has a "black" instead of a bright fringe on-axis. Thus, the fringe pattern is inverted.

R. Bracewell suggested this observing mode for the specific purpose of searching for extra-solar planets [20] when it is required to detect extremely faint objects in the immediate vicinity of a star. The contrast ratio between planet and parent star is as low as 10^{-6}–10^{-10}, depending on wavelength [23], and the angular separation is about 100 mas for Earth-like planets around nearby stars.

How does the inversion of the fringe pattern help us detecting a faint planet? In the monochromatic case, the fringe pattern is just shifted by half a fringe period and the characteristics are the same as without a phase shift. We have to regard observations in polychromatic light to see the effect, as illustrated in Fig. 5.17 for two telescopes. We assume that the telescope diameter is very small compared to the baseline so that the enveloping point-spread function, PSF_a, is rather wide and does not affect the fringe pattern of the central fringes.

Combining the light in the standard way, i.e. without explicit phase shifts in the interferometer arms, the fringe pattern of the on-axis point source has a bright fringe on axis at $\boldsymbol{\alpha} = 0$ (see for instance Fig. 2.3). The first minima are at $1/(2\boldsymbol{R}_B) = \lambda/(2\boldsymbol{B})$ so that for a spectral band like the K-band with $\lambda_0/\Delta\lambda = 5.5$ the positions of the minima vary by $\pm 10\%$ and the resulting polychromatic fringe (shown as a black line) does not go down to zero intensity.

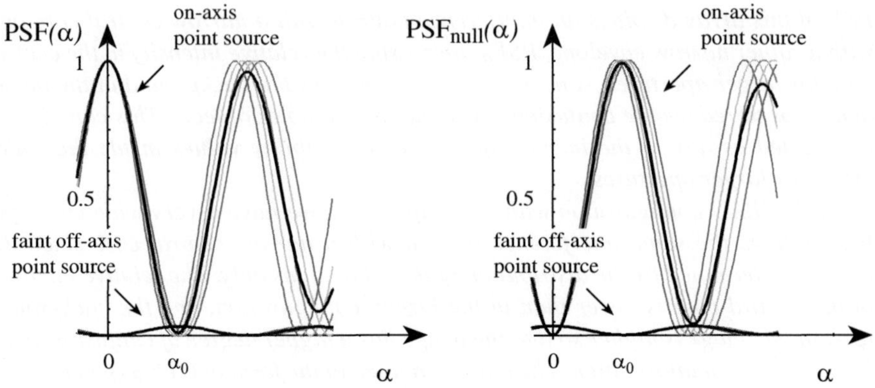

Fig. 5.17 Principle of the Nulling interferometer for two telescopes. *On the left*, the polychromatic PSF for an on-axis source as in Fig. 2.3 is displayed. In this example, five monochromatic fringes add up to the polychromatic PSF (*in black*) with diminishing contrast. In addition, one can see the PSF of an off-axis point source at $\boldsymbol{\alpha}_0$ with 1/30 of the intensity of the on-axis source. The baseline is chosen such that the white-light fringe of the off-axis source at position $\boldsymbol{\alpha}_0$ is exactly at the position of the first minimum of the on-axis object. *On the right*, applying an achromatic phase shift of π in one of the interferometer arms, the on-axis point source produces a PSF with null intensity for all wavelengths at position $\boldsymbol{\alpha} = 0$. The off-axis object displays an achromatic null at $\boldsymbol{\alpha}_0$ but its first maximum is now at $\boldsymbol{\alpha} = 0$

Writing the intensity in the first minimum at $\boldsymbol{\alpha}_{\min} = \lambda_0/(2\boldsymbol{B})$ as

$$I(\boldsymbol{\alpha}_{\min}) = 2\,I_0\big(1 + \cos(2\pi\,\boldsymbol{R}_B \cdot \boldsymbol{\alpha}_{\min})\big) = 2\,I_0\big(1 + \cos(\pi\lambda_0/\lambda)\big) \propto \frac{\pi^2}{4}\frac{(\lambda_0 - \lambda)^2}{\lambda^2},$$

we compute the integrated polychromatic intensity in the first minimum for small $\Delta\lambda$ as about $\frac{\pi^2}{48}(\Delta\lambda/\lambda_0)^2$. Then, for the K-band the minimum intensity is about 1/140 of the intensity of the white-light fringe.

A point-source with 1/30 of the intensity of the on-axis source is also displayed in Fig. 5.17. The position of the off-axis source is chosen to be exactly at the first minimum of the fringe pattern of the on-axis source so that for our example of the K-band, the maximum of the fringe pattern of the faint object is still five times larger than the minimum of the on-axis source, but it is obvious that planets that are many orders of magnitude fainter than their parent star cannot be detected like this.

Introducing the achromatic phase shift of π in one arm, there is an achromatic null in the intensity distribution of an on-axis source at $\boldsymbol{\alpha} = 0$. Consequently, the off-axis source displays a null at position $\boldsymbol{\alpha}_0$ but the first maximum of its fringe pattern is now visible at $\boldsymbol{\alpha} = 0$ as displayed in Fig. 5.17.

We write the point-spread function, the fringe pattern of an on-axis point source, in the Nulling interferometer as

$$\begin{aligned} \mathrm{PSF}_{\mathrm{null}}(\boldsymbol{\alpha}) &= 2\big(1+\cos(2\pi \boldsymbol{R_B}\cdot\boldsymbol{\alpha}-\pi)\big) \\ &= 2\big(1-\cos(2\pi \boldsymbol{R_B}\cdot\boldsymbol{\alpha})\big), \end{aligned} \tag{5.49}$$

using $\mathrm{PSF}_a(\boldsymbol{\alpha}) \approx 1$, since we assume that the individual apertures are small compared to the baseline.

For a point source at position $\boldsymbol{\alpha}_0$ we find

$$\begin{aligned} I(\boldsymbol{\alpha})|_{\alpha_0} &= I_0\delta(\boldsymbol{\alpha}-\boldsymbol{\alpha}_0) * \mathrm{PSF}_{\mathrm{null}}(\boldsymbol{\alpha}) \\ &= 2\,I_0\big(1-\cos(2\pi \boldsymbol{R_B}\cdot(\boldsymbol{\alpha}-\boldsymbol{\alpha}_0))\big). \end{aligned} \tag{5.50}$$

Thus, a source at position $\boldsymbol{\alpha}_0 = \lambda/(2\boldsymbol{B})$ has a fringe pattern with its first maximum at $\boldsymbol{\alpha} = 0$ as displayed in Fig. 5.17. In the ideal case, the light from the star is completely extinguished on axis and we can detect the faint planet. One should note that the baseline is a free parameter and we can choose it such that the angular distance of the planet is such that its first maximum is on axis.

The position of the maximum, $\lambda/(2\boldsymbol{B})$, is wavelength dependent so that the polychromatic intensity at $\boldsymbol{\alpha} = 0$ is slightly reduced. However, the detectability of the faint source is not affected by this slight reduction of the intensity since it is negligible compared to the advantage of completely extinguishing the star light.

Sometimes, the intensity in (5.50) at $\boldsymbol{\alpha} = 0$ is written as a function of the off-axis source position $\boldsymbol{\alpha}_0$, called the ***transmission map*** *of the Nulling interferometer* since we are primarily interested in the signal that an off-axis source provides when the on-axis source is extinguished. It is easy to see that $I(0)_{\alpha_0}$ is identical to $\mathrm{PSF}_{\mathrm{null}}(\boldsymbol{\alpha}_0)$ in (5.49).

Searching for planets around nearby stars, the stars have an extension of several milliarcseconds. Given that the expected angular distance of Earth-like planets is some 100 mas, we need a baseline of 10 m when observing at a wavelength of 10 μm in order to place the planet at the first maximum of the star's fringe pattern, as in Fig. 5.17. If we approximate the transmission map of the Nulling interferometer, normalized to unity, for small $\boldsymbol{\alpha}_0$ by

$$I_{\mathrm{rel,null}}(\boldsymbol{\alpha}_0) = \frac{1}{2}\big(1-\cos(2\pi \boldsymbol{R_B}\cdot\boldsymbol{\alpha}_0)\big) \approx \pi^2\frac{|\boldsymbol{B}|^2}{\lambda^2}|\boldsymbol{\alpha}_0|^2, \tag{5.51}$$

we find that, integrating over the stellar disk, about $I_{\mathrm{rel,null}}(\boldsymbol{\alpha}_0)/4 \approx 6\times10^{-5}$ of the light from the star with 1 mas radius leaks through. Although this signal is likely larger than the planet's signal, the calibration of the interferometer should help since the signal is deterministic [2].

Considering that the product $|\boldsymbol{\alpha}_0||\boldsymbol{B}|$ in (5.51) describes the OPD between the beams, we find that OPD variations at the nm level cause a leak of about 10^{-7}. This gives an impression to which level of accuracy other random error sources – OPD variations due to vibrations or due to atmospheric turbulence, polarization effects with different OPD for s and p polarization, to name a few – have to be controlled in order to obtain a sufficient rejection of the star light [135]. The best experimental results achieved to date in the laboratory show a null of about 10^{-5} in the mid-infrared at 10 μm with a bandwidth of about 4 μm [77, 183].

The signal detection can be improved by rotating the interferometer around the line-of sight, with the star on axis, so that the intensity at $\boldsymbol{\alpha} = 0$ oscillates between a maximum when $\boldsymbol{\alpha}_0$ is parallel to $\boldsymbol{R}_B$ and a minimum of zero when the vectors are orthogonal. Note that for a rotation frequency of ω, the signal from the planet oscillates at 2ω. This was already suggested by Bracewell in his original paper [20] for an interferometer in space.

Combining more than two telescopes, the transmission map can be made wider, for instance proportional to $|\boldsymbol{\alpha}_0|^4$ [5], so that the leaks are smaller, but at the price of increasing the number of optical components and paths to be controlled. Choosing an odd number of telescopes, one can distinguish in the oscillating signal the contribution by the planet on one side of the star – hence, an asymmetric component – from the contribution by symmetric components like zodiacal light, since the signal of the latter oscillates at twice the frequency of that of the planet [137].

Coronagraphy

Apart from applying interferometric techniques to the planet search, there is a wealth of concepts for *coronagraphic* methods on single telescopes [91] rejecting the star light by placing aperture masks for amplitude or phase in the image or in the pupil plane. One concept, the *achromatic interference coronagraph* [78] combines interferometry with single telescope observations by splitting the light from a single telescope into two interferometric arms when in one arm an achromatic phase shift of π is performed by passing through an intermediate focus (see also Sect. 2.2.3).

Having started this section with questions of the visibility in Michelson configuration, we are now at the other extreme, trying to detect the signal in the image intensity distribution that signifies an extra-solar planet. Thus, combining light from separate telescope interferometrically opens many possibilities not necessarily limited to improved angular resolution but also enhancing the detectability of faint sources.

Direct Imaging: Summary

We call N_T the number of telescopes in the array and N_B the number of baselines with $N_B = N_T(N_T - 1)/2$ if the distribution of telescopes is non-redundant.

In Fizeau configuration, we obtain the image intensity distribution as a convolution of the object brightness distribution $O_b(\boldsymbol{\alpha}')$ with the shift-invariant interferometer PSF, writing

$$I(\boldsymbol{\alpha}) = O_b(\boldsymbol{\alpha}/m_p) * \Big(\mathrm{PSF}_a(\boldsymbol{\alpha})\big(N_T + \sum_{j=1}^{N_B} 2\cos(2\pi \boldsymbol{R}_{B_j} \cdot \boldsymbol{\alpha})\big)\Big), \quad (5.29)$$

with $\boldsymbol{\alpha}/m_p = \boldsymbol{\alpha}'$, and, accordingly, the baselines in the exit pupil m_p times smaller than the baselines in the entrance pupil, $\boldsymbol{R}_{B_j} = \boldsymbol{R}'_{B'j}/m_p$. Down-scaling apertures and baselines by the same factor m_p ensures homothetic mapping.

For a large number of baselines (10 and more), the interferometer PSF consists of a narrow central core with diameter $1/\boldsymbol{R}_{B_{\max}}$ that is surrounded by a halo of fringes and sidelobes. The envelope of the interference pattern is given by the Airy disk of an individual aperture in the exit pupil, PSF_a, the primary beam, with an FWHM of λ/D. The field of view is limited by the optical design of the telescope only.

A. Labeyrie pursued the idea of a densified pupil when the baselines and the individual apertures are not reimaged homothetically but when the apertures are relatively larger than in the entrance pupil [122]. In this Hypertelescope, we have the same interference pattern as in the Fizeau configuration but the enveloping PSF is much narrower. The price to pay is the restriction of the field of view, which is identical to that in Michelson configuration discussed in Sect. 5.1.2. We call m_b the demagnification factor of the baselines and m_p that of the apertures obtaining the maximum field of view as $\alpha'_{\max} \ll \lambda/(m_b D'/m_p) = \lambda/(\gamma_d D')$, when $\gamma_d = m_b/m_p$ is the densification factor.

We have to use the small object approximation to write the image intensity distribution as a convolution of the object brightness distribution $O_b(\boldsymbol{\alpha}')$ with the interference pattern,

$$I(\boldsymbol{\alpha}) = \mathrm{PSF}_a(\boldsymbol{\alpha})\Big(O_b(\boldsymbol{\alpha}/m_b) * \big(N_T + \sum_{j=1}^{N_B} 2\cos(2\pi \boldsymbol{R}_{B_j} \cdot \boldsymbol{\alpha})\big)\Big). \quad (5.34)$$

Due to the multiplication with PSF_a, the peak intensity of the central core is reduced for off-axis objects. For instance an 8-m telescope and a pupil densification of $\gamma_d = 10$ show a reduction in peak intensity by a factor of two for an object at 2.8 mas off-axis.

Unlike all other combination schemes forming an interference pattern in the image plane, the IRAN configuration provides an image in the pupil plane. This idea of F. Vakili [245] is based on forming individual images directly behind each aperture in the exit pupil, maintaining the distribution of the apertures (see Fig. 5.14). Adding a large lens behind the image plane, all apertures are reimaged in one common position and an interference pattern forms inside the reimaged aperture. The distribution of fringes and sidelobes in the interference pattern is the same as in Fizeau configuration and in the Hypertelescope but the envelope now has the rectangular form of the aperture function.

The angle coordinate $\boldsymbol{\beta}$ in the pupil plane is related to the angle coordinate in the sky by $\boldsymbol{\alpha}' = \boldsymbol{\beta}/m_b$, and to the physical coordinate by $\boldsymbol{\xi}' = \boldsymbol{\beta} f$, with f the focal length of the lens reimaging the apertures in one common image.

The imaging process is now described by the convolution of the object brightness distribution $O_b(\boldsymbol{\alpha}')$ with $\mathrm{PSF_{IRAN}}$ (5.45) as

$$I(\boldsymbol{\beta}) = |a(\boldsymbol{\beta})|^2 \times \Big(O_b(\boldsymbol{\beta}/m_b) * \big(N_T + \sum_{j=1}^{N_B} 2\cos(2\pi \boldsymbol{R}_{B_j} \cdot \boldsymbol{\beta})\big)\Big). \quad (5.46)$$

The distribution of fringes and sidelobes in the interference pattern is the same as in Fizeau configuration (5.24) and in the Hypertelescope (5.34) but the envelope now has the rectangular form of the aperture function.

The limit on the field of view imposed by the aperture diameter is comparable to the limit in the Hypertelescope.

The Nulling interferometer is a specialised facility to detect the very faint (10^{-6}–10^{-10}) signal of an extra-solar planet, or any faint source, in the immediate vicinity (about 100 mas) of its parent star. It is based on inverting the fringe pattern by introducing an achromatic phase shift of π in one of the interferometer arms, producing an achromatic null instead of a bright fringe on axis. Choosing the baseline such that half of the fringe spacing corresponds to the angular separation of the planet, we find the first maximum of the planet in the null of the bright star.

One can write the image intensity on axis as a function of the angular coordinate $\boldsymbol{\alpha}_0$ of the planet, also called the *transmission map* of a two-telescope Nulling interferometer as

$$I_{\mathrm{rel,null}}(\boldsymbol{\alpha}_0) = \frac{1}{2}\big(1 - \cos(2\pi \boldsymbol{R}_B \cdot \boldsymbol{\alpha}_0)\big) \approx \pi^2 \frac{|\boldsymbol{B}|^2}{\lambda^2} |\boldsymbol{\alpha}_0|^2. \quad (5.51)$$

Given that nearby stars have an angular diameter of a few milliarcseconds, we find that the star is not completely extinguished but that – using a wavelength of 10 μm and baseline of 10 m – about 6×10^{-5} of its light leaks through. While this deterministic leak can be calibrated, other random errors cannot and have to be controlled so that the subsequent residual OPD are at the nm level.

Using more than two telescopes, the transmission map can be made wider (proportional to $|\boldsymbol{\alpha}_0|^4$) so that the leaks are smaller, but at the price of increasing the number of optical components and paths to be controlled.

5.4 Layout of Interferometer Arrays

The design of interferometer arrays is driven by the available number of telescopes, and by the preference either for high angular resolution, preferring long baselines, or for a dense coverage of the uv-plane, preferring a more densely populated array. While a diluted array with long baselines has a point-spread function with a narrow core and a large number of fringes and sidelobes, a densely populated array has fewer sidelobes, making it easier to observe complex objects.

We will distinguish the two cases of many and of few apertures, when the dividing line is at about ten apertures. Here, few apertures means a sparsely populated array, i.e. rather small telescopes on rather long baselines. The number of telescopes has an impact on the beam combination scheme, and we will give examples of direct imaging for the case of many apertures. With few apertures, instantaneous direct imaging is not the preferred solution and the rotation of the Earth is exploited to improve the uv-plane coverage with observations over several hours, as discussed in Sect. 3.4.2.

5.4.1 Many Apertures

In radio interferometry when usually many apertures are available, a lot of research in the last decades went into determining the optimal configuration providing the best uv-plane coverage for a given number of telescopes [104]. We will summarise the main results and compare them to the situation in optical interferometry.

Uniformly Filling the uv-Plane

One goal of the optimisation procedure is to fill the uv-plane uniformly without redundant baselines. In one of the first studies by M. Golay in 1970 [86] an iterative method was used to build up the number of apertures. The result was an apparently irregular distribution of apertures in the array, but a regular periodic grid – either square or hexagonal – of points in the uv-plane. The problem with a periodic grid is that its Fourier transform, the PSF, is a periodic distribution of peaks as well, each peak representing the Fourier transform of the overall support of the array (see Appendix A.1). In addition, an image with this type of PSF consists of replica of the object at each peak in the grid due to the convolution of the object brightness distribution with the PSF. In Fig. 5.18, an example of an interferometer array with periodic but non-uniform coverage of the uv-plane is displayed that – despite the lack of uniformity – illustrates the impact of a periodic grid on the PSF. If we denote by $B_{\rm min}$ the grid period in the uv-plane, which is the shortest baseline in the array, we find the object replica spaced by $\lambda/B_{\rm min}$. With this kind of regular repetition of the object in the image plane, the unambiguous field of view is limited to $\lambda/B_{\rm min}$,

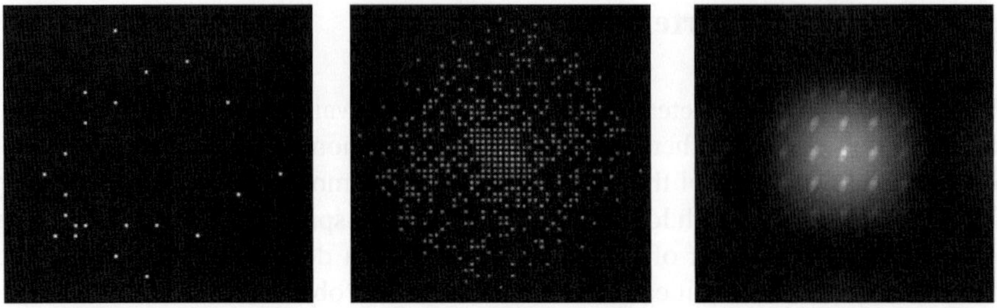

Fig. 5.18 Example of an interferometer array of 22 telescopes with non-redundant baselines. The periodic grid in the uv-plane with grid constant $B_{\min}$, the shortest baseline, converts into a periodic grid in the PSF with grid constant $\lambda / B_{\min}$. Computing the image of a model galaxy we find replica at each peak as displayed *on the right*. The envelope of the image intensity distribution is given by the Airy disk of an individual telescope, the primary beam (Courtesy C. Aime [4])

sometimes called the *clean field of view* [127], so that the replica of the objects are clearly separated.

It should be noted that the clean field of view depends entirely on the distribution of points in the uv-plane. In case of a random instead of a periodic uv-plane coverage as discussed in Sect. 5.3, we find random sidelobes in the PSF (see Fig. 5.12) that would add to random noise but not to an identifiable replica of the object.

Optimising the distribution of telescopes in the array by employing numerical methods, the optimisation procedure showed a tendency to place the telescopes on regular geometric curves, either on the perimeter of a circle, T.J. Cornwell [44], or of a Reuleaux triangle, which is a triangle with rounded sides, E. Keto [113]. The regularity of the distribution of telescopes transforms into a regular and uniform uv-plane coverage so that despite achieving the goal of a good uniformity, sidelobes are also distributed regularly and therefore not suppressed satisfactorily (see Fig. 5.19).

As a remedy, both Cornwell and Keto suggested to slightly perturb the regular distribution of apertures – for instance by modifying their equidistant positions on a circle – so that the sidelobes are scattered irregularly forming an almost homogeneous background. This suppresses the peaks in the halo of the interferometric point-spread function, the primary beam, so that mainly the narrow central core remains with a diameter of about $\lambda/(\sqrt{2}B_{\max})$ and $B_{\max}$ the longest baseline in the array.

Using a fairly large number of telescopes the narrow central core displays diffraction rings like an Airy disk of an annular aperture (see Fig. 2.8). This is not very surprising when placing the telescopes on an approximately circular array. The problem is the rather high intensity level of the first diffraction ring of about 15% of the peak intensity. This can make the image reconstruction more difficult.

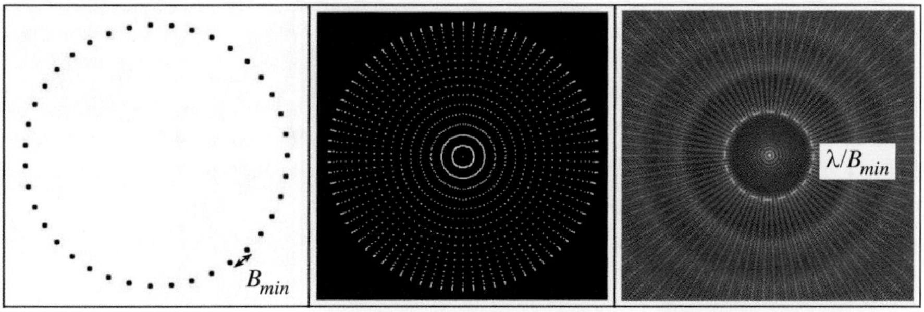

Fig. 5.19 Example of a circular interferometer array with equidistant spacing B_{min} of 39 telescopes. The uv-plane is filled uniformly with points spread over multiple regularly spaced, nested rings. The polychromatic PSF shows a regular pattern that is slightly smeared due to the spectral bandwidth with $\lambda/\Delta\lambda = 5$. The spacing B_{min} determines the radius λ/B_{min} of the central circle with homogeneous and low background, called the clean field of view. The narrow central core has rather prominent diffraction rings like the Airy disk of an annular aperture in Fig. 2.8 (Courtesy O. Lardière [127])

Emphasising Short Baselines

This problem is inherent in the goal of a uniform uv-plane coverage resulting in the approximately circular distribution of telescopes [104]. If, instead, we attempt a uv-plane coverage that is more dense for shorter baselines so that the density of points in the uv-plane decreases with increasing baseline length, we find the uv-plane coverage – more exactly the density of points – to be similar to the OTF of a single aperture. Then the narrow central core has diffraction rings that are lower in intensity, comparable to those of an Airy disk of a circular aperture.

A typical design of this type of array consists of several nested rings providing higher weighting for short baselines, as displayed in Fig. 5.20. The shortest baseline B_{min} determines the radius of the area in the centre of the uv-plane without points, the "hole", where no values of the visibility function can be measured. This converts into the clean field of view with radius λ/B_{min} that has particularly low background before the distribution of sidelobes provides an elevated background. Here, the same perturbation of the perfect symmetry can be applied as for circular and Reuleaux arrays so that sidelobes are distributed irregularly [104].

While the shortest baseline limits the clean field of view it does not impose an absolute limit to the object size as in the case of a periodic array displayed in Fig. 5.18. This was already discussed in the paragraph on the content of information in Sect. 3.3.5. Despite the lack of information on the visibilities for baselines shorter than B_{min}, we can observe large objects as indicated by the PSF in Fig. 5.20, albeit not with the highest fidelity.

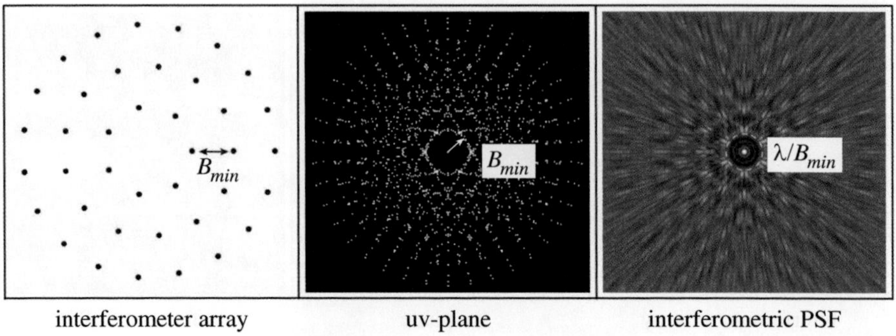

Fig. 5.20 Example of an interferometer array with three nested rings of a total of 39 telescopes. The uv-plane displays a higher concentration of points for short than for long baselines, and a hole in the centre with radius $B_{\rm min}$. The polychromatic PSF with $\lambda/\Delta\lambda = 5$ shows diffraction rings around the narrow central core. Due to the wider spacing $B_{\rm min}$ of the telescopes in this example the clean field of view is smaller than in Fig. 5.19 (Courtesy O. Lardière [127])

Direct Imaging

Having discussed possible layouts of the interferometer array, we now look into the beam combination schemes. Assuming a large number of telescopes, we can apply direct imaging methods. In contrast to the situation in radio interferometry, direct imaging in the optical relies on the instantaneous cophasing of all telescopes with extremely high precision. In particular for a large number of telescopes, this is a daunting task.

We use the interferometer array displayed in Fig. 5.20 to illustrate the different properties of the beam combination schemes – Fizeau, Hypertelescope and IRAN – that were discussed in Sect. 5.3.

The Fizeau configuration, computing the point-spread functions as Fourier transform of the uv-plane coverage, was used in Figs. 5.19 and 5.20. The wide envelope is given by the Airy disk, PSF_a, of an individual telescope, and the clean field of view is depicted by the small dark circle around the narrow central core. The PSF is shift-invariant and is simply reproduced at each object position $\boldsymbol{\alpha}_0$, as is shown in Fig. 5.21. Then, the image forms as a convolution of the object brightness distribution with the PSF.

In Sect. 5.3, we discussed the Hypertelescope and the IRAN configuration for a random telescope distribution. The same formalism as in (5.33) and (5.45) will be applied here for the interferometer array in Fig. 5.20. The instrumental parameters of the configurations are chosen in order to have the same field of view that is given by the central core of the Airy disk, PSF_a, of an individual telescope in the Hypertelescope, and by the pupil diameter in the IRAN configuration. The diameter of the clean field of view, given by $\lambda/B_{\rm min}$, is independent of the type of beam combination. The envelope PSF_a of the Hypertelescope is smaller than that in Fizeau configuration due to the pupil densification with larger apertures in the exit pupil than in Fizeau configuration (compare to Fig. 5.12).

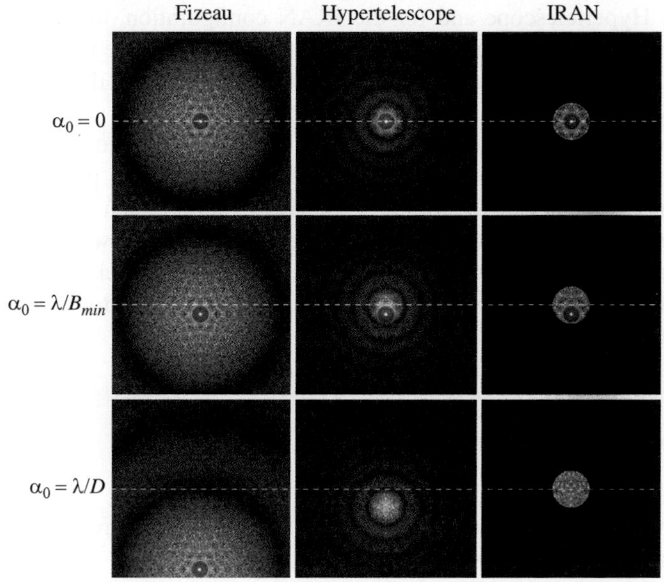

Fig. 5.21 Comparison of the imaging properties of beam combination in Fizeau configuration, in the Hypertelescope and in IRAN configuration, using an interferometer array as in Fig. 5.20. Point-spread functions for three different object positions $\boldsymbol{\alpha}_0$ are displayed. The clean field of view is depicted in each PSF by a small dark circle with diameter $\lambda / B_{\min}$ around the narrow central core. The shift-invariant PSF in Fizeau configuration moves with $\boldsymbol{\alpha}_0$. The shape of the Hypertelescope PSF varies with object position since PSF_a moves much slower than the interference pattern inside with its narrow core. In IRAN configuration, the interference pattern is inside the telescope pupil that is fixed in position (Courtesy O. Lardière [127])

The PSF of both the Hypertelescope and the IRAN configuration depend on the object position $\boldsymbol{\alpha}_0$ as shown in Fig. 5.21 and are not shift-invariant. The PSF of the Hypertelescope is given by (5.33), describing the behaviour that we find in the figure: the enveloping Airy disk, PSF_a, is shifted at a different rate than the interference pattern with its narrow central core. The consequence is that the intensity of the central peak is reduced according to its position inside the Airy disk, as displayed in Fig. 5.13.

In the IRAN configuration, the enveloping pupil function does not move with the object position but the narrow central core does, as described by (5.45), so that the field of view is also limited. However, the central peak inside the field of view has constant intensity unlike in the Hypertelescope.

Although the field of view in Fizeau configuration is unlimited in principle, one could argue that the conditions for good image quality are better when restraining the object size to the small clean field of view. Even a weak companion of a bright star could be detected inside the clean field of view while it might be undetectable outside of it due to the higher background. However, the advantage of a Fizeau configuration is the large field of view and not the sensitivity for weak companions.

For the Hypertelescope and for the IRAN configuration, the size of the clean field of view is much closer to the limit that is imposed by the principle of the two configurations so that it could be interesting to consider it by adjusting the smallest baseline $B_{\min}$ to the other parameters of the configuration.

Masking the aperture of a single large telescope, the required phasing of the sub-apertures can be achieved by an adaptive optics system providing a diffraction limited point-spread function. Then, direct imaging can be done in Fizeau, Hypertelescope or IRAN configuration as described here. However, the advantage of aperture masking to be able to calibrate the visibilities of each individual baseline as discussed in Sect. 5.1.1 would then be lost, so that it seems to be wiser to use the full aperture when aiming at direct imaging.

5.4.2 Few Apertures

All the considerations of uv-plane coverage in Sect. 5.4.1 are valid for instantaneous, snapshot imaging when the Earth's rotation has not moved the effective baselines. For a large number of telescopes, these snapshots provide a sufficiently good uv-plane coverage and a good point-spread function.

In practice, optical interferometers of today combine less than 10 telescopes, with six – in Michelson configuration – being the highest number to date [162], and it is almost inevitable exploiting the rotation of the Earth to improve the uv-plane coverage as discussed in Sect. 3.4. The uv-plane coverage varies with the sky coordinates and with the duration of the observation, and any particular interferometer layout with sophisticated distribution of telescopes will then present different properties than those that drove the design. Therefore, other more practical considerations become more important.

One type of layout that has been used for several interferometers is the so-called Y layout when the telescopes are placed along a three-arm structure like a Y. The practical advantage of this design is the efficient use of the required pipes or underground tunnels in which the light is transported from the telescopes to the delay lines and to the beam combiner since all telescopes in one arm send their light down the same pipe.

The telescopes can be distributed on the Y according to the requirements of the uv-plane coverage, trying to take the Earth rotation into account. An additional boundary condition might be the need to subdivide long baselines into several short, or even redundant, baselines that can be used for fringe tracking. Tracking the fringes on all short baselines with potentially larger visibility values, the fringes on the long baseline are stabilized, too. This technique is called *baseline bootstrapping* and will be discussed in Sect. 6.2.1. Combining very few telescopes only, it is difficult to apply this method.

In Fig. 5.22, the CHARA interferometer is displayed as an example of a Y layout [229]. Being integrated in the existing observatory on Mt. Wilson, California, there were very practical aspects to be considered for the interferometer design. Aiming

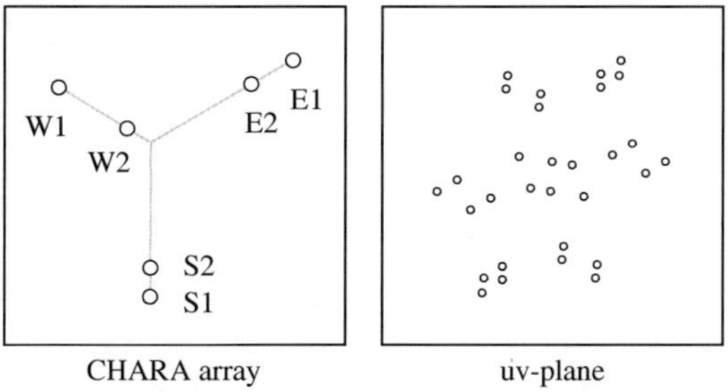

Fig. 5.22 The layout of the CHARA interferometer on Mt. Wilson as an example for a non-redundant *Y* layout with six telescopes. The autocorrelation of the interferometer array provides the instantaneous uv-plane coverage *on the right* that should be compared to Fig. 5.23

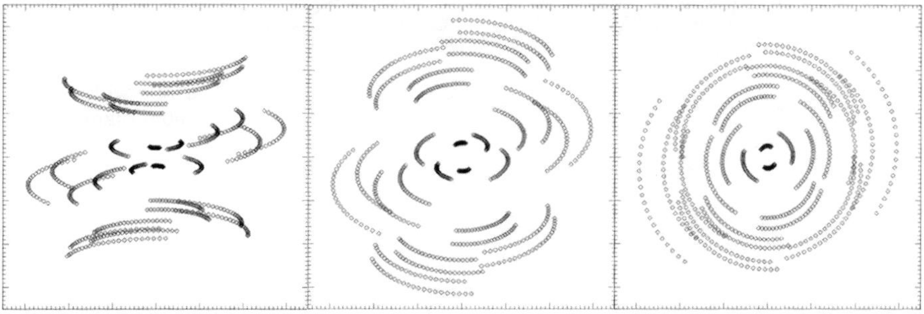

Fig. 5.23 Coverage of the uv-plane of the CHARA array when observing over 6 h for declinations $-15°$ (*left*), $+30°$ (*middle*) and $+75°$ (*right*), displaying the very different characteristics of the uv-plane coverage (Courtesy T.A. ten Brummelaar [229])

at a maximum baseline of 330 m the array had to be compatible with the existing infrastructure keeping also ecological aspects in mind. Thus, the final layout was a compromise between all these aspects.

Six telescopes are combined non-redundantly, with physical baselines varying between 34 and 330 m, resulting in a uv-plane coverage that is rather sparse (see Fig. 5.22) since the emphasis is put on long baselines – and, thus, high angular resolution – instead of on a densely filled array. However, observing over several hours, the uv-plane is covered much better displaying remarkably different distributions for different sky coordinates, as shown in Fig. 5.23.

The *Y* layout is simple and efficient but once the decision on telescope position is taken it is difficult to modify the design. Another approach was pursued at ESO's VLTI, see Fig. 5.24, by building an array of 30 stations for movable 1.8-m telescopes (AT) with a maximum baseline of 200 m, and an array of four 8-m telescopes (UT) with a maximum baseline of 130 m [82]. Placing the observatory in the middle of the Atacama desert on top of Cerro Paranal, there were no particular constraints for the layout of the array.

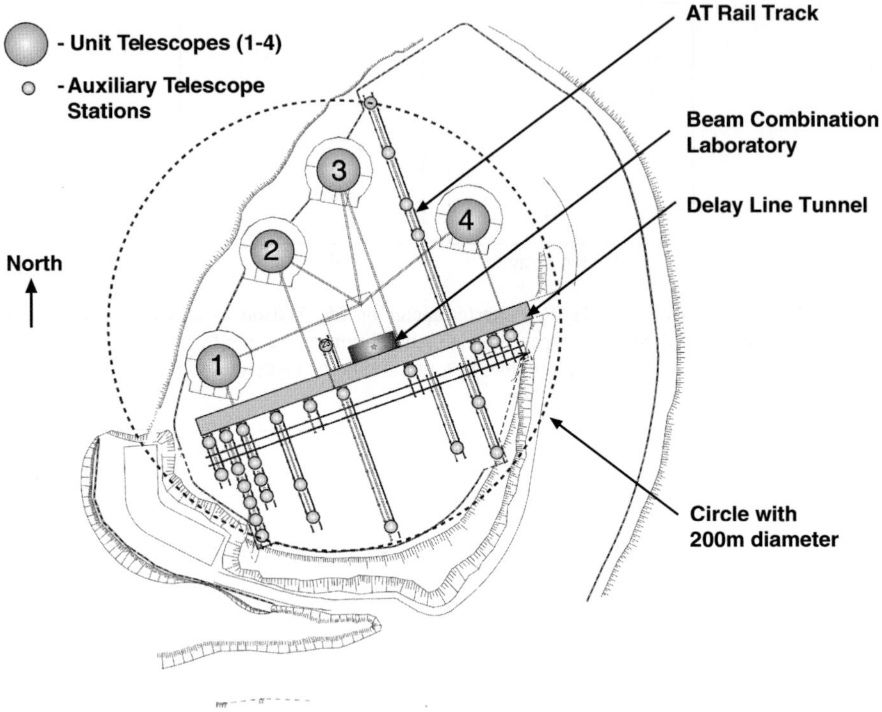

Fig. 5.24 The layout of the VLTI on Cerro Paranal. The four 8-m Unit Telescopes (UT) and the stations for the 1.8-m Auxiliary Telescopes (AT) are displayed. The AT stations are connected by rail tracks on which the AT can be relocated. Also shown are the delay line tunnel and the beam combination laboratory. The delay line tunnel has room for eight delay lines allowing the operation of eight AT and a total of 28 baselines. To date, four AT are in operation. The longest baseline with two AT is 200 m (indicated by *the dotted circle* with 200 m diameter), and the longest baseline with two UT is 130 m

The AT can be moved on rails to individual telescope stations. They can be placed on any telescope station – with no more than one per rail track – and a maximum number of eight AT could in principle be combined simultaneously. The number is limited by the number of delay lines that can be accommodated in the delay line tunnel. This leaves almost complete freedom to adopting the array, and the uv-plane coverage, to the observing situation. To date, six delay lines and four AT are in operation, and the largest number of beams that can be combined in an instrument is three with AMBER. Thus, there is a lot of room to fully exploit this facility with future instrumentation, keeping in mind that the advantage of combining four (and soon six) beams has been demonstrated impressively by CHARA [114].

The 8-m telescopes (UT) are placed on a fixed, rather dense pattern with a maximum baseline of 130 m. In Fig. 5.25 the uv-plane coverage for eight hours of observations is displayed. Due to the large diameter of the individual UT, and their rather wide OTF, the uv-plane seems to be filled rather densely. However, we must

Fig. 5.25 Computed uv-plane coverage of the VLTI combining the four 8-m Unit Telescopes observing an object at $-15°$ declination over eight hours. In Fig. 3.24 a similar result is given for different parameters

not forget that all beam combining instruments are operating in Michelson configuration so that the visibility function inside the OTF is represented by a single value. This is the equivalent of limiting the field of view to the Airy disk of an individual telescope. This was discussed in Sect. 3.4.2 with respect to observing a binary star.

Although the infrastructure of the VLTI permits the installation of a beam combiner for direct imaging in Fizeau configuration there are no plans to implement this mode in the foreseeable future, and the LBT will be the only interferometer operating in Fizeau configuration.

In Sect. 3.4.2 we discussed the impact of Earth rotation synthesis on the image reconstruction process in general. Instantaneous combination in direct imaging provides a PSF that is positive by definition, and image processing methods like those for individual telescopes are suitable. In Michelson configuration, we combine values in the uv-plane that were accumulated by the rotation of the Earth so that the PSF, the dirty beam, is the Fourier transform of the synthesized uv-plane distribution. Then the PSF has different characteristics, and image reconstruction methods as presented in Sect. 3.4.3 can be applied, processing the visibility values in the uv-plane instead of the image intensity.

5.4.3 Delay Lines

So far, we have ignored the fact that with telescopes based on the ground, we have to compensate for an additional optical path when observing objects that are not in the zenith. This optical path can be almost as long as the baseline when observing objects close to the horizon, and its compensation requires a substantial investment in terms of infrastructure and opto-mechanical systems. Figure 5.26 illustrates the situation and shows how the introduction of a delay line compensates for the additional path. Remember that interferometers like the LBT with telescopes on a common mount do not require a delay line since their baseline is always perpendicular to the pointing direction.

The position of the object in the sky is given by the unit vector $\boldsymbol{s}$, defining the line of sight when pointing at the object. In Sect. 3.4, we introduced sky coordinates, the equatorial reference frame, and we defined a local coordinate system $\boldsymbol{\alpha}' = (\Delta RA, \Delta\delta)$ with offset coordinates for right ascension, RA, and declination, δ, centred at the object and orthogonal on $\boldsymbol{s}$.

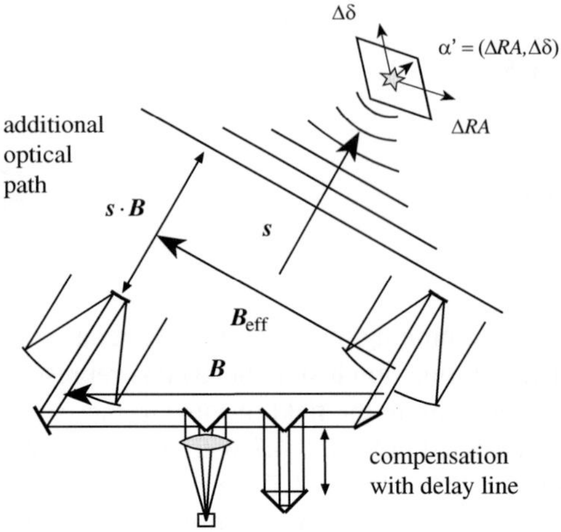

Fig. 5.26 Observing an object at position $\boldsymbol{s}$ with two telescopes on the ground with baseline $\boldsymbol{B}$. The additional optical path, $\boldsymbol{s} \cdot \boldsymbol{B}$, is compensated for by a delay line in the light path of the telescope on the right. Due to the Earth's rotation the position vector $\boldsymbol{s}$ moves with time, and the delay line has to move, too, with up to 10 mm/s for baseline around 130 m

Given that the object position $\boldsymbol{s}$ moves with time due to the Earth's rotation, the additional optical path, $\boldsymbol{s} \cdot \boldsymbol{B}$, is also time dependent so that the delay line has to track the varying path difference, like the telescopes have to track the object by pointing at $\boldsymbol{s}(t)$. Ideally, the delay line perfectly compensates for the additional optical path so that the optical path difference, the OPD, is always zero, and the white-light fringe of a point source is at $\boldsymbol{\alpha} = 0$ in the image plane. This is the case that we have assumed so far when discussing the properties of beam combination in the preceding sections.

Since tracking with the delay lines is never perfect, there will always be an uncompensated OPD affecting the fringe position. This has to be compared to the OPD variations due to atmospheric turbulence that were discussed in Sect. 4.3.1. A straightforward requirement for the performance of the delay line tracking is that the residual OPD should add less than 10% to the atmospheric OPD variance. Using (4.55), we found that under typical atmospheric conditions the rms OPD variations are about 640 nm over 100 ms. Under the same conditions, the atmospheric OPD variations are 360 nm over 50 ms, and 95 nm over 10 ms. Thus, the residual OPD should be smaller than 30 nm over 10 ms, 110 nm over 50 ms and 200 nm over 100 ms.

This gives an idea of the range of performance values that are required for the residual OPD if the delay line shall not add noticeably to the OPD variations of the atmosphere. To assess the maximum rate of the OPD variations, we assume an east-west baseline of 100 m and an object at zenith, and we find that the OPD moves at a maximum rate of $\frac{\partial \boldsymbol{s}}{\partial t} \cdot \boldsymbol{B} = 7.3$ mm/s close to zenith.

Thus, the requirement for the delay line is to move with a speed of $\frac{7.3}{2}$ mm/s and to maintain this motion with an accuracy better than $\frac{110}{2}$ nm over 50 ms. The division by 2 has to be applied to convert the optical path length into mechanical position, as illustrated by Fig. 5.26.

Keeping in mind that the delay line can be almost as long as the baseline, it is clear that these are challenging requirements for a moving optical system [103]. Usually, optical cat's eye retroreflectors on rails are employed and the position of the retroreflectors is controlled by laser metrology [103, 213]. Since this kind of high-technology control equipment was unavailable in the 1920s, stellar interferometry had been abandoned until the 1970s when modern interferometry started [121,215]. The noticeable exception is intensity interferometry when the requirement on OPD accuracy is in the many centimetre range, so that the first successful tests were performed in the early 1960s, as discussed in Sect. 2.5.

Layout of Interferometer Arrays: Summary

The design of interferometer arrays is driven by the number of telescopes and by the preferences for the coverage of the uv-plane. Aiming at high angular resolution with few telescopes inevitably ends up with a sparsely covered uv-plane. And a densely covered uv-plane with a good instantaneous image quality in direct imaging has only limited angular resolution unless a large number of telescopes are available.

In radio interferometry, when the combination of a large number of apertures does not pose particular problems, a lot of research went into investigating how to distribute the apertures for different criteria of uv-plane coverage [104]. Aiming at filling the uv-plane homogeneously avoiding redundant baselines, one can end up with regular periodic grid in the uv-plane [86] and, subsequently, with a periodic grid of object replica in the image plane (see Fig. 5.18).

Another result of the optimisation towards a homogeneous uv-plane was the distribution of the apertures on circles [44] or on the perimeter of the Reuleaux triangle [113]. While this avoided the periodic replica in the image plane the regular distribution of apertures on these geometric curves still produced a less prominent but regular distribution of sidelobes (see Fig. 5.19) disturbing the image reconstruction. Perturbing the regularity in the uv-plane provided an interferometric PSF that has a rather smooth distribution of sidelobes without particular peaks.

Distributing a fairly large number of apertures on a circle, the PSF is similar to an Airy disk of an annular aperture. This PSF has a first diffraction ring that raises up to 15% of the peak intensity (see Fig. 2.8).

Therefore, the optimal solution was a uv-plane coverage that has a higher density of points at short baselines so that the overall shape is more like that of the OTF of a single aperture, resulting in a PSF with a conveniently low first diffraction ring (see Fig. 5.20).

A particular feature of all these designs is an area around the narrow central core that is free of sidelobes and has particularly low background. This area is sometimes called the clean field of view [127] and has a radius of $\lambda / B_{\rm min}$ with $B_{\rm min}$ the shortest baseline in the array. While objects that are confined to the clean field of view would be imaged with very good quality, the permitted field of view is larger although the image quality of large objects would go down due to the peculiar form of the PSF.

Populating an interferometer array with few (less than 10) apertures, the uv-plane coverage and hence the instantaneous PSF is rarely good enough for direct imaging, and aperture synthesis has to be applied. This implies Michelson beam combiners to measure the visibility of each baseline directly. The rotation of the Earth is exploited to fill the uv-plane so that image reconstruction methods as in Sect. 3.4.3 have to be applied.

Amongst the designs of interferometers with few apertures, the Y layout is very popular since – with all telescopes placed on a three-arm structure like a Y – the required facilities transporting the light in pipes or tunnels can be shared in each of the three arms. The CHARA array is an example of an interferometer with six apertures in Y layout. In Fig. 5.23 the uv-plane coverage due to Earth rotation synthesis is shown. The VLTI is an example of an interferometer with less than 10 telescopes abandoning the Y layout. Here, movable 1.8-m telescopes can be placed on 30 stations within a circle of 200 m, with almost unlimited freedom to cover the uv-plane in the preferred distribution. Four 8-m telescopes on a fixed pattern are also available and their uv-plane coverage is displayed in Fig. 5.25.

Interferometers with telescopes on the ground require delay lines to compensate for the additional optical path when observing objects off zenith. Figure 5.26 illustrates the principle. Due to their required lengths – almost as long as the baseline when observing close to the horizon – and due to the required performance for tracking the OPD due to Earth rotation – typically about 50 nm accuracy over 50 ms while tracking at about 4 mm/s – this is high-technology equipment that requires a substantial effort when building interferometers.

Chapter 6
Observing Through Atmospheric Turbulence

Astronomical observations from the ground are limited in sensitivity – not so much in performance – by atmospheric turbulence. Removing the sensitivity limit by observing a reference star close to the object of interest, to freeze the turbulence and to permit long integration times, one is then restricted to a small field of less than 1 arcmin around the reference star.

The statistics of atmospheric turbulence was discussed in great detail in Chap. 4. In this chapter, we will apply the results to the analysis of the performance of interferometers when observing through turbulence. We will first describe the measurement of visibility and phase, investigating accuracy and sensitivity limits in the uncorrected case, and then, in Sect. 6.2, we will discuss the improvements when using a reference star for fringe tracking.

The measurement of the phase of the visibility function suffers in particular from atmospheric turbulence since its reference point, the zero OPD position, cannot be determined without a point-like reference star. Simultaneously observing object and reference star, with a dual-feed system, we can measure the phase of the visibility function of individual baselines, as described in Sect. 6.2.2. Adding the phases of three baselines, from three telescopes forming a closed loop, the sum of the phases that is called the closure phase, is not affected by turbulence, see Sect. 6.2.3. The measurement of the closure phase does not require a dual-feed system but, akin to speckle interferometry, exposure times have to be sufficiently short to freeze the turbulence.

We will restrict the discussion to the Michelson configuration with a field of view much smaller than the Airy disk. This is not only the most common interferometer configuration, but it is also straightforward to extend the main conclusions to any other interferometer configuration.

Interferometers with large telescopes require adaptive optics on each telescope to increase the sensitivity and to improve the measurement accuracy. The formalism is very similar to fringe tracking, and we will discuss the basic principles in the last section.

A. Glindemann, *Principles of Stellar Interferometry*, Astronomy and Astrophysics Library, DOI 10.1007/978-3-642-15028-9_6,

6.1 Visibility Measurement Through Atmospheric Turbulence

Our starting point is the intensity distribution of the fringe pattern formed by two telescopes in quasi-monochromatic approximation, as given by (5.14) for multi-axial combination in Michelson configuration. This means mainly the restriction to small objects with wavelength independent shape.

Observing through turbulence, the fringes are in permanent motion due to random OPD variations between the apertures as discussed in Sect. 4.3.1. The fringe motion is considered by the random phase difference $\Delta\varphi$ at the mean frequency ν_0, that varies with time, writing the instantaneous fringe pattern as

$$I(\boldsymbol{\alpha}) = 2I_0\,\mathrm{PSF}_a(\boldsymbol{\alpha}) \times \Big(1 + g_B(\boldsymbol{\alpha} - \boldsymbol{\alpha}_\varphi)|\mu(\boldsymbol{R}_B)|\cos\big(\phi(\boldsymbol{R}_B) - 2\pi\boldsymbol{R}_B\cdot\boldsymbol{\alpha} + \Delta\varphi\big)\Big), \tag{6.1}$$

when g_B, the Fourier transform of the spectrum G_B, is the envelope function of the fringe package, centred at $\boldsymbol{\alpha}_\varphi$ so that $\boldsymbol{\alpha}_\varphi \cdot \boldsymbol{R}_B = \Delta\varphi/(2\pi)$.

Compared to (5.14), the notation is simplified for clarity, writing $\boldsymbol{R}_B$ instead of $\boldsymbol{R}'_{0,B'}$, but keeping in mind that in quasi-monochromatic approximation the visibility μ is taken at $\boldsymbol{R}'_{0,B'}$, the effective baseline at the mean frequency ν_0 in the entrance pupil.

The reduction of fringe contrast $\Delta\mathcal{V}/\mathcal{V}$ due to turbulence is derived by regarding the propagation of a monochromatic plane wave from a point-like star at zenith. Assuming that the incoming plane wave has the amplitude $V_0 = 1$ and that the interferometer has two apertures $a(\boldsymbol{\xi})$ separated by $\boldsymbol{B}$, we write the turbulence induced random amplitude in the aperture plane as $V_\varphi(\boldsymbol{\xi}) = \exp(\mathrm{i}\varphi(\boldsymbol{\xi}))$, with $\varphi(\boldsymbol{\xi})$ denoting the random phase.

The time averaged intensity of the monochromatic fringe pattern, the interferometer PSF, then reads as

$$\begin{aligned}
\mathrm{PSF}_\varphi(\boldsymbol{\alpha}) &= < \frac{1}{(\lambda F)^2}\left|\int V_\varphi(\boldsymbol{\xi})\big(a(\boldsymbol{\xi} - \boldsymbol{B}/2) + a(\boldsymbol{\xi} + \boldsymbol{B}/2)\big)\mathrm{e}^{-\mathrm{i}\frac{2\pi}{\lambda}\boldsymbol{\xi}\cdot\boldsymbol{\alpha}}\mathrm{d}\boldsymbol{\xi}\right|^2 > \\
&= 2\mathrm{PSF}_a(\boldsymbol{\alpha})\Big(1 + < V_\varphi(\boldsymbol{B}/2)V_\varphi^*(-\boldsymbol{B}/2) > \cos\big(2\pi\frac{\boldsymbol{B}}{\lambda}\cdot\boldsymbol{\alpha}\big)\Big) \\
&= 2\mathrm{PSF}_a(\boldsymbol{\alpha})\Big(1 + \Gamma_\varphi(\boldsymbol{B})\cos\big(2\pi\frac{\boldsymbol{B}}{\lambda}\cdot\boldsymbol{\alpha}\big)\Big),
\end{aligned} \tag{6.2}$$

with $\frac{\boldsymbol{B}}{\lambda} = \boldsymbol{R}_B$.

As in Sect. 4.3.5, discussing the speckle pattern, the time averaged PSF is denoted by the subscript φ and the averaging process is applied to the product of amplitudes V_φ only, since the aperture function $a(\boldsymbol{\xi})$ is time-invariant. Using (4.26), we write the correlation of the amplitude[1] as $\Gamma_\varphi(\boldsymbol{B}) = \exp(-\frac{1}{2}D_\varphi(\boldsymbol{B}))$, with D_φ the phase

[1] Note that the time average is replaced by the ensemble average since the random process is ergodic as discussed in Sect. 4.2.

structure function (4.25). Thus, the contrast of the fringe pattern of the PSF is determined by D_φ. The phase of the averaged fringe package remains unaffected by turbulence since the phase difference $\Delta\varphi$ has zero mean, so that $\Gamma_\varphi(\boldsymbol{B})$ is a real function.

With (4.45), the loss of contrast can also be expressed through the OPD variance, σ^2_{OPD}, yielding

$$\frac{\Delta\mathcal{V}}{\mathcal{V}} = \Gamma_\varphi(\boldsymbol{B}) = \mathrm{e}^{-\frac{1}{2}D_\varphi(\boldsymbol{B})} = \mathrm{e}^{-\frac{1}{2}\left(\frac{2\pi}{\lambda}\right)^2\sigma^2_{\mathrm{OPD}}}. \tag{6.3}$$

Computing σ^2_{OPD} with (4.45), we find rms fluctuations of the order of 5–100 μm, for baselines longer than about 10 m, depending on r_0 and on the outer scale of turbulence (see Fig. 4.4). This reduces the contrast in very long integration times to less than 10^{-3} at observing wavelengths up to 10 μm. That means practically a complete loss of contrast for long time exposures.

For exposure times up to a few hundred milliseconds we use $\sigma^2_{\mathrm{OPD},T}$, (4.55), that was derived in Sect. 4.3.1, writing the average reduction of contrast as

$$\frac{\Delta\mathcal{V}}{\mathcal{V}}_T = \mathrm{e}^{-\frac{1}{2}\left(\frac{2\pi}{\lambda}\right)^2\sigma^2_{\mathrm{OPD},T}} = \mathrm{e}^{-0.71\left(\frac{\bar{v}}{r_0}\right)^{5/3}T^{5/3}}. \tag{6.4}$$

Note that for short exposure times, the contrast loss is approximately independent of baseline.

Taking a sequence of short time exposures, one finds that each exposure shows a different fringe contrast $\mathcal{V}$ – on average reduced by $\Delta\mathcal{V}/\mathcal{V}_T$ – and a different average fringe position $\overline{\Delta\varphi}$.

If on average a contrast loss of 10% is acceptable, i.e. $\Delta\mathcal{V}/\mathcal{V} = 0.9$, we find that the integration time T should be shorter than $0.31 r_0/\bar{v}$ – typically 20 ms in the K-band – which is approximately the atmospheric coherence time τ_0 (4.39) that was defined in a different context in Sect. 4.2.3, discussing the rms fluctuations of the phase in the turbulent layer.

Moving from a point source at zenith to extended objects, we have to convolve $\mathrm{PSF}_\varphi(\boldsymbol{\alpha})$ with the object brightness distribution $O_b(\boldsymbol{\alpha})$ as discussed in Sects. 3.3.2 and 4.4, yielding the time average of the fringe pattern in (6.1) as

$$\begin{aligned} < I(\boldsymbol{\alpha}) > &= O_b(\boldsymbol{\alpha}) * \mathrm{PSF}_\varphi(\boldsymbol{\alpha}) \\ &= 2I_0\,\mathrm{PSF}_a(\boldsymbol{\alpha}) \\ &\quad \times \Big(1 + g_B(\boldsymbol{\alpha})\Gamma_\varphi(\boldsymbol{R}_B\lambda)|\mu(\boldsymbol{R}_B)|\cos\big(\phi(\boldsymbol{R}_B) - 2\pi\boldsymbol{R}\cdot\boldsymbol{\alpha}\big)\Big), \end{aligned} \tag{6.5}$$

when the envelope function $g_B(\boldsymbol{\alpha})$ is centred like the fringe pattern, since the long time average of $\Delta\varphi$ is zero.

Thus, the contrast of the fringe pattern is given by the product of object visibility $|\mu|$ and Γ_φ when the latter can be expressed by σ^2_{OPD} for long integration times, as in (6.3), and for shorter integration times by $\sigma^2_{\mathrm{OPD},T}$, as in (6.4).

Observing through turbulence, short time exposures display speckle patterns in each telescope, which replace the point-spread function PSF_a of the individual apertures in (6.5), as discussed in Sect. 4.3.5. The resulting fringe pattern looks like a speckle pattern with fringes as displayed in Fig. 4.11. The average reduction in fringe contrast would still be given by Γ_φ but it is difficult to measure the visibility with an accuracy better than 10% [49, 164].

Discussing the fringe pattern in quasi-monochromatic approximation, the field of view is limited to less than an Airy disk, so that **spatial filtering** can be applied without any further loss in field of view (see Sect. 3.3.6). Normally, monomode optical fibres are used that do not only filter the intensity but can also serve as waveguides as discussed in Sect. 5.1.3.

Injecting the Airy disk into the monomode fibre, only the fundamental mode is transmitted so that the beam at the exit of the fibre has a wave front without aberrations and usually a Gaussian shape. Injecting a speckle pattern into the fibre, the aberrations due to atmospheric turbulence transform into intensity fluctuations at the exit of the fibre since the speckle pattern is highly variable, and only if the random position of a speckle coincides with the optical fibre, intensity is injected. This does not only mean that the injection is a random event but that only a fraction of the total intensity is injected. Correcting the speckle pattern with adaptive optics, providing a much more stable Airy disk, the intensity fluctuations are largely reduced, and the injected signal is much stronger.

The interference pattern of the light exiting from two fibres is a clean fringe pattern with varying fringe position $\Delta\varphi$ due to OPD fluctuations. But now, the contrast values, $\mathcal{V}$, of a sequence of short time exposures vary more than predicted by the OPD fluctuations alone, due to the intensity fluctuations in each beam, see (2.60). It is easier, however, to determine the visibility with varying intensities after spatial filtering than with random fringes in the speckle image [49]. With spatial filtering, accuracies below 1% for the measurement of μ^2 can be achieved [111, 227].

There is an additional effect, increasing the speed of OPD fluctuations due to the wave front phase of the speckle pattern [239]. Regarding the amplitude instead of the intensity of the speckle pattern, we find that the phase of the image amplitude is not plane but that it has a random distribution, as in the aperture plane, so that each speckle has a different phase as illustrated by Fig. 6.1. These individual values fluctuate faster than the overall phase since the speckle distribution is determined by higher order Zernike modes, and each speckle is due to a random composition of high order modes so that the variances of their fluctuations add, when each individual mode has a variance at high frequencies that is comparable to low orders like piston, as discussed in Sect. 4.3.3. Injecting the light of an individual speckle into the optical fibre we do not only have to account for the overall OPD variance as given by (6.3) but also for the additional random phase variations in the individual speckles.

This effect can reduce the exposure time for a given contrast loss up to a factor of 5 unless adaptive optics is used. Only when employing high order adaptive optics with a Strehl ratio above 30% can this effect be neglected [239].

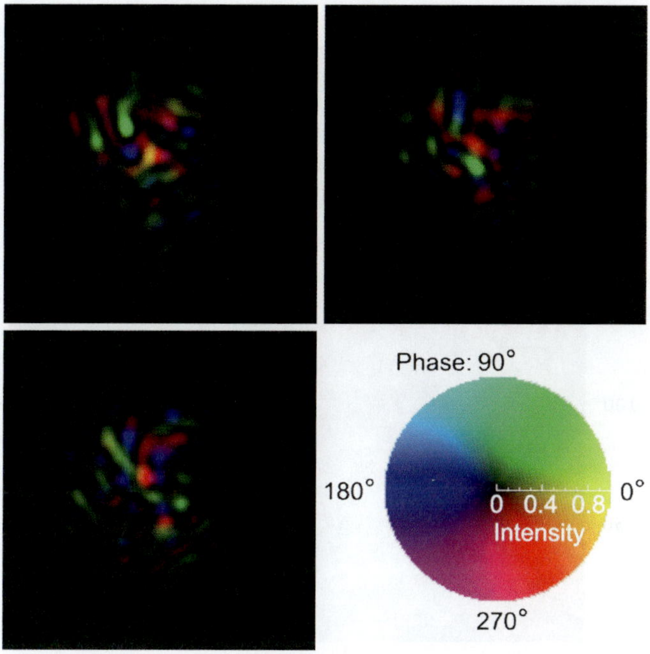

Fig. 6.1 Phase of the light amplitude in monochromatic speckle images. The *circle* in the *lower right* defines the colour scheme with respect to the phase in degrees. The three speckle images show that the phase in a speckle pattern is completely random. (Courtesy R. Tubbs)

Thus, if the telescopes are so large that a speckle pattern develops, adaptive optics is mandatory not only to improve the injected intensity but also to restrict the fringe motion to the level of small telescopes without speckles. We will discuss the basic concept of adaptive optics in Sect. 6.3.

Co-Axial Combination

In co-axial combination, the fringe pattern is a function of time due totemporal OPD modulation, as discussed in Sect. 5.1.3. Then, atmospheric OPD fluctuations do not only cause a loss of contrast as in multi-axial combination but, as we shall see, affect also the fringe spacing.

Using (5.18), we write the instantaneous fringe pattern in co-axial combination as

$$I_{\text{mod}}(t) = 2I_0'\Big(1 + g_p\big(\psi(t) - \Delta\varphi\big)|\mu(\boldsymbol{R_B})|\cos\big(\phi(\boldsymbol{R_B}) + \Delta\varphi - \psi(t)\big)\Big), \quad (6.6)$$

with $I_0' = I_0 \int \text{PSF}_a(\boldsymbol{\alpha})\text{d}\boldsymbol{\alpha}$ and $\Delta\varphi$ the random phase difference. The temporal modulation of the OPD inside the instrument is expressed as a phase modulation $\psi(t)$ at the average wavelength λ_0 through $\psi(t) = \frac{2\pi}{\lambda_0}\text{OPD}(t)$. The envelope

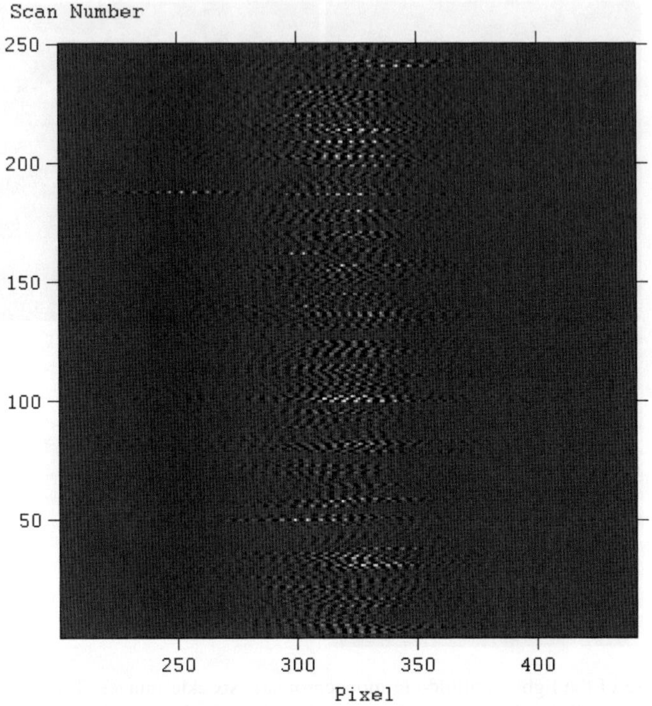

Fig. 6.2 Fringe patterns of Achernar observed with the instrument VINCI at the VLTI in co-axial combination. Each *horizontal line* shows a K-band fringe pattern displaying about 11 fringes as a succession of maxima and minima. The total scan length is about five times as long as the fringe package. The average position, $\overline{\Delta\varphi}_j$, of the white-light fringe is different for each scan j, and the fluctuations of fringe positions are much smaller than the width of the fringe package due to coherencing, i.e. slow fringe tracking. Varying intensities of the scans are due to fluctuations of the injected light into the fibre

function of the fringe package $g_p(\psi(t) - \Delta\varphi)$, centred at $\Delta\varphi$, is written as a function of phase modulation, using the relation $g_p(\psi(t)) = g(\psi(t)/(2\pi\nu)) = g(\tau)$ so that the time delay τ between the interfering beams is replaced by the dependence on the modulated phase.

If the phase is modulated with an OPD amplitude larger than the coherence length l_c then one scan displays the complete fringe package, as shown in Fig. 6.2. Due to the limited integration time, the average fringe position of each fringe package has varying values $\overline{\Delta\varphi}_j$ with j indicating the numbering of the scans.

The average phase of the fringe pattern, which is the position of the white-light fringe with respect to zero OPD, $\psi(t) = 0$, is denoted by

$$\phi'_j = \phi(\boldsymbol{R}_B) + \overline{\Delta\varphi}_j, \tag{6.7}$$

so that, in the absence of atmospheric turbulence, the phase of the object visibility determines the position of the white-light fringe.

Usually the phase modulation is slower than the fringe motion given by the temporal variation of $\Delta\varphi$ and the fringes will move during the integration. One can separate different regimes of the fringe motion: if the phase is practically stable during one scan, with constant value $\Delta\varphi_j$ for scan j, then the phase fluctuations result in different fringe positions for a succession of scans. This is the obvious effect in Fig. 6.2. If the fringe motion is faster, moving steadily during one scan, the resulting fringe package is compressed (or decompressed) displaying a smaller (or wider) fringe spacing. This is equivalent to a fringe pattern of a smaller wavelength so that the Fourier transform of the signal shows the spectrum at a different frequency, and the spectrum is slightly wider if the fringe package is compressed. And if the fringe motion is accelerated during the scan, we even find different fringe spacings in the fringe package, deforming the spectrum [111, 181].

The impact on the fringe contrast can also be classified in these categories: If the fringes are basically frozen during the integration, there is no contrast loss in each individual scan. Linearly or non-linearly moving fringes, however, cause a loss of contrast according to the OPD variance during that time span T when ψ moves by about $\pi/2$ corresponding to moving through one maximum (or minimum) of the fringe.

Observing through turbulence, all these effects mix and affect the measurement of both modulus and phase of the visibility. We write the intensity distribution of the fringe pattern of scan j as

$$I_{\mathrm{mod},j}(t) = I_{1,j}(t) + I_{2,j}(t) + 2\sqrt{I_{1,j}(t)I_{2,j}(t)} \\ \times g_p\big(\psi(t) - \overline{\Delta\varphi_j}\big)\frac{\Delta\mathcal{V}}{\mathcal{V}_{T,j}}|\mu(\boldsymbol{R_B})|\cos\big(\phi(\boldsymbol{R_B}) + \overline{\Delta\varphi_j} - \psi(t)\big), \tag{6.8}$$

when $I_{1,j}$, $I_{2,j}$ are the intensities in each beam (compare to 2.60), and $\Delta\mathcal{V}/\mathcal{V}_{T,j}$ is the contrast reduction during scan j, (6.4), with T the time it takes to scan a distance $\lambda/4$, i.e. one maximum of the fringe pattern. The average position of the white-light fringe is $\phi'_j = \phi(\boldsymbol{R_B}) + \overline{\Delta\varphi_j}$. The fluctuation of the latter usually is a much more prominent effect than the contrast loss as illustrated by Fig. 6.2.

The succession of scans displayed in Fig. 6.2 were taken with the instrument VINCI at the VLTI observing Achernar [111]. The fluctuations of the fringe positions are smaller than the width of the fringe package since after each scan VINCI provides a feed-back of the fringe positions to the delay lines. This slow fringe tracking is also called *coherencing* since the position of the fringe package varies less than the coherence length, which is given by the width of the fringe package. We will come back to fringe tracking in Sect. 6.2.1.

The Need of a Reference Star

Taking a large number of measurements and computing the average of all fringe positions ϕ'_j and of the fringe contrast, it is still impossible to derive absolute values for modulus $|\mu(\boldsymbol{R_B})|$ and phase $\phi(\boldsymbol{R_B})$ of the object visibility function with any useful accuracy. First, the absolute position of zero OPD cannot be identified so that each ϕ'_j is determined with respect to an assumed zero OPD. Thus, the fringe positions are not only affected by phase fluctuations $\overline{\Delta\varphi}_j$ due to atmospheric turbulence – which should average to zero over a long time – but also by a random bias. Dispersing the light into different spectral channels one can only measure the *differential phase* so that despite the lack of a zero OPD reference point, the variation of the phase with wavelength can be determined [184].

Second, there are too many error sources, for instance variation of turbulence parameters, tracking accuracy of delay lines and telescopes, opto-mechanical stability etc, that make it practically impossible to compute the average contrast reduction $\Delta\mathcal{V}/\mathcal{V}_T$ based on (6.4) and then derive $|\mu(\boldsymbol{R_B})|$ from the fringe contrast [24].

Therefore, one has to rely on unresolved reference stars that have a visibility modulus of 1 and a phase of 0. Measuring the fringe contrast of the reference star, any reduction of fringe contrast to values below 1 is only due to turbulence and to instrumental effects so that we can determine the overall contrast loss.

Observing the object of interest – the science object – in quick succession to the observation of the reference star, the statistical parameters of the atmosphere and the instrumental parameters normally vary so little that the visibility of the science object on average suffers the same loss of fringe contrast as the reference star. Since it is only required that the average values are stable, there is no particular requirement for the reference star to be very close to the object. This procedure permits to determine the modulus $|\mu(\boldsymbol{R_B})|$ of the visibility function of the science object, as will be discussed in Sect. 6.1.1.

Regarding the phase $\phi(\boldsymbol{R_B})$ of the visibility function, it is not enough to compare the averages of ϕ'_j of successive observations of science object and reference star since the opto-mechanical stability of the interferometer is not sufficient to maintain the absolute position of zero OPD with adequate accuracy. It is necessary to measure object and reference simultaneously requiring dedicated additional hardware in the interferometer, usually referred to as a *dual-feed system*, so that the position of zero OPD can be determined.

For the simultaneous measurement of the difference of fringe positions, the differential OPD, the requirement is not on the statistical parameters of turbulence (like the average), but on the correlation of the instantaneous random values of $\Delta\varphi$ so that the fringes on both objects move as synchronously as possible. This limits the permitted distance between object and reference star to some 10 arcsec as discussed in Sect. 4.3.1 The rms fluctuations of the differential fringe motion is given in (4.60) as a function of angular distance $\Delta\boldsymbol{s}$ between object and reference, so that the isoplanatic angle as the maximum separation for a sufficient correlation can be quantified. This will be discussed in Sect. 6.2.2.

Having discussed the necessity of a reference star for measurements of the visibility function, we have to note that combining three telescopes and adding the phases of the three fringe patterns, the three phase differences $\Delta\varphi_i$, $i = 1, 2, 3$, cancel and we obtain the closure phase as the sum of three phases $\phi(\boldsymbol{R_{B_i}})$ of the visibility function. Formally identical, this technique is applied in speckle interferometry, see Sect. 4.4, and in intensity interferometry, see Sect. 2.5. We will discuss it for amplitude interferometry in Sect. 6.2.3.

6.1.1 Power Spectrum of the Fringe Pattern

The general principle of the visibility measurement in the Fourier spectrum was introduced in Sect. 2.4.3. Here, we will first discuss the Fourier transform of the multi-axial fringe pattern (6.1), and then we will look in more detail at the fringes in co-axial combination.

The image intensity spectrum of the instantaneous fringe pattern in multi-axial combination is given in (5.13) and is now extended by the influence of atmospheric turbulence as in (6.1), yielding

$$\begin{aligned}\hat{I}(\boldsymbol{R}) = {} & I_0 \mathrm{OTF}_a(\boldsymbol{R}) \\ & * \left(\mu(\boldsymbol{R}_B) G_B(\boldsymbol{R}) \mathrm{e}^{\mathrm{i}2\pi\boldsymbol{\alpha}_\varphi\cdot\boldsymbol{R}} + 2\delta(\boldsymbol{R}) + \mu^*(\boldsymbol{R}_B) G_B(-\boldsymbol{R}) \mathrm{e}^{-\mathrm{i}2\pi\boldsymbol{\alpha}_\varphi\cdot\boldsymbol{R}}\right),\end{aligned} \tag{6.9}$$

when the random phase shift $\Delta\varphi$ transforms into a linear phase[2] in the uv-plane. For $\boldsymbol{R} = \boldsymbol{R}_B = \boldsymbol{B}/\lambda_0$ where the spectrum $G_B(\boldsymbol{R})$ is centred, the phase has the value $2\pi\boldsymbol{\alpha}_\varphi \cdot \boldsymbol{R}_B = \Delta\varphi$.

The long time average that is given by (6.5) for the monochromatic fringe pattern converts to the following Fourier spectrum for polychromatic light:

$$\begin{aligned}< \hat{I}(\boldsymbol{R}) > = {} & I_0 \mathrm{OTF}_a(\boldsymbol{R}) \\ & * \left(\mu(\boldsymbol{R}_B) G_B(\boldsymbol{R}) \Gamma_\varphi(\boldsymbol{B}) + 2\delta(\boldsymbol{R}) + \mu^*(\boldsymbol{R}_B) G_B(-\boldsymbol{R}) \Gamma_\varphi(-\boldsymbol{B})\right),\end{aligned} \tag{6.10}$$

with $\boldsymbol{B} = \boldsymbol{R}_B\lambda$.

Writing the contrast reduction $\Gamma_\varphi(\boldsymbol{B}) = \Delta\mathcal{V}/\mathcal{V}$ as a function of OPD variance in (6.3), we find that the fringe contrast of the long term average practically disappears under normal atmospheric conditions. Accordingly, the peak in the spectrum around $\boldsymbol{R}_B$, the interferometric peak, is reduced to very small values so that only the central OTF remains, representing the spectrum of a PSF without fringes.

Computing the average of the power spectrum $< |\hat{I}(\boldsymbol{R})|^2 >$ instead, the interferometric peak does not disappear and we can derive the squared visibility as the ratio of the integrals over the interferometric peak and over the central OTF as given by (3.74). We will discuss this in more detail for co-axial combination in the following. This procedure can also be applied to interferograms with speckles [164].

[2] Note that the linear increase of $\Delta\varphi$ with $\boldsymbol{R} = \boldsymbol{B}/\lambda$ means that the atmospheric OPD, $\Delta\varphi\frac{\lambda}{2\pi}$, is independent of wavelength, as discussed in Sect. 4.2.

Co-Axial Combination

Processing the fringe pattern in co-axial combination as given in (6.8), including all disturbances, the first step is to remove the influence of the different intensities $I_{1,j}$, $I_{2,j}$ on the fringe pattern [49], yielding the varying part, proportional to cos(.), of the corrected interferogram as

$$\begin{aligned} I_{\mathrm{cor},j}(t) &= \frac{I_{\mathrm{mod},j}(t) - I_{1,j}(t) - I_{2,j}(t)}{2\sqrt{I_{1,j}(t) I_{2,j}(t)}} \\ &= g_p\big(\psi(t) - \overline{\Delta\varphi}_j\big) \mathrm{TF}_j(\boldsymbol{R_B}) |\mu(\boldsymbol{R_B})| \cos\big(\phi(\boldsymbol{R_B}) + \overline{\Delta\varphi}_j - \psi(t)\big), \end{aligned} \tag{6.11}$$

with $\mathrm{TF}_j(\boldsymbol{R_B})$ the *interferometric transfer function* during scan j that combines the contrast reduction $\Delta\mathcal{V}/\mathcal{V}_{T,j}$ due to turbulence, with instrumental effects reducing the contrast, for instance tracking errors of delay lines or telescopes, vibrations, polarization mismatch etc.

Writing down (6.11), this correction looks simple. The intensity measurement could be provided by means of a beam splitter, distributing the light into a photometric and into an interferometric channel. It is mandatory, however, to precisely know the splitting ratio of the beam splitter over the observed wavelength range, requiring an elaborate calibration procedure. Thus, (6.11) would have to be replaced by more complex data processing [49].

Finally, we compute the difference signal between the two outputs A and B, removing the correlated noise and all the background, and amplifying the fringe patterns, which are in anti-phase as discussed in Sect. 5.1.3, yielding

$$I_{\mathrm{rec},j}(t) = I^A_{\mathrm{cor},j}(t) - I^B_{\mathrm{cor},j}(t). \tag{6.12}$$

Figure 6.3 gives an example of a measurement with VINCI, the VLTI test camera, showing the raw fringe patterns in both exits of the beam splitter and the two photometric signals of the two telescopes. The impressive fluctuations of the photometric signals demonstrate the necessity to correct for the individual intensities. The reconstructed interferogram, the top curve in Fig. 6.3, serves as input for the Fourier spectrum that is used to compute the visibility [49, 111, 181].

Sometimes the fringe pattern in co-axial combination is expressed as a function of OPD instead of time t. Then the distance between two maxima, the fringe spacing, corresponds to an OPD of λ_0 and its Fourier transform is a function of wavenumber $1/\lambda$.

We will keep the notation as a function of t, and we compute the **Fourier spectrum** as a function of frequency ν as given by (5.19). Fourier transforming $I_{\mathrm{rec},j}(t)$ only has contributions around $\pm\nu_0$ and not around $\nu = 0$ because we removed the constant term in (6.11). Since the part of the spectrum at $-\nu_0$ is the complex conjugate of the part at positive frequencies we will only consider the latter without losing any information, writing

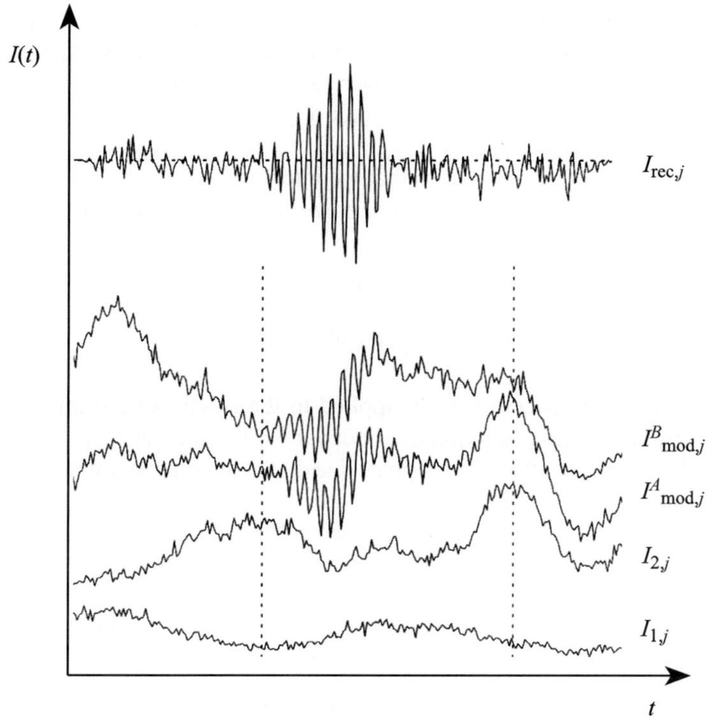

Fig. 6.3 Fringe patterns of an individual scan j in co-axial combination with VINCI, the VLTI test camera. The curves are shifted vertically for clarity only. From *bottom* to *top*, the curves show the photometric signals from the two telescopes, $I_{1,j}$ and $I_{2,j}$, the raw interferograms, $I^A_{\mathrm{mod},j}$ and $I^B_{\mathrm{mod},j}$, from the two exits, A and B, of the beam combiner (therefore in anti phase), and, at the *top*, the corrected interferogram. (Courtesy P. Kervalla [111])

$$\hat{I}_{\mathrm{rec},j}(\nu) = G(\nu)\mathrm{TF}_j(\boldsymbol{R}_B)\mu(\boldsymbol{R}_B)\mathrm{e}^{\mathrm{i}\overline{\Delta\varphi}_j \nu/\nu_0}, \tag{6.13}$$

when, observing only small objects in quasi-monochromatic approximation, the visibility function μ over the spectrum $G(\nu)$ is represented by a single value at the average frequency ν_0 with $\boldsymbol{R}_B = \frac{\boldsymbol{B}}{c}\nu_0$.

This notation of the Fourier spectrum ignores the effects of squeezing the fringes due to fringe motion during the scan that we discussed in the previous section. Considering this effect we would have to introduce varying spectra $G(\nu)$ shifted to higher or lower frequencies depending on the amount of fringe compression or decompression. In good approximation, it can be assumed that this effect adds to the noise only [49], and we maintain our notation.

The average of $< \hat{I}_{\mathrm{rec},j} >$ over many scans is zero due to the random fringe position $\overline{\Delta\varphi}_j$ so that we compute the average of the power spectrum $< |\hat{I}_{\mathrm{rec},j}|^2 >$ as in speckle interferometry, Sect. 4.4.

The measurement of the visibility function in the Fourier spectrum was discussed under general aspects without atmospheric turbulence in Sect. 2.4.3. There the visibility was derived as the ratio between the integral of the Fourier spectrum around ν_0, the interferometric peak, and the integral around $\nu = 0$, the photometric peak, based on the idea of F. Roddier and P. Léna [196]. Here, we proceed directly to the power spectrum computing the average of the squared visibility to obtain an estimator of $\mathcal{V}^2$ that is unbiased by detector or photon noise [224].

Processing the corrected interferogram in (6.13), the constant term and, thus, the photometric peak is removed so that we integrate the interferometric peak only, writing the average of the integral of the power spectrum over many scans as

$$\begin{aligned} < \mathcal{V}^2 > &= < \int \left| G(\nu)\mathrm{TF}_j(\boldsymbol{R_B})\mu(\boldsymbol{R_B})\right|^2 \mathrm{d}\nu > \\ &= < |\mathrm{TF}_j(\boldsymbol{R_B})|^2 > |\mu(\boldsymbol{R_B})|^2, \end{aligned} \quad (6.14)$$

when the averaging process is only applied to the square of the transfer function since the object visibility function does not vary with time. One has to assure, however, that the changes of the effective baseline do not alter the visibility values during the observation.

In case of noisy measurements, the noise variance has to be subtracted from each measured power spectrum to obtain the unbiased estimator for $\mathcal{V}^2$ [49].

Repeating the same procedure with a reference star with $\mu = 1$, or with a precisely known value μ_{ref}, we obtain the squared visibility of the science object as

$$|\mu_{\mathrm{ob}}(\boldsymbol{R_B})|^2 = |\mu_{\mathrm{ref}}(\boldsymbol{R_B})|^2 \frac{< \mathcal{V}^2_{\mathrm{ob}} >}{< \mathcal{V}^2_{\mathrm{ref}} >}. \quad (6.15)$$

We have to remember that this procedure of determining the transfer function with a reference star relies on the temporal stability of the transfer function during both the observations of object and reference star. Observing sequences like: reference star – object – reference star, the transfer function can be monitored, for instance by comparing the reference star observations at the beginning and at the end, improving the accuracy of the measurement.

Faint Objects

Observing faint objects, when the photon noise $\sqrt{N_{ph}}$ is no longer negligible, we then find the variance of the $\mathcal{V}^2$ measurement in one exit of the beam combiner as [35]

$$\sigma^2_{\mathcal{V}^2,ph} = \frac{8}{N^2_{ph}}(1 + N_{ph}\mathcal{V}^2), \quad (6.16)$$

with N_{ph} the number of photons contributing to the measurement, and $\mathcal{V}^2$ the measured squared visibility, with $< \mathcal{V}^2 > = < |\mathrm{TF}|^2 > |\mu_{\mathrm{ob}}|^2$.

Using both exits of the beam combiner, the signal-to-noise ratio, $\mathrm{SNR} = \mathcal{V}^2/\sigma_{\mathcal{V}^2}$, for the visibility measurement after M scans is given by [224]

$$\mathrm{SNR}(\mathcal{V}^2) = \frac{N_{ph}\mathcal{V}^2}{2}\sqrt{M}\left(1 + N_{ph}\mathcal{V}^2\right)^{-1/2}. \tag{6.17}$$

With the variance of the measurement being reduced linearly by M, the SNR improves proportional to $\sqrt{M}$.

Thus, in the limit of very faint objects, in the photon-starved regime with $N_{ph} < 1$ on average, the SNR is proportional to $N_{ph}\mathcal{V}^2$, while for brighter objects, in the photon-rich regime with $N_{ph} > 1$, the SNR goes with $\sqrt{N_{ph}}\mathcal{V}$ as discussed in Sect. 5.2.2. Note that for background limited observations, the SNR in the photon-rich regime is proportional to $N_{ph}\mathcal{V}$ [211].

Since it is the product of number of photons and visibility that determines the SNR we find that an object that is partially resolved with a visibility smaller than 1, needs to be brighter than an unresolved object in order to have the same SNR. Therefore, bright objects might not provide the best SNR if they are fairly large.

In the presence of detector read noise or background noise, σ_r, we can use the following approximation in the photon-starved regime, [38]:

$$\mathrm{SNR}(\mathcal{V}^2) = \frac{N_{ph}^2\mathcal{V}^2}{2(N_{ph} + \sigma_r^2)}\sqrt{M}, \tag{6.18}$$

so that, as expected, in the case of very few photons the read noise is more dominant, while we arrive at the same result as above for faint objects if the read noise is negligible.

6.1.2 ABCD Method

In analogy to Michelson's definition of visibility, the ABCD method relies on intensity measurements at four different points of an individual fringe to determine modulus and phase of the visibility function. The principle was described in Sect. 2.4.2. Here, we will extend the discussion to co-axial combination including atmospheric turbulence.

Discussing the power spectrum in the preceding section, the scan of the fringe pattern was much longer than the coherence length so that the complete fringe package was scanned. Here, it is required to scan an individual fringe only, so that the OPD modulation is reduced to an amplitude of λ. Assuming a sawtooth modulation of the path length with amplitude λ, exactly one fringe is scanned and $I_{\mathrm{mod},j}(t)$, (6.8), describes a complete cosine fringe during the scan interval T as displayed in Fig. 6.4.

The intensity is measured in intervals of $T/4$ as displayed in Fig. 6.4, and the maximum exposure time per measurement is $T/4$ [214]. Applying (2.70), we determine the visibility modulus $\mathcal{V}$ of the fringe pattern which is the product of the modulus of the object visibility $|\mu|$ and the interferometric transfer function TF. The phase ϕ' of the fringe pattern is determined by the sum of the phase of the object visibility function ϕ and the random phase $\overline{\Delta\varphi}$, due to atmospheric turbulence, as given by (6.7).

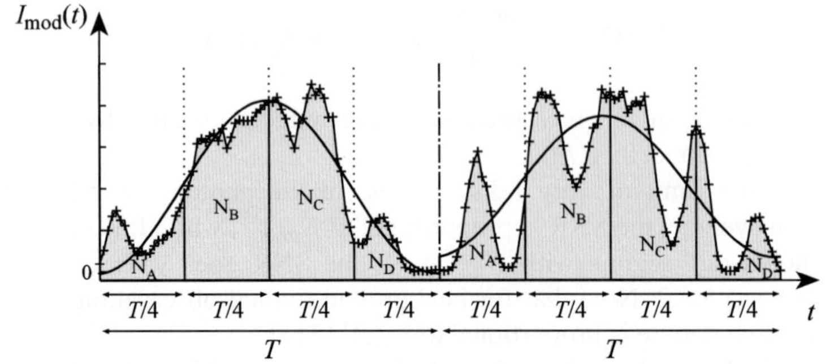

Fig. 6.4 Example of temporal modulation in co-axial combination. The phase is modulated by 2π over time T so that one complete fringe is scanned. Measuring the intensity in four bins, each integrated over $T/4$, provides the four signals that are required for the ABCD method. Individual photon counts are marked by *crosses*, and the fitted cosine fringe is sketched in. (Courtesy R. Wilhelm [253])

If we label the four intensities after integration over $T/4$ by I_A, I_B, I_C and I_D, and their sum by I_{tot} we write

$$\mathcal{V} = \frac{\pi}{\sqrt{2}} \frac{\sqrt{(I_A - I_C)^2 + (I_B - I_D)^2}}{I_{\text{tot}}}$$
$$\phi' = \tan^{-1}\left(\frac{I_A - I_C}{I_B - I_D}\right). \tag{6.19}$$

The factor $\pi/\sqrt{2}$ has to be replaced by a factor 2 if the intensity values are measured at individual points spaced by $\lambda/4$ on the fringe pattern.

Unless observing monochromatic light, the visibility of the fringe depends on its distance from the white-light fringe as discussed in Sect. 3.3.3 and as illustrated for instance by Fig. 6.2. Since the rms fluctuations of the uncorrected fringe motion are of the order of 5–100 μm, it is possible that the scanned fringe is far from the white-light fringe so that the measured visibility is much smaller than the object visibility. This contrast loss is difficult to calibrate so that it has to be assured that the white-light fringe is scanned, usually requiring fringe tracking [167].

In any case, the ABCD method requires the observation of a reference star to derive the object visibility $|\mu_{\text{ob}}|$ from the fringe visibility $\mathcal{V}_{\text{ob}}$. As for the power spectrum, processing the squared visibility $\mathcal{V}_{\text{ob}}^2$ permits to define an unbiased estimator, and the procedure given by (6.15) can be used to compute the squared object visibility $|\mu_{\text{ob}}|^2$.

Even with a reference star, the phase of the visibility ϕ_{ob} cannot simply be derived from the fringe phase ϕ', as we discussed in Sect. 6.1, unless a dual-feed system with a fringe tracker is available. The ABCD method is particularly well suited for fringe tracking since scanning only individual fringes, the phase of the fringe pattern can be measured at a much higher rate than when scanning the complete fringe package to compute the power spectrum. As we shall see in Sect. 6.2.1, this enables stabilizing the fringe motion to a fraction of a fringe.

Spatial Modulation

All issues of temporal modulation, e.g. fringe motion during a scan, that we discussed in Sect. 6.1 have to be considered here, too. These problems are avoided when doing spatial phase modulation as discussed in Sect. 5.1.3 and illustrated by Fig. 5.4. Then, the two outputs of the beam combiner are split into two orthogonal polarization components each, and the four signals display phase shifts of 0, $\pi/2$, π and $3\pi/2$. This is equivalent to measuring the intensities at four points of the fringe pattern and we can apply (6.19) replacing the factor $\pi/\sqrt{2}$ by the factor 2.

Comparing the measured intensities of both temporal and spatial modulation we assume the same time T for measuring one fringe. If we call I_0 the total intensity that is available at the entrance of the beam combiner during the time T, we find that each output in spatial modulation receives $I_0/4$. Applying temporal modulation, the intensity $I_0/2$ in each output of the beam combiner is scanned in bins of $T/4$ so that the signal in each bin is $I_0/8$. Adding the measured signals of both beam combiner outputs we have $I_0/8 + I_0/8$ per bin, thus, the same signal level as in spatial modulation, but we have to add their variances so that the SNR in temporal modulation is $\sqrt{2}$ lower than in spatial modulation.

However, this difference is negligible since there are other issues in spatial modulation that do not appear in temporal modulation, like: injection into four optical fibres vs. one fibre, non ideal phase shifts vs. precisely subdividing the scan in $T/4$, response of four pixels vs. one pixel, all requiring careful calibration [204], so that the two methods are basically equivalent.

Faint Objects

For the ABCD method, the unbiased estimator for the visibility measurement of faint objects reads as [35, 213]

$$< \mathcal{V}^2 >= \frac{\pi^2}{2} \frac{< (N'_A - N'_C)^2 + (N'_B - N'_D)^2 - \sigma_N^2 >}{< N'_{ph} >^2}, \tag{6.20}$$

with σ_N^2 the sum of the noise variances of photon and of read noise, N'_A to N'_D are the photon counts in each bin corrected for background, and N'_{ph} is the total number of photons contributing to the measurement with the background noise subtracted.

The estimator for the phase is similar to (6.19), writing

$$\phi' = \tan^{-1}\left(\frac{N'_A - N'_C}{N'_B - N'_D}\right). \tag{6.21}$$

The signa-to-noise ratio of the ABCD measurement after M scans is as in (6.17) but with a different constant due to the measurement method [35], yielding

$$\mathrm{SNR}(\mathcal{V}^2) = \frac{2N_{ph}\mathcal{V}^2}{\pi^2}\sqrt{M}\left(1 + \frac{4}{\pi^2}N_{ph}\mathcal{V}^2\right)^{-1/2}, \tag{6.22}$$

with N_{ph} the number of photons contributing to the measurement, and $\mathcal{V}^2$ the measured squared visibility, with $< \mathcal{V}^2 > = < |\mathrm{TF}|^2 > |\mu_{\mathrm{ob}}|^2$.

The SNR of visibility and phase measurement are identical [211], so that for faint objects with, $N_{ph} > 1$ we obtain the phase variance for a single measurement as the reciprocal of the SNR as [214]

$$\sigma_{ph}^2 = \frac{\pi^4}{4N_{ph}^2\mathcal{V}^4}\left(1 + \frac{4}{\pi^2}N_{ph}\mathcal{V}^2\right)\ [\mathrm{rad}^2], \tag{6.23}$$

with the limiting cases of $\pi^2/(N_{ph}\mathcal{V}^2)$ in the photon-rich regime [214], and $\pi^4/(4N_{ph}^2\mathcal{V}^4)$ in the photon-starved regime. Thus, the phase variance, like the SNR, improves proportional to the product of number of photons and visibility.

We should not forget that even when perfectly determining the fringe phase ϕ' we still have the influence of atmospheric turbulence so that we cannot not directly derive the phase ϕ_{ob} of the object visibility function. We need a fringe tracker and a dual-feed system that we will discuss in the following section.

Visibility Measurement Through Atmospheric Turbulence – Summary

We discuss the fringe pattern formed by two telescopes in quasi-monochromatic approximation, i.e. for small objects with wavelength independent shape. The same formalism can be applied when combining several telescopes.

Observing through turbulence, the instantaneous fringe pattern in multiaxial combination is given by

$$\begin{aligned} I(\boldsymbol{\alpha}) &= 2I_0\,\mathrm{PSF}_a(\boldsymbol{\alpha}) \qquad (6.1)\\ &\times \Big(1 + g_B(\boldsymbol{\alpha} - \boldsymbol{\alpha}_\varphi)|\mu(\boldsymbol{R_B})|\cos\big(\phi(\boldsymbol{R_B}) - 2\pi\boldsymbol{R_B}\cdot\boldsymbol{\alpha} + \Delta\varphi\big)\Big), \end{aligned}$$

when the fringe motion is considered by the random phase difference $\Delta\varphi$ varying with time. g_B is the envelope function of the fringe package, centred at $\boldsymbol{\alpha}_\varphi$ so that $\boldsymbol{\alpha}_\varphi \cdot \boldsymbol{R_B} = \Delta\varphi/(2\pi)$.

Then, the time averaged fringe pattern is written as

$$\begin{aligned} < I(\boldsymbol{\alpha}) > &= 2I_0\,\mathrm{PSF}_a(\boldsymbol{\alpha}) \qquad (6.5)\\ &\times \Big(1 + g_B(\boldsymbol{\alpha})\Gamma_\varphi(\boldsymbol{R_B}\lambda)|\mu(\boldsymbol{R_B})|\cos\big(\phi(\boldsymbol{R_B}) - 2\pi\boldsymbol{R}\cdot\boldsymbol{\alpha}\big)\Big). \end{aligned}$$

The envelope function $g_B(\boldsymbol{\alpha})$ is now centred like the fringe pattern, since the long time average of $\Delta\varphi$ is zero.

The contrast reduction $\Delta\mathcal{V}/\mathcal{V}$ of the fringe pattern is given Γ_φ when the latter can be expressed by $\sigma^2_{\mathrm{OPD},T}$ for exposure times up to a few 100 ms, writing the average reduction of contrast as

$$\frac{\Delta\mathcal{V}}{\mathcal{V}}_T = \mathrm{e}^{-\frac{1}{2}\left(\frac{2\pi}{\lambda}\right)^2\sigma^2_{\mathrm{OPD},T}} = \mathrm{e}^{-0.71\left(\frac{\bar{v}}{r_0}\right)^{5/3}T^{5/3}}. \tag{6.4}$$

The instantaneous fringe pattern in co-axial combination reads as

$$I_{\mathrm{mod}}(t) = 2I_0'\Big(1 + g_p\big(\psi(t) - \Delta\varphi\big)|\mu(\boldsymbol{R_B})|\cos\big(\phi(\boldsymbol{R_B}) + \Delta\varphi - \psi(t)\big)\Big), \tag{6.6}$$

with $I_0' = I_0 \int \mathrm{PSF}_a(\boldsymbol{\alpha})\mathrm{d}\boldsymbol{\alpha}$.

The temporal modulation of the OPD inside the instrument is expressed as a phase modulation $\psi(t)$ at the average wavelength λ_0 through $\psi(t) = \frac{2\pi}{\lambda_0}\mathrm{OPD}(t)$. The envelope function of the fringe package g_B is now written as a function of phase modulation, $g_p(\psi(t) - \Delta\varphi)$. Modulating the OPD by more than the coherence length, the complete fringe package is scanned.

Taking all effects into account, we write the intensity distribution of the fringe pattern of scan j as

$$I_{\mathrm{mod},j}(t) = I_{1,j}(t) + I_{2,j}(t) + 2\sqrt{I_{1,j}(t)I_{2,j}(t)} \tag{6.8}$$

$$\times\, g_p\big(\psi(t) - \overline{\Delta\varphi_j}\big)\frac{\Delta\mathcal{V}}{\mathcal{V}}_{T,j}|\mu(\boldsymbol{R_B})|\cos\big(\phi(\boldsymbol{R_B}) + \overline{\Delta\varphi_j} - \psi(t)\big),$$

when $I_{1,j}$, $I_{2,j}$ are the varying intensities in each beam. The average position of the white-light fringe is $\phi_j' = \phi(\boldsymbol{R_B}) + \overline{\Delta\varphi_j}$. The fluctuation of the latter usually is more obvious than the contrast loss as illustrated by Fig. 6.2.

Discussing the Fourier spectrum of the fringe pattern, the influence of the varying intensities $I_{1,j}$, $I_{2,j}$ and the constant term are removed from (6.8), before Fourier transforming the fringe pattern, so that only the interferometric peaks remain. We write the contribution at positive frequency as

$$\hat{I}_{\mathrm{rec},j}(\nu) = G(\nu)\mathrm{TF}_j(\boldsymbol{R_B})\mu(\boldsymbol{R_B})\mathrm{e}^{\mathrm{i}\overline{\Delta\varphi_j}\nu/\nu_0}, \tag{6.13}$$

with $\mathrm{TF}_j(\boldsymbol{R_B})$ the *interferometric transfer function* during scan j, which combines the contrast reduction $\Delta\mathcal{V}/\mathcal{V}_{T,j}$ due to turbulence, with instrumental effects.

Observing only small objects in quasi-monochromatic approximation, the visibility function μ over the spectrum $G(\nu)$ is constant, represented by a single value at the average frequency ν_0 with $\boldsymbol{R}_B = \frac{B}{c}\nu_0$.

Processing the power spectrum to obtain an estimator for $\mathcal{V}^2$ that is unbiased by noise we write the average of the integral over the interferometric peak as

$$\begin{aligned} < \mathcal{V}^2 > &=< \int \left|G(\nu)\mathrm{TF}_j(\boldsymbol{R}_B)\mu(\boldsymbol{R}_B)\right|^2 \mathrm{d}\nu > \\ &=< |\mathrm{TF}_j(\boldsymbol{R}_B)|^2 > |\mu(\boldsymbol{R}_B)|^2. \end{aligned} \tag{6.14}$$

If we repeat the same procedure with a reference star with $\mu = \mu_{\mathrm{ref}}$, the squared visibility of the science object is obtained as

$$|\mu_{\mathrm{ob}}(\boldsymbol{R}_B)|^2 = |\mu_{\mathrm{ref}}(\boldsymbol{R}_B)|^2 \frac{< \mathcal{V}_{\mathrm{ob}}^2 >}{< \mathcal{V}_{\mathrm{ref}}^2 >}. \tag{6.15}$$

The ABCD method relies on measuring the intensity of a single fringe at four points separated by $\lambda/4$. If we label the four intensities after integration over one quarter of the fringe, or over $T/4$ in temporal modulation, by I_A, I_B, I_C and I_D, and their sum by I_{tot} we write

$$\begin{aligned} \mathcal{V} &= \frac{\pi}{\sqrt{2}} \frac{\sqrt{(I_A - I_C)^2 + (I_B - I_D)^2}}{I_{\mathrm{tot}}} \\ \phi' &= \tan^{-1}\left(\frac{I_A - I_C}{I_B - I_D}\right). \end{aligned} \tag{6.19}$$

The observation of a reference star is required to derive the object visibility $|\mu_{\mathrm{ob}}|$ from the fringe visibility $\mathcal{V}_{\mathrm{ob}}$. As for the power spectrum, processing the squared visibility $\mathcal{V}_{\mathrm{ob}}^2$ permits to define an unbiased estimator, and the procedure given by (6.15) can be used to compute the squared object visibility $|\mu_{\mathrm{ob}}|^2$.

Observing faint objects, the variance of a single phase measurement with the ABCD method is

$$\sigma_{ph}^2 = \frac{\pi^4}{4N_{ph}^2\mathcal{V}^4}\left(1 + \frac{4}{\pi^2}N_{ph}\mathcal{V}^2\right) \ [\mathrm{rad}^2], \tag{6.23}$$

6.2 Beating Atmospheric Turbulence

The rapid fringe motion does not only wipe out the fringes after some hundred milliseconds, but it also makes it impossible to determine the phase ϕ_{ob} of the object visibility function, as discussed in Sect. 6.1.

These limitations can be removed by using a reference star. Then, faint objects can be observed, reducing the fringe motion by measuring and stabilizing the fringes on a nearby reference star with a fringe tracking system. The residual fringe motion of the faint object is determined by the ability of the fringe tracker to stabilize the fringe pattern on the reference star, and by the isoplanatic angle, i.e. the differential optical path fluctuations between the two objects that we discussed in Sect. 4.3.1.

If the residual fringe motion is much smaller than the fringe spacing, ensuring a good fringe contrast, the integration time on the faint science object can be increased to tens of seconds, instead of some 10 ms, so that the sensitivity is improved by upto 7 stellar magnitudes (see Appendix A.2).

Using a dual-feed system with laser metrology to monitor all internal optical paths, the white-light fringe positions of the reference star and of the science object can be identified so that the phase ϕ_{ob} of the object visibility is determined in addition to its modulus, providing an imaging mode for faint objects.

In this section, we will first investigate principles and requirements of a fringe tracking system and then we will discuss dual-feed systems with their application to phase referenced imaging and to astrometry. Finally, we will also regard the principle of closure phase as a passive method to determine the phase through turbulence.

6.2.1 Fringe Tracking

The principle of fringe tracking relies on stabilizing the fringe motion by rapidly scanning the fringe pattern and feeding the information on the white-light fringe position into a servo loop. After each scan, the OPD is adjusted with the delay line or with a dedicated actuator in the optical path, so that the fringe motion is reduced to a fringe jitter ideally much smaller than the fringe spacing.

In Sect. 6.1, we discussed fringe patterns in co-axial configuration that are displayed in Fig. 6.2. Applying a slow fringe tracking, the fringe package is centred with a residual motion smaller than the coherence length. Therefore, this is called coherencing. The scan rate that was applied was apparently too slow to reduce the residual fringe motion to the level of an individual fringe spacing.

If we want to quantify the scan rate that is required to achieve a given residual motion, we have to regard the power spectrum of OPD fluctuations given in (4.54). We used the power spectrum in Sect. 4.3.1 to determine the variance of the fringe motion during a limited observing time T, by applying a filter function and computing the variance as the integral of the product.

A fringe tracking **servo loop** can also be described by a filter function that is applied to the power spectrum of OPD fluctuations. Given that we reduce the fringe motion by tracking the fringes but that we are left with some high frequency fringe jitter, we expect that the form of the filter function is such that low frequencies are reduced, or *rejected*, considerably but that there is only little effect at high frequencies.

Control theory provides the tools to describe servo loops like the fringe tracker and we find the rejection transfer function, $\mathrm{S}(f)$, as [90]

$$\mathrm{S}(f) = \frac{(f/f_{3\mathrm{dB}})^2}{1 + (f/f_{3\mathrm{dB}})^2}, \tag{6.24}$$

with $f_{3\mathrm{dB}}$ the *3 dB servo bandwidth* which is about ten times smaller than the scan rate, depending on the parameters of the closed loop algorithm. Thus, if a servo bandwidth of 10 Hz is required, the scan rate has to be 100 Hz, scanning a fringe in $T = 1/(10 f_{3\mathrm{dB}}) = 10\,\mathrm{ms}$.

The exact shape of the transfer function depends on many parameters of the control loop and of the hardware. For the ideal case in (6.24) it is assumed for instance that the measured fringe position is transferred without delay to the actuator correcting the OPD. If there is a delay in the signal transport or a limited response time of the actuator, it is possible that there is an amplification of the disturbance, with $\mathrm{S}(f) > 1$ around $f/f_{3\mathrm{dB}} = 1$, as illustrated by the dashed curve in Fig. 6.5. The shape of the transfer function can be tuned by the *gain* of the system. The gain is the fraction of the measured signal that is applied to the actuator. While it seems to be logical to correct the OPD exactly by the measured fringe position – a gain of 1 – the gain has to be reduced to avoid the overshoot around $f/f_{3\mathrm{dB}} = 1$. However, applying a lower gain, the rejection at lower frequencies deteriorates as displayed by the dotted curve in Fig. 6.5. Thus, the servo loop needs to be carefully designed to optimize the performance.

In the following we will provide performance estimates for the transfer function in (6.24) keeping in mind that a precise estimate is only possible when considering all parameters of the system.

Applying the transfer function $\mathrm{S}(f)$ to the power spectrum of OPD fluctuations in (4.54) we obtain the corrected power spectrum. We have to be aware that we consider the atmospheric power spectrum only, ignoring any contribution, like vibrations, from the interferometer. In practice, mechanical vibrations in the optical system often pose a bigger problem than atmospheric turbulence. At sharply defined frequencies, the contribution from vibrations can easily be a factor of 10 larger than the contributions from turbulence especially at frequencies beyond 10 Hz when the rejection by the closed loop system is weak. The parameters of the control system can be tuned to reject individual frequencies but only to a limited extent, and one should take measures to avoid mechanical vibrations as far as possible.

We multiply the temporal power spectrum $\Phi_{\mathrm{OPD},t}(f)$ (4.54) of the OPD fluctuations with the transfer function $\mathrm{S}(f)$ to obtain the corrected power spectrum. It should be emphasised that the corrected power spectrum still has contributions at

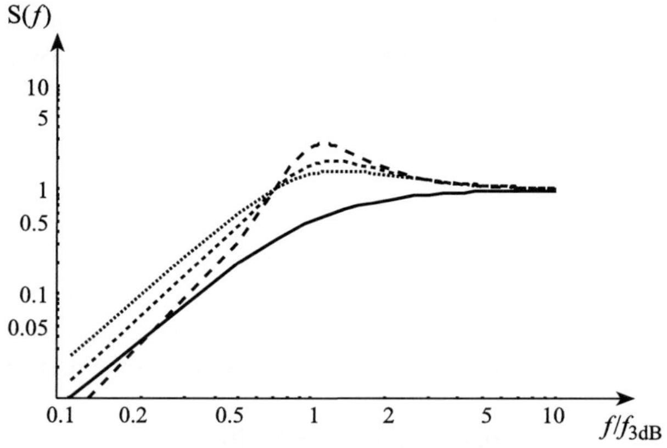

Fig. 6.5 Rejection transfer functions for different parameters of the servo loop. The *solid line* represents the transfer function given in (6.24). The *dashed lines* incorporate the effect of the finite bandwidth of the actuator correcting the OPD, where the curve with the highest peak at $f/f_{3\text{dB}}$ represents the system with the highest gain. Lower gain results in a lower peak, but the properties at low frequencies are deteriorated

frequencies smaller than the servo bandwidth $f_{3\text{dB}}$, as illustrated by the shape of the rejection transfer function, reducing but not eliminating low frequencies.

The **variance of the residual fringe jitter**, $\sigma^2_{\text{OPD},FT}$, is given by the integral of the corrected power spectrum, obtaining the residual rms fringe motion for a bandwidth $f_{3\text{dB}}$ larger than 10 Hz as

$$\sigma_{\text{OPD},FT} = 0.7\,\lambda \left(\frac{\bar{v}}{r_0}\right)^{5/6} T^{5/6}\ [\text{m}], \tag{6.25}$$

with T the exposure time as the reciprocal of the scan rate of $10 f_{3\text{dB}}$, i.e. $T < 10\,\text{ms}$.

Writing $\sigma_{\text{OPD},FT}$ as a function of atmospheric coherence time, we replace $\bar{v}/r_0$ by $0.314/\tau_0$ (4.39) or, if the interferometric coherence time $\tau_{0,2}$ (4.40) is preferred, by $0.207/\tau_{0,2}$.

For a wavelength of 2.2 μm, an r_0 of 0.6 m, typical in the K-band, an effective wind speed of $\bar{v} = 10\,\text{m/s}$, and a scan rate of 250 Hz, corresponding to a servo bandwidth of about $f_{3\text{dB}} = 25\,\text{Hz}$, we find that the residual rms fringe motion are about $\sigma_{\text{OPD},FT} = 160\,\text{nm}$, i.e. about $\lambda/15$ in the K-band. Thus, under typical conditions, scan rates of a few hundred Hertz are required to arrive at a residual fringe motion of a few hundred nanometer.

Scanning the complete fringe package with some hundred Hertz means that each pixel – assuming about 250 pixels per scan as in Fig. 6.2 – has an exposure time of some microsecond. It is clear that there are only very few reference stars that are sufficiently bright to provide a useful signal.

Therefore, the ABCD method is used, taking four measurements per scan, so that the exposure time per bin is about 1 ms. We measure the phase ϕ' of the fringe pattern using (6.19) when the variance σ^2_{ph} of the phase measurement due to photon noise was given in (6.23).

Limiting Magnitude

Determining the residual fringe motion of a fringe tracker one has to take both the variance due to limited bandwidth, $\sigma^2_{\mathrm{OPD},FT}$, and due to photon noise, σ^2_{ph}, into account, since it is the sum of the variances that determines the performance, writing

$$\sigma^2_{\phi'} = \sigma^2_{\mathrm{OPD},FT} + \sigma^2_{\mathrm{ph}} \ [\mathrm{rad}^2]. \tag{6.26}$$

For a complete analysis, other variances, for instance due to detector read-noise, have to be considered as well.

With $\sigma^2_{\mathrm{OPD},FT} \propto T^{5/3}$ and $\sigma^2_{ph} \propto 1/N_{ph}$ in the photon-rich regime, we reduce the bandwidth variance by a factor of 3.17 when doubling the bandwidth and we increase the photon noise variance by a factor of two since we only have half the number of photons if the bandwidth is doubled. Therefore, the sum of the variances decreases – the fringe motion is reduced – when increasing the scan rate. However, when entering the photon-starved regime, the variance due to photon noise goes with $1/N^2_{ph}$ so that σ^2_{ph} increases by a factor of four when doubling the scan rate, and the net effect is an increase of the overall variance, increasing the residual fringe motion.

This permits a simple assessment of the required brightness of the reference star, the limiting magnitude. If we want to reduce the fringe motion to a few hundred nanometer a scan rate of a few hundred Hertz is required and we have to ensure that the number of photons per scan, contributing to the measurement is well above the photon-starved regime, i.e. $N_{ph} > 100$ in order not to increase the overall variance. Assuming that about 2% of the photons in the telescope aperture are fed into the monomode fibre and are registered at the detector – taking transmission and injection losses and quantum efficiency into account – we need about 5,000 photons in the telescope aperture per scan of a few milliseconds. If the telescope diameter is 1.8 m and we observe in the K-band, the limiting magnitude is about $m_K = 9$ (see Appendix A.2) to reduce the fringe motion to a few hundred nanometer. Despite the simple assumptions leading to this result, the magnitude of this value is confirmed by observations [204, 249], and presents a good indication of the requirements.

Scanning an individual fringe with the ABCD method does not provide the information on the distance of the fringe from the white-light fringe since the phase is measured modulo 2π. We discussed the impact on the visibility measurement in Sect. 6.1.2. For fringe tracking, we need to know the position of the white-light fringe – identifying the position of zero OPD when observing a point source – since we want to stabilize the fringes at zero OPD. Only then can we determine the phase of the visibility function of a nearby object as we will discuss in Sect. 6.2.2.

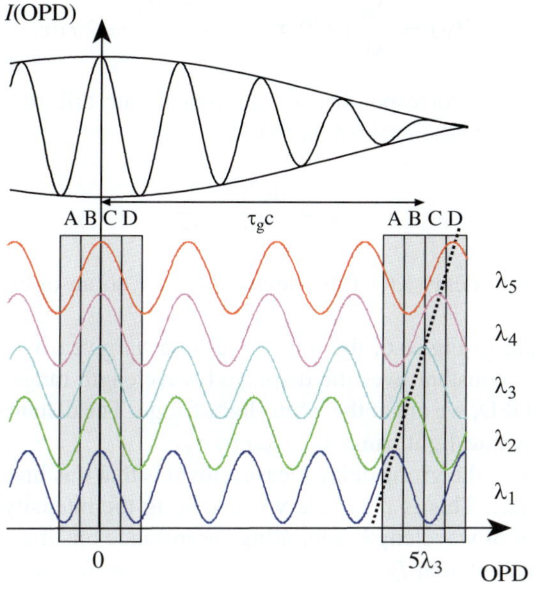

Fig. 6.6 The K-band fringe package *(top)* and its distribution into five spectral channels as a function of OPD, with arbitrary vertical shifts for clarity. For a point source, the white-light fringe, at the centre of the fringe package envelope, is centred at zero OPD. The four stripes at an OPD of 0 and of $5\lambda_3$ indicate two OPD positions where the fringe phase is measured with the ABCD method. At zero OPD, the fringes in all channels have a phase of zero. For an OPD of $5\lambda_3$ the phase is zero only for λ_3. In the other channels, the phases depend linearly on the wavelength, with the *dotted line* marking the measured phase in each channel. The tilt of the *dotted line*, i.e. the variation of phase with wavelength, indicates the group delay τ_g, and $\tau_g c$ is the distance between the white-light fringe and the OPD position of the measurement

Group Delay

Distributing the light into several spectroscopic channels, we can measure the *group delay* providing the information on the distance of the scanned fringe from the white-light fringe position. Figure 6.6 depicts the principle.

If the measurement is taken at zero OPD, the phases at all wavelengths are zero as displayed, unless there is dispersion. Moving to another, unknown OPD, which might be the situation that one encounters when starting the observation, the fringe patterns all show different phase. An OPD larger than λ cannot be determined by individual phase measurements, each limited to $\pm\pi$, but we can measure the differential phase between the channels that is indicated by the dotted line in Fig. 6.6, when the tilt of the dotted line provides the information on the position of the white-light fringe.

The phase ϕ' at each wavelength, $\frac{2\pi}{\lambda j}$ OPD, is determined with respect to the centre of the ABCD measurement, which is at $5\lambda_3$ in Fig. 6.6. Writing ϕ' as a function of frequency,

$$\phi'(\nu) = \frac{2\pi}{\lambda\nu}\nu\,\mathrm{OPD} = 2\pi\nu\frac{\mathrm{OPD}}{c} = 2\pi\nu\tau, \tag{6.27}$$

with τ the time delay corresponding to the optical path difference, also called the *phase delay*, we define the group delay as

$$\tau_g = \frac{\mathrm{d}\phi'(\nu)}{\mathrm{d}\nu}\frac{1}{2\pi}. \tag{6.28}$$

Note that the group delay as a time delay is related to an optical path difference through $\mathrm{OPD} = \tau_g c$.

In our example in Fig. 6.6, the derivative of ϕ' is given by the tilt of the dotted line, which is constant over the displayed wavelength range. If τ_g is zero, for instance at zero OPD, we are at the white-light fringe. The distance of the measured fringes from the white-light fringe is given by $\tau_g c$.

The advantage of the group delay measurement is that it is linear up to the coherence length $\lambda^2/\Delta\lambda$. This is plausible when writing the intensity of the individual bins as a function of wavelength, regarding for instance bin A at an OPD of $5\lambda_3$ in Fig. 6.6. It is apparent, that $I_A(\lambda)$, sometimes called the *channelled spectrum* [218], shows a decline of intensity when moving from λ_1 to λ_5. For an OPD as long as the coherence length we would see that the intensity follows the full cycle of a cosine function, like a fringe pattern with a period length of $\lambda_5 - \lambda_1$. This corresponds to a jump of 2π of the phase at wavelengths λ_1 and λ_5 so that the group delay cannot be computed unambiguously beyond an OPD of the coherence length.

Using the notation $I_A(\lambda)$ for each bin, the group delay can be computed by a somewhat more elaborate procedure, performing a two dimensional Fourier transform of the function $I(\lambda, x)$ with x indicating the OPD corresponding to bins A to D [38, 128].

Comparing the group delay to the measurement of the phase delay, which is defined as $\phi'/(2\pi\nu)$, we find that the variance of the group delay measurement is by a factor of $12(\lambda/\Delta\lambda)^2$ larger than that of the phase delay [130], which means a factor of 360 for the K-band with $\lambda/\Delta\lambda = 5.5$.

Once the white-light fringe is identified we centre the ABCD measurement at this position – by moving the delay line accordingly – and track the fringes around zero OPD by measuring the phase ϕ' in a single spectroscopic channel or on the white-light fringe. In the ideal case, the group delay measurement is no longer required. However, occasionally there are sudden phase jumps of more than 2π in the atmospheric phase so that the fringe tracker may jump by exactly one fringe since this phase jump is invisible for the ABCD measurement of ϕ'. Measuring the group delay, phase jumps would be recognised but the prohibitive variance of the group delay measurement pushes the limiting magnitude to very bright objects.

In practice, one seeks a compromise by reading the fringes at the extreme wavelengths at a slower rate and sending most of the light into the central channel, which provides the phase delay at the required scan rate. The group delay is used only to ensure that fringe jumps are recognised [204]. Then, the larger variance of the group delay measurement does not affect the performance of the fringe tracker.

Dispersion

So far, we have completely ignored the dispersion in air. While this is a reasonable approximation for the effects that we have discussed in the preceding sections, we have to consider the dispersion when discussing astrometry and measurements of the visibility phase [223]. Observing sources off zenith, the optical path difference that we correct with delay lines is due to an additional optical path above the atmosphere in vacuum. In air-filled or partially evacuated delay lines, tiny variations of the refractive index $n(\nu)$ over the spectral band, provide a measurable variation of the optical path $n(\nu)L$, with L being the path length. This is called *longitudinal dispersion.*[3]

Comparing the path, L, in vacuum with the optical path, $n(\nu)L$, in air, we find the OPD as $(n(\nu) - 1)L$. We choose the central wavelength of the observed spectral band to define the position of zero OPD, as displayed in Fig. 6.7 with λ_3 as reference wavelength. For normal dispersion, the refractive index $n(\nu)$ increases with frequency, so that at longer wavelengths the fringe patterns are shifted to negative OPD and at shorter wavelengths to positive OPD. The situation is displayed in Fig. 6.7 assuming a linear increase of n with frequency. Then the shifts are linear with wavelength as indicated by the tilted line.

Measuring the phase at an OPD of $5\lambda_3$, we find that the dotted line, marking the phase in each channel, has a smaller tilt than in the case without dispersion in Fig. 6.6, and, accordingly, the group delay is reduced. This geometrical illustration of the effect of longitudinal dispersion can be replaced by the computation of the derivative of ϕ' in (6.28) when the OPD, $(n(\nu) - 1)L$, is a function of frequency.

According to the smaller group delay, we find that the white-light fringe, more precisely the centre of the fringe package envelope, is now closer to the OPD position, here at $5\lambda_3$, of the phase measurement. This means, that due to dispersion, the envelope of the fringe package has moved from zero OPD closer to the OPD position of the measurement. It should be kept in mind that the group delay is measured with respect to the position where the measurement is performed. Thus, determining the group delay at zero OPD, τ_g would provide the distance of the envelope from zero OPD.

In general, we denote the distance between the position of zero OPD for the central wavelength and the centre of the envelope function by $\mathrm{OPD_{GD}}$, and it is

$$\mathrm{OPD_{GD}} = L\nu \frac{\mathrm{d}n(\nu)}{\mathrm{d}\nu}. \tag{6.29}$$

The group delay depends on the dispersion and on the optical path L in vacuum. To reduce this effect, delay lines can be put in vacuum pipes. Then, the optical path in air is shorter and the formula has to be corrected accordingly.

---

[3] The dispersion also affects the shape of the Airy disk that is elongated when observing objects off-zenith. This is called transversal dispersion and is usually corrected by an atmospheric dispersion corrector.

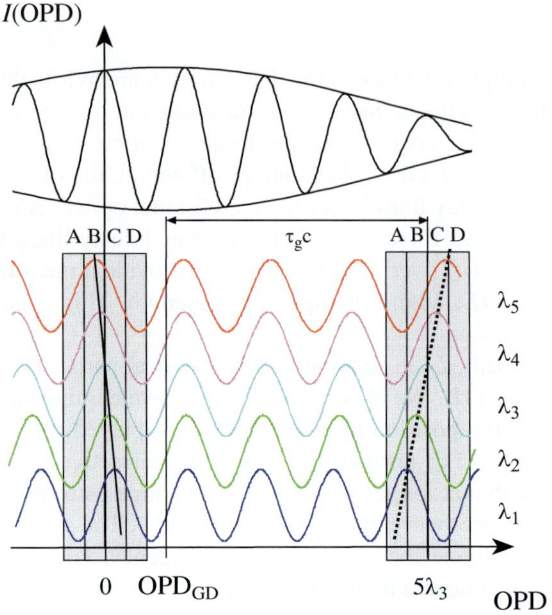

Fig. 6.7 The K-band fringe package *(top)* and its distribution into five spectral channels as a function of OPD, as in Fig. 6.6, but in the presence of dispersion. We use the central wavelength λ_3 to define the position of zero OPD. Due to the dispersion the phases at zero OPD vary with wavelength, and OPD_{GD} indicates the distance between zero OPD and the centre of the envelope function. Measuring the group delay at position $5\lambda_3$, we determine the distance $\tau_g c$ with respect to the centre of the envelope function. As indicated by the tilt of the *dotted line*, this group delay is smaller than without dispersion in Fig. 6.6 and the centre of the envelope function is closer to the position $5\lambda_3$

Regarding the situation when fringes are tracked at zero OPD and the object is at zenith with $L = 0$, we find that the group delay is zero and the fringe package is centred at zero OPD, with $\text{OPD}_{\text{GD}} = 0$. In the course of the observation, L increases and – while the fringes at the central wavelength are kept stable with a fringe tracker – it is the envelope of the fringe package that moves, so that the tracked fringe at zero OPD is no longer at the centre of the envelope.

Thus, having the choice to track either the fringe phase, i.e. the phase delay, or the group delay one has to be aware, that the fringe pattern moves within its envelope so that tracking the group delay, the fringes would smear in a long time integration. In the K-band for example, the refractive index n varies by 10^{-7} so that OPD_{GD} increases by one wavelength for a path length of $L = 4\,\text{m}$, i.e. after about 10 min of observations. This means, tracking the group delay would completely wipe out the fringe pattern after a 10 min exposure, while tracking the phase delay keeps the fringe pattern stabilized.

The dispersion has to be considered in particular when, for instance, tracking fringes in the K-band and observing in the N-band[4] [154]. It is possible, that the envelope function moves so far off the zero OPD position that the measurement of the visibility at zero OPD is affected by the reduced contrast. One has to compensate for this effect in the N-band beam combiner.

6.2.2 Dual-Feed System

The measurement of the phase of the object visibility function requires to observe additionally a point-like reference star with zero visibility phase. Unless an interferometer in Fizeau configuration is available – with a large field of view so that we can pick two stars with arbitrary separation – we face the restrictions in Michelson configuration reducing the field of view to less than an Airy disk.

Measuring the phase of object and reference star, we rely on the correlation of the differential fringe motion that requires an angular distance smaller than some 10 arcsec as discussed in Sect. 4.3.1. Since we are interested in exploiting the maximum separation to increase the chance of finding a sufficiently bright reference star – estimated in Sect. 6.2.1 to have at least a magnitude of $m_K = 9$ – the field of view in Michelson configuration is not sufficient and one has to operate two Michelson beam combiners in parallel. Then, the light from the two objects is fed into separate beam trains via a periscope system in each telescope focus. This is called a dual-feed system.

In general, a fringe tracker runs in one feed of the system so that the integration time for the science object in the other feed can be increased considerably without losing fringe contrast. Thus, we do not only enable the phase measurement with a dual-feed system, but we also increase the sensitivity by several stellar magnitudes.

In Sect. 6.2.1, we discussed how to measure the zero OPD position when observing a point source and how to use this signal to operate a fringe tracker. How do we explore this knowledge to determine the position of zero OPD in a second beam combiner?

Additional Sub-Systems

The calibration of the zero OPD position on the second detector is done by first observing the reference star simultaneously in both beam combiners – hence the periscope is sometimes called a *star separator* – so that the measured phase of the fringe pattern marks the position of zero OPD on both detectors. Then, moving the periscope in the telescope focus from the reference object to the science object at distance Δs, while the reference star in the other feed is still being observed, we

[4] The N-band is the atmospheric band at $\lambda = 10.2 \pm 2.5\,\mu\text{m}$, see Appendix A.2.

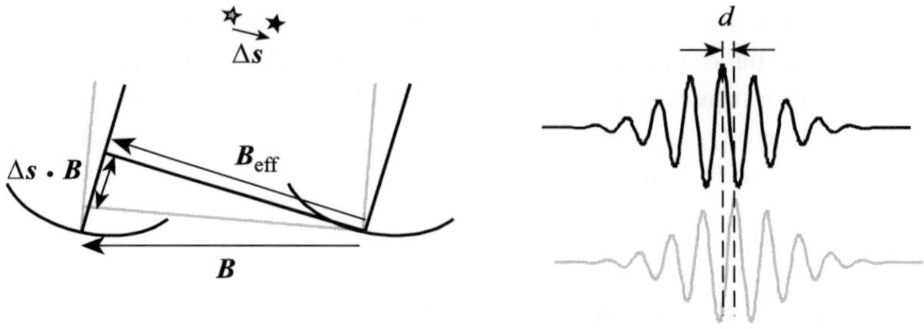

Fig. 6.8 Principle of dual-feed operation. The differential delay d of the white-light fringe positions is determined by the differential OPD $\Delta \boldsymbol{s} \cdot \boldsymbol{B}$, by internal optical path differences, $\mathrm{dOPD}_{\mathrm{int}}$, by the random differential fringe motion due to atmospheric turbulence $\mathrm{dOPD}_{\mathrm{turb}}$, and by the phase of the visibility function $\phi(\boldsymbol{R}_{\mathrm{B}}) = \phi(\boldsymbol{B}_{\mathrm{eff}}/\lambda)$

can relate the phase measurement of the science object to the zero OPD position that was determined with the reference star, as illustrated by Fig. 6.8.

However, this procedure relies on the stability of all optical paths so that any shift of the phase of the science object fringe pattern is due to the object phase and not to any random internal path length variation. It is impossible to guarantee this down to a level of a few hundred nanometer without monitoring the optical paths. A *laser metrology* system is required to measure all internal optical path lengths from the beam combiners through the star separator to the telescope mirrors so that we can determine the differential internal optical path difference, $\mathrm{dOPD}_{\mathrm{int}}$, contributing to the shift of the fringe pattern. This puts an additional constraint on the star separator since the laser beam must not be interrupted when the periscope moves from the reference object to the science object.

For a maximum distance between object and reference star of $\Delta s = 1$ arcmin, and a baseline of $B = 200$ m, we have a differential OPD of about $\Delta \boldsymbol{s} \cdot \boldsymbol{B} = 60$ mm. If the main delay line is used to set the OPD of the reference star to zero, we have to compensate for the differential OPD in the second beam combiner. This requires another, relatively short delay line, the *differential delay line*, that is used to place the fringe package at the zero OPD position that was determined previously with the reference star.

The differential delay line is the third additional sub-system, after star separator and laser metrology system, that is required for a dual-feed facility [38].

Compensating for the additional optical path difference $\Delta \boldsymbol{s} \cdot \boldsymbol{B}$ with the differential delay line, the angular separation $\Delta \boldsymbol{s}$ needs to be known precisely to move the fringe pattern exactly to the zero OPD position of the second beam combiner. Any deviation of the real separation from the nominal separation, which we apply to the differential delay line, shows in the phase measurement of the fringe pattern as an additional shift with respect to zero OPD. Thus, measuring the phase of the fringe pattern, we determine both the precise separation $\Delta \boldsymbol{s}$ and the phase $\phi_{\mathrm{ob}}(\boldsymbol{R}_{\mathrm{B}})$ of the object visibility function, so that the dual-feed system can not only be used for phase referenced imaging but also for astrometry.

Finally, observing through atmospheric turbulence, the fringe patterns are affected also by random differential fringe motion $\mathrm{dOPD_{turb}}$, as discussed in Sect. 4.3.1. We call the sum of all contributions the *differential delay*, d, written as

$$d = \Delta s \cdot \boldsymbol{B} + \mathrm{dOPD_{int}} + \mathrm{dOPD_{turb}} + \frac{\lambda}{2\pi}\phi_{\mathrm{ob}}(\boldsymbol{R}_{\mathrm{B}}) \text{ [m]}. \tag{6.30}$$

Tracking fringes with the reference star, the fringe pattern of the science object shows the differential delay d, with a fast jitter due to $\mathrm{dOPD_{turb}}$ and, additionally, due to the residual fringe motion of the fringe tracker (6.25), causing a contrast loss in long time exposures.

In Sect. 4.3.1, the rms fluctuations of the differential OPD were given by (4.60) as

$$\begin{aligned} \sigma^{d}_{\mathrm{dOPD},T} &= 0.1\,\lambda r_0^{-5/6}\,\bar{v}^{8/6}\bar{h}\,d^{-3/2}\,\Delta s\,T^{4/3} \text{ [m]}. \\ &= 88\,d^{-3/2}\,\Delta s\,T^{4/3} \text{ [nm]}, \end{aligned} \tag{6.31}$$

with the separation Δs in radians in the first line and in arcsec in the second line. The numerical value summarizes the parameters of average atmospheric conditions, $r_0 = 0.6$ m at $\lambda = 2.2\,\mu$m, $\bar{v} = 10$ m/s and $\bar{h} = 2.5$ km. Note that these parameters can vary considerably so that the formula only gives an indication of the order of magnitude of the effect.

The overall OPD variance is given by the sum of differential OPD variance $\sigma^{d\,2}_{\mathrm{dOPD},T}$ and the residual fringe motion $\sigma^2_{\phi'}$, (6.26) in the fringe tracker. For a separation of $\Delta s = 10$ arcsec and an exposure time of $T = 1$ s, we find the rms fluctuations of $\mathrm{dOPD_{turb}}$ to be about 310 nm. The contrast reduction due to OPD fluctuations was given in (6.3) as $\Delta\mathcal{V}/\mathcal{V} = \exp(-2\pi^2/\lambda^2\sigma^2_{\mathrm{OPD}})$ so that, observing in the K-band, we have a contrast reduction to 67%. The additional residual fringe motion of the fringe tracker of 150 nm reduces the contrast additionally by a factor of 0.9 so that in total the contrast is reduced to 60%.

If the contrast loss is the metric to define the isoplanatic angle and if a reduction to 60% is acceptable, the isoplanatic angle for interferometry is 10 arcsec in the K-band. Smaller angular separations reduce the rms fluctuations linearly, and the computation of the contrast loss is straightforward.

The **gain in sensitivity** is given by the increase of exposure time over the uncorrected case. Using (4.55), we find that uncorrected fringe motion shows fluctuations of 300 nm over 40 ms. With a dual-feed system we can integrate 1 s before reaching this level of fringe motion so that we can observe objects, that are 3.5 magnitudes fainter at a distance of 10 arcsec from the reference star.

Reducing the exposure time, the fluctuations are smaller and we could observe at a larger angular distance from the reference star. However, the object has to be brighter to provide the same number of photons during the integration.

In Sect. 4.3.1 we discussed that the OPD fluctuations saturate for very long integration times at about four times the value of a 1-s integration, i.e. at $\sigma^{d}_{\mathrm{dOPD},\infty} \approx 1\,\mu$m. Since this value lowers the contrast to a few percent, we have to reduce the separation $\Delta \boldsymbol{s}$ so that the fluctuations are at least in the range of 500 nm with a contrast of 30%. This means that the angular distance between object and reference star has to be smaller than 10 arcsec for very long exposure times reaching the maximum sensitivity improvement.

Thus, the sensitivity of an interferometer can be improved as a function of the angular distance between object and reference star. For an improvement of 3–4 magnitudes, the science object has to be at about 10 arcsec from the reference star. The maximum improvement of about seven magnitudes in the K-band, corresponding to integration times of about 30 s, requires a reference star within considerably less than 10 arcsec.

For astrometric measurements, with both objects usually sufficiently bright for short exposure times, the maximum theoretical separation of many arcminute is given by the *very narrow angle regime* [212]. We will see in the following, (6.33), that the achievable accuracy is linear with the separation so that, in practice, the maximum angular distance is about 1 arcmin.

Regarding the **phase measurement** of the fringe pattern of the science object, we are interested in the average value, $\bar{d}$, of the difference of fringe positions after a long exposure time in order to determine the angular separation, $\Delta \boldsymbol{s}$ and the phase, $\phi_{\mathrm{ob}}(\boldsymbol{R_B})$, of the visibility function. The accuracy of this measurement depends mainly on the optical path fluctuations due to atmospheric turbulence and we will discuss these first before setting the requirements on the other subsystems.

Ideally, the optical path fluctuations average to zero since atmospheric turbulence has zero mean, as discussed in Sect. 4.2. For finite observing time, however, the average, $\overline{\mathrm{dOPD}_{\mathrm{turb}}}$, is larger than zero. The variance $< \overline{\mathrm{dOPD}}^2_{\mathrm{turb}} >$ for an integration time of T can be computed in frequency space, by applying a filter function $\mathrm{sinc}(\pi T f)$ – the Fourier transform of the rectangular time window – to the power spectrum of differential OPD fluctuations, $\Phi_{\mathrm{dOPD},t}(f)$, (4.58), just like when discussing the fringe motion in Sect. 4.3.1. However, there the variance of the fringe motion during individual measurements was computed while we are now interested in the variance of the average fringe position after time T. Averaging M measurements of duration T, we have to consider the overall integration time $T_M = M \times T$.

The variance is then the integral of the product of $\Phi_{\mathrm{dOPD},t}(f)$ with the sinc function. For long integration times T_M, it is sufficient to consider only the constant part at low frequencies [212], yielding

$$\sigma^2_{\overline{\mathrm{dOPD}_{\mathrm{turb}}},T_M} = 0.628\,\lambda^2 \left(\frac{\bar{v}}{r_0}\right)^{5/3} \left(\frac{\Delta s\,\bar{h}}{\bar{v}}\right)^2 \left(\frac{B}{\bar{v}}\right)^{2/3} T_M^{-1}\ [\mathrm{m}^2]. \quad (6.32)$$

Thus, the variance of the average fringe position decreases with observing time T_M, as expected.

For astrometry, it is more convenient to express the fluctuations for the average separation $\overline{\Delta s} = \overline{\mathrm{dOPD}_{\mathrm{turb}}}/B$, writing the rms fluctuations [212] as

$$\sigma_{\overline{\Delta s},T_M} = \frac{\sigma_{\overline{\mathrm{dOPD}_{\mathrm{turb}}},T_M}}{B} = 0.79\,\lambda r_0^{-5/6}\,\bar{v}^{-1/2}\bar{h}\,B^{-2/3}\,\Delta s\,T_M^{-1/2}\ \text{[arcsec]}.$$
$$= 2100\,B^{-2/3}\Delta s\,T_M^{-1/2}\ [\mu\text{arcsec}], \qquad (6.33)$$

with the baseline B in m and the separation Δs in arcsec. The numerical value, 2100, summarizes the parameters of average atmospheric conditions, $r_0 = 0.6\,\mathrm{m}$ at $\lambda = 2.2\,\mu\mathrm{m}$, $\bar{v} = 10\,\mathrm{m/s}$ and $\bar{h} = 2.5\,\mathrm{km}$, so that we obtain a manageable expression with a numerical value that is within the range of values computed with detailed data for C_n^2 profiles and wind speeds for observatory sites [34, 212].

Assuming integration times of a few hours and a separation of $\Delta s = 10$ arcsec, the rms fluctuations have a value of 10 μarcsec for a 100-m baseline – corresponding to dOPD fluctuations of 5 nm – under average atmospheric conditions.

It should be noted, that the phase of the visibility function is likely to vary when observing over several hours since the effective baseline moves. Thus, depending on object morphology and position, there are restrictions for the duration of the measurement. For an astrometric measurement, the restrictions are less rigid since the phase measurement is determined by $\Delta\boldsymbol{s} \cdot \boldsymbol{B}$, and the sidereal motion shows in the separation vector $\Delta\boldsymbol{s}$ that rotates slowly with respect to the baseline. As a linear effect, its calibration is straightforward.

If we set 5 nm OPD fluctuations as a goal for the accuracy of the measurement of the differential delay d, the errors that are contributed by the other sub-systems have to be of the same order of magnitude. This means that the laser metrology system has to achieve an accuracy better than 5 nm, which is achievable with state-of-the-art systems.

The accuracy for the baseline length can be derived from $\Delta\boldsymbol{s} \cdot \boldsymbol{B}$, and we find that $\delta\Delta\boldsymbol{s}/\Delta\boldsymbol{s} \approx \delta B/B$. If we want to determine the separation vector with an accuracy of 10 μ arcsec for a 10 arcsec separation, the length of the baseline vector, e.g. 100 m, needs to be known to about 100 μm. The average baseline can be determined to this accuracy level by observing calibrator stars with known diameter. However, it is more difficult to ensure the dynamic stability since this requires that the shape of the telescope entrance pupils – their separation defining the baseline – has to be stable to this level while tracking a star.

At this extreme level of accuracy, the spectra of object and reference star also affect the measurement. Discussing dispersion in Sect. 6.2.1, the zero OPD position was defined for the central wavelength of the spectral band, indirectly assuming a rectangular spectrum. Even it the spectrum is not perfectly rectangular this does not affect the measurement as long as object and reference star have the same spectrum. Observing stars of different spectral type, however, the zero OPD positions are different and we have to consider this effect when moving from the reference star to the object.

An astrometric accuracy of 100 μarcsec was achieved with the Palomar Testbed Interferometer (PTI) with a baseline of 110 m, measuring the angular separation of the two components in 61 Cyg with $\Delta s = 31$ arcsec, [125]. Also at PTI, binaries with a separation of some 100 mas – within one Airy disk and not requiring a dual-feed system – were measured with an accuracy of 20 μarcsec, [168].

So far, we have concentrated on the error sources, but how do we disentangle the impact of the separation vector $\Delta \boldsymbol{s}$ from that of the visibility phase, $\phi(\boldsymbol{R_B})$, in the differential delay d? Taking a large number of measurements with different baselines one can separate the distance term that is linear with the baseline, $\Delta \boldsymbol{s} \cdot \boldsymbol{B}$, from the phase of the visibility function that can have almost any shape as a function of baseline $\boldsymbol{R_B} = \boldsymbol{B}_{\text{eff}}/\lambda$.

The measurement of the phase $\phi_{\text{ob}}(\boldsymbol{R_B})$ of the object visibility function provides the last piece of information that is required to form images of celestial objects with an interferometer using the reconstruction procedures described in Sect. 3.4.3.

6.2.3 Closure Phase

Apart from the complex operational scenario of the dual-feed system, the phase of the visibility function can also be determined when combining three telescopes simultaneously eliminating the influence of the turbulence by adding up the phases of all three baselines. This technique has been applied successfully in speckle interferometry, see Sect. 4.4, but it is restricted to rather bright objects.

If we regard the random phase differences $\Delta\varphi_{ij}$ for each baseline $\boldsymbol{R}_{ij}$ as the difference of the random phases φ_i in each telescope, we find that the sum of three measured phases φ'_{ij} around a closed loop of three telescopes amounts to zero, $\boldsymbol{R}_{12} + \boldsymbol{R}_{23} + \boldsymbol{R}_{31} = 0$.

If the observed object has a non-zero phase, the measured phase for each baseline is the sum of the object phase $\phi(\boldsymbol{R}_{ij})$ plus the random phase difference, $\Delta\varphi_{ij}$. Adding up the measured phases of the three baselines then eliminates the random OPD fluctuations, obtaining the closure phase as $\phi(\boldsymbol{R}_{12}) + \phi(\boldsymbol{R}_{23}) + \phi(\boldsymbol{R}_{31})$.

With ϕ'_i the phase of the each fringe pattern, we write

$$
\begin{aligned}
\phi'_{12} &= \phi(\boldsymbol{R}_{12}) + \Delta\varphi_{12} = \phi(\boldsymbol{R}_{12}) + \varphi_1 - \varphi_2, \\
\phi'_{23} &= \phi(\boldsymbol{R}_{23}) + \Delta\varphi_{23} = \phi(\boldsymbol{R}_{23}) + \varphi_2 - \varphi_3, \\
\phi'_{31} &= \phi(\boldsymbol{R}_{31}) + \Delta\varphi_{31} = \phi(\boldsymbol{R}_{31}) + \varphi_3 - \varphi_1, \\
\phi'_{12} + \phi'_{23} + \phi'_{31} &= \phi(\boldsymbol{R}_{12}) + \phi(\boldsymbol{R}_{23}) + \phi(\boldsymbol{R}_{31}).
\end{aligned} \tag{6.34}
$$

That means that the random phase terms disappear in the sum, the closure $\psi(\boldsymbol{R}_{12}, \boldsymbol{R}_{21})$, that is written as

$$
\psi(\boldsymbol{R}_{12}, \boldsymbol{R}_{21}) = \phi(\boldsymbol{R}_{12}) + \phi(\boldsymbol{R}_{23}) - \phi(\boldsymbol{R}_{12} + \boldsymbol{R}_{23}), \tag{6.35}
$$

with $\boldsymbol{R}_{12} + \boldsymbol{R}_{23} = -\boldsymbol{R}_{31}$. Note that the linear term of the phase, determining the source position, is also removed in the closure phase.

The number of closure phases for a different number of telescopes is given in Table 6.1. For a small number of telescopes, the phase measurement with a dual-feed system, determining one visibility phase for each baseline, provides more

Table 6.1 Number of baselines and closure phases for different number of telescopes

No. of telescopes N_T	2	4	6	8
No. of baselines $N_T(N_T-1)/2$	1	6	15	28
No. of closure phases $(N_T-1)(N_T-2)/2$	0	3	10	21

information than the limited number of closure phases. With growing number of telescope, however, the initial disadvantage disappears and the number of baselines and of closure phases are both approximately $\propto N_T^2$.

Combining closure phases of observations with different baselines, the phases of the individual baselines $\phi(\boldsymbol{R}_{ij})$ can be recovered using methods developed for the phase reconstruction from the bispectrum in speckle interferometry [7, 9, 84, 115], see Sect. 4.4. In intensity interferometry (Sect. 2.5) the closure phase can be employed also to recover the object phase.

Although the random phase disappears in the closure phase we are not independent of turbulence since it has to be assured that the exposure times are sufficiently short to measure the phase on each baseline. Using (6.4) we find that after about 100 ms the contrast is reduced to 20%. Increasing the exposure time makes the phase measurement less accurate since, as given by (6.23) it is the product of measured visibility $\mathcal{V}$ and number of photons N_{ph} that determines the variance of the measurement. The signal to noise ratio of the closure phase measurement is given by [211]

$$\mathrm{SNR} = \left(3\left(\frac{2}{N_{ph}\mathcal{V}^2}\right) + 6\left(\frac{2}{N_{ph}\mathcal{V}^2}\right)^2 + 4\left(\frac{2}{N_{ph}\mathcal{V}^2}\right)^3\right)^{-1/2} . \tag{6.36}$$

If the zero OPD position in each phase measurement is not precisely known, a bias is introduced. However, this can be calibrated by observing a point source. Using all-in-one beam combiners as shown in Fig. 5.7, combining three telescopes, the phases can be measured without a bias.

The closure phase technique was first applied to masked apertures [8, 100]. The first demonstration with an interferometer of independent apertures was given by J.E. Baldwin [68] presenting the first image reconstruction with the COAST interferometer.

Beating Atmospheric Turbulence – Summary

A reference star for fringe tracking is required when observing through turbulence, unless, combining three telescopes, the three visibility phases are added up providing the closure phase that is independent of turbulence.

The principle of fringe tracking relies on stabilizing the fringe motion by rapidly scanning the fringe pattern and feeding the information on the white-light fringe position into a servo loop.

The residual rms fringe jitter, $\sigma_{\mathrm{OPD},FT}$, for a servo loop bandwidth $f_{3\mathrm{dB}}$ larger than 10 Hz, is given by

$$\sigma_{\mathrm{OPD},FT} = 0.7\,\lambda\left(\frac{\bar{v}}{r_0}\right)^{5/6} T^{5/6}\ [\mathrm{m}], \tag{6.25}$$

with T the exposure time as the reciprocal of the scan rate of $10 f_{3\mathrm{dB}}$, it is $T < 10\,\mathrm{ms}$.

The limiting magnitude, i.e. the required brightness of the reference star, is about $m_K = 9$ on 2-m class telescopes, so that the phase variance σ^2_{ph}, (6.23), due to photon noise is smaller than $\sigma^2_{\mathrm{OPD},FT}$.

The identification of the white-light fringe of the fast moving fringe pattern requires to measure the group delay defined as

$$\tau_g = \frac{\mathrm{d}\phi'(\nu)}{\mathrm{d}\nu}\frac{1}{2\pi}, \tag{6.28}$$

with $\phi'(\nu)$ the measured phase as function of frequency. The group delay is related to an optical path difference through $\mathrm{OPD} = \tau_g c$.

Unlike the phase measurement that suffers from a 2π ambiguity, and is therefore blind to fringe jumps of λ, the group delay is linear up to the coherence length, i.e. over the complete fringe package.

Observing objects off-zenith, the compensation of the additional optical path L by delay lines in air makes the OPD a function of wavelength owing to the dispersion of air. Computing the group delay, we find that the distance $\mathrm{OPD}_{\mathrm{GD}}$ between the centre of the envelope of the fringe package and the position of zero OPD is given by

$$\mathrm{OPD}_{\mathrm{GD}} = L\nu\frac{\mathrm{d}n(\nu)}{\mathrm{d}\nu}. \tag{6.29}$$

The group delay depends on the dispersion and on the optical path L in vacuum. For evacuated delay lines, the optical path in air is shorter and the formula has to be corrected accordingly.

Using a dual-feed system, an object in the vicinity of a bright reference star benefits from the fringe stabilization on the reference star. The differential delay d, which is the overall OPD difference between object and reference star is given by

$$d = \Delta\boldsymbol{s}\cdot\boldsymbol{B} + \mathrm{dOPD_{int}} + \mathrm{dOPD_{turb}} + \frac{\lambda}{2\pi}\phi_{\mathrm{ob}}(\boldsymbol{R}_{\mathrm{B}})\ [\mathrm{m}], \tag{6.30}$$

when $\Delta\boldsymbol{s} \cdot \boldsymbol{B}$ is due to the angular separation $\Delta\boldsymbol{s}$ between object and reference star, $\mathrm{dOPD_{int}}$ is the differential internal OPD due to imperfections of the opto-mechanics, $\mathrm{dOPD_{turb}}$ is the differential OPD due to turbulence and $\phi_{\mathrm{ob}}(\boldsymbol{R}_{\mathrm{B}})$ is the phase of the object visibility function. The differential delay is the measured quantity that – by monitoring $\mathrm{dOPD_{int}}$ and by averaging (see below) $\mathrm{dOPD_{turb}}$ – is determined by $\Delta\boldsymbol{s}$ and $\phi_{\mathrm{ob}}(\boldsymbol{R}_{\mathrm{B}})$, providing an astrometry and a phase referenced imaging mode.

The differential fringe motion during exposure time T is determined by the differential OPD yielding the rms fluctuations as

$$\begin{aligned}\sigma^{d}_{\mathrm{dOPD},T} &= 0.1\,\lambda r_0^{-5/6}\,\bar{v}^{8/6}\bar{h}\,d^{-3/2}\,\Delta s\,T^{4/3}\ [\mathrm{m}].\\ &= 88\,d^{-3/2}\,\Delta s\,T^{4/3}\ [\mathrm{nm}], \qquad (6.31)\end{aligned}$$

with the separation Δs in the second line in arcsec, and the numerical parameter summarizing the parameters of average atmospheric conditions. Note that these parameters can vary considerably so that the formula only gives an indication of the order of magnitude of the effect.

Measuring the phase, we are interested in the average value $\overline{\mathrm{dOPD_{turb}}}$ after a long time T_M. The variance of the average value is given by the variance of the individual measurement divided by the overall duration T_M of the measurement. For astrometry, it is more convenient to express the fluctuations for the average separation $\overline{\Delta s} = \overline{\mathrm{dOPD_{turb}}}/B$, writing the rms fluctuations [212] as

$$\begin{aligned}\sigma_{\overline{\Delta s},T_M} &= 0.79\,\lambda r_0^{-5/6}\,\bar{v}^{-1/2}\bar{h}\,B^{-2/3}\,\Delta s\,T_M^{-1/2}\ [\mathrm{arcsec}].\\ &= \frac{\sigma_{\overline{\mathrm{dOPD_{turb}}},T_M}}{B} = 2100\,B^{-2/3}\Delta s\,T_M^{-1/2}\ [\mu\mathrm{arcsec}], \qquad (6.33)\end{aligned}$$

with the baseline B in m, the separation Δs in arcsec, and the numerical parameter summarizing the parameters of average atmospheric conditions.

Assuming integration times of a few hours and a separation of $\Delta s =$ 10 arcsec, the rms fluctuations have a value of 10 μ arcsec for a 100-m baseline – corresponding to dOPD fluctuations of 5 nm – under average atmospheric conditions. This sets the accuracy limits for the other contributions to the differential delay in (6.30).

Regarding the random optical path difference fluctuations, $\mathrm{OPD_{turb}}$, on each baseline $\boldsymbol{R}_{ij}$ as the difference of the individual optical paths to each telescope, we find that the sum of the three $\mathrm{OPD_{turb}}(\boldsymbol{R}_{ij})$ around a closed loop of three telescopes, amounts to zero.

If the observed object has a non-zero phase, the measured phase on each baseline is the sum of the object phase $\phi(\boldsymbol{R}_{ij})$ plus a random term, $\mathrm{OPD_{turb}}(\boldsymbol{R}_{ij})$. Adding up the measured phases of the three baselines then eliminates the random OPD fluctuations obtaining the closure phase as

$$\psi(\boldsymbol{R}_{12}, \boldsymbol{R}_{21}) = \phi(\boldsymbol{R}_{12}) + \phi(\boldsymbol{R}_{23}) - \phi(\boldsymbol{R}_{12} + \boldsymbol{R}_{23}). \tag{6.35}$$

Combining closure phases of observations with different baselines permits us to reconstruct the phases of the individual baselines $\phi(\boldsymbol{R}_{ij})$.

6.3 Adaptive Optics

Interferometers with large telescopes require adaptive optics on each telescope to increase the sensitivity and to improve the measurement accuracy, as discussed in Sect. 6.1.

Adaptive optics relies on measuring the aberrations in the telescope pupil with a wave front sensor using a reference star. The signal is then applied to an actuator, a deformable mirror, forming a servo loop like a fringe tracker. The principle is depicted in Fig. 6.9. In contrast to tracking the fringe position and correcting a single parameter, adaptive optics requires measuring optical aberrations in a two dimensional plane and correcting many parameters.

In the following, we will briefly discuss the concept of adaptive optics providing sufficient detail to understand the basic principle, but we will not treat technical developments of deformable mirrors and detectors. More detailed information can be found in the books by F. Roddier [200], by M.C. Roggemann and B.M. Welsh [201], and by R.K. Tyson [244].

Discussing optical aberrations, Zernike polynomials were introduced in Sect. 3.1.2, and we investigated their properties in the case of atmospheric turbulence in Sect. 4.3.3. These properties will be used in the following to determine the performance of adaptive optics systems.

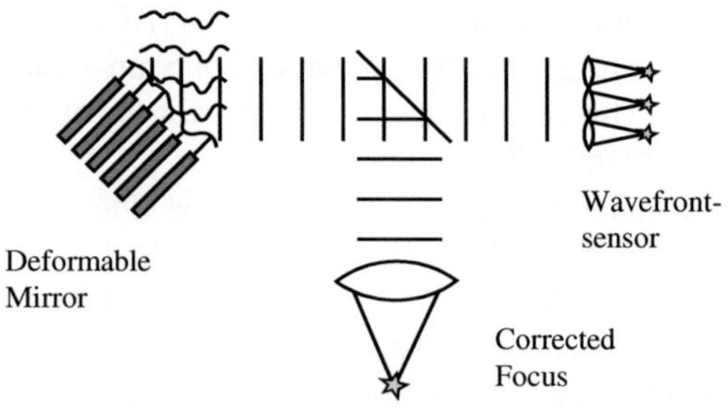

Fig. 6.9 The main elements of an adaptive optics system. The wave front sensor measures the aberrations and sends the information to the deformable mirror to flatten the wave front. A camera in the corrected focus takes the corrected image

We will first introduce the Shack–Hartmann sensor as an example for wave front sensing and then we will discuss the requirements and the performance of an adaptive optics servo loop.

6.3.1 Wave Front Sensing

There are mainly three types of wave front sensors in adaptive optics: the Shack–Hartmann sensor [202], the curvature sensor [195], and the pyramid sensor [190]. The Shack–Hartmann sensor being conceptually the simplest, will serve to discuss general principles of wave front sensing.

The Shack–Hartmann sensor divides the telescope aperture into an array of smaller sub-apertures, and a lenslet array is used to produce multiple images (see Fig. 6.10). The centroid displacement of each of these sub-images gives an estimate of the average wave front gradient over the sub-aperture [187] that can be calculated using (6.37). The Shack–Hartmann sensor is achromatic – the image movement is independent of wavelength – and extended sources can be used as long as they fit into the sub-image boundary.

In practice, a Shack–Hartmann sensor is built by putting a lenslet array in the reimaged telescope pupil. The sub-images from each sub-aperture are imaged onto a detector.

The measurement of the image intensity centroid after the light has passed through a sub-aperture in a Shack–Hartmann sensor provides an estimate of the wave front slope. The centroid, or first-order moment M, of the image intensity

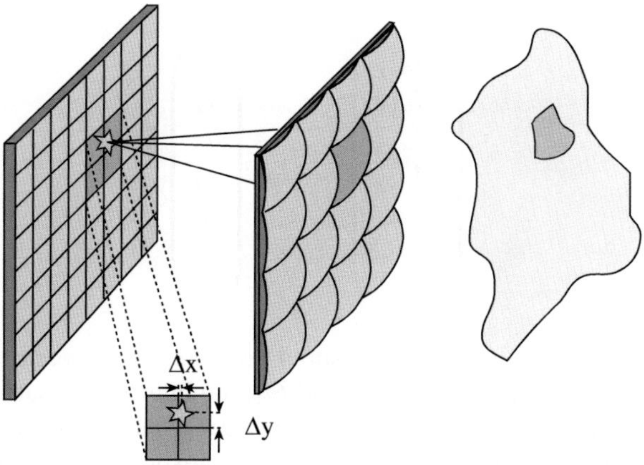

Fig. 6.10 Measurement principle of a Shack–Hartmann sensor. The incoming aberrated wave is subdivided by the lenslet array and the image centroid in every sub-image is shifted according to the average wave front slope over the sub-aperture formed by the lenslet

$I(\alpha_x, \alpha_y)$ with respect to the α_x-direction in the image, is given by the averaged gradient θ_ξ (4.64), the tip-tilt of the wave front in the sub-aperture as [187]

$$M_{\alpha_x} = \frac{\iint_{\text{image}} I(\alpha_x, \alpha_y)\, \alpha_x \, d\alpha_x d\alpha_y}{\iint_{\text{image}} I(\alpha_x, \alpha_y)\, d\alpha_x d\alpha_y}$$

$$= \frac{\lambda}{2\pi} \iint_{\text{sub-aperture}} \frac{\partial \varphi(\xi, \zeta)}{\partial \xi} d\xi d\zeta. \tag{6.37}$$

Zernike polynomials as introduced in Sect. 3.1.2 are a set of orthogonal polynomials defined on the unit circle, here on the telescope aperture (see Sect. 3.1.2). Computing the gradient over a sub-aperture we have to precisely define the position of the sub-aperture within the aperture. Then, we can calculate the interaction matrix Θ_{sh} linking the centroid positions to the modes a_i of the Zernike polynomial. However, this information might be difficult to obtain. Therefore, instead of calculating the sub-image centroid positions for the Zernike modes the deformable mirror is driven to form these modes and the sub-image centroids are then measured.

As an example a 2×2 Shack–Hartmann sensor is modelled, subdividing the aperture into four quarter circles. The integration in (6.37) has to be performed over the sub-apertures, i.e. for $\rho = 0$ to 1 and $\theta = 0$ to $\pi/2$ for the first sub-aperture. The centroid displacement in x for the first sub-aperture therefore becomes

$$M_{\alpha_{x1}} = \frac{\pi}{2} a_2 + \frac{4}{\sqrt{3}} a_4 + \frac{2}{3}\sqrt{6}(a_5 + a_6) + \frac{3}{2}\sqrt{2}(a_7 + a_9) + \frac{\pi}{\sqrt{2}} a_8 + \ldots .$$

As each lenslet yields two measurements, one in α_x and one in α_y, one obtains a total of eight gradients. Usually this system of eight equations is written in matrix form

$$\begin{pmatrix} M_{\alpha_{x1}} \\ M_{\alpha_{y1}} \\ M_{\alpha_{x2}} \\ M_{\alpha_{y2}} \\ \cdot \\ \cdot \\ M_{\alpha_{y8}} \end{pmatrix} = \Theta_{\text{sh}} \begin{pmatrix} a_1 \\ a_2 \\ a_3 \\ a_4 \\ \cdot \\ \cdot \\ a_8 \end{pmatrix}, \tag{6.38}$$

with $\boldsymbol{M}$ the vector containing the measured gradients, $\boldsymbol{a}$ the vector containing the coefficients of the Zernike polynomial and Θ_{sh} the interaction matrix. The optimal number of Zernike coefficients estimated from a 2×2 Shack–Hartmann sensor is not 8 but 6 because the Zernike coefficients are not statistically independent. The non-diagonal covariance matrix (Table 4.2) shows their interdependency. As discussed in Sect. 4.3.3, with Karhunen-Loève functions this increase is less severe, however, with increased computational effort.

Faint Objects

For low light levels, the variance of the tip-tilt measurement in an individual sub-aperture can be found in [187,202]. Assuming that the seeing disk is approximately Gaussian and that it is centred, the related variance of the wave front phase over the sub-aperture is given by

$$\sigma_{\rm ph}^2 = \frac{\pi^2}{2}\frac{1}{N_{ph}}\left(\frac{d}{r_0}\right)^2 \ [\mathrm{rad}^2], \tag{6.39}$$

when N_{ph} is the number of photons contributing to the measurement and d the diameter of the aperture. The connection between the variance of the phase and that of the tip-tilt is given by (4.67).

In the seeing limited case, with $r_0 \leq d$ the phase variance is independent of the sub-aperture size since the number of photons $N_{\rm ph}$ is proportional to the area d^2 of the sub-aperture. In the diffraction limited case, when the sub-apertures are smaller than r_0, the quotient d/r_0 is set equal to 1 and the variance is directly proportional to $1/N_{ph}$.

Measuring the centroid in a large number of sub-apertures, the reconstruction of the phase is computed by a linear process and the noise of each sub-aperture measurement propagates linearly with

$$\sigma_{\rm noise}^2 = P(J)\sigma_{\rm ph}^2, \tag{6.40}$$

where $P(J)$ is the error propagation coefficient as a function of the number J of corrected modes. It depends on the properties of the system, like sensor and mirror geometry, number of sensors and actuators etc. For Zernike modes an approximation is given by [193]

$$P(J) \approx 0.34 \ln(J) + 0.10. \tag{6.41}$$

The error propagation coefficient $P(J)$ of the curvature sensor goes linear with J so that the curvature sensor is particularly suited for low order correction [197].

6.3.2 Closed Loop Operation

In analogy to fringe tracking, we measure the wave front aberrations at a high rate and feed the signal to the deformable mirror, correcting the aberrations in a closed loop. The residual aberrations as function of frame rate can be computed by applying the rejection transfer function, S(f) as given by (6.24), to the power spectra, $\Phi_{Z^n,t}(f)$, of the individual Zernike modes that are displayed in Fig. 4.9.

Since the power spectra are not available in closed form, the integrals have to be computed numerically. For a first estimate, Greenwood [90] presented a simplified approach, applying the rejection transfer function to the power spectrum $\Phi_t(f)$ of the phase fluctuations of the wave front (4.41). This implies a servo loop that, for infinite bandwidth, corrects the wave front perfectly, and all aberrations are caused by the finite bandwidth of the servo system only.

The bandwidth that he defined was subsequently called the *Greenwood frequency* that, by using the single layer approximation, is simplified yielding

$$f_G = 0.43 \frac{\bar{v}}{r_0}. \tag{6.42}$$

The Greenwood frequency is defined such that a servo bandwidth $f_{3\text{dB}}$ equal to f_G provides a residual variance of 1 rad^2, writing.

$$\sigma_G^2 = (f_G/f_{3\text{dB}})^{5/3}\ [\text{rad}^2]. \tag{6.43}$$

For a better performance, the bandwidth has to be increased and other error terms due to the limited number of corrected Zernike modes and due to photon noise have to be considered as well.

The performance of an adaptive optics system is usually given by the Strehl ratio, S, defining the relative peak brightness of the corrected PSF with the diffraction limited Airy disk having a Strehl ratio of 1. A residual variance of 1 rad^2 corresponds to a Strehl ratio of $S = \exp(-\sigma_G^2) = 35\%$, using (3.15).

The performance is not only determined by the bandwidth of the servo loop but also by the residual phase variance σ_J^2, (4.72), after correcting J Zernike modes and by the photon noise σ_{noise}^2, (6.40).

The residual phase variance after correction of J Zernike modes is listed in Table 4.1 for J < 12, and an approximation for large J is given in (4.73), writing

$$\sigma_J^2 = \Delta_J \approx 0.2944 J^{-\sqrt{3}/2}\ (D/r_0)^{5/3}\ [\text{rad}^2] \text{ for large } J\ ,$$

with D the diameter of the telescope aperture. σ_J is also called the fitting error.

We can now write the sum of the variances as

$$\sigma_\varphi^2 = \sigma_G^2 + \sigma_J^2 + P(J)\sigma_{\text{ph}}^2\ [\text{rad}^2], \tag{6.44}$$

listing the fundamental error sources. For the complete analysis of a given system, other error sources of the hardware in use, like detector read noise, have to be considered as well.

Using again average atmospheric conditions with $r_0 = 0.6$ m at $\lambda = 2.2\,\mu$m, and $\bar{v} = 10$ m/s on an 8-m telescope, we sketch the requirements and the limiting magnitude for an adaptive optics system.

Assuming that we correct 100 Zernike modes, we find the residual variance of the uncorrected modes as $\sigma_J^2 = 0.4$ rad^2 corresponding to a Strehl ratio of 67%.

The Greenwood frequency is 7 Hz, under the given atmospheric conditions, corresponding to a residual variance of 1 rad^2, substantially larger than the fitting error σ_J^2. If we want to add not more than 0.15 rad^2 to σ_J^2, we have to set the servo bandwidth $f_{3\text{dB}}$ to $3 f_G = 20$ Hz so that the residual variance is $\sigma_G^2 = (1/3)^{5/3} = 0.15$ rad^2, reducing the Strehl ratio by only 14%, from 67% to

58%. The frame rate of the wave front sensor is about ten times larger than the bandwidth, i.e. we have to read out 200 frames per second, and the exposure time per frame is about 5 ms.

We found a similar bandwidth requirement for fringe tracking. Concluding on the **limiting magnitude**, we assumed that 2% of the photons arrive at the detector. Here, without long optical beam trains and without single mode fibres to feed, we estimate that 10% of the photons contribute to the measurement. Assuming a square grid of 10×10 sub-apertures, i.e. about 80 subapertures inside the circular aperture are sufficient to determine 100 Zernike modes, the diameter of each sub-apertures on an 8-m telescopes is 0.8 m. Then, using (6.39), we find that the read noise per subaperture is $\sigma_{\mathrm{ph}}^2 = 9/N_{ph}$ and, using (6.41), the correcting factor for error propagation is $P(J) = 1.7$, so that we have $\sigma_{\mathrm{noise}}^2 = 15/N_{ph}$.

If less than 0.15 rad^2 shall be added to the overall variance, we need 100 photons on the detector and 1,000 photons per 5 ms in the sub-aperture, corresponding to a magnitude of $m_K = 9$ (see Appendix A.2) for a Strehl ratio of 50%. The same computation for a wave front sensor in the visible shows that the limiting magnitude improves by 1–2 magnitudes to $m_V = 10 - 11$.

One parameter that has not been treated so far is the isoplanatic angle that determines the maximum distance between object and reference star. For adaptive optics, this depends heavily on the degree of correction and on the layer structure of the atmosphere. In general, the requirement is less restrictive than for fringe tracking, and in the near-infrared separations of about 30 arcsec are permitted reducing the Strehl ratio to about 37% of the on-axis value.

A limiting magnitude around $m_V = 10$ and a maximum separation of 30 arcsec means that the fraction of the sky that can be observed, the *sky coverage* is very low. The situation improves drastically if artificial reference stars, *laser guide stars*, are used that can be placed anywhere in the sky [85]. Then, only the tip-tilt, invisible with a laser guide star, needs to be determined. Using the full aperture, the limiting magnitude for the natural tip-tilt reference star on an 8-m telescope improves by $\Delta m = 5$ to $m_V = 15 - 16$.

For interferometry, the concept of a laser guide star does not work, since the fringe motion – being the tip-tilt of the wave front across the array – is invisible with an artificial reference star launched from the ground. Unfortunately, to date, there is no valid concept for an artificial reference star in interferometry.

Not required for interferometry but very useful for single telescope imaging, multi-conjugate adaptive optics (MCAO) increases the size of the corrected field of view to a few arcminutes. Here, several deformable mirrors are placed in planes that are conjugate to the most turbulent layers in the atmosphere, correcting the layers individually [10, 59]. Several wave front sensors are needed to provide the required information and the reference stars have to be distributed over the entire field of view. Ideally, several laser guide stars are used. Objects that are placed off-axis but within the field of view defined by the reference stars are then corrected with equal quality, avoiding the instant drop of image quality in an adaptive optics system with a single deformable mirror. First tests of MCAO systems were very promising [16, 149].

Chapter 7
Modern Interferometers

Following Labeyrie's successful demonstration of stellar interferometry with independent apertures [121], founding the era of modern interferometry, there were several projects in the 1980s and 1990s, at the Observatoire de la Côte d'Azur in France (Grand Interféromètre à 2 Télescopes, GI2T) [124, 164], in the US, Mark I - III [213–215] at Mt. Wilson, California, the Palomar Testbed Interferometer (PTI) [38] at Palomar, California, and the Infrared-Optical Telescope Array (IOTA) at Mt. Hopkins [238], Arizona that, in the meantime, have all been taken out of operation.

In addition, there are the Navy Prototype Optical Interferometer (NPOI) [6] in Flagstaff, Arizona, the Sydney University Stellar Interferometer (SUSI) [53, 54] in Narrabri, Australia, the Cambridge Optical Aperture Synthesis Telescope (COAST) [68] in Cambridge, UK, and the Mitaka optical InfraRed Array (MIRA) [170] at the Mitaka Campus of the National Astronomical Observatory of Japan. All these interferometers, with apertures between 0.4 and 1.5 m and baselines ranging from 2 to 640 m, have operated in the visible or in the near-infrared.

The noticeable exception is the Infrared Spatial Interferometer (ISI) at Mt. Wilson, a three-telescope interferometer using heterodyne detection at an observing wavelength between 9 and 12 μm (N-band) with a maximum baseline of 56 m [92].

The Mark I interferometer was the first to demonstrate active fringe tracking [215], and the PTI had the first dual-feed system in operation [38]. The first closure phase image was produced by COAST [68].

The scientific success of these interferometers was sufficient motivation to equip some of the world's largest telescopes with an interferometric mode. The Keck observatory with two 10-m telescopes at Mauna Kea, Hawaii, and ESO's Very Large Telescope (VLT) observatory with four 8-m Unit Telescopes (UT) at Paranal, Chile, combine the advantage of large apertures with that of long baselines. Since the telescope apertures are substantially larger than r_0, the Fried parameter, they are all equipped with adaptive optics systems.

The Keck Interferometer has a single baseline of 85 m, while the six baselines of the VLT Interferometer (VLTI) range from 47 to 130 m. In addition, the VLTI has four 1.8-m Auxiliary Telescopes (AT) that can be moved on a grid providing baselines from 8 to 203 m. The layout of the VLTI is displayed in Fig. 5.24.

A. Glindemann, *Principles of Stellar Interferometry*, Astronomy and Astrophysics Library, DOI 10.1007/978-3-642-15028-9_7,

Both the Keck Interferometer and the VLTI had first fringes with siderostats within 4 weeks early in 2001 [39, 83]. Very quickly thereafter, a number of instruments were deployed in the near- and mid-infrared producing scientific results [36, 112].

In the same year, the CHARA interferometer on Mt. Wilson, California, also had first fringes [229]. CHARA combines six 1-m telescopes arranged in a *Y* layout, shown in Fig. 5.22. With a maximum baseline of 330 m and the MIRC instrument combining four (and soon six) beams simultaneously [163], CHARA has produced very interesting results despite its rather small apertures [114].

The VLTI has to wait for the second generation of instruments to see the combination of four telescopes. The planning foresees MATISSE in the mid-infrared, [142], and GRAVITY in the near-infrared [67]. GRAVITY will provide narrow angle astrometry and phase referenced imaging with an astrometric accuracy of 10 μ arcsec on the UT for objects separated up to 2 arcsec. The prime scientific goal is the Galactic Center, studying motions to within a few times the radius of the event horizon and potentially testing General Relativity in its strong field limit.

All these interferometers are operated in Michelson configuration so that the field of view is reduced to less than an Airy disk. Only the Large Binocular Telescope (LBT) will use the Fizeau configuration providing 20 arcsec field of view. First Fizeau fringes were achieved in October 2010.

Appendix A

A.1 The Fourier Transform

We call $f(x)$ the Fourier transform of a function $F(s)$, if it is

$$f(x) = \int_{-\infty}^{\infty} F(s)\, \mathrm{e}^{-\mathrm{i}2\pi xs} \mathrm{d}s, \tag{A.1}$$

with $x, s \in \mathbb{R}$, and we write the Fourier back-transform as

$$F(s) = \int_{-\infty}^{\infty} f(x)\, \mathrm{e}^{\mathrm{i}2\pi xs} \mathrm{d}x. \tag{A.2}$$

In our notation the back-transform has a plus sign in the exponential.[1]

$f(x)$ and $F(s)$ form a *Fourier pair*. They can both be complex functions. We denote the Fourier transform by $\mathcal{F}$ using subscript x or s to indicate the coordinate, and the superscript -1 to note the back-transform. Thus, the back-transform of a function of x would be abbreviated by $\mathcal{F}_x^{-1}\left(f(x)\right)$. While this looks a bit artificial for a function of one variable, it is important in the case of several variables.

The translation of a function by s_0, $F(s - s_0)$, yields the Fourier transform $\mathcal{F}_s\left(F(s - s_0)\right) = f(x)\mathrm{e}^{-\mathrm{i}2\pi xs_0}$ (compare for instance to (2.62)), and the derivative of the function $F(s)$, $\frac{\mathrm{d}F(s)}{\mathrm{d}s}$ yields $\mathcal{F}_s\left(\frac{\mathrm{d}F(s)}{\mathrm{d}s}\right) = f(x)\mathrm{i}2\pi x$.

The Fourier transform can be generalised to two and more dimensions, if both f and F are defined on R^2.

If f is a function of time, with $x = t$, then F represents the spectrum of temporal frequencies, $s = \nu$, and we call $f(t)$ the *temporal Fourier transform* of $F(\nu)$. Similarly, going to two dimensions, assuming that $f(\boldsymbol{\alpha})$ is a function of the two dimensional vector $\boldsymbol{\alpha}$, we find that $f(\boldsymbol{\alpha})$ is the *spatial Fourier transform* of $F(\boldsymbol{R})$, representing the spatial frequency spectrum, which is a function of the spatial frequency vector $\boldsymbol{R}$.

[1] This choice of sign is arbitrary since there is no fixed sign convention for Fourier transforms.

A. Glindemann, *Principles of Stellar Interferometry*, Astronomy and Astrophysics Library, DOI 10.1007/978-3-642-15028-9,

Table A.1 Symmetry properties of Fourier pairs. We call even functions symmetric with $f(x) = f(-x)$, and odd functions anti-symmetric with $f(x) = -f(-x)$

$f(\boldsymbol{\alpha})$	$F(\boldsymbol{R})$
real + even	real + even
real + odd	imaginary + odd
imaginary + even	imaginary + even
real + arbitrary	real part even, imaginary part odd
real + positive + arbitrary	modulus even, phase odd

Starting with Sect. 2.3, we denoted two-dimensional integrals by a single integral sign, and integration boundaries are assumed to be $\pm\infty$ unless noted otherwise, so that we write the two-dimensional Fourier transform as

$$f(\boldsymbol{\alpha}) = \int F(\boldsymbol{R})\,\mathrm{e}^{-\mathrm{i}2\pi \boldsymbol{R}\cdot\boldsymbol{\alpha}}\mathrm{d}\boldsymbol{R}. \tag{A.3}$$

$f(\boldsymbol{R})$ is the spatial frequency spectrum that can be computed as Fourier back-transform of $f(\boldsymbol{\alpha})$.

It is interesting to remember that, in Sect. 2.2, replacing $\boldsymbol{\alpha}$ for small angles by x/z one yields $2\pi \boldsymbol{R}\cdot\boldsymbol{\alpha} = 2\pi\boldsymbol{\xi}\cdot\boldsymbol{x}/(\lambda z) = k\boldsymbol{\xi}\cdot\boldsymbol{x}/z$ as argument in the exponential like in (2.22). Thus, the Fourier coordinate pairs are either angle $\boldsymbol{\alpha}$ and spatial frequency $\boldsymbol{R}$, or the spatial coordinates $\boldsymbol{\xi}$ and $\boldsymbol{x}$ weighted by wavelength λ and distance z, respectively.

The symmetry properties in Table A.1 can be derived from (A.3), writing the complex exponential as a sum of cosine and sine function, and expressing the function $F(\boldsymbol{R})$ as the sum of even and odd functions.

The example in the last line is particularly important since for instance the intensity $I(\boldsymbol{\alpha})$ or the point-spread function, PSF, are functions that are real, positive and arbitrary in shape. Their Fourier transforms, the visibility function μ, (3.24), or the OTF, (3.7), have exactly the properties of having an symmetric function in the modulus, $|F(\boldsymbol{R})| = |F(-\boldsymbol{R})|$ and an anti-symmetric function in the phase, $\varphi_F(\boldsymbol{R}) = -\varphi_F(-\boldsymbol{R})$. Functions with these symmetry properties are called *Hermitian*.

Often, we use the Dirac **$\boldsymbol{\delta}$-function** to represent a point-like distribution, for instance a pinhole. The $\boldsymbol{\delta}$-function is defined such that it is $\int\boldsymbol{\delta}(x)\mathrm{d}x = 1$, $\boldsymbol{\delta}(0) \to \infty$ and $\boldsymbol{\delta}(x) = 0$ if $x \neq 0$. The integral of the product of the $\boldsymbol{\delta}$-function with another function $g(x)$ is $\int\boldsymbol{\delta}(x-x_0)g(x)\mathrm{d}x = g(x_0)$, yielding the function value $g(x_0)$ at the position of the peak of the $\boldsymbol{\delta}$-function.

Thus, the Fourier transform of the $\boldsymbol{\delta}$-function is

$$\int \boldsymbol{\delta}(\boldsymbol{R}-\boldsymbol{R}_0)\mathrm{e}^{-\mathrm{i}2\pi \boldsymbol{R}\cdot\boldsymbol{\alpha}}\mathrm{d}\boldsymbol{R} = \mathrm{e}^{-\mathrm{i}2\pi \boldsymbol{R}_0\cdot\boldsymbol{\alpha}},$$

or, for $\boldsymbol{R} = 0$, a constant.

Regarding the sum of two $\boldsymbol{\delta}$-functions as in Young's experiment, we obtain the Fourier transform $f(\boldsymbol{\alpha})$ as

Table A.2 Examples of Fourier pairs

	$f(\boldsymbol{\alpha})$ or $f(\alpha)$ (1D)	$F(\boldsymbol{R})$ or $F(R)$ (1D)
1D:	$\mathrm{sinc}(2\pi\alpha D/2)$	$\mathrm{rect}(\frac{R}{D})$
	$\sum_{j=-\infty}^{\infty}\delta(\alpha - j\alpha_0)$	$\sum_{j=-\infty}^{\infty}\delta(R - 2\pi j/\alpha_0)$
2D:	$\mathrm{Besinc}(2\pi\lvert\boldsymbol{\alpha}\rvert D/2)$	$\mathrm{circ}(\frac{\lvert\boldsymbol{R}\rvert}{D/2})$
	$\int\delta(\boldsymbol{R}-\boldsymbol{R}_0)\mathrm{e}^{i2\pi\boldsymbol{R}\cdot\boldsymbol{\alpha}}\mathrm{d}\boldsymbol{R} = \mathrm{e}^{-i2\pi\boldsymbol{R}_0\cdot\boldsymbol{\alpha}}$	$\delta(\boldsymbol{R}-\boldsymbol{R}_0)$
	$\cos(2\pi\boldsymbol{\alpha}\cdot\boldsymbol{R}_0)$	$\frac{1}{2}(\delta(\boldsymbol{R}-\boldsymbol{R}_0)+\delta(\boldsymbol{R}+\boldsymbol{R}_0))$
	$\sin(2\pi\boldsymbol{\alpha}\cdot\boldsymbol{R}_0)$	$-\mathrm{i}\frac{1}{2}(\delta(\boldsymbol{R}-\boldsymbol{R}_0)-\delta(\boldsymbol{R}+\boldsymbol{R}_0))$
	$\sum_{\boldsymbol{j}=(-\infty,-\infty)}^{(\infty,\infty)}\delta(\boldsymbol{\alpha}-\boldsymbol{j}\alpha_0)$	$\sum_{\boldsymbol{j}=(-\infty,-\infty)}^{(\infty,\infty)}\delta(\boldsymbol{R}-2\pi\boldsymbol{j}/\alpha_0)$
	$\mathrm{e}^{-\lvert\boldsymbol{\alpha}\rvert^2/(2\alpha_0^2)}$	$2\pi\alpha_0^2\,\mathrm{e}^{-2\pi^2\alpha_0^2\lvert\boldsymbol{R}\rvert^2} = \frac{1}{2\pi R_0^2}\mathrm{e}^{-\lvert\boldsymbol{R}\rvert^2/(2R_0^2)}$ with $R_0 = \frac{1}{2\pi\alpha_0}$

$$
\begin{aligned}
f(\boldsymbol{\alpha}) &= \int_{-\infty}^{\infty}(\delta(\boldsymbol{R}-\boldsymbol{R}_0)+\delta(\boldsymbol{R}+\boldsymbol{R}_0))\mathrm{e}^{-\mathrm{i}2\pi\boldsymbol{R}\cdot\boldsymbol{\alpha}}\,\mathrm{d}\boldsymbol{R} \\
&= \cos(2\pi\boldsymbol{\alpha}\cdot\boldsymbol{R}_0) - \mathrm{i}\sin(2\pi\boldsymbol{\alpha}\cdot\boldsymbol{R}_0) + \cos(-2\pi\boldsymbol{\alpha}\cdot\boldsymbol{R}_0) - \mathrm{i}\sin(-2\pi\boldsymbol{\alpha}\cdot\boldsymbol{R}_0) \\
&= 2\cos(2\pi\boldsymbol{\alpha}\cdot\boldsymbol{R}_0),
\end{aligned}
$$

describing the amplitude in the plane of observation if the $\boldsymbol{\delta}$-functions represent the two pinholes.

In order to become familiar with the Fourier transforms of the some typical functions, Table A.2 summarises a variety of one- and two-dimensional functions and their Fourier transforms.

The one-dimensional rectangular function, $\mathrm{rect}(R/D)$, describes a rectangular spectrum with bandwidth $\Delta\nu$, i.e. $G(\nu) = \mathrm{rect}(\nu/\Delta\nu)$. Its Fourier transform is $g(\tau) = \mathrm{sinc}(\pi\tau\Delta\nu)$, with its first zero at $\tau = 1/\Delta\nu$. Shifting the spectrum to the average frequency ν_0, its Fourier transform is $\mathrm{sinc}(\pi\tau\Delta\nu)\mathrm{e}^{-\mathrm{i}2\pi\tau\nu_0}$, see (2.63).

A two-dimensional circular function, $\mathrm{circ}(\frac{|\boldsymbol{R}|}{D/2})$, describes a circular aperture with diameter D, and its Fourier transform is the amplitude $V(\boldsymbol{\alpha})$ in the image plane as given by (2.24).

An infinite distribution of $\boldsymbol{\delta}$-functions, equally spaced by R_0 on a two-dimensional grid, is called a *Dirac comb*. Its Fourier transform is a replica of itself with grid constant $1/\alpha_0$. This was discussed in Sect. 5.4.1.

The last line in Table A.2 displays the Fourier transforms of a two-dimensional Gaussian functions. For a one-dimensional Gaussian, one has to take $\sqrt{2\pi\alpha_0^2}$ as coefficient.

Very often, we discuss the **convolution** product of two functions, defined as

$$
f(x) = \int g(x')h(x-x')\mathrm{d}x' = g(x) * h(x), \tag{A.4}
$$

with $*$ denoting the convolution.

The Fourier back-transform of $f(x)$ is then written as

$$\begin{aligned} F(s) &= \int f(x)\,\mathrm{e}^{\mathrm{i}2\pi xs}\mathrm{d}x \\ &= \int\left[\int g(x')h(x-x')\mathrm{d}x'\right]\mathrm{e}^{\mathrm{i}2\pi xs}\mathrm{d}x \\ &= \iint g(x')h(x-x')\mathrm{e}^{\mathrm{i}2\pi x's}\mathrm{e}^{\mathrm{i}2\pi(x-x')s}\mathrm{d}x'\mathrm{d}x \\ &= \int g(x')\,\mathrm{e}^{\mathrm{i}2\pi x's}\mathrm{d}x \int h(x'')\,\mathrm{e}^{\mathrm{i}2\pi x''s}\mathrm{d}x \quad \text{with } x''=x-x' \\ &= G(s)\times H(s), \end{aligned} \tag{A.5}$$

so that the Fourier transform of a convolution $\mathcal{F}^{-1}\left(g(x) * h(x)\right)$ is the product of the individual Fourier transforms, $G(s)$ and $H(s)$. The imaging process provides an example of a convolution when the object intensity distribution $O(\boldsymbol{\alpha})$ is convolved with the PSF to obtain the image intensity distribution (3.4) as discussed in Sect. 3.1.1. The Fourier transform of the convolution product is the product of the object intensity spectrum with the OTF, $\hat{O}(\boldsymbol{R}) \times \mathrm{OTF}(\boldsymbol{R})$, (3.11).

The convolution of a an aperture function with two $\boldsymbol{\delta}$-functions represents the interferometer aperture. Its Fourier transform is then the product of the Fourier transform of the individual aperture and the Fourier transform of the two $\boldsymbol{\delta}$-functions, yielding the fringe pattern enveloped by an Airy disk.

A convolution with a Dirac comb is particularly interesting since its Fourier transform is again a sum of $\boldsymbol{\delta}$-functions (see Table A.2). The Fourier transform of the product of this function with a visibility function then yields a convolution of the Fourier transform of the visibility function, i.e. the object intensity, with a sum of $\boldsymbol{\delta}$-functions as displayed in Fig. 5.18. Thus, the image shows replica of the object at each peak of the Dirac comb.

Regarding the Fourier transform of an autocorrelation $\int g(x')g^*(x'-x)\mathrm{d}x'$ we apply the same computation as in (A.5), now for $g^*(x'-x)$ instead of $h(x-x')$, yielding the Fourier transform of the autocorrelation as the product of $G(s)$ and its complex conjugate $G^*(s)$. This is also known as the Wiener-Khinchine theorem. We compute the image intensity $I(\boldsymbol{\alpha})$ as the product of the amplitude $V(\boldsymbol{\alpha})$ with its complex conjugate. For a point source this provides the PSF, and its Fourier transform is the OTF (3.6), which is the autocorrelation of the aperture function. This is illustrated by Fig. 3.2.

A.2 Atmospheric Transmission Bands

The atmosphere is transparent only at certain wavelengths. Table A.3 lists frequently used photometric bands from the V-band in the visible to the Q-band in the mid-infrared.

Table A.3 Frequently used astronomical filter passbands, with central wavelength and bandwidth, and reference flux densities ε_0 defining stellar magnitude zero in each band. The number of photons per circular aperture with $D = 1$ m for a magnitude 10 star are given, both per second and nm, and per msec and integrated over the band, assuming a rectangular shape

Band	λ_c [μm]	$\Delta\lambda$ [μm]	ε_0 [W m^{-2} nm^{-1}]	Photons per 1-m aperture [ph s^{-1} nm^{-1}]	Photons per band per 1-m aperture [ph msec^{-1}]
V	0.55	0.09	3.44×10^{-11}	7470	670
R	0.70	0.22	1.76×10^{-11}	4870	1070
I	0.90	0.24	8.3×10^{-12}	2950	710
J	1.25	0.30	3.01×10^{-12}	1490	450
H	1.65	0.35	1.18×10^{-12}	767	270
K	2.20	0.40	4×10^{-13}	350	140
L	3.40	0.55	7.3×10^{-14}	98	54
M	5.00	0.30	2.2×10^{-14}	43	13
N	10.20	5.00	1.23×10^{-15}	5	25
Q	21.00	8.00	6.8×10^{-17}	0.6	5

A scale for the brightness of stars is given by the *stellar magnitude* system providing a relative measure for the brightness. A difference of one magnitude corresponds to a factor of $10^{0.4} = 2.512$. The zero point of the magnitude scale is given by the flux density ε_0 in Table A.3 [138] and http://jcmtarchive.jach.hawaii.edu/UKIRT/astronomy/utils/conver.html. Its value is chosen so that a star of spectral type A0 has the same magnitude in each band. Note that a brighter star has a smaller magnitude with $m = -2.5 \log(\varepsilon/\varepsilon_0)$. An increase by a factor of 100 is then a difference of -5 magnitudes.

The number of photons collected by a telescope is often needed for performance estimates. This number is given in Table A.3 for each band for an $m = 10$ star and for a telescope with a 1-m aperture.

References

1. J.G. Ables, Maximum entropy spectral analysis. Astron. Astrophys. Suppl. Ser. **15**, 383–393 (1974)
2. O. Absil, R. den Hartog, P. Gondoin, P. Fabry, R. Wilhelm et al., Performance study of ground-based infrared bracewell interferometers. Astron. Astrophys. **448**, 787–800 (2006)
3. S. Van Aert, D. Van Dyck, A.J. den Dekker, Resolution of coherent and incoherent imaging systems reconsidered – classical criteria and a statistical alternative. Opt. Express **14**, 3830–3839 (2006)
4. C. Aime, Imaging with hypertelescopes: a simple modal approach. Astron. Astrophys. **483**, 361–364 (2008)
5. J.R.P. Angel, N.J. Woolf, An imaging nulling interferometer to study extrasolar planets. Astrophys. J. **475**, 373–379 (1997)
6. J.T. Armstrong, D. Mozurkewich, L.J. Pickard, D.J. Hutter, J.A. Benson et al., The navy prototype optical interferometer. Astrophys. J. **496**, 550–571 (1998)
7. G.R. Ayers, M.J. Northcott, J.C. Dainty, Knox-Thompson and triple correlation imaging through atmospheric turbulences. J. Opt. Soc. Am. A **5**, 963–985 (1988)
8. J.E. Baldwin, C.A. Haniff, C.D. Mackay, P.J. Warner, Closure phase in high-resolution optical imaging. Nature **320**, 595–597 (1986)
9. H. Bartelt, A.W. Lohmann, B. Wirnitzer, Phase and amplitude recovery from bispectra. Appl. Opt. **23**, 3121–3129 (1984)
10. J.M. Beckers, Increasing the size of the isoplanatic patch with multiconjugate adaptive optics. In *Proceedings of the ESO conference on Very Large Telescopes and their Instrumentation*, vol. 30, ed. by M.-H. Ulrich (Garching, Germany, 1988), pp. 693–703
11. J.M. Beckers, The VLT Interferometer III. Factors affecting wide field-of-view operation. In *Advanced Technology Optical Telescopes IV*, ed. by L.D. Barr. Proceedings of SPIE, vol. 1236, (1990), pp. 379–389
12. M. Benisty, J.-P. Berger, L. Jocou, P. Labeye, F. Malbet, K. Perraut, P. Kern, An integrated optics beam combiner for the second generation VLTI instruments. Astron. Astrophys. **498**, 601–613 (2009)
13. M.J. Beran, G.B. Parrent, *Theory of Partial Coherence* (Prentice-Hall, Englewood Cliffs, NJ, 1964)
14. D.H. Berger, J.D. Monnier, R. Millan-Gabet, T.A. ten Brummelaar, M. Anderson et al., CHARA Michigan phase tracker (CHAMP): a preliminary performance report. In *Optical and Infrared Interferometry*, ed. by W.C. Danchi, M. Schöller, F. Delplancke. Proceedings of SPIE, vol.7013 (2004), pp. 701319–1–701319–10
15. J.P. Berger, P. Haguenauer, P. Kern, K. Perraut, F. Malber et al., Integrated optics for astronomical interferometry. Astron. Astrophys. **376**, L31–L34 (2001)
16. T. Berkefeld, D. Soltau, O. von der Lühe, Multi-conjugate solar adaptive optics with the VTT and GREGOR. In *Advances in Adaptive Optics II*, ed. by B.L. Ellerbroek, D. Bonaccini Calia. Proceedings of SPIE , vol.6272 (2006) p. 627205

17. Y. Biraud, A new approach for increasing the resolving power by data processing. Astron. Astrophys. **1**, 124–127 (1969)
18. K.P. Birch, M.J. Downs, An updated Edlén equation for the refractive index of air. Metrologia **30**, 155–162 (1993)
19. M. Born, E. Wolf, *Principles of Optics,* 6th edn. (Cambridge University Press, Cambridge, 1980)
20. R.N. Bracewell, Detecting nonsolar planets by spinning infrared interferometer. Nature, **274**, 780–781 (1978)
21. A.H. Bridle, F.R. Schwab, Bandwidth and time-average smearing. In *Synthesis Imaging in Radio Astronomy II*, ed. by G.B. Taylor, C.L. Carilli, R.A. Perley. ASP Conference Series, vol. 180 (ASP, San Francisco, 1999), pp. 371–381
22. D.S. Briggs, F.R. Schwab, R.A. Sramek, Imaging. In *Synthesis Imaging in Radio Astronomy II*, ed. by G.B. Taylor, C.L. Carilli, R.A. Perley. ASP Confernce Series, vol.180 (ASP, San Francisco, 1999), pp. 127–149
23. A. Burrows, D. Sudarsky, J.I. Lunine, Beyond the T dwarfs: theoretical spectra, colors, and detectability of the coolest brown dwarfs. Astrophys. J., **596**, 587–596 (2003)
24. D.F. Buscher, Optimizing a ground-based optical interferometer for sensitivity at low light levels. Mon. Not. R. Astron. Soc., **235**, 1203–1226 (1988)
25. D.F. Buscher, *Direct maximum-entropy method image reconstruction from the bispectrum* ed. by J.G. Robertson, W.J. Tango. Very High Angular Resolution Imaging, vol. 158 (Proc. IAU Symposium, 1994), pp. 91–93
26. D.F. Buscher, J.T. Armstrong, C.A. Haniff, A. Quirrenbach, D. Mozurkewich et al., Interferometric seeing measurements on Mt. Wilson: power spectra and outer scales. Appl. Opt. **34**, 1081–1096 (1995)
27. W.H. Carter, E. Wolf, Coherence and radiometry with quasihomogeneous planar sources. J. Opt. Soc. Am. **67**, 785–796 (1977)
28. F. Chassat, Calcul du domaine disoplanétisme dun systéme doptique adaptative fonctionnant á travers la turbulence atmosphérique. J. Opt. (Paris) **20**, 13–23 (1989)
29. P.E. Ciddor, Refractive index of air: new equations for the visible and near infrared. Appl. Opt. **35**, 1566–1573 (1996)
30. P.E. Ciddor, Refractive index of air: 3. the roles of CO_2, H_2O, and refractivity virials. Appl. Opt. **41**, 2292–2298 (2002)
31. B.G. Clark An efficient implementation of the algorithm CLEAN. Astron. Astrophys. **89**, 377–378 (1980)
32. S.F. Clifford, The classical theory of wave propagation in a turbulent medium. In *Laser Beam Propagation in the Atmosphere, Topics in Applied Physics, vol. 25, ed. by J.W. Strobehn* (Springer, Berlin, 1968), pp. 9–43
33. M.M. Colavita, Design considerations for very long baseline fringe-tracking interferometers. In *Amplitude and Intensity Interferometer*, ed. by J. Breckinridge. Proceedings of SPIE, vol. 1237 (1990), pp. 80–86
34. M.M. Colavita, Measurement of the atmospheric limit to narrow-angle interferometric astrometry using the Mark III stellar interferometer. Astron. Astrophys. **283**, 1027–1036 (1994)
35. M.M. Colavita, Fringe visibility estimators for the Palomar Testbed Interferometer. Publ. Astron. Soc. Pac. **111**, 111–117 (1999)
36. M.M. Colavita, R. Akeson, P. Wizinowich et al, Observations of DG Tauri with the Keck interferometer. Astrophys. J. **592**, L83–L86 (2003)
37. M.M. Colavita, M. Shao, D.H. Staelin, Atmospheric phase measurements with the Mark III stellar interferometer. Appl. Opt. **26**, 4106–4112 (1987)
38. M.M. Colavita, J.K. Wallace, B.E. Hines, Y. Gursel, F. Malbet, D.L. Palmer et al., The Palomar testbed interferometer. Astrophys. J. **510**, 505–521 (1999)
39. M.M. Colavita, P. Wizinowich, Keck interferometer update. *Interferometry for Optical Astronomy II*, ed. by W. Traub. Proceedings of SPIE, vol. 4838 (2002), pp. 79–88
40. J.-M. Conan, G. Rousset, P.-Y. Madec, Wave-front temporal spectra in high-resolution imaging through turbulence. J. Opt. Soc. Am. A **12**, 1559–1570 (1995)

41. R. Conan, A. Ziad, J. Borgnino, F. Martin, A. Tokovinin, Measurements of the wave front outer scale at Paranal: influence of this parameter in interferometry. In *Interferometry in Optical Astronomy*, ed. by P.J. Léna, A. Quirrenbac. Proceedings of SPIE, vol. 4006 (2000), pp. 963–973
42. T. Cornwell, R. Braun, D.S. Briggs, Deconvolution. *Synthesis Imaging in Radio Astronomy II*, ed. by G.B. Taylor, C.L. Carilli, R.A. Perley. ASP Conference Series, vol. 180 (ASP, San Francisco, 1999), pp. 151–170
43. T.J. Cornwell, A method of stabilizing the clean algorithm. Astron. Astrophys. **121**, 281–285 (1983)
44. T.J. Cornwell, A novel principle for optimization of the instantaneous fourier plane coverage of correlation arrays. IEEE Trans. Antennas Propag. **36**, 1165–1167 (1988)
45. T.J. Cornwell, K.F. Evans, A simple maximum entropy deconvolution algorithm. Astron. Astrophys. **143**, 77–83 (1985)
46. W. Cotton, J. Monnier, F. Baron et al., 2008 imaging beauty contest. In *Optical and Infrared Interferometry*, ed. by M. Schöller, W.C. Danchi. Proceedings of SPIE, vol. 5491 (2008), p. 70131N
47. W.D. Cotton, Special problems in imaging. In *Synthesis Imaging in Radio Astronomy II*, ed. by G.B. Taylor, C.L. Carilli, R.A. Perley. ASP Confernce Series, vol. 180 (ASP, San Francisco, 1999), pp. 357–370
48. V. Coudé de Foresto, S. Ridgway, FLUOR a stellar interferometer using single-mode infrared fibers. In *ESO Conference on High-resolution imaging by interferometry II*, ed. by J.M. Beckers, F. Merkle (European Southern Observatory, Garching, Germany, 1991), pp. 731–740.
49. V. Coudé de Foresto, S. Ridgway, J.-M. Mariotti, Deriving object visibilities from interferograms obtained with a fiber stellar interferometer. Astron. Astrophys. Suppl. Ser. **121**, 379–392 (1997)
50. J.C. Dainty, F.R. Fienup, Phase retrieval and image reconstruction for astronomy. In *Image Recovery: Theory and Application* (Academic, New York, USA, 1987) pp. 231–274
51. L.A. D'Arcio, Selected aspects of wide-field stellar interferometry. PhD thesis, TU Delft (1999)
52. J. Davis, P.R. Lawson, A.J. Booth, W.J. Tango, E.D. Thorvaldson, Atmospheric path variations for baselines up to 80 m measured with the Sydney University Stellar Interferometer. Mon. Not. R. Astron. Soc. **273**, L53–L58 (1995)
53. J. Davis, W.J. Tango, A.J. Booth, T.A. ten Brummelaar, R.A. Minard, The Sydney University stellar interferometer – I. the instrument. Mon. Not. R. Astron. Soc. **303**, 773–782 (1999)
54. J. Davis, W.J. Tango, A.J. Booth, E.D. Thorvaldson, J. Giovannis, The Sydney University stellar interferometer – II. commissioning observations and results. Mon. Not. R. Astron. Soc. **303**, 783–791 (1999)
55. A. Domiciano de Souza, P. Kervella, S. Jankov, L. Abe, F. Vakili, E. di Folco, F. Paresce, The spinning-top Be star Achernar from VLTI-VINCI. Astron. Astrophys. **407**, L47–L50 (2003)
56. A.J. den Dekker, A. van den Boos, Resolution: a survey. J. Opt. Soc. Am. A **14**, 547–557 (1997)
57. E. di Folco, B. Koehler, P. Kervella, M. Sarazin, V. Coudé du Foresto, M. Schöller, M. Wittkowski, Atmospheric and internal turbulence measured on the Very Large Telescope Interferometer with VINCI. *Interferometry for Optical Astronomy II*, ed. by W.A. Traub. Proceedings of SPIE, vol. 4838 (2002), pp. 1115–1126
58. G.T. di Francia, Super-gain antennas and optical resolving power. Nuovo Cimento Suppl. **9**, 426–435 (1952)
59. R.H. Dicke, Phase-contrast detection of telescope seeing errors and their corrections. Astrophys. J. **198**, 605–615 (1975)
60. S.J. Dobson, A. Cox, Fast image-quality-based optimization of optical systems. Appl. Opt. **37**, 8008–8011 (1998)
61. D. Dravins, L. Lindegren, E. Mezey, A.T. Young, Atmospheric intensity scintillation of stars. I. Statistical distributions and temporal properties. Publ. Astron. Soc. Pac. **109**, 173–207 (1997)

62. D. Dravins, L. Lindegren, E. Mezey, A.T. Young, Atmospheric intensity scintillation of stars. II. Dependence on optical wavelength. Publ. Astron. Soc. Pac. **109**, 725–737 (1997)
63. D. Dravins, L. Lindegren, E. Mezey, A.T. Young, Atmospheric intensity scintillation of stars. III. Effects for different telescope apertures. Publ. Astron. Soc. Pac. **110**, 610–633 (1998)
64. P.M. Duffieux, *L'intégral de Fourier et ses Applications à l'Optique* (Societé Anonyme des Imprimeries Oberthur, Rennes, 1946)
65. P.M. Duffieux, *The Fourier transform and its applications to optics* 2nd edn. (Wiley, New York, 1983)
66. B. Edlén, The refractive index of air. Metrologia **2**, 71–80 (1966)
67. F. Eisenhauer, G. Perrin, W. Brandner, C. Straubmeier et al., GRAVITY: getting to the event horizon of Sgr A*. In *Optical and Infrared Interferometry*, ed. by M. Schöller, W.C. Danchi, F. Delplancke. Proceedings of SPIE, vol. 7013 (2008), pp. 70132A–1–70132A–13
68. J.E. Baldwin et al., The first images from an optical aperture synthesis array: mapping of Capella with COAST at two epochs. Astron. Astrophys. **306**, L13–L16 (1996)
69. H. Fizeau, Prix Borodin: rapport sur le concours de l'année 1867. Comptes Rendus de l'Academie des Sciences **66**, 932–934 (1868)
70. A.T. Forrester, R.A. Gudmundsen, P.O. Johnson, Photoelectric mixing of incoherent light. Phys. Rev. **99**, 1691–1700 (1955)
71. D.L. Fried, Statistics of a geometric representation of wavefront distortion. J. Opt. Soc. Am. **55**, 1427–1435 (1965)
72. D.L. Fried, Optical resolution through a randomly inhomogeneous medium for very long and very short exposures. J. Opt. Soc. Am. **56**, 1372–1379 (1966)
73. D.L. Fried, Angular dependence of the atmospheric turbulence effect in speckle interferometry. Opt. Acta (Lond) **26**, 597–613 (1979)
74. D.L. Fried, Anisoplanatism in adaptive optics. J. Opt. Soc. Am. **72**, 52–61 (1982)
75. C. Froehly, Coherence and interferometry through optical fibers. In *ESO Conference on Scientific Importance of High Angular Resolution at Infrared and Optical Wavelengths*, ed. by M.H. Ulrich, K. Kjaer (European Southern Observatory, Garching, Germany, 1981), pp. 285–293
76. D. Gabor, Theory of communication. Part I. the analysis of information. J. Inst. Elect. Eng. Lond. **93**, 429–441 (1946)
77. R.O. Gappinger, R.T. Diaz, A. Ksendzov, P.R. Lawson, O.P. Lay et al., Experimental evaluation of achromatic phase shifters for mid-infrared starlight suppression. Appl. Opt. **48**, 868–880 (2009)
78. J. Gay, Y. Rabbia, Principe d'un coronographa interférentiel. C. R. Acad. Sci. Paris. Ser. IIB: Mech. Phys. Chim. Astron **322**, 265–271 (1996)
79. E. Gershnik, E.N. Ribak, Light propagation through multilayer atmospheric turbulence. Opt. Commun. **142**, 99–105 (1997)
80. A. Glindemann, Das Bild der Kante bei teilkohärenter Beleuchtung – Rechnungen und Experimente. Dissertation, Technische Universität Berlin, 1989
81. A. Glindemann. Relevant parameters for tip-tilt systems on large telescopes. Publ. Astron. Soc. Pac. **109**, 682–687 (1997)
82. A. Glindemann, M. Albertsen, L. Andolfato, G. Avila, P. Ballester et al., VLTI technical advances – present and future. In *New Frontiers in Stellar Interferometry*, ed. by W.A. Traub. Proceedings of SPIE, vol. 5491 (2004), pp. 447–453
83. A. Glindemann, J. Argomedo, R. Amestica, P. Ballester, B. Bauvir et al., The VLTI – A status report. In *Interferometry for Optical Astronomy II*, ed. by W. Traub. Proceedings of SPIE, vol. 4838 (2002), pp. 89–100
84. A. Glindemann and J.C. Dainty, Object fitting to the bispectral phase using least-squares. J. Opt. Soc. Am. A **10**, 1056–1063 (1993)
85. A. Glindemann, S. Hippler, T. Berkefeld, W. Hackenberg, Adaptive optics on large telescopes. Exp. Astron. **10**, 5–47 (2000)
86. M.J.E. Golay, Point arrays having compact, nonredundant autocorrelations. J. Opt. Soc. Am. **61**, 272–273 (1970)
87. J.W. Goodman, *Statistical Optics* (Wiley, New York, 1985)

88. J.W. Goodman, *Introduction to Fourier optics*, 3rd edn. (Roberts, Englewood, Colorado, 2005)
89. F. Gori, Mode propagation of the field generated by Collett-Wolf Schell-model sources. Opt. Commun. **46**, 149–154 (1983)
90. D.P. Greenwood, Bandwidth specification for adaptive optics systems. J. Opt. Soc. Am. **67**, 390–393 (1977)
91. O. Guyon, E.A. Pluzhnik, M.J. Kuchner, B. Collins, S.T. Ridgway, Theoretical limits on extrasolar terrestrial planet detection with coronagraphs. Astrophys. J. Suppl. **167**, 81–99 (2006)
92. D.D.S. Hale, M. Bester, W.C. Danchi, W. Fitelson et al., The Berkeley infrared spatial interferometer: a heterodyne stellar interferometer for the mid-infrared. Astrophys. J. **537**, 998–1012 (2000)
93. R. Hanbury Brown. *The Intensity Interferometer* (Taylor and Francis, London, 1974)
94. R. Hanbury Brown, R.C. Jennison, M.K. Das Gupta, Apparent angular sizes of discrete radio sources. Nature **170**, 1061 (1952)
95. R. Hanbury Brown, R.Q. Twiss, A new type of interferometer for use in radio-astronomy. Philos. Mag. **45**, 663 (1954)
96. R. Hanbury Brown, R.Q. Twiss, A test of a new type of stellar interferometer on Sirius. Nature **178**, 1046–1048 (1956)
97. R. Hanbury Brown and R.Q. Twiss, Interferometry of the intensity fluctuations in light. part I. Proc. R. Soc. A **242**, 300–324 (1957)
98. C.A. Haniff, J.E. Baldwin, P.J. Warner, T.R. Scott, Atmospheric phase fluctuation measurement: interferometric results from the WHT and COAST telescopes. In *Amplitude and Spatial Interferometry II*, ed. by J. Breckinridge. Proceedings of SPIE, vol. 2200 (1994), pp. 407–417
99. C.A. Haniff, D.F. Buscher, Diffraction-limited imaging with partially redundant masks. I. Infrared imaging of bright objects. J. Opt. Soc. Am. A **9**, 203–218 (1992)
100. C.A. Haniff, C.D. Mackay, D.J. Titterington, D. Sivia, J.E. Baldwin, P.J. Warner, The first images from optical aperture synthesis. Nature **328**, 694–696 (1987)
101. J.M. Hill, R.F. Green, J.H. Slagle, The Large Binocular Telescope. In *Ground-based and Airborne Telescopes*, ed. by L.M. Stepp. Proceedings of SPIE, vol. 6267 (2006), pp. 1–15
102. J.A. Högbom Aperture synthesis with a non-regular distribution of interferometer baselines. Astron. Astrophys. Suppl. **15**, 417–426 (1974)
103. H. Hogenhuis, M. Visser, F. Derie, Commissioning of the VLTI delay lindes on mount paranal. In *Interferometry for Optical Astronomy II*, ed. by W. Traub. Proceedings of SPIE, vol. 4838 (2002), pp. 1148–1154
104. M.A. Holdaway, T.T. Helfer, Interferometric array design. In *Synthesis Imaging in Radio Astronomy II*, ed. by G.B. Taylor, C.L. Carilli, R.A. Perley. ASP Conference Series, vol. 180 (ASP, San Francisco, 1999), pp. 537–563
105. H.H. Hopkins, The use of diffraction-based criteria of image quality in automatic optical design. Opt. Acta(Lond) **13**, 343–369 (1966)
106. R.E. Hufnagel, Variations of atmospheric turbulence. In *Digest of Technical Papers, Topical Meeting on Optical Propagation through Turbulence* (Optical Society of America, Washington, D. C., 1974), pp. WA1/1–WA1/4
107. A. Ishimura, *Wave Propagation and Scattering in Random Media* (Academic, New York, 1978)
108. E.T. Jaynes, Information theory and statistical mechanics. Phys. Rev. **108**, 171–190 (1957)
109. M.A. Johnson, A.L. Betz, C.H. Townes, 10-μm heterodyne stellar interferometer. Phys. Rev. Lett. **33**, 1617–1620 (1974)
110. P. Kern, F. Malbet, I. Schanen-Duport, P. Benech, Integrated optics single-mode interferometric beam combiner for near infrared astronomy. In *Integrated Optics for Astronomical Interferometry* (Bastianelli-Guirimand, Grenoble, France, 1997), pp. 195–203
111. P. Kervella, D. Ségransan, V. Coudé du Foresto. Data reduction methods for single-mode optical interferometry – application to the VLTI two-telescope beam combiner VINCI. Astron. Astrophys. **425**, 1161–1174 (2004)

112. P. Kervella, F. Thévenin, D. Ségransan, G. Berthomieu, B. Lopez et al., The diameters of α Centauri A and B. Astron. Astrophys., **404**, 1087–1097 (2003)
113. E. Keto, The shapes of cross-correlation interferometers. Astrophys. J. **475**, 843–852 (1997)
114. B. Kloppenborg, R. Stencel, J.D. Monnier, G. Schaefer, M. Zhao et al., Infrared images of the transiting disk in the ϵ Aurigae system. Nature **464**, 870–872 (2010)
115. K.T. Knox, B.J. Thompson, Recovery of atmospherically degraded short-exposure images. Astrophys. J. **193**, 45–48 (1974)
116. L. Koechlin and J.-P. Pérez, Limit in field-resolution ratio for interferometric arrays. In *Interferometry for Optical Astronomy II*, ed. by W.A. Traub. Proceedings of SPIE, vol. 4838 (2003), pp. 411–415
117. A.N. Kolmogorov Dissipation of energy in the locally isotropic turbulence. *C. R. (Dok.) Acad. Sci. URSS* **32**, 16–18 (1941)
118. A.N. Kolmogorov, The local structure of turbulence in incompressible viscous uid for very large Reynolds numbers. C. R. (Dok.) Acad. Sci. URSS **30**, 301–305 (1941)
119. M.M. Komesaroff, R. Narayan, R. Nityananda, The maximum entropy method of image restoration – properties and limitations. Astron. Astrophys. **93**, 269–281 (1981)
120. A. Labeyrie, Attainment of diffraction limited resolution in large telescopes by Fourer analysing speckle patterns in star images. Astron. Astrophys. **6**, 85–87 (1970)
121. A. Labeyrie, Interference fringes obtained with Vega with two optical telescopes. Astrophys. J. **196**, L71–L75 (1975)
122. A. Labeyrie, Resolved imaging of extra-solar planets with future 10-100 km optical interferometric arrays. Astron. Astrophys. Suppl. Ser. **118**, 517–524 (1996)
123. A. Labeyrie, S.G. Lipson, P. Nisenson, *An Introduction to Optical Stellar Interferometry* (Cambridge University Press, Cambridge, 2006)
124. A. Labeyrie, G. Schumacher, M. Dugué, C. Thom, P. Bourlon et al., Fringes obtained with the large boules interferometer at CERGA. Astron. Astrophys. **162**, 359–364 (1986)
125. B.F. Lane, M.M. Colavita, A.F. Boden, P.R. Lawson, Palomar testbed interferometer – update. In *Interferometry in Optical Astronomy*, ed. by P.J. Léna, A. Quirrenbac. Proceedings of SPIE, vol. 4006 (2000), pp. 453–458
126. R.G. Lane, Blind deconvolution of speckle images. J. Opt. Soc. Am. A **9**, 1508–1541 (1992)
127. O. Lardière, F. Martinache, F. Patru, Direct imaging with highly diluted apertures I. Field-of-view limitations. Mon. Not. R. Astron. Soc. **375**, 977–988 (2007)
128. P.R. Lawson, Group-delay tracking in optical stellar interferometry with the fast Fourier transform. J. Opt. Soc. Am. A **12**, 366–374 (1995)
129. P.R. Lawson, Notes on the history of stellar interferometry. In *Principles of Long Baseline Stellar Interferometry*, ed. by P.R. Lawson. NASA-JPL (Publication 00-009) (2000), pp. 325–332
130. P.R. Lawson, M.M. Colavita, P.J. Dumont, B.F. Lane, Least-squares estimation and group delay in astrometric interferometers. In *Interferometry in Optical Astronomy*, ed. by P.J. Léna, A. Quirrenbac. Proceedings of SPIE, vol. 4006 (2000), pp. 397–406
131. P.R. Lawson, W.D. Cotton, C.A. Hummel et al., An interferometry imaging beauty contest. In *New Frontiers in Stellar Interferometry*, ed. by W.A. Traub. Proceedings of SPIE, vol. 5491 (2004), pp. 886–899
132. P.R. Lawson, W.D. Cotton, C.A. Hummel et al., The 2006 interferometry imaging beauty contest. In *Advances in Stellar Interferometry*, ed. byJ.D. Monnier, M. Schöller, W.C. Danchi. Proceedings of SPIE, vol. 6268 (2006), pp. 62681U
133. P.R. Lawson, D.M.A. Wilson, J.E. Baldwin, A desktop interferometer for optical synthesis imaging. In *Interferometry for Optical Astronomy II*, ed. by W.A. Traub. Proceedings of SPIE, vol. 4838 (2002), pp. 404–410
134. P.R. Lawson (ed.), *Selected Papers on Long Baseline Stellar Interferometry* SPIE MS139 (SPIE, Bellingham, 1997)
135. O.P. Lay, Systematic errors in nulling interferometers. Appl. Opt. **43**, 6100–6123 (2004)
136. G. Le Besnerais, S. Lacour, L.M. Mugnier, E. Thiébaut, G. Perrin, S. Meimon, Advanced imaging methods for long-baseline optical interferometry. *IEEE J. Sel. Top. Signal Process.* **2**, 767–780 (2008)

137. A. Léger, M. Mariotti, B. Mennesson, M. Ollivier, J.L. Puget, D. Rouan, J. Schneider The DARWIN project. Astrophys. Space Sci. **241**, 135–146 (1996)
138. P. Léna, F. Lebrun, F. Mignard, *Observational Astrophysics*, 2nd edn. (Springer, Berlin, 1997)
139. L. Lindegren, Atmospheric limitations of narrow-field optical astrometry. Astron. Astrophys. **89**, 41–47 (1980)
140. A.W. Lohmann, G. Weigelt, B. Wirnitzer, Speckle masking in astronomy: triple correlation theory and applications. Appl. Opt. **22**, 4028–4037 (1983)
141. G. Lombardi, J. Navarrete, M. Sarazin, Combining turbulence profiles from MASS and SLODAR. In *Ground-based and Airborne Telescopes II*, ed. by L.M. Stepp, R. Gilmozzi. Proceedings of SPIE, vol. 7012 (2008), pp. 701221–11
142. B. Lopez, P. Antonelli, S. Wolf, S. Lagarde, W. Jaffe et al., MATISSE: perspective of imaging in the mid infrared at the VLTI. In *Optical and Infrared Interferometry*, ed. by M. Schöller, W.C. Danchi, F. Delplancke. Proceedings of SPIE, vol. 7013 (2008), pp. 70132B–1–70132B–10
143. A. Luiz, L.L. Sanchez-Soto, A quantum description of the beam splitter. Quantum Semiclassical Opt. **7**, 153–160 (1995)
144. W. Lukosz, Optical systems with resolving powers exceeding the classical limit. J. Opt. Soc. Am. **56**, 1463–1472 (1966)
145. R.F. Lutomirski, H.T. Yura, Wave structure function and mutual coherence function of an optical wave in a turbulent atmosphere. J. Opt. Soc. Am. **61**, 482–487 (1971)
146. L. Mandel, Fluctuations of light beams. In *Progress in Optics II*, ed. by E. Wolf. (North-Holland, Amsterdam, 1963), pp. 181–248
147. L. Mandel, E. Wolf, *Optical Coherence and Quantum Optics* (Cambridge University Press, Cambridge, 1995)
148. A.S. Marathay, *Elements of Optical Coherence Theory* (Wiley, New York, 1982)
149. E. Marchetti, R. Brast, B. Delabre, R. Donaldson, E. Fedrigo et al., MAD on sky results in star oriented mode. In *Adaptive Optics Systems*, ed. by N. Hubin, C.E. Max, P.L. Wizinowich. Proceedings of SPIE, vol. 7015 (2008), p. 70150F
150. J.-M. Mariotti, S.T. Ridgway, Double Fourier spatio-spectral interferometry: combining high spectral and high spatial resolution in the near infrared. Astron. Astrophys. **195**, 350–363 (1988)
151. H.M. Martin, Image motion as a measure of seeing quality. Publ. Astron. Soc. Pac. **99**, 1360–1370 (1987)
152. R.J. Mathar, Calculated refractivity of water vapor and moist air in the atmospheric window at 10 μm. Appl. Opt. **43**, 928–932 (2004)
153. P. Mege, F. Malbet, A. Chelli, Interferometry with single mode waveguide. In *Interferometry for Optical Astronomy II*, ed. by W.A. Traub. Proceedings of SPIE, vol. 4838 (2003), pp. 329–337
154. J. Meisner, R. Le Poole, Dispersion affecting the VLTI and 10 micron interferometry using MIDI. In *Interferometry for Optical Astronomy II*, ed. by W.A. Traub. Proceedings of SPIE, vol. 4838 (2002), pp. 609–624
155. E. Menzel, Optik **8**, 295–301 (1951)
156. A.A. Michelson. On the application of interference methods to astronomical measurements. Philos. Mag. **30**, 1–20 (1890)
157. A.A. Michelson, E.W. Morley, On the relative motion of the Earth and the luminiferous ether. Am. J. Sci. **34**, 333–345 (1887)
158. A.A. Michelson, F.G. Pease, Measurement of the diameter of α Orionis with the interferometer. Astrophys. J. **53**, 249–259 (1921)
159. F. Millour, Interférométrie différentielle avec AMBER. Thesis, Université Joseph Fourier, Grenoble (2006)
160. J. Monnier, M. Zhao, E. Pedretti et al, Imaging the surface of Altair. Science **317**, 342–345 (2007)
161. J.D. Monnier, Optical interferometry in astronomy. Rep. Prog. Phys. **66**, 789–857 (2003)

162. J.D. Monnier, J.-P. Berger, R. Millan-Gabet, T. ten Brummelaar, The Michigan infrared combiner (MIRC). In *New Frontiers in Stellar Interferometry*, ed. by W.A. Traub. Proceedings of SPIE, vol. 5491 (2004) pp. 1370–1378
163. J.D. Monnier, E. Pedretti, N. Thureau, J.-P. Berger, R. Millan-Gabet et al., Michigan infrared combiner (MIRC): commissioning results at the CHARA array. In *Advances in Stellar Interferometry*, ed. by M. Schöller, J.D. Monnier, W.C. Danchi. Proceedings of SPIE, vol. 6268 (2004), pp. 62681P–1–62681P–11
164. D. Mourard, I. Tallon-Bosc, A. Blazit, D. Bonneau, G. Merlin et al., The GI2T interferometer on Plateau de Calern. Astron. Astrophys. **283**, 705–713 (1994)
165. D. Mozurkewich, A hybrid design for a six way beam combiner. In *Amplitude and Intensity Spatial Interferometry II*, ed. by J.B. Breckinridge. Proceedings of SPIE, vol. 2200 (1994), pp. 76–80
166. D. Mozurkewich, Beam combination with a large number of apertures. In *Optical and Infrared Interferometry*, ed. by M. Schöller, W.C. Danchi, F. Delplancke. Proceedings of SPIE, vol. 7013 (2008), pp. 70132D–1–70132D–9
167. D. Mozurkewich, K.J. Johnston, R.S. Simon, P.F. Bowers, R. Gaume, Angular diameter measurements of stars. Astron. J. **101**, 2207–2219 (1991)
168. M.W. Muterspaugh, B.F. Lane, M. Konacki, B.F. Burke, M.M. Colavita et al., Scientific results from high-precision astrometry at the palomar testbed interferometer. In *Advances in Stellar Interferometry*, ed. by J.D. Monnier, M. Schöller, W.C. Danchi. Proceedings of SPIE, vol. 6268 (2006), pp. 62680F–1–62680F–8
169. R. Narayan, R. Nityananda, Maximum entropy image restoration in astronomy. Annu. Rev. Astron. Astrophys. **24**, 127–170 (1986)
170. J. Nishikawa, K. Sato, M. Yoshizawa, T. Fukushima, Y. Machida et al, Mitaka optical and infrared array first stage (MIRA-I.1) instruments. In *Interferometry in Optical Astronomy*, ed. by P.J. Léna, A. Quirrenbac. Proceedings of SPIE, vol. 4006 (2000), pp. 681–687
171. R.J. Noll, Zernike polynomials and atmospheric turbulence. J. Opt. Soc. Am. **66**, 207–211 (1976)
172. A.M. Obukhov, Izv. Akad. Nauk S.S.S.R. Ser. Geogr. Geofiz. **13**, 58 (1949)
173. A. Ofir, E. Ribak, Offline, multidetector intensity interferometers – I. Theory. Mon. Not. R. Astron. Soc. **368**, 1646–1651 (2006)
174. A. Ofir, E. Ribak, Offline, multidetector intensity interferometers – II. Implications and applications. Mon. Not. R. Astron. Soc. **368**, 1652–1656 (2006)
175. J.C. Owens, Optical refractive index of air: dependence on pressure, temperature and composition. Appl. Opt. **6**, 51–59 (1967)
176. A. Papoulis, *Probability, Random Variables, and Stochastic Processes* (McGraw-Hill, New York, 1965)
177. A. Papoulis, *Systems and Transforms with Applications in Optics* (McGraw-Hill, New York, 1968)
178. R.R. Parenti, R.J. Sasiela, Laser-guide-star systems for astronomical applications. J. Opt. Soc. Am. A **11**, 288–309 (1994)
179. T. Paumard, G. Perrin, A. Eckart, R. Genzel, P. Lena, R. Schoedel, F. Eisenhauer, T. Mueller, S. Gillessen, Scientific prospects for VLTI in the Galactic Centre: Getting to the Schwarzschild radius. Astron. Nachr. **326**, 568–572 (August 2005)
180. T.J. Pearson, A.C.S. Readhead, Image formation by self-calibration in radio astronomy. Astron. Astrophys. **22**, 97–130 (1984)
181. G. Perrin, Correction of the piston effect in optical astronomicalinterferometry - I. Modulus and phase gradient of the visibility function restoration. Astron. Astrophys. Suppl. Ser. **121**, 553–568 (1997)
182. G. Perrin, S. Lacour, J. Woillez, E. Thiébaut, High dynamic range imaging by pupil single-mode filtering and remapping. Mon. Not. R. Astron. Soc. **373**, 747–751 (2006)
183. R.D. Peters, O.P. Lay, M. Jeganathan, Broadband phase and intensity compensation with a deformable mirror for an interferometric nuller. Appl. Opt. **47**, 3920–3926 (2008)
184. R.G. Petrov, F. Malbet, G. Weigelt et al., AMBER, the near-infrared spectro-interferometric three-telescope VLT instrument. Astron. Astrophys. **464**, 1–12 (2007)

185. R.K. Pina, R.C. Puetter, Incorporation of spatial information in Bayesian image reconstruction: The maximum residual likelihood criterion. Publ. Astron. Soc. Pac. **104**, 1096–1103 (1992)
186. S. Prasad, G. Loos, Spatial filtering of atmoshperic decorrelation from wavefronts. In *ESO Conference on High-Resolution Imaging by Interferometry II*, ed. by F. Merkle (European Southern Observatory, Garching, Germany, 1991), pp. 1057–1065
187. J. Primot, G. Rousset, J.C. Fontanella, Deconvolution from wave-front sensing: a new technique for compensating turbulence-degraded images. J. Opt. Soc. Am. A **7**, 1598–1608 (1990)
188. A. Quirrenbach, Optical interferometry. Annu. Rev. Astron. Astrophys. **39**, 353–401 (2001)
189. A. Quirrenbach, The development of astronomical interferometry. Exp. Astron. **26**, 49–63 (2009)
190. R. Ragazzoni, Pupil plane wavefront sensing with an oscillating prism. J. Mod. Optics **43**, 289–293 (1996)
191. S.H. Reiger, Starlight scintillation and atmospheric turbulence. Astron. J. **68**, 395 (1963)
192. E.N. Ribak, M. Gai, D. Loreggia, S.G. Lipson, Simple beam combination for stellar interferometry. Opt. Lett. **32**, 1075–1077 (2007)
193. F. Rigautm, E. Gendron, Laser guide star in adaptive optics: the tilt determination problem. Astron. Astrophys. **261**, 677–684 (1992)
194. F. Roddier, The effects of atmospheric turbulence in optical astronomy. In *Progress in Optics XIX*, ed. by E. Wolf (North-Holland, Amsterdam, 1981), pp. 281–376
195. F. Roddier, Curvature sensing and compensation: a new concept in adaptive optics. Appl. Opt. **27**, 1223–1225 (1988)
196. F. Roddier, P. Lena, Long-baseline Michelson interferometry with large ground-based telescopes operating at optical wavelengths – I. General formalism. Interferometry at visible wavelengths. J. Opt. (Paris) **15**, 171–182 (1984)
197. F. Roddier, M. Northcott, J.E. Graves, A simple low-order adaptive optics system for near-infrared applications. Publ. Astron. Soc. Pac. **103**, 131–149 (1991)
198. F. Roddier, M.J. Northcott, J.E. Graves, D.L. McKenna, D. Roddier, One-dimensional spectra of turbulence-induced zernike aberrations: time-delay and isoplanicity error in partial adaptive compensation. J. Opt. Soc. Am. A **10**, 957–965 (1993)
199. N. Roddier, Atmospheric wavefront simulation using Zernike polynomials. Opt. Eng. **29**, 1174–1180 (1990)
200. F. Roddier (ed.), *Adaptive Optics in Astronomy* (Cambridge University Press, Cambridge, 2004)
201. M.C. Roggemann, B.M. Welsh, *Imaging Through the Atmosphere* (CRC, Boca Raton, 1996)
202. G. Rousset, Wavefront sensing. In *Adaptive Optics for Astronomy*, ed. by D.M. Alloin, J.M. Mariotti. NATO Advanced Study Institute Series, vol. C423 (Kluwer, New York, 1994), pp. 115–137
203. M. Ryle, A. Hewish, The synthesis of large radio telescopes. Mon. Not. R. Astron. Soc. **120**, 220–230 (1960)
204. J. Sahlmann, S.Ménardi, R. Abuter, M. Accordo, S. Mottini, F. Delplancke, The PRIMA fringe sensor unit. Astron. Astrophys. **507**, 1739–1757 (2009)
205. T.R.M. Sales, G.M. Morris, Fundamental limits of optical superresolution. Opt. Lett. **22**, 582–584 (1997)
206. M. Sarazin, F. Roddier, The ESO differential image motion monitor. Astron. Astrophys. **227**, 294–300 (1990)
207. R.J. Sasiela, J.D. Shelton, Transverse spectral filtering and mellin transform techniques applied to the effect of outer scale on tilt and tilt anisoplanatism. J. Opt. Soc. Am. A **10**, 646–660 (1993)
208. O.H. Schade, Electro-optical characteristics of television systems. RCA Rev. IX:5 (Part I), 245 (Part II), 490 (Part III), 653 (Part IV) (1948)
209. M. Schöller, R. Wilhelm, B. Koehler, Modeling the imaging process in optical stellar interferometers. Astron. Astrophys. Suppl. Ser. **144**, 541–552 (2000)

210. S.B. Shaklan, F. Roddier, Coupling starlight into single-mode fiber optics. Appl. Opt. **27**, 2334–2338 (1988)
211. M. Shao, M.M. Colavita, Long-baseline optical and infrared stellar interferometry. Annu. Rev. Astron. Astropys. **30**, 457–498 (1992)
212. M. Shao, M.M. Colavita, Potential of long-baseline infrared interferometry for narrow-angle astrometry. Astron. Astrophys. **262**, 353–358 (1992)
213. M. Shao, M.M. Colavita, B.E. Hines, D.H. Staelin, D.J. Hutter et al., The Mark III stellar interferometer. Astron. Astrophys. **193**, 357–371 (1988)
214. M. Shao, D.H. Staelin, Long-baseline optical interferometer for astrometry. J. Opt. Soc. Am. **67**, 81–86 (1977)
215. M. Shao, D.H. Staelin, First fringe measurements with a phase-tracking stellar interferometer. Appl. Opt. **19**, 1519–1522 (1980)
216. J. Skilling, R.K. Bryan, Maximum entropy image reconstruction: general algorithm. Mon. Not. R. Astron. Soc. **211**, 111–124 (1984)
217. A. Starikov, E. Wolf, Coherent-mode representation of Gaussian Schell-model sources and of their radiation fields. J. Opt. Soc. Am. **72**, 923–928 (1982)
218. W.H. Steel, *Interferometry* (Cambridge University Press, Cambridge, 1987)
219. M. Stéphan, C. R. Acad. Sci. **78**, 1008 (1874)
220. D.R. Stinebring, M.A. McLaughlin, J.M. Cordes, K.M. Beckers, J.E.E. Goodman, M.A. Kramer, J.L. Sheckard, C.T. Smith, Faint scattering around pulsars: probing the interstellar medium on solar system size scales. Astrophys. J. **549**, L97–L100 (2001)
221. M. Tallon, I. Tallon-Bosc, The object-image relationship in Michelson stellar interferometry. Astron. Astrophys. **253**, 641–645 (1992)
222. I. Tallon-Bosc, M. Tallon, Imaging an extended object witha Michelson stellar interferometer. In *ESO Conference on High-Resolution Imaging by Interferometry II*, ed. by F. Merkle, J.M. Beckers (European Southern Observatory, Garching, Germany, 1991), pp. 805–813
223. W.J. Tango, Dispersion in stellar interferometry. Appl. Opt. **29**, 516–521 (1990)
224. W.J. Tango, R.Q. Twiss, Michelson stellar interferometry. In *Progress in Optics XVII*, ed. by E. Wolf (North-Holland, Amsterdam, 1980), pp. 241–276
225. V.I. Tatarski, *Wave propagation in a turbulent medium* (McGraw-Hill, New York, 1961)
226. V.I. Tatarski, V.U. Zavorotny, Atmospheric turbulence and the resolution limits of large ground-based telescopes: comment. J. Opt. Soc. Am. A **10**, 2410–2417 (1993)
227. E. Tatulli, P. Mege, A. Chelli, Single-mode versus multimode interferometry: a performance study. Astron. Astrophys. **418**, 1179–1186 (2004)
228. G.B. Taylor, C.L. Carilli, R.A. Perley (ed.), *Synthesis Imaging in Radio Astronomy II, A collection of Lectures from the Sixth NRAO/NMIMT Synthesis Imaging Summer School*. ASP Conference Series, vol. 180 (ASP, San Francisco, 1999)
229. T.A. ten Brummelaar, H.A. McAllister, S.T. Ridgway, W.G. Bagnuolo Jr., N.H. Turner et al., First results from the CHARA array. II. A description of the instrument. Astrophys. J. **628**, 453–465 (2005)
230. E. Thiébaut, Optimization issues in blind deconvolution algorithms. In *Astronomical Data Analysis II*, ed. by J.-L. Starck, F.D. Murtagh. Proceedings of SPIE, vol. 4847 (2002), pp. 174–183
231. E. Thiébaut and J.-M. Conan, Strict a-priori constraints for maximum-likelihood blind deconvolution. J. Opt. Soc. Am. A **12**, 485–492 (1995)
232. A.R. Thompson, J.M. Moran, G.W. Swenson Jr, *Interferometry and Synthesis in Radio Astronomy* (Wiley, New York, 1986)
233. A. Tokovinin, From differential image motion to seeing. Publ. Astron. Soc. Pac. **114**, 1156–1166 (2002)
234. A. Tokovinin, M. Le Louarn, Isoplanatism in a multiconjugate adaptive optics system. J. Opt. Soc. Am. A **17**, 1819–1827 (2000)
235. A. Tokovinin, M. Sarazin, A. Smette, Testing turbulence model at metric scales with mid-infrared VISIR images at the vlt. Mon. Not. R. Astron. Soc. **378**, 701–708 (2007)
236. A. Tokovinin, T. Travouillon, Model of turbulence profile at Cerro Pachón. Mon. Not. R. Astron. Soc. **365**, 1235–1242 (2006)

237. C.H. Townes, Spatial interferometry in the mid-infrared region. J. Astrophys. Astron. **5**, 111–130 (1984)
238. W.A. Traub, Recent results from the IOTA interferometer. In *Astronomical Interferometry*, ed. by R.D. Reasenberg. Proceedings of SPIE, vol. 3350 (1998), pp. 848–855
239. R. Tubbs, Effect of wavefront corrugations on fringe motion in an astronomical interferometer with spatial lters. Appl. Opt. **44**, 6253–6257 (2006)
240. J. Turunen, E. Tervonen, A.T. Friberg, Coherence theoretic algorithm to determine the transverse-mode structure of lasers. Opt. Lett. **14**, 627–629 (1989)
241. P.G. Tuthill, J.D. Monnier, W.C. Danchi, A dusty pinwheel nebula around the massive star WR104. Nature **398**, 487–489 (1999)
242. P.G. Tuthill, J.D. Monnier, W.C. Danchi, E.H. Wishnow, C.A. Haniff, Michelson interferometry with the Keck I telescope. Publ. Astron. Soc. Pac. **112**, 555–565 (2000)
243. G.A. Tyler, Bandwidth considerations for tracking through turbulence. J. Opt. Soc. Am. A **11**, 358–367 (1994)
244. R.K. Tyson, *Principles of adaptive optics* (Academic, Boston, 1998)
245. F. Vakili, E. Aristidi, L. Abe, B. Lopez, Interferometric remapped array nulling. Astron. Astrophys. **421**, 147–156 (2004)
246. G.C. Valley, Long- and short-term Strehl ratios for turbulence with finite inner and outer scales. Appl. Opt. **18**, 984–987 (1979)
247. G.C. Valley, Isoplanatic degradation of tilt correction and short-term imaging systems. Appl. Opt. **19**, 574–577 (1980)
248. P.H. van Cittert, Die wahrscheinliche Schwingungsverteilung in einer von einer Lichtquelle direkt oder mittels einer Linse beleuchteten Ebene. Physica **1**, 201–210 (1934)
249. G. Vasisht, A.J. Booth, M.M. Colavita, R.L. Johnson, E.R. Ligon et al., Performance and verification of the Keck interferometer fringe detection and tracking system. In *Interferometry for Optical Astronomy II*, ed. by W.A. Traub. Proceedings of SPIE, vol. 4838 (2002), pp. 824–834.
250. H. Vazquez Ramio, M. Reyes, J. Miguel Delgado, E. Hernandez, M. Nunez Cagigal et al., Cute-SCIDAR at paranal for E-ELT site characterisation. ESO Messenger **132**, 29–31 (2008)
251. A. Walther, Radiometry and coherence. J. Opt. Soc. Am. **58**, 1256–1259 (1968)
252. G. Weigelt, Triple-correlation imaging in optical astronomy. In *Progress in Optics*, vol. 29, ed. by E. Wolf (Elsevier, Amsterdam, 1991), pp. 293–319
253. R. Wilhelm. Novel numerical model for dynamic simulations of optical stellar interferometers. PhD Thesis, TU Berlin (2000)
254. J.C. Wyant. Use of an ac heterodyne lateral shear interferometer with real-time wavefront correction systems. Appl. Opt. **14**, 2622–2626 (1975)
255. A.T. Young. Seeing: its cause and cure. Astrophys. J. **8**, 587–604 (1974)
256. T. Young. On the theory of light and colors. Phil. Trans. R. Soc. Lond. **387**, 12–48 (1802)
257. A. Zeilinger. General properties of lossless beam splitters in interferometry. Am. J. Phys. **49**, 882–883 (1981)
258. F. Zernike. Beugungstheorie des Schneidenverfahrens und seiner verbesserten Form, der Phasenkontrastmethode. Physica **1**, 689 (1934)
259. F. Zernike. The concept of degree of coherence and its application to optical problems. Physica **5**, 785–795 (1938)
260. A. Ziad, R. Conan, A. Tokovinin, F. Martin, J. Borgnino, From the grating scale monitor to the generalized seeing monitor. Appl. Opt. **39**, 5415–5425 (2000)

Index

ASTRONOMY AND ASTROPHYSICS LIBRARY

The Stars By E. L. Schatzman and F. Praderie

Modern Astrometry 2nd Edition
By J. Kovalevsky

The Physics and Dynamics of Planetary Nebulae By G. A. Gurzadyan

Galaxies and Cosmology By F. Combes, P. Boissé, A. Mazure and A. Blanchard

Observational Astrophysics 2nd Edition
By P. Léna, F. Lebrun and F. Mignard

Physics of Planetary Rings Celestial Mechanics of Continuous Media
By A. M. Fridman and N. N. Gorkavyi

Tools of Radio Astronomy 4th Edition, Corr. 2nd printing
By K. Rohlfs and T. L. Wilson

Tools of Radio Astronomy Problems and Solutions 1st Edition, Corr. 2nd printing
By T. L. Wilson and S. Huttemeister

Astrophysical Formulae 3rd Edition (2 volumes)
Volume I: Radiation, Gas Processes and High Energy Astrophysics
Volume II: Space, Time, Matter and Cosmology
By K. R. Lang

Galaxy Formation 2nd Edition
By M. S. Longair

Astrophysical Concepts 4th Edition
By M. Harwit

Astrometry of Fundamental Catalogues
The Evolution from Optical to Radio Reference Frames
By H. G. Walter and O. J. Sovers

Compact Stars. Nuclear Physics, Particle Physics and General Relativity 2nd Edition
By N. K. Glendenning

The Sun from Space By K. R. Lang

Stellar Physics (2 volumes)
Volume 1: Fundamental Concepts and Stellar Equilibrium
By G. S. Bisnovatyi-Kogan

Stellar Physics (2 volumes)
Volume 2: Stellar Evolution and Stability
By G. S. Bisnovatyi-Kogan

Theory of Orbits (2 volumes)
Volume 1: Integrable Systems and Non-perturbative Methods
Volume 2: Perturbative and Geometrical Methods
By D. Boccaletti and G. Pucacco

Black Hole Gravitohydromagnetics
By B. Punsly

Stellar Structure and Evolution
By R. Kippenhahn and A. Weigert

Gravitational Lenses By P. Schneider, J. Ehlers and E. E. Falco

Reflecting Telescope Optics (2 volumes)
Volume I: Basic Design Theory and its Historical Development. 2nd Edition
Volume II: Manufacture, Testing, Alignment, Modern Techniques
By R. N. Wilson

Interplanetary Dust
By E. Grün, B. Å. S. Gustafson, S. Dermott and H. Fechtig (Eds.)

The Universe in Gamma Rays
By V. Schönfelder

Astrophysics. A New Approach 2nd Edition
By W. Kundt

Cosmic Ray Astrophysics
By R. Schlickeiser

Astrophysics of the Diffuse Universe
By M. A. Dopita and R. S. Sutherland
The Sun An Introduction. 2nd Edition
By M. Stix

Order and Chaos in Dynamical Astronomy
By G. J. Contopoulos

Astronomical Image and Data Analysis
2nd Edition By J.-L. Starck and F. Murtagh

The Early Universe Facts and Fiction
4th Edition By G. Börner

ASTRONOMY AND ASTROPHYSICS LIBRARY

The Design and Construction of Large Optical Telescopes By P.Y. Bely

The Solar System 4th Edition
By T. Encrenaz, J.-P. Bibring, M. Blanc, M. A. Barucci, F. Roques, Ph. Zarka

General Relativity, Astrophysics, and Cosmology By A. K. Raychaudhuri, S. Banerji, and A. Banerjee

Stellar Interiors Physical Principles, Structure, and Evolution 2nd Edition
By C. J. Hansen, S.D. Kawaler, and V. Trimble

Asymptotic Giant Branch Stars
By H. J. Habing and H. Olofsson

The Interstellar Medium
By J. Lequeux

Methods of Celestial Mechanics (2 volumes)
Volume I: Physical, Mathematical, and Numerical Principles
Volume II: Application to Planetary System, Geodynamics and Satellite Geodesy
By G. Beutler

Solar-Type Activity in Main-Sequence Stars
By R. E. Gershberg

Relativistic Astrophysics and Cosmology
A Primer By P. Hoyng

Magneto-Fluid Dynamics
Fundamentals and Case Studies
By P. Lorrain

Compact Objects in Astrophysics
White Dwarfs, Neutron Stars and Black Holes
By Max Camenzind

Special and General Relativity
With Applications to White Dwarfs, Neutron Stars and Black Holes
By Norman K. Glendenning

Stellar Physics
2: Stellar Evolution and Stability
By G. S. Bisnovatyi-Kogan

Principles of Stellar Interferometry
By A. Glindemann